Predictive Maintenance in Dynamic Systems

Edwin Lughofer • Moamar Sayed-Mouchaweh
Editors

Predictive Maintenance in Dynamic Systems

Advanced Methods, Decision Support Tools and Real-World Applications

Editors
Edwin Lughofer
Fuzzy Logic Laboratorium Linz-Hagenberg
Department of Knowledge-Based
Mathematical Systems
Johannes Kepler University Linz
Linz, Austria

Moamar Sayed-Mouchaweh
Institute Mines-Telecom Lille Douai
Douai, France

ISBN 978-3-030-05644-5   ISBN 978-3-030-05645-2 (eBook)
https://doi.org/10.1007/978-3-030-05645-2

Library of Congress Control Number: 2019931901

© Springer Nature Switzerland AG 2019
This work is subject to copyright. All rights are reserved by the Publisher, whether the whole or part of the material is concerned, specifically the rights of translation, reprinting, reuse of illustrations, recitation, broadcasting, reproduction on microfilms or in any other physical way, and transmission or information storage and retrieval, electronic adaptation, computer software, or by similar or dissimilar methodology now known or hereafter developed.
The use of general descriptive names, registered names, trademarks, service marks, etc. in this publication does not imply, even in the absence of a specific statement, that such names are exempt from the relevant protective laws and regulations and therefore free for general use.
The publisher, the authors and the editors are safe to assume that the advice and information in this book are believed to be true and accurate at the date of publication. Neither the publisher nor the authors or the editors give a warranty, express or implied, with respect to the material contained herein or for any errors or omissions that may have been made. The publisher remains neutral with regard to jurisdictional claims in published maps and institutional affiliations.

This Springer imprint is published by the registered company Springer Nature Switzerland AG.
The registered company address is: Gewerbestrasse 11, 6330 Cham, Switzerland

To our families and friends

Preface

During recent years, rapid technological developments and breakthroughs in various industrial automatization processes with the support of modern machines, big data storages, and clouds of computers operating in parallel led to a significant increase in system complexity and dynamic processes. This makes a manual supervision and maintenance of machines, system components, and production chains more and more unaffordable and thus unrealistic to be conducted in a reasonable amount of time with reasonable efforts and costs for companies.

Therefore, automated predictive maintenance (APdM) has more and more become a central cornerstone in today's industrial applications and systems ranging from online manufacturing rails and production lines through (cyber) security problems and infrastructure management to energy fabrication, maritime systems, and exploitation facilities. This is because APdM addresses not only strategies for the early detection and prediction of significant machine wearing towards component failures, degraded performance of the system, undesired situations and occurrences, or downtrends in product quality but also for taking appropriate actions upon the recognition and prediction of such occasions. Such actions are indispensable for reducing waste, repair, and production costs, or even customer complaints, and thus, in the long run, also for increasing the income of companies; for guaranteeing higher quality of production items, network functionality, software, and user front ends; and finally for reducing the pollution of the environment. In the extreme case of catastrophic system failures, any severe damage to the infrastructure, machine, or system and severe risks for operators working with the system can be avoided by predictive maintenance.

The necessity of APdM in theory and practice is reflected in several methodological and application-oriented developments during the last 15–20 years, where, according to the ISI Web of Knowledge/Science database (Thomson Reuters)—www.webofknowledge.com—the number of publications permanently grew from around 100 per year around the beginning of the 2000s up to more than 500 per year around 2017, and the number of citations grew even more intensively. Furthermore, APdM became an essential component in today's Industry 4.0 environments and applications, and several objectives with associated calls have been established during the recent years under the umbrella of the Horizon 2020 Framework Programme "Nanotechnologies, Advanced Materials, Advanced Manufacturing and Processing, and Biotechnology."

In a typical predictive maintenance framework, embedded system models play a key role for producing (quality) forecasts, for indicating arising problems and faults at an early stage, or for conducting any deeper diagnosis about upcoming expected (as predicted) anomalous process behaviors in various forms. The high dynamics in today's processes or parts of processes often has the effect that already modeled/learned dependencies become outdated, which requires system models to self-adapt over time in order to maintain their predictive performance and to expand their "knowledge" and "validation range." This is hardly considered in the current state of the art of predictive maintenance; therefore, it is a central aspect in this book to show new trends in this direction—in fact, most of the chapters are dealing with (data-driven) modeling, optimization, and control (MOC) strategies, which possess the ability to be trainable and adaptable on the fly based on changing system behavior and nonstationary environmental influences.

Apart from this, several new applications in the context of predictive maintenance as well as combinations of MOC methodologies to successfully establish predictive maintenance are demonstrated in this book. According to the essential steps in predictive maintenance systems from early anomaly and fault detection during the process through the prognostics of eventually arising problems in the (near) future to their diagnosis and proper reactions on these (through optimization, control for repair, and self-healing), the book is structured into three main parts, where in each of them, important real-world systems and application scenarios are discussed:

- Anomaly detection and localization
- Prognostics and forecasting
- Diagnosis, optimization, and control

Furthermore, the first three chapters round off the whole book by discussing important aspects, principal concepts, and requirements and which are of general relevance in predictive maintenance systems and thus can be of significant importance in any of the three methodological steps (book parts).

Finally, the editors are very grateful to all authors and reviewers for contributing with substantial and very valuable material to make this volume become alive and to set another cornerstone in the research and publication history of predictive maintenance methodologies and applications. The first editor acknowledges the support by the "LCM—K2 Center for Symbiotic Mechatronics" within the framework of the Austrian COMET-K2 program. We also acknowledge Mary E. James and Menas Donald Kiran for establishing the contract with Springer and supporting us in any organizational aspects. We hope that the volume will be a useful basis for further fruitful investigations and fresh ideas as well as a motivation and inspiration for newcomers to join this important and still emerging field of research.

Linz, Austria Edwin Lughofer
Douai, France Moamar Sayed-Mouchaweh
October 2018

Contents

Prologue: Predictive Maintenance in Dynamic Systems 1
Edwin Lughofer and Moamar Sayed-Mouchaweh

Smart Devices in Production System Maintenance 25
Eike Permin, Florian Lindner, Kevin Kostyszyn, Dennis Grunert,
Karl Lossie, Robert Schmitt, and Martin Plutz

**On the Relevance of Preprocessing in Predictive Maintenance
for Dynamic Systems** 53
Carlos Cernuda

Part I Anomaly Detection and Localization

**A Context-Sensitive Framework for Mining Concept Drifting Data
Streams** 97
Chamari I. Kithulgoda and Russel Pears

**Online Time Series Changes Detection Based on Neuro-Fuzzy
Approach** 131
Yevgeniy Bodyanskiy, Artem Dolotov, Dmytro Peleshko,
Yuriy Rashkevych, and Olena Vynokurova

**Early Fault Detection in Reciprocating Compressor Valves
by Means of Vibration and pV Diagram Analysis** 167
Kurt Pichler

**A New Hilbert-Huang Transform Technique for Fault Detection in
Rolling Element Bearings** 207
Shazali Osman and Wilson Wang

**Comparison of Genetic and Incremental Learning Methods
for Neural Network-Based Electrical Machine Fault Detection** 231
Daniel Leite

Evolving Fuzzy Model for Fault Detection and Fault Identification of Dynamic Processes 269
Goran Andonovski, Sašo Blažič, and Igor Škrjanc

An Online RFID Localization in the Manufacturing Shopfloor 287
Andri Ashfahani, Mahardhika Pratama, Edwin Lughofer, Qing Cai, and Huang Sheng

Part II Prognostics and Forecasting

Physical Model-Based Prognostics and Health Monitoring to Enable Predictive Maintenance 313
Tiedo Tinga and Richard Loendersloot

On Prognostic Algorithm Design and Fundamental Precision Limits in Long-Term Prediction 355
Marcos E. Orchard and David E. Acuña

Performance Degradation Monitoring and Quantification: A Wastewater Treatment Plant Case Study 381
Iñigo Lecuona, Rosa Basagoiti, Gorka Urchegui, Luka Eciolaza, Urko Zurutuza, and Peter Craamer

Fuzzy Rule-Based Modeling for Interval-Valued Data: An Application to High and Low Stock Prices Forecasting 403
Leandro Maciel and Rosangela Ballini

Part III Diagnosis, Optimization and Control

Reasoning from First Principles for Self-adaptive and Autonomous Systems 427
Franz Wotawa

Decentralized Modular Approach for Fault Diagnosis of a Class of Hybrid Dynamic Systems: Application to a Multicellular Converter ... 461
Moamar Sayed-Mouchaweh

Automated Process Optimization in Manufacturing Systems Based on Static and Dynamic Prediction Models 485
Edwin Lughofer, Alexandru-Ciprian Zavoianu, Mahardhika Pratama, and Thomas Radauer

Distributed Chance-Constrained Model Predictive Control for Condition-Based Maintenance Planning for Railway Infrastructures 533
Zhou Su, Ali Jamshidi, Alfredo Núñez, Simone Baldi, and Bart De Schutter

Index 555

Contributors

David E. Acuña Department of Electrical Engineering, Faculty of Mathematical and Physical Sciences, University of Chile, Santiago, Chile

Goran Andonovski Faculty of Electrical Engineering, University of Ljubljana, Ljubljana, Slovenia

Andri Ashfahani Nanyang Technological University, Singapore, Singapore

Simone Baldi Delft Center for Systems and Control, Delft, The Netherlands

Rosangela Ballini Institute of Economics, University of Campinas, Campinas, SP, Brazil

Rosa Basagoiti Faculty of Engineering, Mondragon Unibertsitatea, Arrasate - Mondragon, Spain

Sašo Blažič Faculty of Electrical Engineering, University of Ljubljana, Ljubljana, Slovenia

Yevgeniy Bodyanskiy Kharkiv National University of Radio Electronics, Kharkiv, Ukraine

Qing Cai Nanyang Technological University, Singapore, Singapore

Carlos Cernuda BCAM - Basque Center for Applied Mathematics, Bilbao, Spain
Faculty of Engineering (MU-ENG), Mondragon Unibertsitatea, Arrasate, Spain

Peter Craamer MSI Grupo, Andoain, Spain

Bart De Schutter Delft Center for Systems and Control, Delft, The Netherlands

Artem Dolotov Kharkiv National University of Radio Electronics, Kharkiv, Ukraine

Luka Eciolaza Faculty of Engineering, Mondragon Unibertsitatea, Arrasate - Mondragon, Spain

Dennis Grunert Fraunhofer Institute for Production Technology IPT, Aachen, Germany

Ali Jamshidi Section of Railway Engineering, Delft, The Netherlands

Chamari I. Kithulgoda Department of Computer Science, Auckland University of Technology, Auckland, New Zealand

Kevin Kostyszyn Fraunhofer Institute for Production Technology IPT, Aachen, Germany

Iñigo Lecuona Faculty of Engineering, Mondragon Unibertsitatea, Arrasate - Mondragon, Spain

Daniel Leite Department of Engineering, Federal University of Lavras, Lavras, Minas Gerais, Brazil

Florian Lindner Fraunhofer Institute for Production Technology IPT, Aachen, Germany

Richard Loendersloot Dynamics Based Maintenance, University of Twente, Enschede, The Netherlands

Karl Lossie Fraunhofer Institute for Production Technology IPT, Aachen, Germany

Edwin Lughofer Fuzzy Logic Laboratorium Linz-Hagenberg, Department of Knowledge-Based Mathematical Systems, Johannes Kepler University Linz, Linz, Austria

Leandro Maciel Sao Paulo School of Politics, Economics and Business, Federal University of Sao Paulo, Osasco, SP, Brazil

Alfredo Núñez Section of Railway Engineering, Delft, The Netherlands

Marcos E. Orchard Department of Electrical Engineering, Faculty of Mathematical and Physical Sciences, University of Chile, Santiago, Chile

Shazali Osman Department of Electrical and Computer Engineering, Lakehead University, Thunder Bay, ON, Canada

Russel Pears Department of Computer Science, Auckland University of Technology, Auckland, New Zealand

Dmytro Peleshko IT Step University, Lviv, Ukraine

Eike Permin Fraunhofer Institute for Production Technology IPT, Aachen, Germany

Kurt Pichler Linz Center of Mechatronics GmbH, Linz, Austria

Martin Plutz oculavis GmbH, Aachen, Germany

Mahardhika Pratama School of Computer Science and Engineering, Nanyang Technological University, Singapore, Singapore

Thomas Radauer Stratec Consumables, Anif, Austria

Yuriy Rashkevych Ministry of Education and Science of Ukraine, Kyiv, Ukraine

Moamar Sayed-Mouchaweh Institute Mines-Telecom Lille Douai, Douai, France

Robert Schmitt Fraunhofer Institute for Production Technology IPT, Aachen, Germany

Laboratory for Machine Tools and Production Engineering (WZL) of RWTH Aachen University, Aachen, Germany

Huang Sheng Singapore Institute of Manufacturing Technology, Singapore, Singapore

Igor Škrjanc Faculty of Electrical Engineering, University of Ljubljana, Ljubljana, Slovenia

Zhou Su Delft Center for Systems and Control, Delft, The Netherlands

Tiedo Tinga Dynamics Based Maintenance, University of Twente, Enschede, The Netherlands

Gorka Urchegui MSI Grupo, Andoain, Spain

Olena Vynokurova Kharkiv National University of Radio Electronics, Kharkiv, Ukraine

IT Step University, Lviv, Ukraine

Wilson Wang Department of Mechanical Engineering, Lakehead University, Thunder Bay, ON, Canada

Franz Wotawa Technische Universität Graz, Institute for Software Technology, Graz, Austria

Alexandru-Ciprian Zavoianu Department of Knowledge-Based Mathematical Systems, Johannes Kepler University Linz, Linz, Austria

Urko Zurutuza Faculty of Engineering, Mondragon Unibertsitatea, Arrasate - Mondragon, Spain

Prologue: Predictive Maintenance in Dynamic Systems

Edwin Lughofer and Moamar Sayed-Mouchaweh

1 From Predictive to Preventive Maintenance in Dynamic Systems: Motivation, Requirements, and Challenges

Predictive maintenance (PdM) relies on real-time monitoring and diagnosis of system components, processes, and production chains [35]. The primary strategy is to take action when items or parts show certain behaviors that usually result in machine failure, degraded performance, or a downtrend in product quality. Originally, predictive maintenance was motivated by the execution of system checks at predetermined intervals to analyze the health of equipment, machines, or components in machineries [46]. During recent years, predictive maintenance also has been more and more applied in (cyber) security problems, infrastructure management, energy fabrication and power plants, maritime systems, exploitation facilities as well as in production chains or in factories of the future, see [35] and several chapters in this book below. In this sense, it became an essential component in today's Industry 4.0 environments and applications [23, 68], and several objectives with associated calls have been established under the umbrella of the Horizon 2020 framework programme "Nanotechnologies, Advanced Materials, Advanced Manufacturing and Processing, and Biotechnology"—there, it is typically addressed in the context of *zero-defect manufacturing* [48], where predictive maintenance aspects place strong contributions in various prognostics tasks as well as in the

---

E. Lughofer (✉)
Fuzzy Logic Laboratorium Linz-Hagenberg, Department of Knowledge-Based Mathematical Systems, Johannes Kepler University Linz, Linz, Austria
e-mail: edwin.lughofer@jku.at

M. Sayed-Mouchaweh
Institute Mines-Telecom Lille Douai, Douai, France
e-mail: moamar.sayed-mouchaweh@mines-douai.fr

© Springer Nature Switzerland AG 2019
E. Lughofer, M. Sayed-Mouchaweh (eds.), *Predictive Maintenance in Dynamic Systems*, https://doi.org/10.1007/978-3-030-05645-2_1

context of *refurbishments and remanufacturing of industrial equipments*, where predictive maintenance plays a strong role in performing functional diagnosis and in providing appropriate repairing actions.

Compared to classical quality control and to condition monitoring [41, 47], which basically operate in a kind of retrospective and reactive manner [70], i.e., problems such as severe wearings and failures of machines [12] and tools [54] or downtrends in production quality (e.g., production parts not meeting the quality boundaries) [57, 72] are recognized *after* they have de facto happened, a central aspect of predictive maintenance in all kinds is to recognize untypical system behavior [5, 46] or to identify undesired trends [45, 69] *at an early stage*. Ideally, this should be achieved as early as possible (by long-term predictions), in order to have enough time for appropriate reactions to avoid bad quality, to decrease the likelihood of machine (components) failures, and to even reduce risks for operators in subsequent processing stages [26]. In the extreme case of catastrophic system failures, any severe damage to the infrastructure, machine(s), or to the whole system and thus any severe risks for operators working with the system can be avoided by predictive maintenance. This makes predictive maintenance extremely important for reducing waste, repair and production costs, finally even for omitting customer complaints and thus, in the long run, also for increasing the income of companies, for guaranteeing higher quality of production items, network functionality, software and user front-ends, and finally for reducing the pollution of the environment.

Apart from work-programmes within the scope of Horizon 2020 specifically dedicated to predictive maintenance aspects, the necessity of predictive maintenance in theory and practice is thus also reflected in several developments during the last 15 to 20 years, where, according to the Isi Web of Knowledge/Science data base (Thomson Reuters)—www.webofknowledge.com—the number of publications permanently grew from around 100 per year around the beginning of the 00's up to more than 500 per year around 2017, and the number of citations grew even more intensively. Figure 1 depicts this tremendous explosion of both over the years.

Automatization in predictive maintenance (APM) is an essential aspect to deal with the ever-increasing system complexity and open-loop characteristics of components and installations, which more and more makes a manual supervision and maintenance of machines, system components, and production chains unaffordable and thus unrealistic to be conducted in a reasonable amount of time with reasonable efforts and costs for companies. Manual supervision thus becomes more and more a bottleneck, also because of being affected by human inconsistencies occurring over time subject to fatigue, boredom, uncertainty, or different cognitive abilities during different daytimes and workloads [1]. In some cases, different operators/experts may have even different modes of operation and especially different experience levels [43, 44], such that contradictory monitoring and reaction behavior among them could severely affect the consistency of the manual supervision over time. This may result in irregular and infrequent machining and system cycles.

In order to increase the level of automatization and the consistency of operation (modes), the so-called *system models* are typically required, which are able to automatically and permanently produce (quality) forecasts, for indicating arising

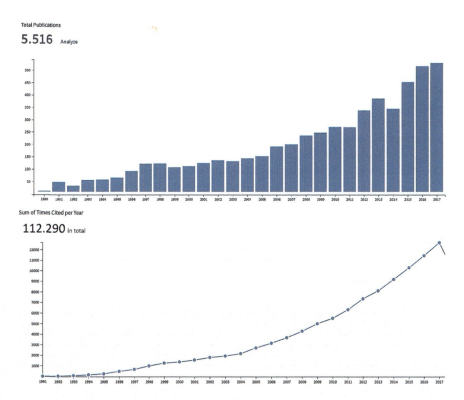

Fig. 1 Upper: the development of the number of papers appearing in the field of predictive maintenance per year from 1990 on; lower: the same for the development in terms of the number of cites; based on the Isi Web of Knowledge/Science data base, www.webofknowledge.com/

problems and faults at an early stage or for conducting any deeper diagnosis about upcoming expected (as predicted) anomal process behaviors in various forms (see also Sect. 2). With such models, it is possible to provide consistent outputs over time, i.e., the same (or very similar) current process trend(s) (reflected in states (e.g., samples) gathered from the process) will always produce the same (or very similar) model outputs. This is because, models do not get tired or are not affected by any "vague" knowledge. On the other hand, models may also suffer by low experience, e.g., when being established for only very particular system modes or based on an insufficient (historic) data basis; furthermore, high dynamics in today's processes or parts of processes often has the effect that already modeled/learned dependencies become outdated (even when being established based on enriched data). Both situations require system models to self-adapt and self-evolve over time in order to expand their "knowledge" and "validation range" properly with new data (reflecting dynamics, changes, etc.) [3, 27, 60] and to refine their parameters and structures for increasing their "significance" with more data—which is necessary for the models to maintain or even improve their predictive performance. This has been hardly addressed in the current state-of-the-art approaches of predictive maintenance (systems) so far, and therefore it is a central aspect in this book to show new trends in this direction.

Another important aspect is the fuzzy transition between predictive maintenance and *preventive maintenance* [20, 35, 74], the latter dealing with the prevention of faults, severe wearings, and quality downtrends before they occur or at least before they develop into major defects. In a predictive maintenance system, upon the prediction or early detection of any sort of problems, typically there exist tools for conducting diagnosis, optimization, and control in order to rebalance the process [52]. These mostly avoid severe defects and thus contribute also to preventive maintenance. The third part of the book will be dedicated to these more extended aspects of predictive maintenance.

2 Components and Methodologies for Predictive Maintenance

An example of a realization of a predictive maintenance framework is shown in Fig. 2. This should serve for readers to provide a more clear picture what a predictive maintenance system is about and especially which components are required and can be expected and how these interplay, from an early detection of a problem to a

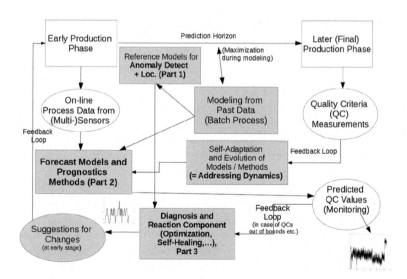

Fig. 2 An example of a predictive maintenance framework and the interplay of components therein; essential components to realize a well-functioning and high-performance system and which are contained in the three parts of this book (Parts 1 to 3) are highlighted in bold font

fully automated reaction by optimization and self-healing strategies. It is based on the past experience of the editors developing for and working with several such systems within applied research projects. In case when there are no data recordings performed or when analytical, physical-based models are sufficient and available, the modeling component has expert knowledge as source instead of a data base containing sensor measurements, etc.

Addressing significant system dynamics, in the example shown in Fig. 2 realized by feeding back the current process states and achieved targets (QCs) to a self-adaptation component for forecast models/methods, will be a central aspect of this book going beyond state-of-the-art. The operational environments of the majority of dynamic systems are very tough (e.g., offshore and far shore wind farms). This is because a lot of interferences and noises are added to the collected data entailing to decrease significantly its quality. This impacts the decision quality and its efficiency for scheduling the predictive maintenance actions. Therefore, it is important to develop advanced signal processing techniques that can improve significantly the quality of data before using them to feed up the decision model or step. Finally, the time-varying conditions of dynamic systems entail a variation in the degradation trends. The latter are accelerated in some operation conditions and significantly reduced in other operation conditions. Therefore, it is important to link the degradation trends to the system operation conditions. It is worth to mention that for some complicated components, the degradation trend may be reversed. This behavior is known to be a self-healing phenomenon. The latter impacts significantly the forecasting precision of the time to failure. Another challenge or requirement for the predictive maintenance in complex dynamic systems is the multi-degradation process. Indeed, since a complex system is composed of a set of interconnected stages/components (also termed as *multistage* processes [56]), a degradation in one stage/component propagates to other stages/components. This entails a multi-degradation process that needs to be taken into account adequately in order to establish methods and models for predicting quality and health states across multiple stages and in order to be able to efficiently schedule the maintenance actions.

In the following subsections, we discuss the essential components in a predictive maintenance system in more detail.

2.1 Models as Backbone Component

To establish APdM sufficiently well, system models are required serving as backbone for either the explicit detection of problems, the forecast of health and quality indicators, or for the automatic and in-line optimization and control of processes. These models can rely on analytical-based, knowledge-based, or purely data-driven models, which are describing the natural behavior of a system under normal, problem-free operation conditions or which are able to forecast arising problems and faults in the (near) future based on the current process trends.

In this book, one strong focus (in most of the chapters) will be placed on the usage of data-driven models, i.e., models generated fully from or operating on data, mostly with the usage of machine learning, soft computing, and statistical techniques which are more or less automatized (subject to some parametrization and final model selection efforts, for which typically an expert is needed). It is already well-known and widely accepted that data-driven models provide a complementary approach to maintenance planning by analyzing significant data sets of individual machine performance and environment variables [11]. They have been used in various PdM applications ranging from power plants [19] and turbines [71] through railway infrastructures [25] and cyber-physical systems [58] to manufacturing and production systems [49].

In the case of data-driven models, a wide variety and thus a great support of methods and techniques for self-adaptation and self-evolution of parameters and structures over time exist, either originating from the fields of data-stream mining [21, 30], incremental (machine) learning [60], or evolving (intelligent) systems [3, 27, 39] (see also Wikipedia for their definitions): three lines of research which started to develop in parallel during the 00's and emerged to wider research communities since 2010. These methods open up the possibility to follow dynamical changes of the system (also termed as drifts [28, 42]) or to integrate new operation modes not respected so far on demand and on the fly [33, 34, 78] (the latter is a special strength of evolving methods and systems [4]), and thus to maintain model performance and furthermore the performance of the whole predictive maintenance chain. This may become in particular essential in the case of dynamic systems, where process setups or measurement settings may change over time [2, 62] or where nonstationary environmental influences may happen [32, 60], such that older (input/output) relations/dependencies established within the models become outdated (thus leading to deteriorations of model performance). Furthermore, in particular systems it may be time consuming and expensive to capture data in advance for a whole run-to-failure process (including also undesired, "bad" states). A possibility to improve this situation is to build an online forecasting model (indicator) that can update (correct) and self-evolve its structure and parameters over time and on the fly in response to the reception of new data during the system run time. The point is that such dynamic, self-adaptive, and self-evolving modeling issues have been rarely addressed within the scope of predictive maintenance systems so far and is thus a sophisticated issue, which is addressed in various chapters in this book (thus significantly going beyond state-of-the-art).

Additionally, knowledge-based models, established through the formulation and coding of the (long-term, cognitive) expertise of experts (expert systems) [7, 20], and analytical, physical-based models may be used as supportive tools (also in a hybrid context with data-driven models [37]) to improve the reliability and precision of the remaining useful life (RUL) of machines and components [61] and to optimize the scheduled operations of maintenance [36]—e.g., by rule-based models for root-cause analysis of quality and machine functionality downtrends [73] or by model-based reasoning concepts [64].

In physical model-based approaches [16], such as unknown-input observers, algebraic methods, parameter estimation, parity space, and analytical redundancy, the physical laws describing the system dynamics in normal operation conditions are used to estimate the remaining useful life or to forecast health/quality indicators. Since, due to cost and security concerns, describing the degradation dynamics by physical laws for the majority of industrial systems is often unfeasible, reliability tests are performed in order to provide this information. Reliability tests [76] during the design step aim at characterizing a component's life cycle in each of the following phases: the break-in phase, the normal operating phase, and the aging or degradation phase. The latter [31] is often modeled by a power law, Gamma process, the Wiener Process, and their variants in order to calculate the mean time to failure when designing the system and mean time before failures after maintenance actions. However, reliability tests are time consuming and very costly. In addition, they are generated under laboratory conditions and not in real conditions. This may impact the precision of the predictions in particular when the real operation conditions are much different from the laboratory ones. Hence, data-driven models (as discussed in the paragraphs above) are often a promising or, depending on the efforts for establishing the models or the reliability tests, even a necessary alternative to physical-based models.

2.2 Methods and Strategies to Realize Predictive Maintenance

Predictive maintenance is mainly based on the following methodological modules: data acquisition and preprocessing, health indicators construction, (early) anomaly detection and localization, forecasting and prognostics of system/machine health state or potentially arising problems/downtrends (including remaining useful life prediction or fault prognostics), and corrective actions when undesired systems states are detected or predicted.

Data Acquisition, Pre-processing, and Health Indicator Construction
Data acquisition aims at capturing different kinds of measurements that can characterize the degradation process of the system components. To this end, several sensors are used to record various environmental, physical, electrical, acoustic, or mechanical signals. There are two additional data sources: logs data and warning events that provide information about the status of the different components, and service and inspection reports that give details about the performed inspection and repair actions. The data transmission and storage equipment belong to data acquisition. The data pre-processing aims at extracting fault indicators using signal processing (time, frequency, time–frequency domains, etc.) and artificial intelligence (residuals, virtual sensors, etc.) techniques. These techniques include: vibration monitoring, thermographic inspection, oil analysis, visual inspection, X-ray radiography, electrical insulation analysis, ultrasonic and acoustic emission analysis, nondestructive analysis, performance testing, etc.

Health indicators construction [31] aims at defining degradation indicators that monitor the performance (health) evolution of a system or one of its component over time. It can be based on the use of one feature (fault indicator) defined through the pre-processing step or the combination of several features. In both cases, the goal of health indicators is to decide when a degradation occurs in order to trigger the fault prognostics and diagnosis modules. The latter aims at estimating, starting from the current time, the time left before a system or one of its components becomes unable to perform its expected task or mission.

Anomaly Detection and Localization

It concerns the detection and localization of any form of atypical appearances in the system or at machines during on-line processing modes. Often, such anomalies are reflected in real faults actually happening at the system (and thus requiring an immediate reaction), but they could be also early indicators for arising problems in various forms (downtrend in production quality, weakening in machine functionality, weakening in component lifetime, etc.). In a data-driven context, anomalies or faults are typically reflected by measurement recordings which deviate from the past regular behavior or do not fit into the characteristics (density, shape, spread, etc.) of past sample occurrences [8, 13, 34, 40]. Anomaly detection often operates in a fully *unsupervised manner*, which means that no quality information about the current process/system/product states needs to be available to establish appropriate predictors or forecast models. This makes it often attractive in predictive maintenance systems, since:

1. The automatization capability of such an approach is expected to be very high: the on-going, regular measurement recordings can be immediately taken as representatives for the anomaly-free production process—and a characterization can be built upon these, which can be further used as an anomaly-free reference situation.
2. Annotation effort in terms of labeling costs for historic data samples [43] can be completely avoided—as is required in case of when establishing classifiers [17] within a decision support tool [66, 75] for actively classifying states into normal and abnormal through direct binary classification.
3. There is no necessity to have a kind of product quality index or even a failure index permanently measured over time, which is often costly to obtain (especially if manually taken), and which thus ends up in very small data set sizes for model training [45], or which in some systems is not really profitable or possible to install at all [31, 63].

On the other hand, there is no real target in the prediction of the current system state, just a hypothesis about an anomal behavior is provided, which makes it typically difficult to identify the problem and especially to find its reason. Furthermore, the training of (anomaly-free) reference models and/or statistical representatives from training data, especially finding the optimal training parameters, is challenging, as no classical statistical evaluation procedures such as cross-validation or bootstrapping [24] can be used (as these are requiring target values to be able to calculate errors and to perform parameter and model selection). Indeed, evaluation of the models may be based on data recorded during anomal or fault phases, but such data is also often rarely available and/or costly to collect [8, 10].

Anomaly/fault localization is an essential aspect in large-scale and distributed systems, where it is hard to find the detected problem by manual checking through all parts of the systems. In fact, due to the localization of the problem, it makes it much easier to identify the type of problem and its reason, so it is an important prerequisite of fault diagnosis aspects and further optimization cycles, see below.

Prognostics and Forecasting

It concerns the prognosis of potentially arising problems in the (near) future. The trade-off between the prediction horizon versus the forecast quality plays an essential role to establish meaningful prognosis [6]. The model should be still accurate enough to deliver reliable predictions, while the prediction and thus reaction time should be still long enough to properly conduct meaningful interventions in the process or in the runtime of machines before some components/parts get broken or something runs completely out of the rudder. Opposed to anomaly detection operating in a more or less unsupervised manner, prognostics has a clear target goal under examination (such as a machine health indicator, a quality criterion for production parts, or an indicator for the level of component wearings), whose concrete values are predicted into the future by an established forecast model [19, 38]. The monitoring of the predicted values is then typically achieved by comparing them with upper and lower allowed limits of the respective indicator, but can be also in form of checking for drifting states or other atypical trends of the indicators. Upon the prediction of undesired values (as an outcome of the monitoring module), operators may be informed and certain actions may be triggered (either in manual or automatic way, the latter being addressed in the subsequent paragraph). In a data-driven context, the forecast models are either established through fixed parameter settings (e.g., for machines in production systems) [22] or by current process values trends which are dynamically changing and recorded over time and which reflect the actual system (machine) status [45, 69]. From these trends, it is a challenge to predict the lifetime of a component or the health state of a machine or the expected quality of production parts. As important trends may last over a longer time frame, the modeling is often confronted with severe curse of dimensionality, which can be adequately addressed with time-series-based transformation or compression techniques [14].

A specific variant of prognostics and forecast is the estimation of the *remaining useful life (RUL)* [61] (e.g., often applied in supervising the charge-state of batteries [53]). To perform RUL in precise and reliable manner is a very challenging task, in particular when the system is complex formed by multiple interacted and nonlinear dynamic components working in strongly nonstationary environments. In general, there are two techniques used to estimate the RUL: model-based and data-driven approaches. (Analytical) Model-based approaches [31, 67] exploit physical knowledge about the system dynamics in order to model the degradation trends, often at component level. They model degradations caused by tear and wearing, crack, fissures, clogging, and corrosion phenomena. Data-driven approaches comprise statistical [80] and artificial intelligence [50, 65] techniques as well as Auto Regressive Moving Average (ARMA), particle filter, artificial neural networks, support vector machines, hidden Markov modes, etc.

Corrective Actions through Diagnosis, Optimization, and Control

When a degradation or a fault is detected or predicted, the corrective actions must be defined and planned. These actions are defined in order to maximize the system availability (breakdown of equipment, components, etc.) and to reduce its maintenance costs (labor cost, energy cost, etc.). In order to schedule these actions, several requirements must be defined such as the maintainability, availability, and severity. Maintainability measures the effort and cost for the maintenance according to skill level required and the availability of getting spare parts and service. Availability defines the time that a component is actually available to perform a task according to the time that it should be available. In this context, also the type and severity of a fault/problem plays a central role—faults are typically classified according to its severity into four categories [55]: catastrophic, critical, marginal, and negligible. A catastrophic or critical fault has much more priority for maintenance actions than a marginal fault. This is due to the fact that their impacts on availability and maintenance costs are more important than in the case of a marginal fault. To find out the level of fault intensity and severity and also the type of the fault, fault diagnosis methodologies are required, see [15, 29, 51, 59].

According to the fault severity, the system can still be operational but may lose of its efficiency or performance (e.g., a slower response or becoming less effective). In this case, a fault-tolerant control system (FTC) can be activated in order to alleviate or accommodate the fault consequences. FTC systems are defined in [79] as control systems that possess the ability to accommodate system component failures automatically. They are capable of maintaining overall system stability and acceptable performance in the event of failures. Further control strategies can be achieved through formulating a model-based predictive control (MPC) problem as a chance-constrained problem, which ensures that the constraints, e.g., bounds on the degradation level, are satisfied subject to some confidence level (see also Chap. 18). In this sense, any severe degradation or even system failures can be avoided in advance.

Optimization also may play a central role in order to balance out undesired situations in the system [52]. Often, optimization is part of the whole control-loop (as, e.g., in MPC) where it typically has to be conducted numerically in a fast manner (to meet on-line reaction demands), but optimization may be also used to provide suggestions to operators/experts for an improved behavior of the system process (e.g., improved production quality by adjusting machine parameters [9], see also Chap. 17 in this book). In the latter case, this often results in general multi-objective or even many-objective optimization problems, where heuristics-based evolutionary algorithms [18] can be used for providing adequate parameter settings/changes which are non-dominated solutions [77].

3 Beyond State-of-the-Art—Contents of the Book

The book basically addresses the data-driven autonomous aspect and the self-adaptation and self-evolution aspect in prediction maintenance systems in order to prevent manual operations (such as expert-based corrective actions) and high development times for system models and in order to be able to follow dynamical changes of the system, to integrate new operation modes and systems states, and to properly respect nonstationary environmental influences on demand and on the fly.

According to the essential steps in predictive maintenance systems as discussed in the previous section, the book is structured into three main parts, where in each of them important real-worlds systems and application scenarios are discussed:

- Anomaly detection and localization
- Prognostics and forecasting
- Diagnosis, optimization, and control

Additionally, there is a general section, which, apart from this introductory chapter, comprises the next two chapters and addresses aspects which are of general relevance and importance in predictive maintenance systems and do not directly fall into any of the aforementioned steps, but can be of significant importance (to be applied) in any of these. Their content can be summarized in the following way:

Chapter 2: Smart Devices in Production System Maintenance
Chapter 2 discusses the requirements and applicability of *smart devices*—which are electronic, mobile devices, which provide functionalities via sensor-based information processing and communication—in (general) production systems, with a specific focus on particular maintenance actions in these (Sect. 3). These include: (1) local data analysis and communication for condition monitoring, (2) remote expert solutions to accelerate machine failure handlings (with the usage of mobile devices), and (3) process data visualization for process monitoring which is interconnected with the productions software landscape through the usage of smart devices. After discussing several variants for smart devices, the authors also provide insights into the market view of smart devices (in order to give a clue to readers about support and popularity of the different variants) and demonstrate a standard implementation approach for smart devices. The latter consists of four essential steps, namely: (1) demonstration (to find out whether the combination of hardware and software is working in general), (2) proof of concept (to find out the functionality within the concrete maintenance use case), (3) pilot project (functionality of the business case temporarily), and (4) rollout (functionality of the business case in the long term). The authors clearly underline the necessity and usage of smart devices in different maintenance actions and also provide limitations of the current realizations and installations (hardware limitations, user acceptance, and information compression issues) as well as resulting challenges to be addressed in the future in order to make smart devices better applicable and acceptable in production maintenance. All in all, the chapter provides a round picture how an advanced and elegant communication with users and operators working with predictive maintenance systems can be established through smart devices.

Chapter 3: On the Relevance of Preprocessing in Predictive Maintenance for Dynamic Systems

Chapter 3 treats the problem of data preprocessing in order to make it meaningful and usable for any type of predictive maintenance system. It studies all the steps involved in data preprocessing required to build robust, accurate, and long-lasting models for highly dynamic systems. More precisely, this chapter presents the most known techniques, and tools used in six preprocessing steps: data cleaning, data normalization, data transformation, missing values treatment, feature engineering (including feature selection and extraction), and imbalanced data treatment. The chapter discusses the links between the preprocessing techniques used and the characteristics of available data about the dynamic system behavior. Finally, this chapter uses two public data sets (PHM Data Challenge 2014 and PHM Data Challenge 2016) in order to evaluate the presented preprocessing techniques for a fault detection problem (classification) and a remaining useful life estimation (regression). The evaluation of these techniques is performed in both, off-line and dynamic on-line learning scenarios.

The first part of the book after the general section comprises seven chapters, where the first five chapters are solely dealing with anomaly/fault detection and the latter two also with identification and localization issues. Their content can be summarized in the following way:

Chapter 4: A Context-Sensitive Framework for Mining Concept Drifting Data Streams

Chapter 4 demonstrates a context-sensitive framework for an appropriate handling of *drifts* in data streams. The latter either denotes changes in the underlying data distribution (input space drift), in the underlying relationship between model inputs and targets (joint drift), or in the prior probabilities of the target class resp. in the distribution of the target vector (target concept drift) [28]. Thus, in a more general context, a drift can be seen as a (significant) change in the system, which furthermore may point to an anomaly or not, depending on the intensity of the drift, its outlook and characteristics, and its effect on the reference base model (e.g., whether its accuracy is deteriorating or not). The authors propose three components: an incremental classifier, a concept drift detector, and an online repository of past concepts. The intensity of the learning process (by using an ensemble of classifiers) changes significantly according to the intensity of changes over time. This allows to speed up the learning time. The changes are monitored using the drift detector SeqDrift2. The latter uses the sample variance during a time window and assumes that the data samples follow a normal distribution. The online repository is based on the use of decision tree forest and Fourier spectra classifiers. These classifiers are updated or new classifiers are created in response to drifts (changes) in order to properly handle significant dynamics in the (PdM) system. In order to ensure that the memory does not overflow in the repository, newly created spectra are aggregated with the most similar existing ones. The proposed scheme is evaluated according to the classification accuracy and memory consumption using synthetic (rotating hyperplane with noises, noisy RBF) and real data sets (stemming from various application scenarios such as electricity, flight, cover type, and occupancy detection). The interest of this chapter is to observe the link between the drift detection and the problem of predictive maintenance. Indeed, a drift can be considered as a degradation, and the actions required for the classifier's creation and updating can be used to understand the degradation dynamics and its evolution. Therefore, the proposed generic scheme for drift detection and handling can be used for the aid of predictive maintenance.

Chapter 5: Online Time Series Changes Detection Based on Neuro-Fuzzy Approach

Chapter 5 handles the problem of online change detection in multidimensional data series. The change is considered to be a consequence of a fault occurrence and its detection is interpreted as an anomaly or fault detection. The chapter presents an online adaptive fuzzy clustering approach that allows monitoring changes in the data structure and adapting the system parameters to these changes. It therefore is able to account for system dynamics and is able to handle large volumes of data through online sequential processing of incoming observations. It is guided by fuzzy sequential clustering of time series with the use of probabilistic and possibilistic procedures as well as of a wavelet-neural network. The latter is learned by a robust algorithm conducting synaptic weights adjustments, which enables suppressing abnormal outliers present in real time series. The evaluation of this approach on data streams shows the simplicity of its numerical implementation and its high performance.

Chapter 6: Early Fault Detection in Reciprocating Compressor Valves by Means of Vibration and pV Diagram Analysis

Chapter 6 handles the problem of condition-based maintenance of reciprocating compressors, in particular broken valves. The goal is to perform an automated fault detection procedure allowing to reduce, and even to eliminate, the unscheduled shutdowns and the frequency of on-site inspections. The chapter presents two independent fault detection methods. The first one is based on vibration analysis, while the second method is based on the analysis of pressure–volume diagram (pV). The goal of this analysis is to extract appropriate features which allow a good separation between faultless and faulty conditions in the feature space. The crux of the whole approach is that it is robust with respect to compressors equipped with different sensors and with different types of valves. It can be also robustly applied to compressors working in varying load and pressure conditions, and reference models can be trained using only faultless conditions, i.e., in a fully unsupervised manner. Thus, time-intensive and costly collection of faulty data samples can be avoided. The obtained results show that even small fissures or leaks in the valve can be detected by the presented approaches.

Chapter 7: A New HHT Technique for Fault Detection in Rolling Element Bearings

Chapter 7 discusses the problem of the detection of faults entailing imperfections in rotating machinery. These imperfections are related to defects in rolling element bearings subject to dynamic loadings. Bearing defects are caused by fatigue damage resulting in micro-cracks and localized defects that lead to cracks. One important challenge that this chapter addresses is to detect the bearing damage in early stage because when the bearing damage propagates, the statistical indicators extracted by signal processing techniques may generate confusing diagnostic results. This chapter proposes an enhanced Hilbert–Huang transform technique in order to detect incipient bearing damages under variable load and speed conditions. Hilbert–Huang transform is a time–frequency domain technique that applies both, time and frequency information, to investigate transient feature properties. The proposed approach is evaluated using tests of different controlled bearing conditions. The obtained results demonstrate the potential of the proposed approach to perform bearing fault detection in real applications as gearboxes.

Chapter 8: Comparison of Genetic and Incremental Learning Methods for Neural Network-Based Electrical Machine Fault Detection

Chapter 8 discusses the problem of condition monitoring and predictive maintenance of induction motors, in particular the inter-turns short-circuit in the stator windings. These faults are primary faults that happen after insulation breakdown. The early detection of these faults, in particular in incipient stage, may lead to significant improvements of availability, quality, and productivity of production lines. This chapter proposes an approach based on a combination of a genetic algorithm and incremental feed-forward neural networks (NNs) that learn parameters in a stream-like dynamic manner. The purpose of the neural networks is to detect and determine the number of shorted turns in the stator windings of induction machines. The aim of the genetic algorithm is to find a suitable number of hidden layers and neurons per layer, which basically determine the neural network generalization ability. The proposed approach is evaluated in a real dynamic environment subject to mechanical asymmetries, voltage unbalance, and measurement noise. The obtained results demonstrate that the proposed approach is able to detect shorted turns successfully.

Chapter 9: Evolving Fuzzy Model for Fault Detection and Fault Identification of Dynamic Processes

Chapter 9 presents an evolving fuzzy modeling approach, which relies on the recursive calculation of local densities of data clouds, which are further associated with fuzzy rules. The local density is actually a measure which determines the closeness and the membership degree of the data to the existing data clouds. Based on this measure, new fuzzy rules are evolved when the partial densities between new samples and existing clouds are lower than a predefined threshold. In this sense, the evolving fuzzy model has an open structure to add new rules (modeling local regions) on demand and on the fly. As the learning procedure of the model starts with known/labeled data for the normal process operation and for (different types of) anomalies (faults), the model is capable of identifying the same types of operation modes the next time they appear. Anomaly/fault identification is thus automatically achieved according to the maximal global density of the current sample in those various data clouds which represent the different anomaly (fault) modes. Due to the incrementally adaptive and evolving characteristics of the fuzzy models over time, significant system dynamics in the form of drifting normal (or anomal) system states can be properly handled, as is successfully verified based on data (including four different types of faults) from a heating, ventilation, and air condition (HVAC) process (achieving high true positive fault recognition rates with low false positive rates).

Chapter 10: An Online RFID Localization in the Manufacturing Shopfloor
Chapter 10 presents an approach for locating equipments and trolleys manually over large manufacturing shopfloor areas (based on radio frequency identification (RFID) technology), which, when being conducted manually, results in time-consuming activities and significantly increased workloads for operators. Location of such equipments is important in the maintenance, repair, and overhaul (MRO) industry, especially, when faults or failures of the equipment are elicited or prognosticated and thus should be repaired or exchanged as early as possible. The authors develop an evolving model based on a novel evolving intelligent system, namely evolving Type-2 Quantum Fuzzy Neural Network (eT2QFNN), which features an interval type-2 quantum fuzzy set with uncertain jump positions and which is used to address the nonstationary, dynamic characteristics of manufacturing shopfloor, which makes a location of manufacturing objects and equipments difficult. The eT2QFNN is equipped with a rule growing mechanism, so it can generate its rules automatically in a single-pass learning mode, and thus it automatically expands its knowledge whenever new location states appear. The effectiveness of the approach has been experimentally validated using real-world RFID localization data from a real-world manufacturing shopfloor environment in Singapore using four RFID tags as references.

The second part of the book comprises four chapters, which are dealing with prognostics and forecast aspects, basically based on time-series data (except Chap. 11, which demonstrates a physical model-based approach) and which are applicable in predictive maintenance systems. Their content can be summarized in the following way:

Chapter 11: Physical Model-Based Prognostics and Health Monitoring to Enable Predictive Maintenance
Chapter 11 presents an overview of methods, techniques and tools used to treat the problem of fault prognostics and its challenges, in particular for complex systems. The latter are represented by complex dynamic responses that evolve in nonstationary environments. The chapter focuses on the use of fault prognostics in order to develop predictive maintenance concepts. It divides the state-of-the-art methods into data-driven and physical model-based approaches. Then, the chapter compares their advantages and drawbacks according to the system complexity (dynamics, size, environmental conditions, etc.), the precision of the estimation of

remaining useful life (RUL), and the capacity of interpretation of the degradation situation and development dynamics. The chapter gives hints for the selection of the method to be used to perform the fault prognostics by showing clearly the relationship to the system characteristics and to the available data (usage, load, and condition) about its functioning. The chapter highlights also the benefit to combine diagnostics and prognostics in particular for critical applications and to bridge the gap between the system highest level (asset, functions, etc.) and its lowest level (components). Finally, the chapter discusses the problem of fault prognostics to enable predictive maintenance using three case studies: maritime systems, railway infrastructure, and wind turbines. The equipment and structure health monitoring aspects are treated using these three case studies. The interest of these case studies is related to their complexity, to their cost and exploitation high costs, and to their varying dynamic environments.

Chapter 12: On Prognostic Algorithm Design and Fundamental Precision Limits in Long-Term Prediction

Chapter 12 deals with the long-term prediction of health condition indicators, which in predictive maintenance systems could appear as various performance metrics such as health indices of system components (indicating the remaining useful life (RUL)), indicators measuring the degree of wearing of machines (or particular parts therein), or as quality criteria of production parts in manufacturing systems. The authors provide a formal mathematical definition of the prognostic problem and a rigorous analysis for performance metrics based on the concept of Bayesian Cramer–Rao lower bounds (BCRLBs) for the predicted state mean square error (MSE) in prognostic algorithms. Furthermore, a step-by-step design methodology to tune prognostic algorithm hyper-parameters is explored, allowing to guarantee that the obtained results do not violate fundamental precision bounds for Time-of-Failure estimates. It is shown how this design procedure allows to detect situations in which the prognostic algorithm implementation generates results that violate the fundamental precision boundaries (and thus requiring actions by operators or by an automatized optimization component). The concepts are applied to the problem of end-of-discharge (EoD) time prognostics in lithium-ion batteries as an illustrative example.

Chapter 13: Performance Degradation Monitoring and Quantification: A Wastewater Treatment Plant Case Study

Chapter 13 presents an approach to perform the fault diagnosis of wastewater treatment plants, in particular its blowers and pumps. The challenge that the presented approach addressed is to detect early and in reliable manner nonlinear changes in the performance of the wastewater treatment plants generated by wear, tear, and clogging situations. The early and reliable detection of changes and further on their analysis contributes to the establishment of a predictive maintenance plan and is crucial in order to increase the availability, to reduce the operational costs, and to improve the maintenance task scheduling of the wastewater treatment plants. The presented approach uses key performance indicators in order to recognize or even prognose a degradation in the pumping systems and the blowers of the aeration process. The presented approach is based on a divide-and-conquer concept by breaking the system into individual components (blowers and pumps). This decomposition of the system into individual components allows to break a complex problem into smaller and easier-to-solve problems and to take into account the dependence between a component and its degradation dynamics. The key performance indicator used in this chapter is the relative mean error between expected and real observations over one day. The observations represent the wastewater flow generated by the external recirculation pump, its frequency, blower diffuser position, and its current. The obtained results show that the presented approach is able to detect and quantify the effects of clogging situations.

Chapter 14: Fuzzy Rule-Based Modeling for Interval-Valued Data: An application to High and Low Stock Prices Forecasting

Chapter 14 proposes an interval fuzzy rule-based model (iFRB) for financial interval time series forecasting and volatility estimation, which plays an important role in many maintenance-related applications in the area of financial markets, such as risk management, derivatives pricing and portfolio selection, as well as supplements the information by the time series of the closing price values. The iFRB is similar to a classical Takagi–Sugeno fuzzy model but by applying affine interval consequents. The learning of the models is achieved through the usage of evolving participatory learning (ePL) paradigm, which is able to process the data in incremental sample-wise manner, and thus the parameters of the model are successively updated, and which is also able to evolve new clusters (representing new states/modes) whenever the arousal index exceeds a predefined threshold. Hence, the approach is able to properly compensate significant system dynamics due to possible changing financial states (such as varying psychological factors, and several political occasions) over time, and it is able to predict a range of possible future values (interval). It thus can be seen as a promising forecast model not only for stock prices but also for other types of health and quality indicators, especially when uncertainties in model outputs should be expressed in the form of prediction intervals. The approach is empirically evaluated on the prediction of the main index of the Brazilian stock market, the IBOVESPA. The results indicate that the iFRB method appears as a promising alternative to traditional univariate and multivariate time series benchmark models as well as to interval multilayer perceptron neural networks for interval-valued financial time series forecasting.

The third part of the book comprises again four chapters, which are dealing with deeper analysis concepts (in terms of problem/fault diagnosis and reasoning aspects) and possibilities how to early react on (based on optimization and repair) and/or even prevent problems/faults (due to control strategies) in predictive maintenance systems. Their content can be summarized in the following way:

Chapter 15: Reasoning from First Principles for Self-Adaptive and Autonomous Systems

Chapter 15 provides concepts for automated model-based reasoning, which is to use a model of a system directly to reason about the system. In a certain instance, i.e., model-based diagnosis, the systems model can be used for identifying root causes in case of an observed behavior that contradicts the expected one (e.g., a fault or machine failure). A model in model-based diagnosis (MBD) comprises the systems structure including its components and interconnections, as well as the component models. The health state of components, i.e., a predicate indicating whether a component is working as expected or not, is used to indicate a root cause. The reasoning concepts can be further used for the intrinsic diagnosis of a problem (leading to a not properly working system)—the authors suggest two different variants, model-based diagnosis and abductive diagnosis and formulate them in a mathematical way. The authors also discuss the use of model-based reasoning and diagnosis for self-adaptation, which is understood as the ability of a system to adapt to dynamic and changing operating conditions autonomously, i.e., without requiring human intervention. In the context of reasoning and diagnosis, self-adaptation can be seen as a form of self-optimization and—healing behavior, which is also known as autonomic computing where a system can detect, diagnose, and repair localized faults originating from software or hardware. The concepts of repair and self-repair are again mathematically formalized by the authors in a generic sense, and thus they could be applied to any predictive maintenance system.

Chapter 16: Decentralized Modular Approach for Fault Diagnosis of a Class of Hybrid Dynamic Systems: Application to a Multicellular Converter

Chapter 16 proposes an approach to perform the diagnosis of faults in discretely controlled continuous systems. The latter is a class of hybrid dynamic systems in which the discrete and continuous dynamics cohabit. The discrete dynamics are described by discrete state variables, while the continuous dynamics are described by continuous state variables. Discretely controlled continuous systems are composed by a plant with continuous dynamics and supervisory discrete control. The presented approach exploits the modularity of the system by dividing it into several discrete components. Then, the local model for each of the latter is built. This local model includes the normal discrete modes as well as the ones reached in response to the occurrence of faults impacting the discrete behavior of this component. The continuous dynamic behavior is defined by a set of analytical redundancy relations (ARRs). The latter are used to generate residuals. The abstraction of these residuals generates events that are used to enrich the local discrete models. Then, a local diagnoser is built for each discrete component based on the use of the corresponding enriched local model. The proposed approach is illustrated and evaluated using a three-cellular power converter. The latter is based on the combination of three switches (cells of commutation) allowing the current flowing from the voltage source toward the output load. The obtained results show the ability of the presented approach to diagnose faults impacting the switches' normal behaviors (stuck close/stuck open).

Chapter 17: Automated Process Optimization in Manufacturing Systems Based on Static and Dynamic Prediction Models

Chapter 17 proposes an approach for the automated optimization of process parameters in manufacturing systems in order to automatically compensate possible downtrends in product quality at an early stage. The approach relies on the combination of: (1) static predictive mappings which are able to predict expected quality criteria at the end of a production stage based on machine parameter settings, (2) dynamic time-series-based forecast models which are able to forecast quality criteria during the runtime of production based on process values trends, and (3) a many-objective optimization strategy (using the predictive mappings and the forecast models as surrogate models) in order to automatically balance out undesired process behavior/trends that lead to a deterioration of product quality, which in turn may even lead to customer complaints, etc. The time-series-based forecast models possess the ability to self-adapt their parameters and to evolve their structures on demand and on the fly. Thus, they are able to compensate certain system dynamics due to new production charges and types, nonstationary environmental influences, or other unexpectedly arising system drifts. This is achieved with different levels of flexibility due to forgetting concepts, incremental updates of the latent variables space (required for reducing dimensionality of the high-dimensional time-series space), and incremental splitting of model components (rules). Several results are included from a micro-fluidic chip production process, where a reduction of the many-objective optimization problem to a (three-dimensional) multi-objective one could be achieved and heuristics-based solvers (using co-evolution strategy) could be successfully applied to produce parameter settings and process values trends leading to significantly improved product quality (omitting further waste and unnecessary machine wearings), compared to standard settings as having been used by operators for months and years.

Chapter 18: Distributed Chance-Constrained Model Predictive Control for Condition-Based Maintenance Planning for Railway Infrastructures

Chapter 18 develops a model predictive control (MPC) approach for condition-based maintenance planning under uncertainty for railway infrastructure systems composed of multiple components. To keep the balance between robustness and optimality, the authors formulate the MPC optimization problem as a chance-constrained problem, which ensures that the constraints, e.g., bounds on the degradation level, are satisfied with a given probabilistic guarantee. In this sense, degradation or even system failures can be avoided in advance, which abandons the necessity of problem detection and reaction (by reasoning and healing) at all. By comparing the chance-constrained MPC approaches with a deterministic approach (considering the mean values of the uncertain parameters) and a traditional time-based maintenance approach (performing grinding and replacing at a predetermined optimal interval), on a particular real-world example dealing with the optimal treatment of squats (a type of rolling contact fatigue), the authors show that despite their high computational requirements, chance-constrained MPC approaches are cost efficient and robust in the presence of uncertainties.

Acknowledgements The first author acknowledges the support by the LCM–K2 Center within the framework of the Austrian COMET-K2 program.

References

1. Akerkar, R., Sajja, P.: Knowledge-Based Systems. Jones & Bartlett Learning, Sudbury (2009)
2. Alippi, C., Roveri, M.: Just-in-time adaptive classifiers Part I: Detecting nonstationary changes. IEEE. Trans. Neural Netw. **19**(7), 1145–1153 (2008)
3. Angelov, P., Filev, D., Kasabov, N.: Evolving Intelligent Systems—Methodology and Applications. Wiley, New York (2010)
4. Angelov, P., Kasabov, N.: Evolving computational intelligence systems. In: Proceedings of the 1st International Workshop on Genetic Fuzzy Systems, pp. 76–82. Granada (2005)
5. Aumi, S., Corbett, B., Mhaskary, P.: Model predictive quality control of batch processes. In: 2012 American Control Conference, pp. 5646–5651. IEEE, Montreal (2012)
6. Box, G., Jenkins, G., Reinsel, G.: Time Series Analysis, Forecasting and Control. Prentice Hall, Engelwood Cliffs (1994)
7. Castillo, E., Alvarez, E.: Expert Systems: Uncertainty and Learning. Computational Mechanics Publications, Boston (2007)
8. Chandola, V., Banerjee, A., Kumar, V.: Anomaly detection: a survey. ACM Comput. Surv. **41**(3), 15 (2009)
9. Chandrasekaran, M., Muralidhar, M., Krishna, M., Dixit, U.: Application of soft computing techniques in machining performance prediction and optimization: a literature review. Int. J. Adv. Manuf. Technol. **46**, 445–464 (2010)
10. Chiang, L., Russell, E., Braatz, R.: Fault Detection and Diagnosis in Industrial Systems. Springer, London (2001)
11. Cline, B., Niculescu, R., Huffman, D., Deckel, B.: Predictive maintenance applications for machine learning. In: Proceedings of the 2017 Annual Reliability and Maintainability Symposium (RAMS). IEEE, Orlando (2017)
12. Collins, J., Busby, H., Staab, G.: Mechanical Design of Machine Elements and Machines. Wiley, Danvers (2010)
13. Costa, B., Angelov, P., Guedes, L.: Fully unsupervised fault detection and identification based on recursive density estimation and self-evolving cloud-based classifier. Neurocomputing **150**(A), 289–303 (2015)
14. Cunha, C., Soares, C.: On the choice of data transformation for modelling time series of significant wave height. Ocean Eng. **26**(6), 489–506 (1999)
15. Ding, S.: Model-based Fault Diagnosis Techniques: Design Schemes, Algorithms, and Tools. Springer, Berlin (2008)
16. Djeziri, M., Nguyen, V., Benmoussa, S., Msirdi, N.: Fault prognosis based on physical and stochastic models. In: Proceedings of the 2016 European Control Conference, pp. 2269–2274. IEEE, Aalborg (2016)
17. Dou, D., Zhou, S.: Comparison of four direct classification methods for intelligent fault diagnosis of rotating machinery. Appl. Soft Comput. **46**, 459–468 (2016)
18. Eiben, A.E., Smith, J.E.: Introduction to Evolutionary Computing, 2nd edn. Springer, Berlin (2015)
19. Ekwaro-Osire, S., Gonçalves, A., Alemayehu, F.: Probabilistic Prognostics and Health Management of Energy Systems. Springer, New York (2017)
20. Fonseca, D.: A knowledge-based system for preventive maintenance. Expert Syst. **17**(5), 241–247 (2000)
21. Gama, J.: Knowledge Discovery from Data Streams. Chapman & Hall/CRC, Boca Raton (2010)
22. García, V., Sánchez, J., Rodríguez-Picón, L., Méndez-Gónzalez, L., de Jesús Ochoa-Domínguez, H.: Using regression models for predicting the product quality in a tubing extrusion process. J. Int. Manag. (2018). https://doi.org/10.1007/s10845-018-1418-7
23. Gilchrist, A.: Industry 4.0: The Industrial Internet of Things. Springer, New York (2016)
24. Hastie, T., Tibshirani, R., Friedman, J.: The Elements of Statistical Learning: Data Mining, Inference and Prediction. Springer, New York (2001)

25. Jamshidi, A., Hajizadeh, S., Su, Z., Naeimi, M., Nunez, A., Dollevoet, R., Schutter, B.D., Li, Z.: A decision support approach for condition-based maintenance of rails based on big data analysis. Transp. Res. C **95**, 185–206 (2018)
26. Kano, M., Nakagawa, Y.: Data-based process monitoring, process control, and quality improvement: recent developments and applications in steel industry. Comput. Chem. Eng. **32**, 12–24 (2008)
27. Kasabov, N.: Evolving Connectionist Systems: The Knowledge Engineering Approach, 2nd edn. Springer, London (2007)
28. Khamassi, I., Sayed-Mouchaweh, M., Hammami, M., Ghedira, K.: Discussion and review on evolving data streams and concept drift adapting. Evol. Syst. **9**(1), 1–23 (2017)
29. Korbicz, J., Koscielny, J., Kowalczuk, Z., Cholewa, W.: Fault Diagnosis—Models, Artificial Intelligence and Applications. Springer, Berlin (2004)
30. Last, M., B., H., Kandel, A.: Data Mining in Time Series and Streaming Databases. World Scientific, Singapore (2017)
31. Lei, Y., Li, N., Guo, L., Li, N., Yan, T., Lin, J.: Machinery health prognostics: a systematic review from data acquisition to RUL prediction. Mech. Syst. Signal Process. **104**, 799–834 (2018)
32. Leite, D., Ballini, R., Costa, P., Gomide, F.: Evolving fuzzy granular modeling from nonstationary fuzzy data streams. Evol. Syst. **3**(2), 65–79 (2012)
33. Leite, D., Palhares, R., Campos, C.S., Gomide, F.: Evolving granular fuzzy model-based control of nonlinear dynamic systems. IEEE Trans. Fuzzy Syst. **23**(4), 923–938 (2015)
34. Lemos, A., Caminhas, W., Gomide, F.: Adaptive fault detection and diagnosis using an evolving fuzzy classifier. Inform. Sci. **220**, 64–85 (2013)
35. Levitt, J.: Complete Guide to Preventive and Predictive Maintenance. Industrial Press, New York (2011)
36. Li, Z., Guo, Z., Zhou, R.: Maintenance scheduling optimization based on reliability and prognostics information. In: Proceedings of the 2016 Annual Reliability and Maintainability Symposium (RAMS), pp. 1–8, IEEE, Tucson (2011)
37. Liao, L., Köttig, F.: A hybrid framework combining data-driven and model-based methods for system remaining useful life prediction. Appl. Soft Comput. **44**, 191–199 (2014)
38. Liao, W., Wang, Y.: Data-driven machinery prognostics approach using in a predictive maintenance model. J. Comput. **8**(1), 225–231 (2013)
39. Lughofer, E.: Evolving fuzzy systems—fundamentals, reliability, interpretability and useability. In: Angelov, P. (ed.) Handbook of Computational Intelligence, pp. 67–135. World Scientific, New York (2016)
40. Lughofer, E.: Robust data-driven fault detection in dynamic process environments using discrete event systems. In: Sayed-Mouchaweh, M. (ed.) Diagnosability, Security and Safety of Hybrid Dynamic and Cyber-Physical Systems, pp. 73–116. Springer, New York (2018)
41. Lughofer, E., Eitzinger, C., Guardiola, C.: On-Line Quality Control with Flexible Evolving Fuzzy Systems. In: Sayed-Mouchaweh, M., Lughofer, E. (eds.) Learning in Non-Stationary Environments: Methods and Applications, pp. 375–406. Springer, New York (2012)
42. Lughofer, E., Pratama, M., Skrjanc, I.: Incremental rule splitting in generalized evolving fuzzy systems for autonomous drift compensation. IEEE Trans. Fuzzy Syst. **26**(4), 1854–1865 (2018)
43. Lughofer, E., Richter, R., Neissl, U., Heidl, W., Eitzinger, C., Radauer, T.: Explaining classifier decisions linguistically for stimulating and improving operators labeling behavior. Inf. Sci. **420**, 16–36 (2017)
44. Lughofer, E., Smith, J.E., Caleb-Solly, P., Tahir, M., Eitzinger, C., Sannen, D., Nuttin, M.: Human-machine interaction issues in quality control based on on-line image classification. IEEE Trans. Syst. Man Cybern. A Syst. Hum. **39**(5), 960–971 (2009)
45. Lughofer, E., Zavoianu, A.C., Pollak, R., Pratama, M., Meyer-Heye, P., Zörrer, H., Eitzinger, C., Haim, J., Radauer, T.: Self-adaptive evolving forecast models with incremental PLS space updating for on-line prediction of micro-fluidic chip quality. Eng. Appl. Artif. Intell. **68**, 131–151 (2018)
46. Mobley, R.: An Introduction to Predictive Maintenance, 2nd edn. Elsevier, Woburn (2002)

47. Montgomery, D.: Introduction to Statistical Quality Control, 6th edn. Wiley, Hoboken (2008)
48. Myklebust, O.: Zero defect manufacturing: a product and plant oriented lifecycle approach. Procedia CIRP **12**, 246–251 (2013)
49. Nikzad-Langerodi, R., Lughofer, E., Cernuda, C., Reischer, T., Kantner, W., Pawliczek, M., Brandstetter, M.: Calibration model maintenance in melamine resin production: integrating drift detection, smart sample selection and model adaptation. Anal. Chim. Acta **1013**, 1–12 (2018)
50. Niu, G., Yang, B.: Intelligent condition monitoring and prognostics system based on data-fusion strategy. Expert Syst. Appl. **37**(12), 8831–8840 (2010)
51. Palade, V., Bocaniala, C.: Computational Intelligence in Fault Diagnosis. Springer, London (2010)
52. Permin, E., Bertelsmeier, F., Blum, M., Bützler, J., Haag, S., Kuz, S., Özdemir, D., Stemmler, S., Thombansen, U., Schmitt, R., Brecher, C., Schlick, C., Abel, D., Popraw, R., Loosen, P., Schulz, W., Schuh, G.: Self-optimizing production systems. Procedia CIRP **41**, 417–422 (2016)
53. Pola, D., Navarrete, H., Orchard, M., Rabie, R., Munoz, M.C., Olivares, B., Silva, J., Espinoza, P., Perez, A.: Particle-filtering-based discharge time prognosis for lithium-ion batteries with a statistical characterization of use profiles. IEEE Trans. Reliab. **64**(2), 710–720 (2015)
54. Pratama, M., Dimla, E., Tjahjowidodo, T., Lughofer, E., Pedrycz, W.: Online tool condition monitoring based on parsimonious ensemble. IEEE Trans. Cybern. (2018). https://doi.org/10.1109/TCYB.2018.2871120
55. Precup, R.E., Angelov, P., Costa, B.S.J., Sayed-Mouchaweh, M.: An overview on fault diagnosis and nature-inspired optimal control of industrial process applications. Comput. Ind. **74**, 75–94 (2015)
56. Renna, P.: Influence of maintenance policies on multi-stage manufacturing systems in dynamic conditions. Int. J. Prod. Res. **50**(2), 345–357 (2011)
57. Sannen, D., van Brussel, H.: A multilevel information fusion approach for visual quality inspection. Inf. Fusion **13**(1), 48–59 (2012)
58. Sayed-Mouchaweh, M.: Diagnosability, Security and Safety of Hybrid Dynamic and Cyber-Physical Systems. Springer, New York (2018)
59. Sayed-Mouchaweh, M.: Fault Diagnosis of Hybrid Dynamic and Complex Systems. Springer, New York (2018)
60. Sayed-Mouchaweh, M., Lughofer, E.: Learning in Non-Stationary Environments: Methods and Applications. Springer, New York (2012)
61. Si, X.S., Wang, W., Hu, C.H., Zhou, D.H.: Remaining useful life estimation—a review on the statistical data driven approaches. Eur. J. Oper. Res. **213**(1), 1–14 (2011)
62. Skrjanc, I.: Evolving fuzzy-model-based design of experiments with supervised hierarchical clustering. IEEE Trans. Fuzzy Syst. **23**(4), 861–871 (2015)
63. Srivastava, A., Han, J.: Machine Learning and Knowledge Discovery for Engineering Systems Health Management. CRC Data Mining and Knowledge Discovery. Chapman & Hall, Boca Raton (2011)
64. Steinbauer, G., Wotawa, F.: Model-based reasoning for self-adaptive systems—theory and practice. In: Camara, J., de Lemos, R., Ghezzi, C., Lopes, A. (eds.) Assurances for Self-Adaptive Systems, LNCS, vol. 7740, pp. 187–213. Springer, Berlin (2013)
65. Toubakh, H., Sayed-Mouchaweh, M.: Hybrid dynamic data-driven approach for drift-like fault detection in wind turbines. Evol. Syst. **6**(2), 115–129 (2015)
66. Turban, E., Aronson, J., Liang, T.P.: Decision Support Systems and Intelligent Systems, 7th edn. Prentice Hall, Upper Saddle River (2004)
67. Uluyol, O., Parthasarathy, G., Foslien, W., Kim, K.: Power curve analytic for wind turbine performance monitoring and prognostics. In: Proceedings of the Annual Conference of the Prognostics and Health Management Society, pp. 1–8 (2011)
68. Ustundag, A., Cevikcan, E.: Industry 4.0: Managing The Digital Transformation. Springer, Cham (2017)

69. Viharos, Z.J., Csanaki, J., Nacsa, J., Edelenyi, M., Pentek, C., Kis, K.B., Fodor, A., Csempesz, J.: Production trend identification and forecast for shop-floor business intelligence. Acta Imeko **5**(4), 49–55 (2016)
70. Wang, L., Gao, R.: Condition Monitoring and Control for Intelligent Manufacturing. Springer, London (2006)
71. Wang, S., Wang, K., Li, Z.: A review on data-driven predictive maintenance approach for hydro turbines/generators. In: Proceedings of the 6th International Workshop of Advanced Manufacturing and Automation (IWAMA 2016), pp. 30–35. Atlantis Press (2016)
72. Weigl, E., Heidl, W., Lughofer, E., Eitzinger, C., Radauer, T.: On improving performance of surface inspection systems by on-line active learning and flexible classifier updates. Mach. Vis. Appl. **27**(1), 103–127 (2016)
73. Wilson, F., Larry, D., Anderson, G.: Root Cause Analysis: A Tool for Total Quality Management, pp. 8–17. ASQ Quality Press, Milwaukee (1993)
74. Wu, S., Zuo, M.: Linear and nonlinear preventive maintenance. IEEE Trans. Reliab. **59**(1), 242–249 (2010)
75. Yam, R., Tse, P., Li, L., Tu, P.: Intelligent predictive decision support system for condition-based maintenance. The Int. J. Adv. Manuf. Technol. **17**(5), 383–391 (2001)
76. Yang, G.: Life Cycle Reliability Engineering. Wiley, New York (2007)
77. Yusup, N., Zain, A., Hashim, S.: Evolutionary techniques in optimizing machining parameters: review and recent applications. Expert Syst. Appl. **39**, 9909–9927 (2012)
78. Zdsar, A., Dovzan, D., Skrjanc, I.: Self-tuning of 2 DOF control based on evolving fuzzy model. Appl. Soft Comput. **19**, 403–418 (2014)
79. Zhang, Y.M., Jiang, J.: Bibliographical review on reconfigurable fault-tolerant control systems. Annu. Rev. Control **32**(2), 229–252 (2008)
80. Zhu, J., Yoon, J., He, D., Qiu, B., Bechhoefer, E.: Online condition monitoring and remaining useful life prediction of particle contaminated lubrication oil. In: Proceedings of the IEEE Conference on Prognostics and Health Management (PHM), pp. 1–14. IEEE, Gaithersburg (2013)

Smart Devices in Production System Maintenance

Eike Permin, Florian Lindner, Kevin Kostyszyn, Dennis Grunert, Karl Lossie, Robert Schmitt, and Martin Plutz

1 Introduction

The introduction of the iPhone about 10 years ago radically changed the market for mobile phones. Featuring a large screen with a touch display, it combined several functions and features that all required separate devices before. Suddenly, a camera, an MP3 player, a telephone, an internet-ready small computer, and many more could be held in one's hand. It was not the first smartphone to enter the market, but the iPhone was the first successful one, thus kick-starting a market turnover.

Today, more actively sending mobile devices than people can be found in most industrialized countries. The maturity as well as the saturation for these devices can be described as quite high in the majority of these markets. Still, smart devices are mostly used in a private environment—as personal organizer and device for surfing the web, etc. On the other hand, current studies are predicting a high growth potential for such devices in the industrial environment, mostly in automation and factory control: by the year 2025, according to a study from PricewaterhouseCoopers, 75% of all smart devices will be found in the area of industrial automation [1]. This is a major turnover in the market for these devices.

E. Permin · F. Lindner · K. Kostyszyn · D. Grunert (✉) · K. Lossie
Fraunhofer Institute for Production Technology IPT, Aachen, Germany
e-mail: dennis.grunert@ipt.fraunhofer.de

R. Schmitt
Fraunhofer Institute for Production Technology IPT, Aachen, Germany

Laboratory for Machine Tools and Production Engineering (WZL) of RWTH Aachen University, Aachen, Germany

M. Plutz
oculavis GmbH, Aachen, Germany

© Springer Nature Switzerland AG 2019
E. Lughofer, M. Sayed-Mouchaweh (eds.), *Predictive Maintenance in Dynamic Systems*, https://doi.org/10.1007/978-3-030-05645-2_2

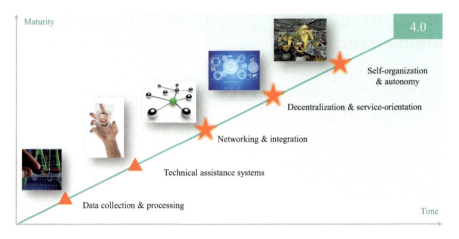

Fig. 1 The stepwise approach towards Industry 4.0

Smart devices are very often perceived as key enablers for company digitalization, as they are rather cheap and provide simple ways to introduce smart capabilities into an industrial environment. In a wider sense, they play a key role as technical assistance systems for the integration of workers in a digitalized factory (see Fig. 1).

Typically, the first step of a producing company towards Industry 4.0—a term, that was introduced in 2011 to describe the endeavors of the federal government and the industry to enable German industry to be prepared for the future of production—or Smart Manufacturing lies in the collection and processing of data, thus turning them into information. Smart devices as technical assistance systems depict the next logical step in the usage of this information on the shop floor and in real time. Integrating and connecting all machinery is typically addressed as a third step, as this means higher efforts regarding machine control, interfaces, networks, and many more. Only then, technically more sophisticated topics such as autonomy or decentralized control can be addressed [2].

When looking at the Industrial Internet of Production, smart devices represent the communication and information exchange layer, thus clustering, evaluating, and aggregating all data coming from the different software layers, machines, and sensors, as depicted in Fig. 2 [3]. Today, most of them are commonly used as representation layer, while storage intensive calculations, etc., are run on external servers and computer. With increasing calculation power and storage, more and more data integration, analytics, and modelling can be achieved locally on these devices.

Four major fields have been identified for the initial industrial application of smart devices: logistics, assembly instructions, quality control, and maintenance. In logistics, glasses can be used as a hands-free option to display real-time information

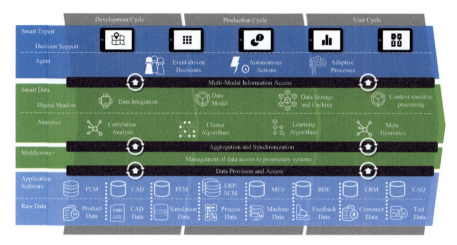

Fig. 2 Internet of production

regarding, e.g., which parts to pick up, where to ship them, etc. First successful applications have already been introduced to the market. In manual assembly, a shift towards more customized or individual products leads to a higher complexity and variety for the workers, which again depicts an interesting field of action for smartglasses: Assembly instructions can be provided locally, with a direct view on the final product. First studies show savings potentials of up to 30% in assembly time when compared to classic, paper-based descriptions [4]. In quality control, smart devices can display evaluation instructions directly to the quality personnel. Through guided processes and direct feedback of pictures or videos, this process again can be digitalized and thus upgraded efficiently.

For maintenance, smart devices and especially glasses depict a promising technical solution to provide instructions and historical information as well as close the feedback loop directly. When it comes to data sources, the maintenance process depends heavily on everything that happened to the specific equipment over time. Starting from first engineering drawings to production information and the service history, the digital twin of the equipment to be maintained plays a major role for the personnel involved, see Fig. 3. The three major distinctions towards the machine or equipment to be maintained are: equipment as planned (design phase), equipment as built (after manufacturing and assembly), and equipment as serviced (history of earlier repairs and services). All of these together constitute the current status of the system to be maintained, while at the same time might be documented in totally different systems [5]. The role of the digital twin as storage tank for all information from the history of a system has thus been stressed extensively in the scientific literature, as, e.g., in [6].

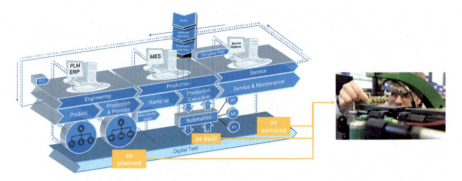

Fig. 3 Different databases and lifecycle steps for a system under maintenance

The following chapter provides an introduction of the possibilities and challenges for smart devices in maintenance processes. The chapter is structured as follows: after the introduction, an overview over the state of the art is given in the second section, including a definition of terms and descriptions of the individual smart devices as well as market shares of the devices and potentials each smart device offers. In the third section, application examples in maintenance are given, including local data analytics and communication for condition monitoring, remote expert solutions, and process data visualization for process monitoring. The fourth section focuses on limitations and challenges smart devices face, including hardware limitations, user acceptance, information compression on smart devices, and legal aspects. The fifth section briefly summarizes the content of the entire chapter.

In the context of the book "Predictive Maintenance in Dynamic Systems," the chapter at hand introduces smart devices as mobile user interfaces, which provide possibilities to integrate humans into modern IT infrastructures in manufacturing companies and thus help humans to take on new roles in maintenance processes. The chapter shows that local data analysis and condition monitoring, process monitoring, and remote expert solutions for maintenance are among the benefits that smart devices provide in the field of predictive maintenance.

2 State of the Art

This section provides a general introduction to smart devices. First, important terms and concepts are presented. Different devices are then categorized and characterized. A view on the market introduces the different vendors and operating systems as well as their importance based on market share. Based on the hardware properties, device selection criteria for different applications and boundary conditions are derived.

2.1 Definition of Terms

In general, smart devices are electronic, mobile devices, which provide functionalities via sensor-based information processing and communication. Smart devices can run applications, programmed for various use cases. With cameras, microphones, and other sensors, they connect humans to the environment and the digital world [7].

For presenting information to humans, Milgram has defined a reality–virtuality continuum, which characterizes different levels of integrating virtual content into the real world. Different devices can be classified on this continuum as shown in Fig. 4 [8].

In virtual reality, the content is separated from the real world by using head-mounted displays [9]. The headset's position is tracked and movements are translated into the virtual reality. The user can interact and manipulate the virtual world with position-tracked controllers, which often represent the users' hands or tools [10]. Assisted reality overlays information, e.g., user manuals or process information with the real world. Mixed reality merges virtual and real world even more than assisted reality. The visual elements are augmented in such a way that they appear to be part of the real world. This requires tracking of the headset's position, which can be either visually or sensor based (e.g., gyroscopes or accelerometers). Visual tracking is often marker based. Without markers, methods of computer vision are used to identify objects and their position. For augmented and mixed reality applications, a mix of these tracking methods is usually used. Augmented reality has the highest degree of merging real and virtual content.

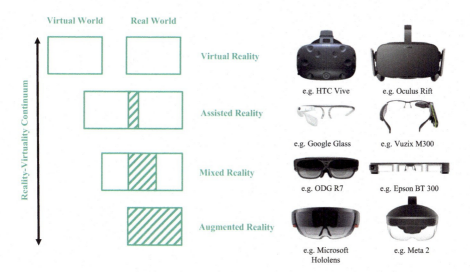

Fig. 4 Different degrees of merging virtual and real world

Devices like Microsoft HoloLens project holograms in the viewer's field of view. Their position can be fixed in real environments, and the user can move freely around the objects.

2.2 Physical Devices/Hardware

There are different types of smart devices, which comply with the definition above. They differ in hardware design but also in functionality. Popular devices used today are smartphones, tablets, smartglasses, and smartwatches. Smartphones and tablets are quite similar and are therefore described together.

2.2.1 Smartphones and Tablets

Smartphones are handheld computer devices, which feature wireless network connectivity (via WLAN and cellular networks) and other wireless technology-like location services (GPS, GLONASS, and Galileo), Bluetooth, and NFC [11, 12]. The telephone function is becoming more and more of a minor matter, in view of the large range of functions provided by smartphones. Smartphone operating systems (i.e., Android and iOS) can run applications programmed for a wide range of industrial use cases. Smartphone CPU and GPU performance has multiplied in recent years. Therefore, they are increasingly capable of running compute-intensive applications. For user interaction, current smartphones generally have large (high-resolution) touchscreens on the front. Most smartphones integrate cameras (front and/or back) and other sensors, like gyroscopes, accelerometers, barometers, proximity sensors, and ambient light sensors [13]. Depending on the CPU and GPU usage, modern smartphones provide between 1 and 2 days of battery life.

The technological innovations stagnated in recent years and improvements are mainly limited to ever-faster processors and better cameras. It can be concluded that smartphones as device category are commonly used in private sectors and characterized by high technological maturity levels.

From a technological perspective, tablets mainly have the same functionalities as smartphones. Usually, they do not provide telephone features. The touchscreen size typically varies between 7 and 13 in. Regarding sensors and wireless connectivity, they are on a par with smartphones. Tablets usually deliver greater performance and battery life due to their larger size, enabling manufacturers to pack larger batteries.

2.2.2 Smartglasses

Apart from their shape, smartglasses are relatively similar to smartphones. They provide similar functionalities in a different construction. Smartglasses (optical head-mounted displays) project information into the user's field of vision through mainly three technologies: optical see-through displays, video see-through displays [14], and retinal projection [15]. It can be distinguished between monocular and binocular smartglasses, whereas monocular smartglasses use a single display unit. In addition, smartglasses are characterized by different levels of combining real and virtual world as described above [16].

Most smartglasses also feature a wide variety of sensors like gyroscopes, accelerometers, microphones, and cameras. They can use the sensors to track their position and orientation in space, which is necessary for augmented and mixed reality applications.

Smartglasses in general suffer from short battery life between 1 and 6 h. Extended battery life can be achieved using wired external batteries [15]. The field of view of smartglasses is still rather small, compared to the human eye (20–60° horizontally vs. 180°), which causes a limited area where the virtual information can be placed without turning the head [17]. Compared to smartphones and tablets, smartglasses are characterized by lower technological maturity. This is mainly due to low battery life, less hardware robustness, and poor ergonomics (low resolution, low field of view, and mostly high weight; see sect. 4).

2.2.3 Smartwatches

Smartwatches are another hardware category used in manufacturing environments. In general, smartwatches are wrist-worn devices, featuring computational power, integrated sensors, and can connect to other devices through the internet [18].

Concerning the hardware, smartwatches mostly integrate touchscreens for information display and interaction. They include sensors, which are, e.g., gyroscopes, accelerometers, barometers, light sensors, and heart rate sensors. Bluetooth and WLAN are used for wireless connectivity and GPS and/or GLONASS for localization [19]. In addition, many watches are waterproof or water resistant. Battery life ranges from one to multiple days, depending on usage and processing requirements of the running applications [18].

Smartwatches are mostly designed to function in interaction with a smartphone. They can relay notifications and alarms from the smartphone to the user's wrist. The smartphone acts as an interface between the watch and external systems such as MES or CAQ-systems [20].

Compared to smartphones and tablets, smartwatches have some key advantages for providing information. Because they are wrist worn, information can be accessed quicker, with less obstruction and "hands-free" [19]. In addition, haptic feedback can be more reliable than acoustic or vibration feedback coming from a smartphone in a person's pocket (e.g., in loud manufacturing environments) [20].

Disadvantages are the low processing power and the small screen. However, smartwatches are still a rather young device category, and huge improvements have been made in the last few years.

2.3 Market View

For assessing vendors and operating systems, it is useful to differentiate between the different device categories. Especially, the smartglasses vendors have little to no overlap to the well-established leaders on the smartphone and tablet market.

Smartphones and Tablets The single biggest vendors for smartphones and tablets are Samsung and Apple with 34.2% in the third quarter 2017 on the smartphone, respectively, 40.8% on the tablet market, see Fig. 5 [21, 22]. Regarding operating systems (OS), there are only two with significant market share: Android and iOS. Android is based on Linux and is developed by the Open Handset Alliance, which is led by Google. In the first quarter 2017, it had a market share of 85.0%. During the same period, iOS, which is Apple's smartphone and tablet OS, had a market share of 14.7% [23]. Most other smartphone and tablet vendors are using Android, which they customize to provide features not available by default.

Smartglasses The market for smartglasses is much smaller, compared to the smartphone and tablet market. In 2016, only 16 million head-mounted displays were sold [24]. In the same time, 1.5 billion smartphones and 175 million tablets were sold [22, 25]. Vendors in the smartglasses market are new startups mixed with traditional electronic manufacturers. Notable manufacturers are, e.g., Atheer, Epson, Google, Meta, Microsoft, ODG, and Vuzix. There are no reliable sales numbers indicating the market leader. What can be said though is that Android is the leading operating system for smartglasses [15]. One important exception is the Microsoft HoloLens, which runs a version of Windows 10. Microsoft has also presented a platform called "Windows Mixed Reality" which provides a framework for AR apps and hardware.

Smartwatches Their sales numbers are projected to double from 2017 to 2021, see Table 1. Apple's WatchOS, Samsung's Tizen, and Google's Wear OS are the relevant operating systems. Whereas WatchOS and Tizen are exclusive to Apple and Samsung, respectively, Wear OS can be used by every interested hardware vendor.

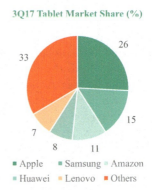

Fig. 5 Smartphone and tablet market share for the third quarter 2017 [21, 22]

Table 1 Sales of wearables as forecasted 2017 (in million units) [24]

Device	2016	2017	2018	2021
Smartwatch	34.80	41.50	48.20	80.96
Head-mounted display	16.09	22.01	28.28	67.17
Total	50.89	63.51	76.48	148.13

Fig. 6 Implementation approach for smart devices

2.4 Device Selection and Potentials

Different categories and smart devices have been established in the previous section. For their application, it is important to match the expected operating conditions with the specific suitability of the device. Choosing a smart device that does not meet the requirements of a specific use case is one of the biggest threats in application projects with wearable devices—even if the use case per se might have big potentials regarding productivity gains.

In order to prevent such mistakes, a guideline will be introduced to specifically guide stakeholders in the process of device selection as well as potential and effort estimation. Figure 6 shows a general approach for an implementation procedure.

The approach follows the phases "Demo," "Proof of Concept," "Pilot project," and "Rollout." Within the first phase, decision makers should find out, whether the targeted combination of software and hardware works in general. This can easily be found out during a short demo that might only take some minutes, therefore has almost no effort but will result in no measurable benefit except for the fact that afterwards the solution has been falsified or verified under laboratory conditions. When this demo phase has been passed, a proof of concept should follow to analyze if the targeted solution also works in the use case's boundary conditions. While the demo might have proved to work under laboratory conditions, realistic conditions might result in the opposite. In case, the solutions also pass this phase of the implementation procedure, the assumed business case should be verified within a proof of concept project. That means that the solution should temporarily be implemented to evaluate if its application over weeks or months results in the desired productivity gains. After the business case has been proven, a rollout is the final step to constantly gather positive productivity effects.

The following criteria are important aspects for device selection but do not claim to be exhaustive. In maintenance, many processes are manual and require the worker to use both hands. In general, devices, which allow hands-free operation, are better suited for these processes. A higher degree of automation usually requires less human interaction, which makes them less suited for smart device support. In addition, the duration of processes is an important factor. Because humans need time (200 ms) to react to visual stimuli [26], longer processes, like maintenance (which normally is not bound to strict cycle times), are better suited for smart device applications. The longest process duration is limited to the device's battery life. Processes which require many and/or very complex steps to perform are better suited for smart device application. The device can provide a detailed explanation and visualization for every step. This is especially useful for processes, which are not performed regularly, like special repair tasks. It is also possible to support workers on how to use the required tools to perform these tasks.

The environment conditions in which the smart devices will be used are an important factor regarding device selection. First, there is temperature, which the manufacturers only guarantee a specific window of operation for the devices to work properly in. The same applies to humidity. This makes some devices unsuitable for rough environment conditions, e.g., maintenance applications in very humid parts of the world. Excessive noise can impede the use of, e.g., voice commands or acoustic feedback, like alarms. When using smart devices, dirt can also have a negative impact, especially for touchscreen usage. The same applies to vibrations. From an organizational perspective, processes with a lot of necessary documentation are better suited for smart device usage. The documentation can be done right on the device, accessing the company's databases, providing seamless integration. Also, if additional data is required for performing the required tasks, smart devices can easily provide with the information. Therefore, processes with a lot of additionally necessary information are better suited for smart device usage.

In the early phases of the introduced implementation approach, device selection can be supported by tools using the described criteria. The tools can prevent decision makers from being already stopped in a demo phase of a solution (see Fig. 7).

Besides an overview of different smart devices that are available at the market and their detailed technical specifications, the tools can offer a questionnaire that allows users to systematically describe their use cases. After sending the questionnaire, a knowledge- and experience-based matching algorithm is applied that gives recommendations about preferred hardware for the entered use case. In addition, a rough estimation of implementation efforts is made which depends on several factors, but especially depends on the integration level into existing IT infrastructure, which usually requires customization and integration programming efforts.

Finally, knowledge exchange between users of the platform is offered to complement the systematic guideline approach with human interactions like commenting the guideline results in order to continuously improve the guideline's underlying heuristics [27].

Fig. 7 Evaluation tool for smartglasses selection [27]

To give an example of the evaluation, two different application scenarios will be presented. The first example is a maintenance application in an indoor automotive assembly environment (climate controlled). It is an unplanned manual repair and therefore requires guidance on the system's components and their interaction. The worker is supported by providing manuals for the equipment. The general process duration is high, compared to, e.g., assembly lines with a fixed cycle time. For the second scenario, a repair task of a construction machine will be considered. The maintenance takes place in hot, humid conditions in the field. There is less routine of the mechanic, because he is not specially trained for the task. Figure 8 shows a comparison of the two applications and their suitability for smart device usage derived from the criteria described above.

These radar charts can provide a decision-making basis for assessing smart device potentials in maintenance and other industry-related use cases. They can be used for a preselection of suitable processes. However, the specific suitability of a process must be examined in detail, as there are always new devices on the market and this classification can only provide an orientation.

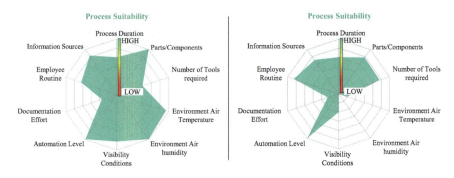

Fig. 8 Machine maintenance in climate-controlled assembly shop (left) compared to maintenance in a construction environment with hot and humid conditions (right)

3 Application Examples in Maintenance

This section provides examples for different applications for the use of smart devices in production system maintenance. First, an example for local data analysis and communication for condition monitoring is given. This includes the presentation of a real-time worker information system as the core of solutions for worker assistance in condition monitoring tasks. Second, the application of smart devices for remote expert solutions is presented. Remote expert solutions enable maintenance engineers to communicate with machine experts via video live stream, to collaborate on fixing problems. Finally, an example is presented which shows how smart devices have changed the way information is displayed to workers in case of process data visualization for process monitoring.

Critical to the successful use of smart devices in the industrial environment is the integration and linking into the relevant system landscape. Instead of a stand-alone solution, planning systems such as enterprise resource planning (ERP), manufacturing execution systems (MES), computer-aided quality (CAQ), and the machine itself exchange data. This is the prerequisite to use smart devices as an integral part in the different applications. Common use cases for smart devices in maintenance applications are found in the area of condition monitoring, remote expert solutions as well as process monitoring [28, 29].

Condition monitoring describes the process of recording machine data for checking the current machine status. This allows the identification of irregularities or errors in the system. Moreover, a condition monitoring system can make predictions about the future system behavior by means of a combination of the analysis of the current system state and historical data. Thus, the monitoring system can detect faulty states early or plan maintenance activities and intervals. Remote expert solutions help to accelerate and improve the maintenance process. Those systems enable engineers to communicate with experts via live streams. Being able to look into the machine while simultaneously displaying all the relevant information on the spot also makes it possible to predict the future state of the machine. Finally, systems connected to other systems such as ERP provide an overview of the state of the entire system on the shop floor. The combination of these three use cases enables companies to use smart devices on a large scale for predictive maintenance.

3.1 Local Data Analysis and Communication for Condition Monitoring

Condition monitoring enables the maintenance engineer to identify the current system status and allows the derivation of future recommendations for action. Prerequisite is the use of real-time machine data in order to interpret and analyze it and to derive actions subsequently. This is useful, e.g., for troubleshooting, maintenance, or predicting future system states. For applications, providing real-time information of the machine to the worker, a direct information exchange between the machine and smart devices is necessary. Information here is often time critical and requires short-term action and intervention options (e.g., in the event of sudden malfunctions or tool changes). Therefore, in the following a real-time worker information system is presented, see Fig. 9. It enables direct communication between smart devices (e.g., tablets and smartglasses) and the machine control. The system supports the machine operator in planned and unplanned maintenance activities.

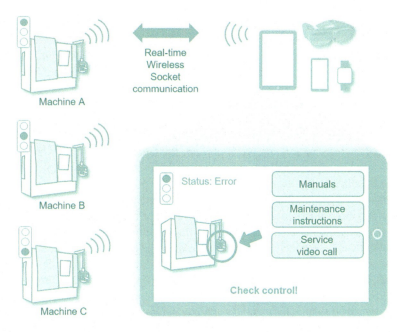

Fig. 9 Communication between machine and device

As stated, data exchange between mobile terminal and machine control is a prerequisite for real-time systems. Direct socket communication enables the data exchange and eliminates a separate arithmetic unit. Smartphones and tablets are used as hardware. Communication with the machine control system takes place, for example, via OPC-UA. For this purpose, the control technologic PLC interface (programmable logic controller) is implemented device-specific. By this, transmission is enabled for many different devices. Other (manufacturer-specific) protocols can also be implemented in the system. A communication protocol ultimately defines the rules and syntax of how data of specific inputs and outputs can be got or set (read and write functions). By constantly retrieving actual data from the controller, real-time information such as machine status or tool condition is transferred and further processed in the information system. The data transport is done wirelessly, e.g., over Wi-Fi.

The information system forms the core of solutions for worker assistance in condition monitoring tasks. In addition to the real-time machine data, the information system also provides a library with specific video manuals and documentation, e.g., manuals in PDF file format. Particularly in the field of "training," users can use video manuals, for example how to replace a tool during maintenance. This is displayed directly on the tablet or the glasses conveniently and location-independent. It gives the worker the information he or she needs without being dependent on paper-based instructions or PC terminals. When using smartglasses, the operator can open these manuals parallel to her or his work, since the glasses provide the information via an integrated, semitransparent display.

The direct communication between the device and machine enables real-time data analysis within the information system. For monitoring reasons, the operator can see the latest machine and order information, such as progress, remaining time, machine status, or overall system effectiveness (OEE). This promises real-time transparency for the employee, because the controller transfers raw data continuously to the smart device. The smart devices then further process and visualize the data locally, condense it into key performance indicators (KPI), and perform automatic updates.

In case of unexpected disturbances, interruptions, or errors, the user automatically receives information, e.g., as a pop-up. Predefined error libraries and codes give the machine operator direct messages via the smart device. Examples of such error and fault information are opened safety devices such as doors or necessary tool changes in the mechanical machining of components. The interpretation of the raw data for a possible incident, the derivation of instructions for the user, and their communication also take place directly in the information system on the smart device. A manual error analysis by error retrieval at the machine terminal is obsolete. This reduces the reaction times in case of unexpected disturbances and can increase the OEE. So, all the required information, such as manuals or repair information for the specific error, are available directly during maintenance.

The described system for machine-related operator support represents a tangible extension of the classical interface between machine control and worker. In the age of Industry 4.0, the system enables direct retrieval, local processing of data as well as the visualization of the correspondingly condensed information via smart devices. In this way, the system supplements the classical human–machine interface (HMI) on the machine terminal by using a flexible and real-time-capable information system directly at the shop floor.

Current challenges and research requirements lie in particular in the integration of different control systems and corresponding communication protocols. In addition to actual data exchange strategies, this also includes the interpretation of the control-specific raw data and the subsequent information compression. In addition, current systems are often limited to communication between a device and a single machine control. In industrial use cases, direct communication from one device to multiple controllers is desirable. For this, it is necessary to define meaningful access routines. For example, a pairing of device and machine could be done via a scan of a machine ID (e.g., QR code) or the automatic recognition of surrounding production machines via Bluetooth or NFC. In the field of predictive maintenance in particular, the predictive models must be further improved and generalized. These improved and generalized models can provide more accurate predictions for wider use cases.

3.2 Remote Expert Solutions

In manufacturing companies, machine downtimes can considerably influence the productivity and result in high costs. New studies show that 82% of surveyed companies have been confronted with unplanned machine failures within the last few years. Most typical reasons were hardware- and software-based malfunctions followed by human errors. On average, machine failures lasted for 4 h and involved costs of two million dollars. Almost 50% of companies believe that downtimes can be decreased when machines are able to request for help by themselves and when they use cloud-based functions to support failure diagnostics [30].

To accelerate machine failure handlings supported by machine experts, mobile devices with remote expert systems can be applied. Those systems enable maintenance engineers to communicate with machine experts via internet connection and provide a live video stream showing the failed machine. Thus, the machine expert can immediately support the troubleshooting and provide professional instructions for a proper failure handling. For maintenance engineers, smartglasses serve as a practical platform for remote expert systems. Their mobility allows them to stay on the shop floor and to use the integrated camera module for sharing their own perspective of the failed machine. The integrated headset enables speech-based communication while being hands-free. Since the machine expert can be consulted immediately without regard to the current location, travel costs can be saved. The machine expert can use a computer or a tablet PC. Similar to classical applications with video conference functionalities, remote expert systems have to be installed on the devices of both conversation partners. As shown in Fig. 10, those systems provide different functions and options depending on the specific device and role of the user.

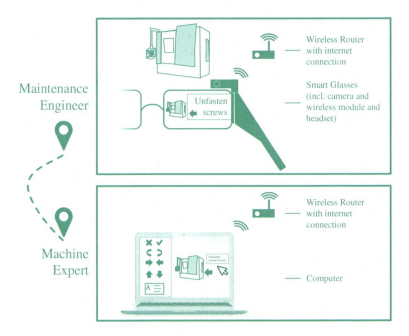

Fig. 10 Connection of maintenance engineer and machine expert via remote expert system

Since most available smartglasses on the market use popular operating systems such as Android, the access to the camera module is standardized. Moreover, remote expert systems connect to a local wireless router with the device-integrated wireless module. The external machine expert, who receives the camera image, can easily guide the maintenance engineer through the troubleshooting and failure handling processes. To support these processes, remote expert systems provide different useful functionalities. For example, the machine expert can add and remove different elements such as symbols, textboxes, images, or checklists to the camera image via drag and drop. Due to automatic synchronization with the smartglasses of the maintenance engineer, those elements will also show up on their screens. This way, the machine expert is able to guide the maintenance engineer to the right spot of the machine and to write comments that can include information about the next steps.

Remote expert systems allow immediate failure handling guided by machine experts and therefore, downtimes and resulting costs can be decreased. For machine suppliers, those systems provide new opportunities to create profitable business models. Up to now, 81% of companies state that aftersales services do not significantly contribute to the profitability due to limited capacities, which result into long reaction times [31]. The ad hoc connection of service employees and customers via a remote expert system can be considered as one feasible approach for an efficient use of personnel capacities.

3.3 Process Data Visualization for Process Monitoring

One essential target of Industry 4.0 is to provide employees with the information they need at the right time to carry out their processes efficiently and to fulfill current quality requirements. This can be realized with modern process data visualization systems that are interconnected with the production's software landscape. Available systems providing CAD, CAM, ERP, MES, MDA, and CAQ functionalities are used as different data sources [4]. Instead of classical computer terminals, smart devices are used, providing a high grade of mobility. With their application, information does not need to be actively requested at a fixed location on the shop floor. Smart devices can display information at any time when it is needed without considering the employee's current location. Audio or vibration signals are typical instruments to gain the employee's attention. Figure 11 illustrates how process visualization systems combine different data sources of the production's software environment. Employees who are equipped with such a system can be provided with various information such as technical product specifications, process and quality data as well as machine condition data. The final choice of information that is visualized on the screen is adapted to the individual needs and functions of the specific system user. For example, machine operators receive information about single processes, while maintenance engineers mainly receive condition data of machines that are under their responsibility.

In case of an occurring machine failure, maintenance engineers can use the integrated menu structure to get access to digital machine handbooks (machine as planned), to specific information about machine components (machine as built),

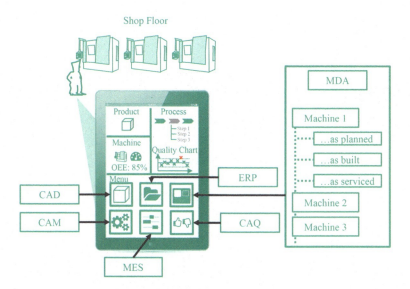

Fig. 11 Process monitoring through mobile devices

and to machine-related maintenance histories (machine as serviced). Such a high availability of information promotes time-efficient troubleshooting and failure handling. With the help of data processing algorithms, various key performance indicators such as the overall equipment effectiveness (OEE) of machines can be calculated and displayed. Predictive algorithms inform machine operators and maintenance engineers about future quality outcomes and machine states. In case of predicted qualities that are outside the tolerances, process chains and parameters can be adapted. Maintenance engineers can be informed about possible machine failures before their occurrence. To provide an impression of modern process monitoring applications, an example is given in the following. Figure 12 shows a system that visualizes a precision glass molding process.

In contrast to grinding and polishing processes for production of optical lenses, precision glass molding describes a replicative molding process. With short cycle times and its ability for production of complex lens geometries with stable qualities, this technology depicts an optimal approach for mass production. In addition, due to the fact that every lens geometry requires a specific and expensive mold, precision glass molding can mostly be found in productions with high output rates of the same products. During the molding process, force and temperature sensors deliver data from different positions within the mold. Since the molding of the glass blank cannot be observed visually, the data acquired during the process is combined with a simulation that visualizes the molding based on a three-dimensional model. Thus, besides the monitoring of the current process, this system supports gaining new process knowledge because it visualizes the correlations between the geometrical specifications of the mold and the resulting forming, forces, and temperatures. This knowledge enables the optimization of process parameters and mold designs. From a maintenance perspective, the acquired data can help to derive the current and future wear state of the tool. This can support maintenance engineers to initiate tool repairs or changes before the output quality is considerably influenced [32, 33].

Fig. 12 Process monitoring during precision glass molding [32]

Process data visualization is an important component of Industry 4.0, which increases process and machine state transparency and therefore, promotes process knowledge building and an increase of experience regarding machine behaviors. From a short-term perspective, visible process data supports operators and maintenance engineers to fulfill predefined quality requirements and to decrease machine downtimes. The long-term application allows continuous optimizations of process parameters and maintenance strategies, e.g., event-based maintenance, that enables preventions of discard, rework, and machine downtimes.

4 Limitations and Challenges

This chapter provides and discusses current limitations and challenges related to the use of smart devices for industrial maintenance applications. First, hardware limitations are discussed. These hardware limitations include human-related limitations like wearing comfort, application-related limitations, which are set by limited accuracies of sensors and cameras, and environment-related limitations like high temperature or dust. The second subsection focusses on user acceptance and emphasizes that an appropriate and practical system design is required to achieve general user acceptance. After that, information compression, which is necessary due to the compact design and reduced possibilities of user interaction on smart devices, is discussed. Finally, legal aspects are considered by describing legislative requirements originating from EU directions referring to safety and health requirements for the workplace, work equipment, and data protection.

4.1 Hardware Limitations

In addition to the variety of possibilities, using smartglasses in industrial environments also leads to certain limitations and challenges. These challenges can be divided into three categories:

- Human-related limitations
- Application-related limitations
- Environment-related limitations

Human-related limitations, for instance, include wearing comfort. Smartglasses should not affect the user's comfort, even if the user wears them over a longer period. Weight of the smartglasses as well as the glasses' fit to the user's head are crucial factors in terms of wearing comfort and will also contribute to the user's acceptance for the smartglasses.

Application-related limitations are limits that result from the hardware sets in terms of accuracy of sensors and cameras. Since sensors have a defined range of measurement inaccuracy, not all glasses are suited for all applications. Smartglasses can only be utilized, if their sensors and cameras fulfill the requirements, which the application defines. For example, if smartglasses are used for technical service through remote expert software, the smartglasses' camera has to be capable of recording high-resolution videos, even when the caller is in motion, so that the receiver of the call can also identify small details (e.g., a tool identification plate) in the streamed video. Another application-related limitation is the battery. The battery life has to be suitable for the applications. As bigger and heavier smartglasses tend to have a bigger battery, the trade-off between comfort and battery life has to be evaluated for every application case.

Environment-related limitations are limits that result from environmental influences on smartglasses. Environmental influences include water, dust, temperature (hot and cold), and atmospheric corrosion. The IP Code provides information on the degree of the device's solid particle protection and liquid ingress protection. Dust and water can damage smartglasses, if they are not selected according to their IP Code. High or low temperatures can also damage smartglasses. As smartphones tend to have a less efficient battery during very low temperatures, the same applies to smartglasses, since they use the same type of lithium-ion battery. Many smartglasses are approved for temperatures near to room temperature. The temperatures of many production facilities exceed these temperature values, though. Another environmental aspect is explosion protection. Electrical devices can possibly become a source of ignition and only the use of certified, intrinsically safe devices is allowed in explosion-hazardous areas. As many smartglasses lack an approval for explosion-hazardous areas, their usage within these areas is strictly limited.

4.2 User Acceptance

Nowadays, working population represents a cross section of different generations and corresponding backgrounds regarding the use of and the familiarity with digital solutions in their daily work. Future digitalized production systems—Smart Factories—require workers to operate with and within the world of data. Here, smart devices represent important interfaces between worker and interconnected production machines and software systems. The worker's specific technical affinity is strongly related to his or her generation and level of training. However, as the smartphone shows, the professional or generation background does not prevent a widespread use of smart devices across almost all sections of population in the private sector.

The industrial sector faces similar characteristics for its future. However, this requires a broad user acceptance by appropriate and practical system design. General requirements for the system design and its interfaces can be derived from ISO 9241—Ergonomics of Human-System Interaction. This series of standards defines boundary conditions and design rules/guidelines for physical aspects such as workplace design and posture. Furthermore, major topics of those standards are related to software ergonomics. Here, aspects such as dialogue management, user interfaces, or interactive system features are considered [34].

Besides generalized guidelines such as ISO 9241, specific end-user requirements need to be taken into account when designing smart device applications for maintenance purposes. A participative approach represents a key success factor. Therefore, maintenance personnel and experts should be actively included during the system development by structured gathering and incorporating their requirements and feedbacks. Figure 13 provides a recommendation of methods to systematically include end-user requirements and feedbacks during different phases of the development, implementation, and rollout of smart maintenance systems-based mobile devices.

Besides the general system design, the user acceptance regarding the implementation of new systems correlates with its level of adaption during the rollout period. The introduction of digital applications in operational processes such as maintenance activities is always related to a change to the employee's way of working. Therefore, change management is a crucial and central aspect for the rollout of smart device applications in maintenance. Experienced maintenance worker might feel left behind or less valued in case their established and proven procedures are replaced—respectively adapted—by new technologies such as interactive failure documentation using augmented reality or guided repair procedures via remote expert solutions. This aspect can be described according to the worker-specific perception of its own competence. A generalized development of this perception during the implementation of changes is visualized in Fig. 14.

It can be seen that the worker's reaction to those changes develops from an initial shock, refusal, stepwise acceptance of the new technologies and a related perception of decreased competence towards a learning curve characterized by improved knowledge and integration. According to this model, the process results in a perception of increased competence [35]. However, it needs to be pointed out that it is in

Planning & Feasibility	Requirements	Design	Implementation	Test & Measure	Post Release
Getting started	User Surveys	Design guidelines	Style guides	Diagnostic evaluation	Post release testing
Stakeholder meeting	Interviews	Paper prototyping	Rapid prototyping	Performance testing	Subjective assessment
Analyse context	Contextual inquiry	Heuristic evaluation		Subjective evaluation	User surveys
ISO 13407	User Observation	Parallel design		Heuristic evaluation	Remote evaluation
Planning	Context	Storyboarding		Critical incidence technique	
Competitor Analysis	Focus Groups	Evaluate prototype		Pleasure	
	Brainstorming	Wizard of Oz			
	Evaluation of existing systems	Interface design patterns			
	Card sorting				
	Affinity diagramming				
	Scenarios of use				
	Task analysis				
	Requirements meeting				

Fig. 13 Recommended methods and approaches towards participative system design

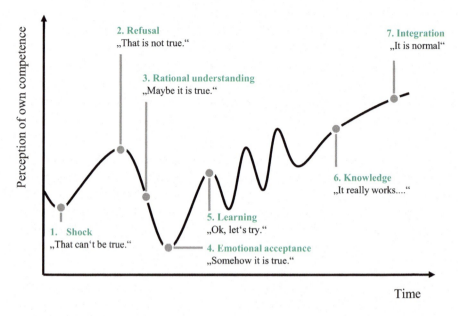

Fig. 14 Perception of competences during the rollout of changes [31]

the nature of things that employees might be skeptical of changes and corresponding implications on their daily work and perception of own competences. Thus, it is all the more important to include the end-user in the development of smart device applications for maintenance following a participative approach as outlined before.

4.3 Information Compression on Smart Devices

Smart devices provide several advantages such as capability for mobile applications. However, due to their compact design and reduced possibilities for user inputs and interactions, the provision of information is not comparable to classic methods such as printed documentation or PC terminals. Documents such as drawings, quality plans or working instructions (e.g., pdfs) are generally provided as extensive information containing all details. Consequently, workers might be overstrained, as documentation needs to be reviewed in terms of relevant information for the very specific task. This aspect represents a major potential for improvement when using smart devices, as the provided content is limited to the relevant information. The direct connection of smart device maintenance applications to superior software and planning systems (e.g., for providing relevant maintenance instructions and checklists) allows to query the very specific and didactically prepared information, rather than entire manuals or process documentation. However, this requires information compression on the chosen device.

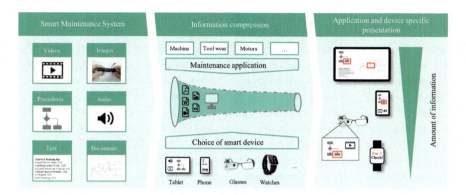

Fig. 15 Scheme for information compression on different devices

Depending on the device technology, display sizes, resolution as well as general features such as audio recording, playback, or vibration capability vary. A current research focus is therefore related to the device- and user-dependent compression of information for industrial applications such as maintenance support via digital repair plans or ad hoc documentation via mobile devices. The general scheme of this information compression is outlined in Fig. 15.

It is necessary to identify the trade-off between loss of information and mental overload through unnecessary or redundant information. While detailed process descriptions or system plans can be provided and interpreted via tablets through intuitive operations such as scrolling or zooming, the use of smartglasses or smartwatches for the same information could lead to confusion rather than support. Consequently, maintenance procedure might even be delayed and more complicated. Here, short and concise requests for subtasks as text instructions, videos, or schematic sketches could be used instead of extensive documentation.

4.4 Legal Aspects

Further challenges for the implementation of smart devices for maintenance activities are also related to legal conditions. Some of those legislative requirements originating from EU directions shall be presented at this stage. Those directions particularly refer to safety and health requirements for the workplace and work equipment as well as data protection.

For countries of the European Union, 89/654/EEC defines minimum safety and health requirements for the workplace [36]. This document also defines specific requirements for mobile virtual display units. Therefore, this regulation is also applicable for maintenance activities supported by smartglasses or tablets. As an example, this document requires a temporally limited use of those mobile display units, except there are no other technical solutions available for the specific tasks.

However, in case they are required and need to be used to execute specific maintenance operations as work equipment, they are covered by 2009/104/EC, which defines minimum safety and health requirements for the use of work equipment [37]. According to this regulation, the employer is required to assess the functional safety of those mobile devices on a regular basis to avoid any hazards for the workers. Defect devices, e.g., indicated by hotter battery systems, shall not be used anymore as they represent potential hazards. In general, employers are required to measure and to achieve improvements in the safety and health of workers at work. For the European Union, this aspect is regulated by 89/391/EEC [38]. The employer has to assess the working conditions for maintenance activities and to introduce countermeasures in case of any hazards. Therefore, personal protective equipment (PPE) is mandatory—especially in the field of maintenance. However, in case smart devices are used for maintenance purposes, the compatibility of smart devices and PPE needs to be guaranteed. Due to this reason, there are different smartglasses on the market that can be easily combined with PPE such as helmets.

In the age of Industry 4.0 and fully connected digitalized manufacturing systems, data protection plays a crucial role. Especially, from employee and work council perspective, the protection of individuals with regard to the processing of personal data is of major interest. For the European Union, this aspect is regulated in 95/46/EC, respectively, 2016/679 regulation. Personal data can clearly be related to a specific person, respectively, worker [39, 40]. Personally identifiable data acquisition, data processing, and data use can only be considered as legal if the affected person authorizes it or if it is legally required or allowed. The acquisition of this kind of data—e.g., for assessing the employee's performance or working speed by use of mobile devices—is critical and should be punished according to this regulation. Even the recording of other (uninvolved) persons when using smartglasses during maintenance documentation requires the explicit permission of the specific person. Therefore, measures need to be executed to guarantee anonymity of data and a limitation of recorded data that is coherent with legislative boundary requirements when introducing smart devices for maintenance purposes.

It can be concluded that several legislative requirements are formulated for work equipment, working conditions, and data acquisition and processing to guarantee safety standards as well as sufficient data protection. Some of those regulations are applicable to the use of smart devices for maintenance applications. To be coherent with those regulations, measures and data acquisition and processing strategies need to be implemented to ensure a lawful implementation of smart devices.

5 Summary

This chapter described the role of smart devices in production system maintenance. With their specific features such as mobility, interconnectivity, and processing performance, they serve as technical assistance systems and promote the integration of workers into the digital factory.

In a definition of terms and a subsequent market view, different types and technologies of smartphones, tablets, smartglasses, and smartwatches were introduced. Besides technical differences, leading providers of different operating systems and hardware were named.

To underline the key role of smart devices in terms of modern maintenance, three application examples were described. The first one described the interface between smart devices and machines through direct socket communication. Specific benefits of local data analysis and condition monitoring were pointed out. Occurring machine downtimes can be decreased with the application of remote expert system that connects maintenance engineers with machine experts through the internet by using smart devices. Their advantages and functionalities were described in the second example. The last example introduced data visualization systems for process monitoring on smart devices as mobile solutions to provide maintenance engineers with machine information without regarding their current location on the shop floor.

In the last part of this chapter, limitations and challenges that are connected to the integration of smart devices into the shop floor were discussed. For each specific application, a different technology can be seen as most suitable. Technical differences and limitations of smart devices are defined by their specific operating systems, technical interfaces, calculation of power, storage, and battery capacities, and by their environmental working conditions. Moreover, the application of smart devices and software can be restricted regarding user acceptance, information compression, and legal aspects.

Considering all existing limitations and requirements, smart devices are forward-looking technologies that promote location-independent and need-based information exchanges on the shop floor. For maintenance, those advantages are crucial to enable quick reactions to machine failures and to unacceptable changes of machine conditions.

References

1. PricewaterhouseCoopers: The internet of things: what it means for US manufacturing. https://www.pwc.se/sv/publikationer/verkstad/the-internet-of-things.html (2015). Accessed 19 Mar 2018
2. Kagermann, H., Wahlster, W., Helbig, J.: Recommendations for implementing the strategic initiative INDUSTRIE 4.0: final report of the Industrie 4.0 Working Group, Berlin (2013)
3. Schuh, G., Brecher, C., Klocke, F., et al. (eds.): Engineering Valley - Internet of Production Auf Dem RWTH Aachen Campus, 1st edn. Aachen, Apprimus Verlag (2017)
4. Schmitt, R., Permin, E., Kerkhoff, J., et al.: Enhancing resiliency in production facilities through cyber physical systems. In: Jeschke, S., Brecher, C., Song, H., et al. (eds.) Industrial Internet of Things: Cybermanufacturing Systems, pp. 287–313. Springer, Cham (2017)
5. Schmitt, R., Bihler, S., Bork, H., et al.: Agile, data-based process design. In: Schuh, G., Brecher, C., Klocke, F., et al. (eds.) AWK Aachen Werkzeugmaschinen-Kolloquium 2017 Internet of Production für agile Unternehmen, 1st edn, pp. 389–407. Apprimus Verlag, Aachen (2017)

6. Tao, F., Cheng, J., Qi, Q., et al.: Digital twin-driven product design, manufacturing and service with big data. Int. J. Adv. Manuf. Technol. **94**(9–12), 3563–3576 (2018). https://doi.org/10.1007/s00170-017-0233-1
7. Niehues, M., Reinhart, G., Schmitt, R.H., et al.: Organisation, Qualität und IT-Systeme für Planung und Betrieb. In: Reinhart, G. (ed.) Handbuch Industrie 4.0: Geschäftsmodelle, Prozesse, Technik, pp. 137–168. Hanser, München (2017)
8. Milgram, P., Takemura, H., Utsumi, A., et al.: Augmented reality: a class of displays on the reality-virtuality continuum. Proc. SPIE. **2351**, 282–292 (1995). https://doi.org/10.1117/12.197321
9. Sicaru, I.A., Ciocianu, C.G., Boiangiu, C.A.: A survey on augmented reality. J. Inf. Syst. Oper. Manag. **11**(2), 263–279 (2017)
10. Gavish, N., Gutiérrez, T., Webel, S., et al.: Evaluating virtual reality and augmented reality training for industrial maintenance and assembly tasks. Interact. Learn. Environ. **23**(6), 778–798 (2013). https://doi.org/10.1080/10494820.2013.815221
11. Apple Inc: iPhone X - technical specifications. https://www.apple.com/iphone-x/specs/ (2018). Accessed 12 Mar 2018
12. Samsung: Samsung Galaxy Note 8 specifications. https://www.samsung.com/us/galaxy/note8/specs/ (2018). Accessed 12 Mar 2018
13. Agu, E., Pedersen, P., Strong, D., et al.: The smartphone as a medical device: assessing enablers, benefits and challenges. In: Knightly, E.W. (ed.) 2013 10th Annual IEEE Communications Society Conference on Sensor, Mesh and Ad Hoc Communications and Networks (SECON): 24–27 June 2013, New Orleans, LA, USA, pp. 76–80. IEEE, Piscataway (2013)
14. Azuma, R., Baillot, Y., Behringer, R., et al.: Recent advances in augmented reality. IEEE Comput. Grap. Appl. **21**(6), 34–47 (2001). https://doi.org/10.1109/38.963459
15. Syberfeldt, A., Danielsson, O., Gustavsson, P.: Augmented reality smart glasses in the smart factory: product evaluation guidelines and review of available products. IEEE Access. **5**, 9118–9130 (2017). https://doi.org/10.1109/ACCESS.2017.2703952
16. Palmarini, R., Erkoyuncu, J.A., Roy, R., et al.: A systematic review of augmented reality applications in maintenance. Robot. Comput. Integr. Manuf. **49**, 215–228 (2018). https://doi.org/10.1016/j.rcim.2017.06.002
17. Kishishita, N., Kiyokawa, K., Orlosky, J., et al.: Analysing the effects of a wide field of view augmented reality display on search performance in divided attention tasks. In: Julier, S. (ed.) IEEE International Symposium on Mixed and Augmented Reality (ISMAR), 2014: 10–12 Sept. 2014, Munich, Germany, pp. 177–186. IEEE, Piscataway (2014)
18. Bieber, G., Kirste, T., Urban, B.: Ambient interaction by smart watches. In: Makedon, F. (ed.) Proceedings of the 5th International Conference on PErvasive Technologies Related to Assistive Environments. ACM, New York (2012)
19. Rawassizadeh, R., Price, B.A., Petre, M.: Wearables: has the age of smartwatches finally arrived? Commun. ACM. **58**(1), 45–47 (2014). https://doi.org/10.1145/2629633
20. Lee, J.: Bosch and Industry 4.0: smartwatches on assembly lines. https://blog.bosch-si.com/industry40/smartwatches-assembly-lines/ (2015). Accessed 16 Mar 2018
21. Gartner: Gartner Says Top Five Smartphone Vendors Achieved Growth in the Third Quarter of 2017. https://www.gartner.com/newsroom/id/3833964 (2017). Accessed 31 Jan 2018
22. IDC: Tablet market declines 5.4% in third quarter despite 4 of top 5 vendors showing positive year-over-year growth, according to IDC. https://www.idc.com/getdoc.jsp?containerId=prUS43193717 (2017). Accessed 31 Jan 2018
23. IDC: IDC: smartphone OS market share. https://www.idc.com/promo/smartphone-market-share/os (2017). Accessed 31 Jan 2018
24. Gartner: Gartner says worldwide wearable device sales to grow 17 percent in 2017. https://www.gartner.com/newsroom/id/3790965 (2017). Accessed 31 Jan 2018
25. Gartner: Gartner says worldwide sales of smartphones grew 7 percent in the fourth quarter of 2016. https://www.gartner.com/newsroom/id/3609817 (2017). Accessed 31 Jan 2018

26. Jain, A., Bansal, R., Kumar, A., et al.: A comparative study of visual and auditory reaction times on the basis of gender and physical activity levels of medical first year students. Int. J. Appl. Basic Med. Res. **5**(2), 124–127 (2015). https://doi.org/10.4103/2229-516X.157168
27. oculavis GmbH: The smart glasses guide. http://smartglasses.guide/ (2018) Accessed 21 Mar 2018
28. Lindner, F., Kostyszyn, K., Grunert, D., et al.: Smart Devices in der Fertigung. ZWF. **112**(10), 662–665 (2017). https://doi.org/10.3139/104.111803
29. Lindner, F., Permin, E., Grunert, D., et al.: Smarte Informationssysteme für den Maschinenbediener. ZWF. **112**(7–8), 515–517 (2017). https://doi.org/10.3139/104.111756
30. Vanson Bourne: After the fall: the costs, causes and consequences of unplanned downtime: full report. http://lp.servicemax.com/Vanson-Bourne-Whitepaper-Unplanned-Downtime-LP.html (2017). Accessed 19 Mar 2018
31. McKinsey & Company: The future of German mechanical engineering operating successfully in a dynamic environment: full report. https://www.mckinsey.com/industries/automotive-and-assembly/our-insights/the-future-of-german-mechanical-engineering-operating-successfully-in-a-dynamic-environment (2014). Accessed 19 Mar 2018
32. Fraunhofer-Institute for Production Technologie: Industrie 4.0 erlaubt Blick in die Präzisionsblankpresse. https://www.fraunhofer.de/de/presse/presseinformationen/2016/Juni/industrie40-erlaubt-blick-in-die-praezisionsblankpresse.html (2016). Accessed 19 Mar 2018
33. Georgiadis, K.: The Failure Mechanisms of Coated Precision Glass Molding Tools, 1. Auflage. Prozesstechnologie, Band 41/2015. Apprimus Verlag, Aachen (2015)
34. International Organization for Standardization: Ergonomic requirements for office work with visual display terminals (VDTs) – part 11: guidance on usability (ISO 9241-11). (1998)
35. Streich, R.K.: Fit for Leadership. Springer Fachmedien Wiesbaden, Wiesbaden (2016)
36. The Council of the European Communities: Council Directive 89/654/EEC of 30 November 1989 concerning the minimum safety and health requirements for the workplace (first individual directive within the meaning of Article 16 (1) of Directive 89/391/EEC)(89/654/EEC). (1989)
37. The European Parliament and the Council of the European Union: Directive 2009/104/EC – use of work equipment of 16 September 2009 concerning the minimum safety and health requirements for the use of work equipment by workers at work (second individual Directive within the meaning of Article 16(1) of Directive 89/391/EEC)(2009/104/EC). (2009)
38. The Council of the European Communities: Directive 89/391/EEC - OSH "Framework Directive" of 12 June 1989 on the introduction of measures to encourage improvements in the safety and health of workers at work – "Framework Directive"(89/391/EEC). (1989)
39. The European Parliament and the Council of the European Union: Directive 95/46/EC of the European parliament and of the council of 24 October 1995 on the protection of individuals with regard to the processing of personal data and on the free movement of such data(95/46/EC). (1995)
40. The European Parliament and the Council of the European Union: Regulation (EU) 2016/679 of the European parliament and of the council of 27 April 2016 on the protection of natural persons with regard to the processing of personal data and on the free movement of such data, and repealing Directive 95/46/EC (General Data Protection Regulation) (2016/679). (2016)

On the Relevance of Preprocessing in Predictive Maintenance for Dynamic Systems

Carlos Cernuda

1 Introduction

Nowadays the volume of data is exploding, and the costs of collecting, storing, and treating them are affordable for many, making big data solutions more science and less fiction. In this world submerged by a data tsunami, predictive maintenance is not an exception. In fact the advances in cheaper, smaller, and much more accurate sensors development, together with highly sophisticated communication protocols, have widely contributed to a continuous rise of data-driven approaches in predictive maintenance.

In any data-driven application in general, thus for predictive maintenance in particular, preprocessing [132] is of uppermost importance in order to make the data meaningful and usable, driving the path from potential to real information. Depending on the author, preprocessing can take different meanings. Some separate, for instance, data compression approaches, such as feature selection, from preprocessing. We will consider any treatment performed to the data before training a model as *preprocessing*. Then, data cleaning, noise filtering, normalizing, and feature selection are part of it, among others.

Therefore, we can think of preprocessing as a step formed by several steps, each of them with a particular purpose, whose order could be sometimes interchanged but in which the commutative property is in general not fulfilled. Considering the amount of possible steps, the variety of possible approaches per step, and the non-commutativity between them, the amount of options explodes existing no

---

C. Cernuda (✉)
BCAM - Basque Center for Applied Mathematics, Bilbao, Spain

Faculty of Engineering (MU-ENG), Mondragon Unibertsitatea, Arrasate, Spain
e-mail: ccernuda@bcamath.org; ccernuda@mondragon.edu

© Springer Nature Switzerland AG 2019
E. Lughofer, M. Sayed-Mouchaweh (eds.), *Predictive Maintenance in Dynamic Systems*, https://doi.org/10.1007/978-3-030-05645-2_3

guaranty that a combination of preprocessing actions would behave better than no preprocessing the raw data at all [39].

Data involved in each problem related to predictive maintenance have specific properties. For instance, data related to fault detection tend to be highly imbalanced because the information regarding faulty situations is much less frequent than the one regarding fault-free situations. In general, the properties of the data should be taken into account when choosing a preprocessing strategy. Unfortunately the task does not provide enough information, meaning that not all datasets used for a task have the same properties. For example, not all datasets for remain useful life (RUL) prediction problems are the same. The properties of each dataset have to be determined. Moreover, sometimes, with the same properties, a preprocessing scheme works for one problem and not for another. Some general hints are provided in the definitions of the different strategies.

In predictive maintenance accurate models are necessary, but accurate today could become inaccurate tomorrow, making robust long-lasting models also a requirement, especially in highly dynamic systems. Proper preprocessing strategies are the foundation of the construction of a robust accurate model.

The rest of the work is as follows. Section 2 establishes a taxonomy, provides brief but beyond a mere citation descriptions of several techniques for each of the preprocessing steps following the previously provided taxonomy, and presents several modeling techniques meant for system monitoring in predictive maintenance. Section 3 fully describes the datasets that define the different scenarios, the complete experimental setup, as well as the evaluation schemata that would allow for a fair comparison of the proposed pretreatment configurations, and comments about the results achieved. Finally, Sect. 4 concludes the study.

2 Preprocessing

We define *preprocessing* as the set of actions performed to raw data prior to a subsequent modeling performance, with the aim of improving the modeling capabilities. The improvement could be understood in several ways, such as increasing accuracy, increasing robustness, shortening computational time, decreasing memory and/or computational power requirements, or reducing monetary costs.

The perfect result would be a combination of several of those (usually conflicting) objectives, leading to multi- and many-objective solutions (in which an algorithm is trained in order to find the best preprocessing strategy) that are far beyond the scope of this work. Generally, the objectives are dependent on the problem and the final user requirements. Therefore, we will focus separately on accuracy and robustness, assuming that the methods are fast enough for our requirements as well as affordable in time, technical resources, and money.

2.1 Taxonomy

The taxonomy we are presenting here is an ordered taxonomy, meaning that the steps, if included in our strategy, should be performed in the given order. Since some of those procedures deal with some calculations using the data (e.g., averages), then any transformation made would affect those calculations in the subsequent steps, which could lead to different resulting actions. We first present the six preprocessing steps, and then we develop in detail the most relevant approaches in each one of them.

1. **Data cleansing**. Most data-driven techniques rely on the supposition of complete, reliable noise-free data. But real-world data are not such ideal clean data, being necessary to define strategies to deal with outliers and noise. Moreover, due to the nature of the data or due to a lack of an adequate data acquisition strategy, redundant or irrelevant features could be considered in the dataset, which could be treated both in the data cleansing step or later in the feature engineering step.[1] Despite expert knowledge could be extremely helpful for data cleansing, we assume a lack of it so that we focus on data-based strategies. Besides, some of the parts of the taxonomy are interconnected. For instance, noise treatment is usually attempted through filtering (data transformation) or compression (data engineering), as well as redundancy and irrelevancy, which are usually overcome through data engineering. Therefore, those cases will be treated in their corresponding steps, being the link mentioned.
2. **Data normalization**. Data coming from diverse heterogeneous origins is collected with ease, which makes actual datasets a compendium of datasets obtained in different parts of the system in different manners. This datasets fusion, known as data integration, is not considered by many authors (including us) as part of preprocessing, but as part of data collection. Some algorithms are highly sensitive to the variety of scales and ranges of the variables, which could lead to a performance degradation if no homogenization is performed in the data.
3. **Data transformation**. Despite the previous steps and some of the posterior ones imply indeed transformations of the data, we reserve this name for transformations in the data by means of certain functions, motivated by knowledge about the system. For instance, if we are performing predictive maintenance of certain industrial machinery by using information about the chemical composition of residual wastes by spectroscopic data (named chemometric multivariate calibration), we can use Beer–Lambert law to realize that the relationship between the chemical composition and the absorbance spectroscopic data (obtained by a

---

[1] It is not irrelevant when the treatment happens, because there are several steps between data cleansing and feature engineering that could be very sensitive to redundancies or heavily affected by features that in the end are irrelevant.

logarithmic transformation) is linear. Therefore, the transformation is beneficial for the posterior use of a linear monitoring technique.
4. **Missing values treatment**. Due to several possible causes some values of certain variables could be missing. A naïve approach is to ignore any sample containing a missing value, but sometimes the amount of samples is small or, in case of imbalanced data, the minority class could become more minor even if we adopt such a destructive approach. The obvious alternative is filling the holes, but how? Depending on the size and intrinsic characteristics of the data, the filling strategy could be tricky.
5. **Feature engineering**. There is not a standard definition of feature engineering. By it, we mean the employment of one or more of: *feature selection* (determination of the most important features according to certain quality criteria), *feature extraction* (creation of new features from some or all of the original ones), and *discretization* (transforming continuous features into discrete ones by using bins).
6. **Imbalanced data treatment**. If our predictive maintenance problem is supervised so that certain type of samples are extremely rare compare to the others (minority class), then we are facing an imbalanced learning problem. There are two logical ways to proceed: (1) balancing somehow the data, and (2) compensate giving somehow more importance to the samples from the minority class. The former is related to *sampling* techniques, and the latter to *weighting* techniques.

2.2 Data Cleansing

Data cleansing is a complicated task in which we frequently have to make strong assumptions. Some of those assumptions might hold theoretically but not in real-world data. Therefore, sometime we walk on quicksand. An example we will show right afterwards is the implicit assumption of Gaussian behavior when applying outlier detection based on Mahalanobis distance. As aforementioned, data cleansing deals with several data artifacts, such as outliers, noise, redundancy, or irrelevancy.

The detection of *outliers* understood as feature values that are too far from the general acceptable trend, and the posterior action on those identified outliers is a tricky task. First, how do we identify the *general acceptable trend*? Second, how do we quantify what *too far* means? Most of the approaches are based on thresholds from distances in certain representation of the feature space.

We will consider two important approaches, which relevance comes not only because they are widely used but also because they can be updated incrementally for data streams. They are based on Mahalanobis distance [77], and on chi-square approximations of the orthogonal (Q) and score (T^2) distances from principal components analysis (PCA) [61]. As it is indeed an orthogonal transformation, PCA will be briefly described in Sect. 2.4.

2.2.1 Outlier Detection Based on Mahalanobis Distance

Mahalanobis distance [77] is defined for two vectors $\mathbf{x}_i$ and $\mathbf{x}_j$ as

$$d_M(\mathbf{x}_i, \mathbf{x}_j) = \sqrt{(\mathbf{x}_i - \mathbf{x}_j)^T \Sigma^{-1} (\mathbf{x}_i - \mathbf{x}_j)} \qquad (1)$$

It takes into account the covariance matrix Σ, where Σ_{ij} is the covariance between $\mathbf{x}_i$ and $\mathbf{x}_j$ and Σ_{ii} is the variance of $\mathbf{x}_i$. Thus we are considering elliptic regions, instead of circular ones, of equidistant points. Figure 1 shows a 2-D example where the point marked with the red square would not be considered as an outlier according to Euclidean distance, but it would be in terms of Mahalanobis distance, which seems to be more reasonable.

The outlier identification procedure consists in calculating the Mahalanobis distance from each sample to a central point and checks whether it exceeds certain threshold. The mean is the classical central measure, but it is not robust against outliers. Also the covariance matrix is not a robust dispersion measure. The robustness can be assumed if the number of samples is quite big, that is usually the case in predictive maintenance. Robust alternatives to the mean and the covariance matrix are, respectively, the *robust location estimator* and the *minimum covariance determinant*, which are the mean and covariance matrix of a subset of the original dataset. For further information, see [116].

If we denote by $\mathbf{x}_c$ the chosen center and by Σ_c the chosen dispersion matrix, then Mahalanobis distance from a sample $\mathbf{x}_i$ to the center is given by

$$d_M(\mathbf{x}_i) = \sqrt{(\mathbf{x}_i - \mathbf{x}_c)^T \Sigma_c^{-1} (\mathbf{x}_i - \mathbf{x}_c)} \qquad (2)$$

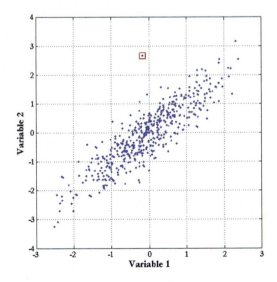

Fig. 1 Example of outlier according to Mahalanobis distance that would not be so according to Euclidean. Considering Euclidean distance, the lines of points with a constant distance to a central point form a circumference. But considering Mahalanobis distance, the shape of those line is elliptical, and adapted to the overall shape of the cloud of points

Assuming that the multivariate data follows a multivariate normal distribution, then the squared Mahalanobis distance follows a χ_N^2 distribution, with N the number of variables. Then a sample would be considered as an outlier if its distance to the mean is higher than the threshold given by a α quantile, $\chi_{N,\alpha}^2$.

For the incremental case, we just need to be able to incrementally update the inverse of the covariance matrix, which is defined as

$$\Sigma_N = \frac{1}{N} \sum_{i=1}^{N} (x_i - \overline{X}_N) \cdot (x_i - \overline{X}_N)^T \tag{3}$$

Then, for the extended data stream considering an extension with one single sample,

$$\Sigma_{N+1} = \frac{1}{N+1} \sum_{i=1}^{N+1} (x_i - \overline{X}_{N+1}) \cdot (x_i - \overline{X}_{N+1})^T$$

If we split the sum in two parts, from 1 to N and $N+1$, we get

$$\Sigma_{N+1} = \frac{1}{N+M} \sum_{i=1}^{N} (x_i - \overline{X}_{N+1}) \cdot (x_i - \overline{X}_{N+1})^T + \frac{1}{N+1} (x_i - \overline{X}_{N+1}) \cdot (x_i - \overline{X}_{N+1})^T$$

We denote both addends as A_1 and A_2, respectively, and expand them separately.

Firstly,

$$A_1 = \frac{1}{N+1} \sum_{i=1}^{N} (x_i - \overline{X}_{N+1}) \cdot (x_i - \overline{X}_{N+1})^T$$

Taking into account that the incremental update of the mean is given by

$$\overline{X}_{N+1} = \frac{N \overline{X}_N + x_{N+1}}{N+1} \tag{4}$$

Then,

$$-\overline{X}_{N+1} = -\overline{X}_N - \frac{1}{N+1} (x_{N+1} - \overline{X}_N)$$

Denoting $\overline{C} := x_{N+1} - \overline{X}_N$, and substituting,

$$A_1 = \frac{1}{N+1} \sum_{i=1}^{N} \left(x_i - \overline{X}_N - \frac{1}{N+1} \overline{C} \right) \cdot \left(x_i - \overline{X}_N - \frac{1}{N+1} \overline{C} \right)^T$$

As $(A - B) \cdot (A - B)^T = A \cdot A^T - A \cdot B^T - B \cdot A^T + B \cdot B^T$, and $\overline{C}$ is constant, then

$$A_1 = \frac{1}{N+1} \sum_{i=1}^{N} (x_i - \overline{X}_N) \cdot (x_i - \overline{X}_N)^T$$

$$- \frac{1}{(N+1)^2} \sum_{i=1}^{N} (x_i - \overline{X}_N) \cdot \overline{C}^T$$

$$- \frac{1}{(N+1)^2} \sum_{i=1}^{N} \overline{C} \cdot (x_i - \overline{X}_N)^T$$

$$+ \frac{1}{(N+1)^3} \sum_{i=1}^{N} \overline{C} \cdot \overline{C}^T$$

$$= \frac{N}{N+1} \Sigma_N + \frac{N}{(N+1)^3} \overline{C} \cdot \overline{C}^T$$

Secondly,

$$A_2 = \frac{1}{N+1} (x_{N+1} - \overline{X}_{N+1}) \cdot (x_{N+1} - \overline{X}_{N+1})^T$$

From Eq. (4), we know that

$$-\overline{X}_{N+1} = -x_{N+1} + \frac{N}{N+1} (x_{N+1} - \overline{X}_N)$$

Therefore,

$$A_2 = \frac{N^2}{(N+1)^3} \overline{C} \cdot \overline{C}^T$$

Then, since $\overline{C} := x_{N+1} - \overline{X}_N$,

$$\Sigma_{N+1} = \frac{N}{N+1} \Sigma_N + \frac{N}{(N+1)^2} (x_{N+1} - \overline{X}_N) \cdot (x_{N+1} - \overline{X}_N)^T \qquad (5)$$

In order to obtain the inverse of the covariance matrix, one option is to update the covariance matrix and calculate its inverse. This requires a huge computational effort unless the number of variables is very low, which is not usually the case. Therefore, a direct update of the inverse covariance matrix is preferable.

The properties of the matrices involved in Eq. (5) allow us to compute the new inverse as a perturbation of the old one by using the following Lemma [1].

Lemma 1 (General Sherman–Morrison Formula) *Suppose $A \in \mathcal{M}_n$ is an invertible matrix, and v and w are vectors of length n so that $1 + w^T A^{-1} v \neq 0$. Then,*

$$\left(A + v \cdot w^T\right)^{-1} = A^{-1} - \frac{A^{-1} v \cdot w^T A^{-1}}{1 + w^T A^{-1} v} \tag{6}$$

where $v \cdot w^T$ is the outer product of v and w.

If we identify $A := \frac{N}{N+1} \Sigma_N$, $v := \frac{N}{(N+1)^2}(x_{N+1} - \overline{X}_N)$, and $w := x_{N+1} - \overline{X}_N$, then we have

$$1 + w^T A^{-1} v = 1 + \frac{1}{N+1}(x_{N+1} - \overline{X}_N)^T \Sigma_N^{-1}(x_{N+1} - \overline{X}_N)$$

that is never null because Σ_N is positive semi-definite, then so its inverse.

In Eq. (5), inverting both sides

$$\Sigma_{N+M}^{-1} = \left(\frac{N}{N+1}\Sigma_N + \frac{N}{(N+1)^2}(x_{N+1} - \overline{X}_N) \cdot (x_{N+1} - \overline{X}_N)^T\right)^{-1} \tag{7}$$

that corresponds to the left part of (6) in the Lemma, with the previous identifications of A, v, and w.

By Sherman–Morrison formula,

$$\Sigma_{N+1}^{-1} = \frac{N+1}{N} \Sigma_N^{-1} - \frac{\frac{N+1}{N}\Sigma_N^{-1} \cdot \frac{N}{(N+1)^2}(x_{N+1} - \overline{X}_N) \cdot (x_{N+1} - \overline{X}_N)^T \cdot \frac{N+1}{N}\Sigma_N^{-1}}{1 + (x_{N+1} - \overline{X}_N)^T \cdot \frac{N+1}{N}\Sigma_N^{-1} \cdot \frac{N}{(N+1)^2}(x_{N+1} - \overline{X}_N)} \tag{8}$$

Therefore, taking common factor $\frac{N+1}{N}$, we get

$$\Sigma_{N+1}^{-1} = \frac{N+1}{N} \cdot \left(\Sigma_N^{-1} - \frac{\Sigma_N^{-1}(x_{N+1} - \overline{X}_N) \cdot (x_{N+1} - \overline{X}_N)^T \Sigma_N^{-1}}{(N+1) + (x_{N+1} - \overline{X}_N)^T \Sigma_N^{-1}(x_{N+1} - \overline{X}_N)}\right) \tag{9}$$

Now, taking the square in Eq. (2), the square Mahalanobis distance from a sample x_i to the center x_c is

$$d_M^2(x_i) = (x_i - x_c)^T S_c^{-1}(x_i - x_c)$$

Suppose we have prefixed a confidence level α, the threshold for the outlier detection is $\chi_{m,\alpha}^2$, where m is the number of variables. Thus it is independent of the number of samples and, then, fixed during the whole online process.

Let us suppose that we have the updated center and inverse dispersion matrix at a time t. Once the next sample, x_{t+1}, from the data stream arrives, its $d_M^2(x_{t+1})$ value is calculated in order to decide whether it is an outlier or not, according to the current center and inverse dispersion matrix. If it is not an outlier, the previous center (obtained by averaging) and inverse dispersion matrix can be incrementally updated as shown.

2.2.2 Outlier Detection Based on χ^2 Approximations of Q and T^2 Statistics

Suppose that we have a centered data matrix $X \in \mathcal{M}_{M,N}$ where the columns correspond to the predictor variables. Therefore we can consider that we are working in an N-dimensional space E. Once selected a number a of principal components, principal components analysis algorithm projects the data onto an a-dimensional subspace V, defined by the a first principal components. Then we can consider the orthogonal supplementary subspace of V, $U = V^\perp$, that is a $(N - a)$-dimensional, meaning that $V \oplus U = E$. Consequently, any element x in E has unique projections in both V and U so that their sum equals x. The selection of a is crucial for the final result. Nevertheless, the way to determine it is out of the scope of this section, and has been widely treated in the literature.

We are interested in two distance measures: (1) the Mahalanobis distance from the projection of x onto V to the center of the cloud of projections of all the data onto V, called *score distance*, and (2) the Euclidean distance from x to V, called *orthogonal distance*, which is related with the Euclidean distance to U. Figure 2 provides the geometric interpretation of both statistics for an original three-dimensional data example projected onto a two-dimensional subspace.

The former distance indicates the variation of each sample within the model. It is also known as Hotelling's T^2 statistic, and can be calculated as [76]

$$T_i^2 = x_i P_a \Lambda^{-1} P_a^T x_i^T = \sum_{j=1}^{a} \frac{t_{ij}^2}{\lambda_j} \tag{10}$$

where $\Lambda = \{\lambda_j\}_{j=1}^{a}$ is a diagonal matrix containing the biggest a eigenvalues and P_a is the loadings matrix.

For a fixed number a of principal components, on the basis of the fact that the data are centered, we can model the score distance, since all random variables t_{ia} have null expectation and variance λ_a/M, as [8]

$$\text{DoF} \cdot \frac{T^2}{\overline{T^2}} \sim \chi^2(\text{DoF}) \tag{11}$$

where DoF and $\overline{T^2}$ are the *degrees of freedom* and the average Hotelling's statistic, respectively. DoF could be estimated by

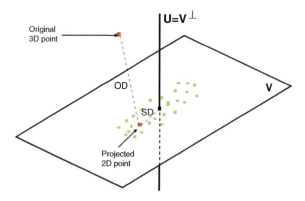

Fig. 2 Geometric interpretation of the score distance *SD* and orthogonal distance *OD* for a three-dimensional example projected onto a two-dimensional subspace *V*. For visual purposes, we have shown an original 3D point with a huge *OD*. Due to the way the principal components (PCs) are selected, this is not usually the case, and, unless the point is an outlier, the *OD* is commonly small

$$\widehat{\mathrm{DoF}} = \frac{2\overline{T^2}^2}{S_{T^2}} \qquad (12)$$

where S_{T^2} is an estimation of the standard deviation of T^2. A robust option, based on the interquartile range (IQR), is obtained by solving wrt $\widehat{\mathrm{DoF}}$ the equation

$$\frac{1}{\widehat{\mathrm{DoF}}}\left[\chi^{-2}(\widehat{\mathrm{DoF}}, 0.75) - \chi^{-2}(\widehat{\mathrm{DoF}}, 0.25)\right] = \frac{1}{\overline{T^2}}\mathrm{IQR}(T_1^2, \ldots, T_M^2) \qquad (13)$$

The latter distance, also known as Q statistic, indicates how well each sample conforms to the model, and it can be defined for a given sample x_i as

$$Q_i = \sum_{j=a+1}^{k} t_{ij}^2 \qquad (14)$$

where $(t_{i1}, \ldots, t_{iN})$ is the ith row of T, and k is the rank of X.

A similar formula to Eq. (11) can be proposed

$$C \cdot \frac{Q}{\overline{Q}} \sim \chi^2(C) \qquad (15)$$

It depends only in one parameter C that can be estimated in an analogous way as in Eq. (12).

Now that we have totally determined the distributions of both distances in terms of χ^2 distributions, p-values can be calculated, for a certain chosen critical level α, which are the probability of occurrence of each T^2 and Q. Considering c_i as any of

T_i^2 or Q_i, the corresponding p-value is

$$P(c_i) = 1 - [1 - \text{CDF}(c_i)]^M \qquad (16)$$

where CDF is the cumulative distribution function of the corresponding distribution. If any of the p-values is below the fixed critical level, then the corresponding input is considered as an outlier.

Assuming we keep the principal components fixed, at a time t we can suppose that we have the updated estimated distributions for Q and T^2. Once the next sample, x_{t+1}, from the data stream arrives, its Q_{t+1} and T_{t+1}^2 values are calculated in order to decide whether it is an outlier or not, according to the current estimations of the distributions of Q and T^2. If it is not an outlier, the mean values for Q and T^2 can be incrementally updated.

Besides, the new estimations of C and DoF can be done just by incrementally estimate the updated IQR. The calculation of the real IQR requires to store all data in memory. Nevertheless, the estimation could be done based on a window [38, 82, 89] (requiring memory for the samples in the window only), or based on quantile approximations [114]. All this allows us to incrementally extend the outlier detection based on Q and T^2 to data streams.

2.3 Data Normalization

Assuming that preprocessing is a preliminary task prior to a subsequent modeling phase using certain method, it is important to understand the characteristics of that method in order to perform a proper data preprocessing.

The most used technique is *mean centering*, consisting on subtract the mean value of every feature (thus column-wise). Some methods, like principal components regression (PCR) or partial least squares (PLS) have connections to distances to a central location of the distribution of the data. Therefore, if the data is not centered, they suffer from certain bias due to the distance to the origin of the raw data points.

Another fundamental normalization technique is *standardization*. Standardization comes from the transformation of a general Gaussian distribution into a standard Gaussian distribution (with null expectation and unitary variance), obtained by mean centering plus dividing column-wise by the standard deviation of every feature. By standardizing we make our data centered and unitary spread, thus correcting differences in the scales and ranges of the features. When employing any monitoring algorithm in which distance calculations are somehow involved, standardization is recommended unless the nature of the features is similar. In such cases, the differences in the ranges of the features are relevant for the process we are monitoring. An example of an algorithm involving distances is support vector machines, in which the widths (distances) between the data groups determined by the support vectors are maximized.

The third normalization approach we will consider is *scaling*. The motivation behind is gaining robustness against tiny feature variances, as well as to avoid zero entries in case of sparse data. In scaling we choose an interval and our data will be scaled so that it fits into that interval. The usual intervals are [0, 1], obtained by subtracting the minimum value and dividing by the range, and [−1, 1], obtained by dividing mean centered data by the value with largest absolute value in each of the features. The latter is the preferred one for sparse data. Both approaches are highly sensitive in the presence of outliers, thus either a proper outlier detection strategy or the use of robust alternatives to the range and standard deviation are recommended.

2.4 Data Transformation

The versatility of the data employed in predictive maintenance opens plenty of possibilities when it comes to transformations. There are two main branches in data transformation for predictive maintenance, which we identify as *statistical transformations* and *signal processing*.

2.4.1 Statistical Transformations

The *statistical transformations* are inspired in those transformations historically used in statistical inference [60]. The use of one or another type depends on the application and the type of data.

In Statistics, data transformations are applied when some prior information motivating them is available. Some of the most famous ones are *logit transformation*, from logistic regression, being related to neural networks and deep learning methods; *square root transformation*, from quadratic regression; and *reciprocal transformation*, obtaining similar scaling transformations as logit but also applicable to negative values.

In general, all those transformations can be generalized by means of the *power transformation* [49] that depends on a parameter λ, being all the aforementioned particular cases for certain λs. As the *identity* is also a particular case, it is possible to infer the most adequate transformation for some given data (by optimizing λ) including not transforming at all (identity). This technique is known as *Box-Cox* [5, 95]. Box-Cox has been successfully employed in fault detection [100].

Another family of transformations with statistical background are the projection on latent subspaces, like PCA and partial least squares (PLS). PCA is easily understandable if we approach it as an iterative procedure. Assuming we have centered data, the first PC will be the single direction on which the variance of the projection of the data is maximum. This direction is always obtainable as a linear combination of all the original features. Once fixed the first principal component

PC_1, we consider the orthogonal supplementary subspace of the subspace defined by PC_1, that is a line. As an example, in 3D the orthogonal supplementary subspace of a line is the plane that is orthogonal to it. In Fig. 2 the plane V is the orthogonal supplementary subspace of the line U. In this supplementary subspace we can also look for the single direction on which the variance of the projection of the data is maximum, getting PC_2. Notice that, as any direction in the subspace, PC_2 is orthogonal to PC_1. As each supplementary subspace we obtain has one dimension less than the previous one, we can continue with the same process until we end up with one last single line, that is the last principal component (PC_N if we had originally N features). Also in Fig. 2, V would be the plane defined by PC_1 and PC_2 (where PC_1 and PC_2 have respectively the direction of the large and small axes of the ellipse formed by the green points), and U would be the line defined by the last component $PC_3 = PC_N$.

PLS could be seen as a supervised equivalent to PCA. It becomes clear when we point out that the procedure for the calculation of the components in PLS (called *latent variables*) is similar to the case of PCA, but the objective is to maximize variance of the projection plus correlation with the target simultaneously. There is also a relevant difference from the algebraic point of view. In PCA, the supplementary space considered is the orthogonal one. Nevertheless, aiming for some flexibility required by the double objective of maximizing not only the variance of the projection but also the correlation with the target, PLS considers a supplementary subspace not necessarily orthogonal. The need of the target makes PLS unfeasible for online outlier detection. The application of a PCA variant is usually referred as performing an orthogonal transformation [101].

Both PCA and PLS are linear transformations, unless we opt for one of their multiple nonlinear extensions. There are several recent nonlinear transformations that are meant for exploiting the relations among the features. By relevance and usage, the most important ones are locally linear embedding (LLE) [91], isomap [110], and derivatives. They rely on the transformation of the original set of features into a smaller amount of projections taking into consideration the geometrical properties of clusters formed by instances, or patches of the underlying manifolds. Therefore, these methods could also fit into Sect. 2.6, because they could be understood as dimensionality reduction approaches.

2.4.2 Signal Processing

The heterogeneity in the properties of the data samples, also called *signals*, leaves margin for transformations coming from many sources. We have seen statistical transformations, but they also could arise from Mathematics, Physics, or Computer Science. It is a matter of semantics, but usually the word signal is reserved for certain type of data that can be ordered in time. Concretely we will focus on waveform data, because most of the predictive maintenance data are based upon this type. Waveform data can be observed from two related domains: time domain and frequency domain, being possible to move from one to the other and back. Depending on the domain we will distinguish three types of techniques in signal processing [109], which are (1) *time domain*, (2) *frequency domain*, and (3) *time–frequency domain techniques*.

The analysis of the *time domain* is the analysis of the original waveform data, which is, from a mathematical point of view, a chronological sequence of the value of certain random variable, having certain expectation, variance, skewness, and kurtosis which calculation could be part of the analysis, helping to characterize the signals. An example of time domain analysis is time series analysis [23], being *autoregressive* models one of the most employed ones. By autoregressive we understand that the feature values depend linearly on the previous ones [94, 99], so we would be assuming independence between features. If we think that it is not the case, then fractal time series take into account dependencies between two waveforms in different ways, such as local or global self-similarity, or short-range or long-range dependency [69].

The consideration of the *frequency domain* has several motivations. One of them is the fact that noise is usually affecting our signals, being recommended to use denoiser filters. These filters, when applied in the time domain, have huge computational costs, as they imply the application of convolution operations. Meanwhile, in the frequency domain they are just multiplications, as they transform differential equations into algebraic ones. Therefore, it is computationally cheaper to transform the data into the frequency domain, apply a filter there, and transform the filtered data back to the time domain in order to perform any posterior analysis there. There are many possible filters to be applied, even designed, depending on the components of the data we need to filter out [109]. Just as an example, a famous digital filter for smoothing the data is Savitzky–Golay filter [81, 84, 96], which is based on a local low-order polynomial interpolation using for each point a window containing some of its neighbor points. Some filters are also suitable for incremental online application on a streaming context [102].

The use of signals for modeling the state of real dynamic systems needs indeed information available in both the time and the frequency domain. For this reason, it is common to use both domains at the same time, moving from one to the other on demand. This use is called *time–frequency domain* analysis.

There are several ways to transform the time domain signals into frequency domain signals [6]. We highlight (1) the Fourier transform [120], (2) the Laplace transform, and (3) the Z transform [109] (known as the discrete version of the Laplace transform), since they are the most relevant ones. There are efficient algorithms to calculate them as well as their inverses. For instance, the fast Fourier transform (FFT) [29, 35, 122] is an efficient algorithm for calculating the Fourier transform. It suffers from a problem because it considers the whole signal. If we are facing, for instance, a fault detection problem trying to identify faults by changes in the signal, we could miss true faults (camouflaged as noise) unless the changes are significantly big wrt the whole signal. A way to overcome this effect is considering the short-time Fourier transform (STFT) that considers a fixed-width time window [25, 109].

Nevertheless there is another issue with STFT, coming from the fact that a good resolution in one domain implies a bad resolution in the other. This forces us to choose the width of the window so that there is a fine trade-off between the resolution in both domains. Another solution consists in employing a wavelet transformation, which provides us with the same effect as having dynamic resolutions in time and frequency. There are continuous and discrete wavelet transformations [26, 78], being the latter more computationally efficient.

Some more sophisticated newer approaches were developed afterwards. The *Hilbert Huang transformation* [56], a two-step method consisting on (1) *empirical mode decomposition*, i.e., the decomposition of the signal into a finite number of intrinsic mode functions, and (2) *Hilbert transform* of the intrinsic mode functions. The fact that those functions are orthogonal [104] implies that they can be understood as having physical meaning, thus being applicable in predictive maintenance [125].

Finally, the *Wigner Ville distribution* [24] was adapted by Ville [118] from Wigner's work in the field of quantum mechanics. It is a quadratic integral transformation in the form of a two-dimensional Fourier transform of a time–frequency autocorrelation function related to both time and frequency. It is not a window-based method, and it provides with the best resolution. Nevertheless, when a signal is a composition of two signals, there appear cross terms that could interfere (by distortion) the result of the analysis [63]. Otherwise, the study of the differences in the cross terms could be used in predictive maintenance problems [119].

2.5 Missing Values Treatment

The appearance of missing values is common in real-world data collected remotely and sent synchronously to a central database. In the same way as in the case of outliers, an obvious approach is to ignore samples in which one or more features presents a hole. As discussed in the case of outliers, sometimes we cannot afford ignoring data. Then there is an obvious alternative, *missing value imputation*. The *what* is obvious, but the *how* is really hard.

Naïve logical options, such as imputing the mean, or median as robust alternative, in numerical features, or the mode, in categorical ones, could be very risky. For instance, in imbalanced data situations the minority class gets great importance in the modeling, thus erroneous imputation could significantly influence model behavior.

In case we have a methodology to compare different samples and check whether they are similar or not just by looking at a subset of the features that defines them, then we could compare a sample with a hole with the samples without holes, and choose for the imputation the value of the hole-free sample. There are alternatives in which the sample to be used as imputer, such as systematically use one sample (*cold deck*) or randomly select from a pool of candidates (*hot deck*). As an example, if the methodology is based on distances using all features and the most similar one (i.e., the closest), then the approach is the same as K-nearest neighbors with $K = 1$. The main drawback of this approach is the difficulty of finding a proper way to compare samples. For instance, if we employ distances and the amount of features is big, we would suffer from the *curse of dimensionality* effect [85], being all distances huge and comparable in terms of magnitude.

As an alternative, an option could be a *double-model* strategy, in which a model is created using the fully available hole-free data in order to be used for imputation only. The estimated value could be directly used (*regression imputation*), or it could be slightly modified by adding a random residual (*stochastic regression imputation*). Once the missing values have been imputed, the main model is trained. There are several options depending on how to consider the imputed samples. Some approaches consider them as regular legitimate data samples, and some others underweight them, making them less influential in the main model. In case of computational and/or time expensive models, the imputing model employed is different from the main one, such that it is cheaper and/or faster. For instance, common algorithms used for this purpose are K-nearest neighbors [113], fuzzy K-means [70], Bayesian PCA [83], and multiple imputations by chained equations (MICE) [92].

In special cases in which the features are related, we could use *extrapolation* or *interpolation* methods for imputation. For instance, in data coming from a spectrometer, the different features consist of measurements made at different but close sequence of wavelengths, thus features in close wavelengths should present similar values.

2.6 Data Engineering

In data engineering we include three approaches (feature selection, feature extraction, and discretization) that could be performed alone or combined. Nevertheless it is not usual to combine them because they actually result in severely modified data as they are deeply invasive procedures.

2.6.1 Feature Selection

By *feature selection* we mean feature subset selection. Some authors consider both concepts as different because there are approaches in which the output is a ranking of all features instead of a subset of them. Nevertheless, we will say just feature selection since the common action is to use rankings to get a subset by truncation.

Feature selection can be understood as an optimization process in which the aim is to find a collection of features that makes certain quality criteria optimum. The simplest approaches, in which one single criterion is optimized, e.g., minimizing the root mean squared error (RMSE) of prediction in a regression problem, can be considered as single-objective optimization problems in which the objective function to be optimized is the quality criterion.

There are two types of feature selection (FS) approaches [44]:

- **Filters** [93]. They ignore the posterior task and focus only on the characteristics of the data to perform the selection, i.e., the criteria to be optimized are intrinsic to the data (e.g., mutual information between the features and the target). They could be understood as some kind of preprocessing selection. They are fast, but they usually ignore the possible redundancy in the data because most of the approaches evaluate the features independently from each other.
- **Wrappers** [65]. The modeling task (e.g., classification or regression) is understood as a black-box, whose performance using the subset of selected features is the goodness of the selection (performance optimization). They can deal with the redundancy, but they are usually computationally expensive, and they tend to overfit if the amount of available data is not big enough.

Some taxonomies include a third type, *embedded methods*, that are those methods in which the selection is internal to the model. As then the feature selection cannot be decoupled from the training, we cannot consider them as preprocessing, thus we keep the two-type taxonomy.

As filter methods rely on the characteristics of the data, the most renowned methods are based on statistical measures suitable for establishing dependencies and/or relationships between inputs and outputs, e.g., sensors information and machinery condition. Perhaps the most important filter method is *correlation-based feature selection* [46], in which the correlation between the features and the target is used.

There are plenty of ways, some employing problem-specific information, for defining what we understand by correlation, leading to different versions of the algorithm. Any way, specific to the predictive maintenance task, to establish a quantifiable relation between a feature (or a subset of them) and the output of the task that is capable of comparing/ordering different features (resp. subsets) could be used as a measure of correlation.

Recently, Brown [11] has found a generalization framework of some of the most extended families of filter methods that facilitates their understanding, given by

$$J_{\text{Brown}} = I(X_n; Y) - \beta \sum_{k=1}^{n-1} I(X_n; X_k) + \gamma \sum_{k=1}^{n-1} I(X_n; X_k|Y) \tag{17}$$

where n is the number of features, X_i the ith feature, Y the output, and $I(X; Y)$ is the mutual information shared by X and Y [103].

The approaches subsumed in the framework, just by playing with β and γ, are *mutual information-based feature selection* [3], *maximum-relevance minimum-redundancy criterion* [86], *joint mutual information* [127], *mutual information uniformly distributed* [67], *conditional info-max* [71], *conditional mutual information maximization* [32], and *informative fragments* [117]. Moreover, it becomes easier to compare the sensitivity of such families of methods with respect to redundancy and noise.

2.6.2 Feature Extraction

We define *feature extraction* [45] as the generation of new features by combining all or some of the existing ones. A common way to extract features is based on expert knowledge, but we will not consider it as it is subjective to the problem and is not fully data-driven.

The most relevant feature extraction approaches are based on the already mentioned projections on latent subspaces. The core methods are PCA and its many variants [57, 61]. In the original PCA the number of extracted features is the same as in the original data, since each principal component is just a linear combination of all the original features and there are as many linear combinations as the original number of features, thus PCA is a linear method. Assuming mean centered data, from a linear algebra viewpoint it consists just on a rotation of the coordinate axes.

The gain when applying PCA is that the new features (principal components) are ordered from higher to lower amount of captured variance in the set of features in the original data (ignoring the target, i.e., unsupervised). The cumulative variance captured by nested subsets of PCs can be easily computed, allowing to set a cut threshold in the number of PCs, leading to a reduction in the number of features (data compression) in such a way that the variance that is left out is small and controlled, possible colinearities between features are overcome, and, theoretically, noise is filtered.

Even when it contradicts intuition, compression is not always a goal in pre-processing when applying PCA. A situation in which it is not worthy, even counterproductive, to compress is noise-filtered data to be used afterwards by an algorithm that internally includes an embedded feature selection, such as random forests [10]. It is not a rare situation in predictive maintenance applications because noise filtering through transformations is well-established.

Many variants are motivated by nonlinear nature of some data. For instance, if there are certain known/intuited nonlinear relations between samples somehow having similar consequences as colinearity, we could model them by means of a specific kernel function and apply KernelPCA [97]. In this way, by using the kernel trick, we transform our feature space into a space where those relations look linear, applying there PCA.

As an alternative to the philosophy behind PCA-based approaches, we can consider neighborhood embedding approaches that try to preserve local neighborhood structures in the data on lower dimensionality spaces. A well-known algorithm for neighborhood embedding is *stochastic neighborhood embedding* [53] (SNE) in which a Gaussian probability distribution describes the potential neighborhood of each original sample in the high-dimensional space. A variation of SNE, with a simpler optimization process and comparable performance, is *t-distributed SNE* [75] (t-SNE). Despite it was originally developed for visualization purposes, it is perfectly applicable for data compression.

Both SNE and tSNE are nonlinear algorithms. A linear method also meant for neighborhood preservation is *locality preserving projections* [52] (LPP). In the same paper, the authors propose a nonlinear extension, named Kernel LPP, just by applying the kernel trick before LPP.

In highly dynamical systems, as is the case here, it is an adequate strategy to perform several local linear models covering the zones of influence of the data as a way to obtain the behavior of a nonlinear global model by aggregation/ensemble [130] of the local linear ones. A use of this technique, in which the aggregation of the local linear models is achieved by means of a fuzzy inference system [74], can be found in [13]. The authors applied the strategy for regression purposes, but it could be adapted for feature extraction using local information. In case of favorable properties in the aggregation algorithm, this strategy is also suitable for online monitoring [14, 16].

2.6.3 Feature Discretization

Some of the most famous algorithms employed in machine learning in general, thus also in monitoring in predictive maintenance, are meant for categorical values (e.g., decision trees). Moreover, sometimes they can only handle such type of data. Besides, the type of data in predictive maintenance applications consists of numerical continuous features with an order relationship, e.g., sensor data. Therefore, it makes sense to think of ways to transform such features into categorical ones, so that those algorithms could be used. This procedure is called *discretization*, and it is performed feature by feature independently.

Assuming we have a feature X_i whose values are numerical values with an order relationship. If we denote the minimum value by m, and the maximum value by M, then a discretization process consists in the definition of K intervals

$$I_1 = [a_0, a_1), I_2 = [a_1, a_2), \ldots, I_{K-1} = [a_{K-2}, a_{K-1}), I_K = [a_{K-1}, a_K]$$

where $a_0 = m$ and $a_K = M$. Notice that the cutpoints for the intervals define the partition of the range unambiguously.

A naïve approach would be to prefix K and split the range $[m, M]$ into K equal-length intervals. There are several drawbacks with this method. First of all, which is the right value for K? If the data is sparse or some extreme values (outliers or not) are present, then the range is huge. In such situation it could happen that certain intervals are empty and some crowded. Therefore, unless our data is uniformly spread and we have a proper way to choose K, it is not a good option. Nevertheless, there are plenty of estimators for the width (W) of the bins, thus for K. One robust option is the *Freedman–Diaconis rule* [33], given by

$$W = 2 \cdot \frac{\mathrm{IQR}(X)}{\sqrt[3]{N}} \qquad (18)$$

where X is the feature under consideration and N the number of samples.

Hence, it is preferable to have a clever way to proceed that, if possible, does not force us to prefix K. The widest used method is a supervised top-down algorithm called *minimum description length principle (MDLP)* [30]. By top-down we mean that it begins with an empty partition and the cutpoints are added on the fly, thus no need to prefix K. It decides whether a new cutpoint is needed and where to locate it by means of information theory, concretely the mutual information with the target [103].

There are many other ways to define discretizations. An exhaustive survey including several taxonomies according to the properties of the methods and the data is available in [36]. The authors present 87 methods, tested on many datasets with different properties, so by comparison of types of data we could try to guess which methods would fit better to our data.

2.7 Imbalanced Data Treatment

In such cases when the data show a lack of balance between the classes of the samples, it is usually the case that the class we are more interested in is the minority class, e.g., faulty and fault-free samples. Despite we have commented in Sect. 2.1 on two ways to deal with imbalanced data, named as sampling- and weighting-based, the latter is more related to the modeling phase instead of the preprocessing phase because the weights are actually introduced in the model creation or the model validation steps, depending on the characteristics of the algorithm that is being employed.

It is also of uppermost importance the metrics employed in the validation. For instance, accuracy is not a valid choice because if the imbalance is 99%–1%, then predicting always the majority class leads to a 99% accuracy. Thinking on the example of faulty and fault-free samples, we would predict that faults never happen, being almost always right. But it is obvious that not all errors in our prediction have the same cost. In order to mitigate this without a need to assign a cost per error it is common to use ROC curves [31]. Nevertheless, this is out of the scope of this chapter, as it does not correspond to preprocessing. Therefore, we focus on imbalance treatment approaches based on sampling techniques.

There are two obvious ways of compensating the imbalance, which are adding samples from the minority class (*oversampling*), and removing samples from the majority class (*undersampling*) [59, 87, 108, 121]. Because both have pros and cons [18, 27, 79], it is also common to opt for a hybrid approach (*mixed sampling*) [2, 66] combining them.

Recently, the importance of ensemble methods has been shown in many applications including imbalance treatment [19]. Both ensembles of repetitions of stochastic techniques or ensembles of diverse deterministic techniques usually overcome the application of single techniques, ensuring robustness by reducing the variance while fixing the bias.

2.7.1 Oversampling

If we think on how to perform oversampling, the first intuitive approach is *random oversampling* (with or without replacement). In [58] the authors consider two random oversampling possibilities: a pure random one and a focused one in which only samples close to the boundary between classes are considered as selectable; both used until parity in the classes is reached. As not all the samples from one class influence the monitoring algorithm in the same way, the samples we replicate could be so influential that we suffer from an overfitting effect. More sophisticated approaches opt for creating new samples by interpolation of some of the existing ones.

There are two main methods, existing several variants for each of them. Those relevant methods are *synthetic minority oversampling technique (SMOTE)* [20] and *adaptive synthetic sampling method (ADASYN)* [51]. In both methods the algorithm to generate new samples is the same. A sample x_i from the minority class is considered. Then the K-nearest neighbors from the minority class are located. One of them x_j is randomly chosen, and the new synthetic sample is a convex combination of them

$$x_{new} = \lambda x_i + (1 - \lambda) x_j$$

where $\lambda \in [0, 1]$ is randomly selected. Graphically, the convex combination of two points is a point located in the segment that joins them. Figure 3(a) shows the generation of a new minority class sample x_{new} (marked as a green cross). The difference between SMOTE and ADASYN is only in the way the neighbor points are taken. The latter uses a prefixed K, and the former chooses it depending on the density of the minority class inside a neighborhood obtained by K'-nearest neighbors.

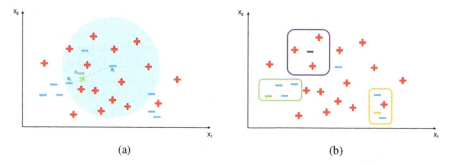

Fig. 3 Examples of (**a**) the generation of a new minority class sample x_{new} from an existing sample x_i and one of its 3-NN x_j (the 3-NN neighborhood of x_i appears as a pale blue circle), and (**b**) the determination of noisy (purple), borderline (orange), and safe (green) minority class samples in borderline extensions of SMOTE (the 3-NN is inside colored rounded squares). Plus and minus symbols represent majority and minority class samples, respectively

The influence of extreme values (or outliers, if not detected) is really high in both SMOTE and ADASYN, being higher in ADASYN. Then SMOTE is usually employed in some of its variants. The most famous ones are *borderline-1 SMOTE*, *borderline-2 SMOTE*, and *SVM SMOTE*. In borderline versions also an auxiliary K' neighborhood is used, where the samples x_i from the minority class are labeled as *noisy* (all nearest neighbors are not from the minority class), *in danger* (at least half of the neighbors are from minority class), or *safe* (all are from the same class as x_i). Then the only samples chosen as initial samples are *in danger* samples. See Figure 3(b) for an example.

The difference between borderline-1 and borderline-2 happens when selecting x_j. Borderline-2 allows to select a sample from any class, not necessary majority class (as borderline-1 does). In SVM SMOTE the support vectors are used to generate the new sample x_{new}.

2.7.2 Undersampling

The major risk when ignoring majority class samples is to potentially ignore really relevant informative samples, leading to a degradation of the general quality of the model. As in oversampling, there are methods that select (sample selection) prototypes in the majority class (most of the approaches) and methods that generate (sample extraction) a smaller set of prototypes from the original bigger set of samples. The only relevant approach in *prototype generation methods* is called *cluster centroids undersampling*, which is based on clustering using representatives (CURE) [42], a famous clustering algorithm in which relevant points of the identified clusters (e.g., the centers) substitute the points inside those clusters, reducing the amount of points but keeping the underlying cluster structure. When it comes to *prototype selection methods*, we can identify two subgroup of methods depending on the possibility by the user of controlling the number of samples after undersampling (*controlled undersampling techniques*) or not (*cleaning undersampling techniques*).

The simplest *controlled undersampling* technique is *random undersampling*, which is the riskiest one as all samples are equiprobably deleted ignoring their potential informativeness/relevance. The most representative approach is called *NearMiss* [131], which includes some heuristic rules in order to select the samples. The authors presented three NearMiss versions, differing in the way the heuristics are defined. *NearMiss-1* selects the majority class samples with minimum average distance to the N closest minority class samples. *NearMiss-2* selects the majority class samples with minimum average distance to the N farthest minority class samples. Finally, *NearMiss-3* has two steps: first, the M nearest neighbors for each minority class sample are kept, then the majority class samples with maximum average distance to the N closest minority class samples are selected.

Also in [131], the authors define another approach, named *MostDistant*, in which the selected majority class samples are those presenting largest average distances to the N closest minority class samples. In the original paper the authors select $N = 3$. Figure 4 shows examples of the three versions of NearMiss in a two-dimensional space with $N = 3$ and $M = 5$.

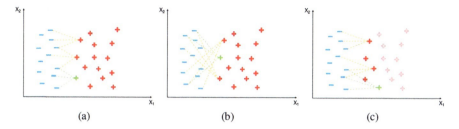

Fig. 4 Examples of the selections performed respectively by all three NearMiss versions. Plus and minus symbols represent majority and minority class samples, respectively. Distances to the 3-NN of some majority class samples are depicted using colored dashed lines. In green we can see the distances corresponding to the selected majority class sample, as well as the sample itself in each version. In (**c**), the samples out of the 5-NN neighborhood are represented

The family of *cleaning undersampling* techniques is bigger. The name comes from the fact that the part of the dataset corresponding to the majority class is cleaned by deleting certain samples considered as dispensable according to certain heuristic algorithm. We describe them in no particular order.

A popular method is based on the so-called *Tomek's links* [112]. We say that two samples from different classes form a Tomek's link if they are nearest neighbors to each other. Mathematically,

$$d(x, z) \geq d(x, y) \text{ and } d(y, z) \geq d(x, y), \forall z \quad (19)$$

The undersampling procedure associated with them has two variants. We can remove (1) only the sample in the Tomek's link corresponding to the majority class, or (2) both samples. It is clear that such pairs of samples are some sort of contradiction. The safest choice would be to remove both, but this could be sometime not an option as it would decrease the size of the minority class. An example of a Tomek's link can be seen in Fig. 3(b), formed by the purple minority class sample and its nearest neighbor.

Inspired by Wilson's studies on the nearest neighbors rules [123], *edited nearest neighbors* edits the dataset by removing those samples which do not *agree enough* with their neighborhood. Different agreement criteria provide different versions. Given one sample, the most restrictive version demands all the samples in the neighborhood to be from the same class of the sample under study. A more relaxed version demands only a majority of samples from the same class. There is also the possibility to run the edition procedure several times iteratively (with the same or different K), so that more samples are removed.

We can define *instance hardness* [106] as the level of difficulty to predict the class of a sample due to the sample characteristics. The usual way to calculate it is by means of an algorithm that assigns to each sample, using cross-validation, a probability of being well classified, thus the lower the probability the harder the instance. The instance hardness undersampling technique consists in establishing a threshold for the probability of the majority class samples, removing those that are below the threshold.

Last but not least, we have the family of *condensed nearest neighbors*, based on the homonymous rule [47]. The undersampling methods that are inspired on it condense the space by removing samples that are far from the decision boundaries. The original method is based on an iterative process with the following steps:

1. Construct a *condensed set* C containing the minority class samples.
2. Add one majority class sample to C, and create a *potential set* P with the rest.
3. Classify each sample in P using 1-NN. If misclassified, move it to C. Otherwise, do nothing.
4. Reiterate until no samples can be added to C.

As this original approach is very sensitive to noisy samples, keeping them in C, some variants were proposed. The variant named *one-sided selection* [66] removes noise by applying Tomek's links first, and then the steps 1–3 of the original approach, thus no iteration over P.

In [68], the authors propose *neighborhood cleaning rule* that proceeds as follows:

1. Get one sample x_i and classify it using 3-NN.
2. If x_i is misclassified, go to next step. If classified go to the first step.
3. If x_i is a majority class sample, then remove x_i. If x_i is a minority class sample, then remove the 3-NN corresponding to the majority class. Go to the first step.

This approach is computationally expensive, and could suffer in case of very large heavily imbalanced datasets. Even when its philosophy is based on cleaning, the result is usually a condensed subset of the original one.

All these condensed family techniques depend on some randomness, when taking samples to begin. Moreover, the order of the samples is relevant for the final undersampled set. Therefore, we cannot expect the same result when repeating them over the same dataset. It is recommended to perform the methods several times and ensemble the results by certain aggregation procedure.

2.7.3 Mixed Sampling

The naïve approach, consisting on combining both random oversampling and random undersampling, was proposed in [72]. The authors used lift analysis instead of accuracy as performance score measurement in their experiments, without obtaining relevant improvements.

A deep study on the mixture of oversampling and undersampling techniques can be found in [2]. The authors point out the good results of mixing SMOTE with both Tomek's links and edited nearest neighbors.

2.8 Models

Our aim is not checking which modeling technique behaves better, but comparing different preprocessing schemata by means of the posterior performance in a regression or classification task. Therefore, we present only a few techniques just to check whether using different algorithms is also relevant in the selection of the right preprocessing scheme apart from the data.

Here we briefly describe some state-of-the-art algorithms suitable for confronting predictive maintenance problems. We distinguish two types of algorithms for two classical problems: classification algorithms for fault detection problems and regression algorithms for remaining useful life prediction problems.

2.8.1 Classification

A regular fault detection problem is a binary *classification* problem in which the aim is to predict whether a concrete system state (sample) corresponds to a faulty or to a fault-free situation. The simplest but still widely used classification methods are naïve Bayes and K-nearest neighbors. *Naïve Bayes* (NB) algorithm [34] is a probabilistic method based on the application of Bayes theorem under strong feature independence assumptions. K-*nearest neighbors* algorithm [105], as all methods based on distance calculations, can suffer from huge distances of some of the neighbors due to the sparseness enforced by a habitual high dimensionality. The attempts to mitigate such problem are the motivation behind *distance-weighted K-nearest neighbor* algorithm [28], that is the variation of K-NN we will consider, consisting in regulating the importance of the votes of the neighbors by means of weights that depend on the distance, so that the closer the more important. Since it is the only variant we will consider, we denote it by *K-NN*.

Support vector machines (SVM) [98, 115] is a well-known nonlinear classification method, based on separating the classes employing hyper-planes is such way that the separation is maximized. This separation is not performed in the original input space but in a kernel-transformed space, i.e., the kernel trick [55]. The samples that are closest to the decision boundary, thus defining the hyper-planes, are called *support vectors*. In [62] the authors compare several classifiers for fault detection, including distance-weighted K-nearest neighbors and support vector machines, among others.

The *random forests* (RF) algorithm [10] is a stochastic ensemble method that performs a bagging strategy (a combination of bootstrapping and aggregation [9]) of weak learners, concretely decision trees. The procedure is simple. Given a prefixed number of trees, for each tree a subset of the original features is randomly selected (*weakness*). Then the tree is trained using those features and a set of samples obtained by random selection with replacement (*bootstrapping*). The decision is obtained by combining all individual tree decisions (*aggregation*). The magic behind RF is that the bias of the full ensemble is equivalent to the bias of each single tree, whereas the variance is much smaller. This robustness, together with its low computational cost and high parallelization and distribution capabilities, makes RF an algorithm to be taken into consideration in predictive maintenance [12, 43, 128].

2.8.2 Regression

Despite the original purpose of RF and SVM is classification, there are versions of both of them for regression purposes. In the case of random forests, it is quite straightforward to substitute decision trees by regression trees, and the voting aggregation by an average prediction [10]. The insights in the case of support vector regression (SVR) are a bit more complex and too long to be commented here [107]. Some applications to RUL prediction can be seen in [4, 73, 90, 126].

Basic linear regression approaches, such as multiple linear regression, suffer from the arising of singularities because of the effect of colinearities between features when calculating the inverse of $X^T X$, required by the least squares solution, being X the input data matrix. In such situations, shrinkage (regularization) methods avoid singularity by perturbing the matrix before it is inverted. The two main approaches in the family of shrinkage methods are *Lasso* [111] and *ridge regression* [49], obtained by introducing ℓ_1 and ℓ_2 penalties, respectively. The *elastic net* [133] includes a penalty based on a combination of both ℓ_1 and ℓ_2 penalties, looking for some elasticity in the regularization, being Lasso and ridge regression particular cases of the elastic net.

Generalized linear models [49] is a generalization of ordinary linear regression that provides flexibility in the sense that the distribution of the errors is not necessarily supposed to be normal, as happens in ordinary linear regression. The combination of the elastic net with generalized linear models (GLMnet) is a regression algorithm based on generalized least squares that uses cyclical coordinate descent [50] in a path-wise fashion [48] in order to select the optimum elasticity in the regularization via the elastic net. The elasticity provided by the possibility of controlling how close we are to Lasso or ridge regression by means of a single parameter allows an efficient exploitation of the regularization benefits.

Up to our knowledge, this approach has not been used in predictive maintenance yet. Nevertheless it has been considered here because of its outstanding results in monitoring dynamic chemical systems in process analytic technology (PAT) [15, 17], that behave quite similarly to regression problems in predictive maintenance with dynamic systems.

Deep learning (DL) is the way to call the use of a complex artificial neural networks. A neuron is a single computation unit that receives an input value (from a data source or another neuron), performs a simple operation consisting on applying certain simple function (*activation function*) over the product of the input by a numerical parameter (*weight*), and outputs the result (towards an output interface or another neuron).

Different types of neurons connected in different ways lead to different network architectures. These neural networks are designed by means of layers of neurons conceived for specific subtasks.

For stream-like data, such as the data usually involved in monitoring tasks in predictive maintenance, the most used networks are *recurrent networks* (RNN), in which the neurons are also connected to themselves. This provides the network with some *memory* in the form of persistence of the information. In general, they suffer when the ideal persistence time grows.

There is a family of RNNs meant for handling long-term information dependencies called *long short term memory networks* (LSTM) [54] that contain an internal mechanism (*cell state*) to filter/retain part of the information as long as necessary. There are several ways of handling the remembering/forgetting part of the learning process, leading to different variants of LSTMs. The most relevant ones are, among others, *vanilla LSTM* [40], *gated recurrent unit* (GRU) [21], *depth gated LSTM* (DG) [129], or *grid LSTM* [64]. In principle, LSTM networks are the most adequate network architectures in predictive maintenance.

Even when the natural output of the network is a number, they could be adapted for classification purposes by linking the classes to certain numerical output ranges.

3 Experimentation

The philosophy derived from non-free-lunch theorem [124], which states that the average performance of all algorithms over all possible problems is asymptotically the same, is that there is not a single universal algorithm that is the best. Therefore, there is always margin for improvement and every particular problem (correspondingly dataset) is better suit for a different method. This applies also to the preprocessing schemata, in the sense that there is not a universal preprocessing schema that is always the best, being the goodness problem/data dependent.

Consequently, providing successful stories for concrete scenarios is perhaps not the best option. It would be more relevant to provide the reader with direct or literature referenced details of the available choices in the market, as well as hints about possible decisions depending on the characteristics of the problems or the data. For such reason we will just employ the already presented classification and regression techniques on some of public available real-world datasets from competitions in the Annual Conference of the Prognostics and Health Management (PHM) Society. We will use them (both the original data and some modified subsets, e.g., for missing values treatment or outlier detection) to compare several preprocessing schemata on different algorithms. Furthermore, we will provide some clues about which preprocessing methods might be more reasonable depending on the particularities of data and problems based on successful applications.

It is obvious that the combination of all possible methods in all steps in different orders would end up in thousands of preprocessing schemata. Moreover, if a schema consisting on seven steps works very well, we would not be able to decide which of them contributed more to that behavior. Therefore, just some schemata involving only a few steps will be tested, and compared also with, we should not forget that it is always a possibility, not preprocessing at all.

3.1 Datasets

In order to have a classification and a regression problem, we have considered the data corresponding to the *PHM Data Challenge 2014* and the *PHM Data Challenge 2016*. The former is transformed into a fault detection problem (classification), and the latter is a RUL estimation problem (regression) in which the average removal rate of material in a polishing process. The lack of exact environmental information about the origin of the datasets impedes us to infer cause–effect reasons for the results. Hence, we focus on the goodness of the application of the preprocessing schemata instead of the underlying reasons.

3.1.1 PHM Challenge 2014

The information about the domain and the data of the *PHM challenge 2014* is not provided due to proprietary concerns. We know that it consists of six datasets, half for training and half for testing with information about (1) *part consumption* (i.e., the replacement of some parts), (2) *usage* (similar to the lines of an odometer), and (3) *failures* (time of failure). The target information for the test files is unknown, so we focus only on training data. By crossing the failure information with the rest we could build by merging a dataset in which the target is binary: faulty or non-faulty, thus it consists on a binary classification problem. For further information on the data, check the call for participation in [37].

In this dataset there is almost no information about the nature of the features. The original aim in the challenge was to predict the health level of the components in a certain time by classifying them into low risk and high risk of failure, equivalent to fault save and faulty in a short future. The variables are numerical and discrete. Nevertheless, the amount of different values is so big that we can employ any method suitable for continuous variables.

The data is heavily imbalanced, belonging to the high risk class (faulty) only 4% of the samples. The modeling algorithms do not take into account the level of imbalance during training. Only some of them, based on iterative optimization processes, are capable of weighting the errors according to the class densities so that they favor avoiding mistakes in the minority class. The main problem of such approaches is the price to pay in the prediction of the majority class. Therefore it is recommended to treat imbalanced in advance as part of the preprocessing.

3.1.2 PHM Challenge 2016

The system under investigation is a wafer chemical-mechanical planarization (CMP) tool that removes material from the surface of the wafer through a polishing process. Figure 5 depicts the CMP process components and operation. The CMP tool is composed of the following components: (1) a rotating table used to hold a polishing pad, (2) a replaceable polishing pad which is attached to the table, (3) a translating and rotating wafer carrier used to hold the wafer, (4) a slurry dispenser, and (5) a translating and rotating dresser used to condition a polishing pad.

During the polishing process, the polishing pad's ability to remove material is diminished. Over time, the polishing pad has to be replaced with a new pad. Similarly, the dresser's capability to roughen the polishing pads is also reduced after successive conditioning operations and after a while the dresser must be replaced. The objective is to predict polishing removal rate of material from a wafer, thus it is a regression problem. For further details, check the call for participation in [88].

A deeper look at the data allows us to infer some characteristics of the data. All variables are numeric (float), with different ranges and dynamics. Some fluctuate up and down approximately in a cyclic way while some others show a continuous increase or decrease that is apparently linear. These differences force us to be careful when selecting the way to apply the preprocessing techniques.

For instance, if we apply a technique that involves mixing the features, such as PCA, then standardization is recommended. On the contrary, in approaches acting on the features individually, such as discretization, it could be counterproductive. Due to the size of the data we are limited to visualization techniques based on certain information summary/compression such as tSNE plots, and scores and loadings plots in PCA. Nevertheless there is not an obvious relationship between the visualizations and the adequate preprocessing techniques.

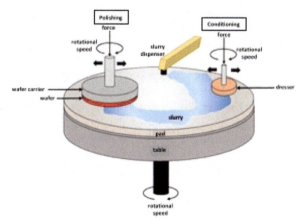

Fig. 5 Chemical-mechanical planarization (polishing) of wafer. This process removes material from wafer surface. This image is the property of the Prognostics and Health Management Society and was taken from the online information about the challenge

3.2 Experimental Schema

The algorithms we consider, whether take advantage of centered data or are translation invariant, thus we have mean centered all the data in advance. All the experiments were made using ten-fold cross-validation because it is known to be a good approximation of the expected prediction error on separate future unseen test samples. We evaluate the performance in classification by means of the area under the ROC curve [7], and in regression with the root mean square error.

Since our intention is not to beat the winners of the competitions, but check whether preprocessing is beneficial or not (and how much), then our comparisons are against not preprocessing. The reason for including several modeling algorithms is not to determine which one is better, but to try to check if that diversity of models is relevant or not for the benefit of preprocessing.

In case we suspect that the best preprocessing strategy is independent of the posterior modeling technique, then we could try the simplest ones in order to guess the right preprocessing scheme. For statistical significance of the differences, we have employed the *Mann–Whitney–Wilcoxon* test [80].

In this study both the outliers and missing values have been artificially introduced, thus we have the chance to check the performance of the methods, as we know the truth. This is not the case in real-world applications. With respect to the rest of preprocessing steps, we have performed the test with the full dataset. The realist approach in an application would be to extract a representative subset of the data in order to perform some preliminary tests and determine a full preprocessing strategy.

When it comes to the study of the approaches for missing values, we have modified the PHM2016 dataset by randomly erasing 1% of the values in 10% variables. Taking into account that there are 21 variables and 346,015 samples, we have introduced in 2 variables 3460 holes per variable. The approaches employed were imputation with the mean value, imputation by averaging using 5-NN, as well as removing the samples (deletion strategy).

For outlier detection, also using PHM2016, we have modified 1% of the total amount of single numerical values by distancing them from the mean of the feature they correspond to. The amount of variation is proportional to their distance to the mean, with factors corresponding to 20%, 50%, 100%, and 200%, meaning 865 variations per level. Each of them has been applied to the one fourth of the modified values, i.e., 0.25% of the total amount of values. In this way we can evaluate the sensitivity to the amount of variation. The approaches employed were the Mahalanobis distance and the Q and T^2 approximations in their offline versions.

As some potential detected outliers could be out of the list of the artificial modifications (false positives), it makes sense also to check the posterior performance, after cleaning, in modeling. For this comparison, we have also included the original modified data, i.e., without looking for outliers.

When it comes to feature engineering, we have designed experiments separately for feature selection (on PHM2016), feature extraction (on both datasets), and discretization (on PHM2016). The algorithms we have employed are:

- **Feature selection**. We have chosen two filter methods (correlation-based feature selection and conditional info-max), and a wrapper approach (using K-NN).
- **Feature extraction**. We have selected PCA, PLS, Kernel-PCA, t-SNE, and Kernel-LPP, so we have two linear and three nonlinear methods. Notice that most of the features in PHM2014 are numerical discrete variables containing natural numbers. Nevertheless, the amount of different values is so big that we can consider them as continuous numerical variables, suitable for feature extraction by PCA or PLS. The adequate number of PCs and LVs has been selected by grid search.
- **Discretization**. We have opted for two approaches, in order to consider one that prefix the number of bins (equal-width intervals using Freedman–Diaconis rule), and another one that does not prefix it (MDLP).

For *imbalanced data treatment* we need a classification problem, thus we use our PHM2014 version, which has a minority class (faulty) represented approximately by a 4% of the data samples. We have not applied all the methods in Sect. 2.7, but some of the most popular ones. Classified according to the provided taxonomy, they are

- **Oversampling**. Random oversampling, SMOTE borderline-2, and ADASYN.
- **Undersampling**. Random undersampling, cluster centroids, NearMiss-2, and Tomek's links.
- **Mixed sampling**. Random oversampling and undersampling combination, and SMOTE with Tomek's links.

3.3 Results

The results are presented by means of tables, whose formats depend on the experiments. As general facts,

- we consider not preprocessing as baseline, and we present the percentage of improvement (positive number) or deterioration (negative number). An exception occurs in the case of missing values, because the usual baseline does not exist. In that case we consider the deletion strategy as the baseline approach. If the shown variation from the baseline is significant, according to the Mann–Whitney–Wilcoxon test, it will be indicated with a ‡ mark. The best results are highlighted in **bold** font. There is also another exception when studying the detection of outliers. In that situation there is not any baseline because there is not any modeling step, but just checking the performance in the detection of the outliers

for different perturbation levels. In this case we just show the detection rates per method and per level, and the † mark means significantly better than the other method;
- in all nearest neighbor related approaches in which we have the chance of choosing K, our choice will be $K = 5$;
- the kernel function used in both SVM and Kernel-PCA is radial basis function (RBF);
- the network architecture used for DL is GRU because it has a simple effective joined input/forgetting mechanism by using the so-called *update gates*, proved to behave similarly to much more complex architectures [41];
- unless explicitly indicated otherwise, the learning parameters of the algorithms are set by default as in the literature. For GRU, the default arguments in Keras [22] have been used.

Table 1 shows the results for missing values. Notice that in this situation all columns are independent because we are comparing, for each modeling technique, the performance of imputation versus deletion for that concrete technique. For instance, the values +1.35 and +2.57 corresponding to RF algorithm mean that imputation is preferred (both are positive values) and the performance when using K-NN method is almost doubly beneficial than mean.

We can see that it is slightly beneficial to use imputation, being a bit better the imputation by means of K-NN. Nevertheless, none of the imputations are statistically significantly better than deletion, for any algorithm except for RF and GRU. This, together with the fact that K-NN requires huge computational and memory resources, shows doubts about its suitability.

The reason for using several models is to check whether the model to be applied after preprocessing has an impact in the right preprocessing scheme. Luckily we can see that the results are similar for all modeling algorithms, thus it seems that the data is more relevant than the algorithm. Nevertheless, we should notice that there are big differences in performance between deletion, mean, and K-NN for the various modeling techniques even when the general trend remains stable.

Table 2 shows the results for the detection of outliers for different deviations. The percentage of outliers is constantly 1%, but the amount of deviation from the original values, artificially introduced, varies from low intensity (20%) to high intensity (200%). The higher the intensity the simpler the detection because the values are much more different from the real ones. In the case of the Q and T^2 approximations method, the number of principal components has been determined by establishing a threshold of the total amount of variance captured, set in 90%.

Table 1 Missing values

Method[a]	RF	SVR	GLMnet	GRU
Mean	+1.35	+0.15	+1.20	+1.57 ‡
K-NN	+2.57 ‡	+0.33	+1.28	+2.04 ‡

[a]Deletion strategy is considered as the baseline

Table 2 Outlier detection accuracy

Method	20%	50%	100%	200%
Mahalanobis	1.04	13.87	53.29	**92.37**
Q and T^2 [a]	**1.62**	**26.82** †	**69.71**	91.91

[a] The number of PCs is 4

Table 3 Outlier detection effect on modeling

Method	RF	SVR	GLMnet	GRU
Mahalanobis	+0.66	+2.09	+4.10 ‡	+3.96
Q and T^2 [a]	**+0.92**	**+2.33**	**+5.01** ‡	**+5.14** ‡

[a] The number of PCs is 4

Table 4 Feature selection

Method	RF	SVR	GLMnet	GRU
CFS	−0.77	+0.03	+0.38	+0.20
Conditional info-max	−0.07	**+0.05**	+0.25	+0.16
K-NN [a]	**+0.02**	+0.03	**+0.41**	**+0.22**

[a] Only wrapper method. The rest are filters

In general, the approximation approach behaves better than Mahalanobis, being that difference higher in the intermediate levels. For the biggest distortions (easier to detect) both methods perform very well.

Table 3 shows the results for outlier detection effect in modeling. Looking at RF column we can see that the advantage is much lower than for the other two algorithms. A possible reason is the fact that RF uses for each tree a reduced dataset, both in the features and in the samples part. Theoretically the expected percentage of the samples from the original set considered for training each tree is indeed 63.2%, thus errors in the detection could be somehow partially neglected.

Besides, we could suspect the difference between SVM (nonlinear) and GLMnet (linear) to be due to the fact that the transformation used for generating the outliers is a linear mapping. Nevertheless, the suspicion is not right because GRU is also nonlinear and behaves almost the same as GLMnet. In the end, GLMnet and GRU have suffered less than SVM from the not detected outliers or the false positives. The latter are very few, almost zero compared to the true outliers.

Table 4 shows the results for feature selection. The most plausible reason for the total lack of advantage in this feature selection process is that the variables are quite independent, containing a similar amount of complementary information. Therefore, selecting features in any way enforces certain information loss. The effect is magnified in RF, as it has an internal tree-wise feature selection step.

Tables 5 and 6 show the results for feature extraction in the classification and regression tasks, respectively. We can say that (1) the data seem to be quite nonlinear, as the nonlinear methods are the best in all algorithms except SVM in classification (probably due to the fact that in that case we are applying twice an equivalent kernel trick), (2) PLS (supervised) behaves better than PCA (unsupervised) because of the possibility of using the target information, and (3) it makes sense to use these feature extraction methods, even when the improvement

Table 5 Feature extraction in fault detection (classification)

Method	NB	K-NN	SVM	RF
PCA[a]	+1.03	−1.11	+1.48	+1.27
PLS[b]	+1.24	−0.22	**+2.17** ‡	+1.33
Kernel-PCA[a]	**+1.36**	+1.04	+2.04 ‡	+2.16 ‡
t-SNE	+1.25	**+1.53** ‡	+2.11 ‡	**+2.35** ‡
Kernel-LPP	+1.21	+1.15	+2.16 ‡	+2.08 ‡

[a]The number of PCs is 3
[b]The number of LVs is 2

Table 6 Feature extraction in RUL estimation (regression)

Method	RF	SVR	GLMnet	GRU
PCA[a]	+2.20 ‡	+1.26	+1.52	+2.15 ‡
PLS[a]	+2.31 ‡	+1.53	+1.59	+3.02 ‡
Kernel-PCA[a]	+3.33 ‡	+1.54	+2.60 ‡	+3.48 ‡
t-SNE	**+4.22** ‡	**+1.78** ‡	**+2.85** ‡	+4.39 ‡
Kernel-LPP	+3.97 ‡	+1.60	+2.71 ‡	**+4.81** ‡

[a]The number of PCs and LVs is 4

Table 7 Discretization

Method	RF	SVR	GLMnet	GRU
Equal-width[a]	−1.61 ‡	−0.32	−0.95	−2.33 ‡
MDLP	**+2.66** ‡	**+0.98**	**+2.76** ‡	**+4.19** ‡

[a]Using Freedman–Diaconis rule

is not statistically significantly better (with one single exception), because they are not much computationally expensive.

Table 7 shows the results for discretization. In this case the comparison between equal-width and MDLP is not totally fair, as the former is unsupervised and the latter supervised. Also, in general, methods that do not need to prefix the number of bins achieve results at least as good as the restrictive ones. In this case, according to the significance tests, it is clearly not an exception, especially in RF, GLMnet, and GRU. Equal-width only makes sense if the density of the features is homogeneous in their ranges, which is rare and not happening here.

In general, RF use to behave better when discretizing. Besides, some algorithms involving complex/computationally expensive optimization processes (like SVR and GRU) could suffer from numerical instabilities that are less likely with discrete features. Nevertheless, discretization is not necessary beneficial always. Also notice that the process affects the features independently, which makes possible to discretize only a subset of the continuous features.

Table 8 shows the results for imbalance data treatment. First, we can compare these results with the ones in Table 5, and point out that imbalanced data treatment schema seems to be a better choice than feature extraction schema, hence, as we have mentioned before that it was worthy to use feature extraction, it is even worthier to use imbalanced data treatment.

Table 8 Imbalanced data treatment

Method	NB	K-NN	SVM	RF
RandOver	−1.83 ‡	−2.01 ‡	−0.92	−1.24
SMOTE[a]	+1.39	+1.06	+2.43 ‡	+1.40
ADASYN	+0.23	−0.82	+1.49	+0.48
RandUnder	−1.71	−1.69	−1.03	−1.43
ClustCentr	−0.02	−0.85	+1.04	+0.62
NearMiss[b]	+1.22	+1.03	+1.15	+1.55
Tomek	+1.42	+0.99	+1.24	+2.51 ‡
RandOverUnder	−1.75	−2.25 ‡	−1.20	−1.37
SMOTE+Tomek	**+1.81**	**+1.13**	**+4.61 ‡**	**+3.49 ‡**

[a] The version is SMOTE borderline-2
[b] The version is NearMiss-2

Looking only at these imbalanced methods, it is clear that all three random approaches are a bad choice, independently of the algorithm employed. Maybe the flexibility provided by its nonlinear nature makes SVM be the least bad. It seems logical that SMOTE+Tomek is the best when SMOTE was the best among the oversampling methods and Tomek's links among the undersampling methods. In this case it has occurred, but it is not always necessary the case.

4 Conclusions

We have presented in detail methods covering all steps involved in preprocessing in predictive maintenance, both for offline and online learning scenarios when the latter was feasible, as well as provided the reader with exhaustive bibliographic references.

We have performed several experiments on public available real-world data from the PHM Data Challenges 2014 and 2016, so that we could empirically test some of the presented approaches.

We have seen that the data seem to have higher relevance than the posterior modeling technique in order to determine the preprocessing schema, both for regression and classification problems.

As possible extensions, some online tests could be performed, in order to check the online versions of the methods. Besides, more preprocessing strategies, modeling algorithms, and datasets could be considered in order to extend the study and check with higher certainty whether the modeling algorithm is much less relevant than the data for the adequate preprocessing scheme.

Acknowledgements This research is supported by the Basque Government through the BERC 2018–2021 and ELKARTEK programs and through project KK-2018/00071; and by Spanish Ministry of Economy and Competitiveness MINECO through BCAM Severo Ochoa excellence accreditation SEV-2017-0718, and through project TIN2017-82626-R.

References

1. Bartlett, M.: An inverse matrix adjustment arising in discriminant analysis. Ann. Math. Stat. **22**(1), 107–111 (1951)
2. Batista, G.E.A.P.A., Prati, R.C., Monard, M.C.: A study of the behavior of several methods for balancing machine learning training data. ACM SIGKDD Explor. Newsl. **6**(1), 20–29 (2004)
3. Battiti, R.: Using mutual information for selecting features in supervised neural net learning. IEEE Trans. Neural Netw. **5**(4), 537–550 (1994)
4. Benkedjouh, T., Medjaher, K., Zerhouni, N., Rechak, S.: Remaining useful life estimation based on nonlinear feature reduction and support vector regression. Eng. Appl. Artif. Intell. **26**(7), 1751–1760 (2013)
5. Box, G.E.P., Cox, D.R.: An analysis of transformations. J. R. Stat. Soc. Ser. B **26**(2), 211–252 (1964)
6. Bracewell, R.N.: The Fourier Transform and Its Applications, 3rd edn. McGraw-Hill, Boston (2000). ISBN 0-07-116043-4
7. Bradley, A.P.: The use of the area under the ROC curve in the evaluation of machine learning algorithms. Pattern Recogn. **30**(7), 1145–1159 (1997)
8. Branden, K.V., Hubert, M.: Robust classification in high dimensions based on the SIMCA method. Chemom. Intell. Lab. Syst. **79**, 10–21 (2005)
9. Breiman, L.: Bagging predictors. Mach. Learn. **24**(2), 123–140 (1996)
10. Breiman, L.: Random forests. Mach. Learn. **45**(1), 5–32 (2001)
11. Brown, G.: A new perspective for information theoretic feature selection. J. Mach. Learn. Res. **13**, 27–66 (2012)
12. Cabrera, D., Sancho, F., Sánchez, R.V., Zurita, G., Cerrada, M., Li, C., Vásquez, R.E.: Fault diagnosis of spur gearbox based on random forest and wavelet packet decomposition. Front. Mech. Eng. **10**(3), 277–286 (2015)
13. Cernuda, C., Lughofer, E., Märzinger, W., Kasberger, J.: NIR-based quantification of process parameters in polyetheracrylat (PEA) production using flexible non-linear fuzzy systems. Chemom. Intell. Lab. Syst. **109**(1), 22–33 (2011)
14. Cernuda, C., Lughofer, E., Suppan, L., Röder, T., Schmuck, R., Hintenaus, P., Märzinger, W., Kasberger, J.: Evolving chemometric models for predicting dynamic process parameters in viscose production. Anal. Chim. Acta **725**, 22–38 (2012)
15. Cernuda, C., Lughofer, E., Hintenaus, P., Märzinger, W., Reischer, T., Pawliczek, M., W., Kasberger, J.: Hybrid adaptive calibration methods and ensemble strategy for prediction of cloud point in melamine resin production. Chemom. Intell. Lab. Syst. **126**, 60–75 (2013)
16. Cernuda, C., Lughofer, E., Mayr, G., Röder, T., Hintenaus, P., Märzinger, W., Kasberger, J.: Incremental and decremental active learning for optimized self-adaptive calibration in viscose production. Chemom. Intell. Lab. Syst. **138**, 14–29 (2014)
17. Cernuda, C., Lughofer, E., Klein, H., Forster, C., Pawliczek, M., Brandstetter, M.: Improved quantification of important beer quality parameters based on nonlinear calibration methods applied to FT-MIR spectra. Anal. Bioanal. Chem. **409**(3), 841–857 (2017)
18. Chawla, N.V.: C4.5 and imbalanced data sets: investigating the effect of sampling method, probabilistic estimate, and decision tree structure. In: Proceedings of the ICML'03 Workshop on Learning from Imbalanced Data sets, Washington, DC, USA (2003)
19. Chawla, N.V.: Data mining for imbalanced datasets: an overview. In: Maimon, O., Rokach, L. (eds.) Data Mining and Knowledge Discovery Handbook, 2nd edn., pp. 875–886. Springer, New York (2010)
20. Chawla, N.V., Bowyer, K.W., Hall, L.O., Kegelmeyer, W.P.: SMOTE: synthetic minority oversampling technique. J. Artif. Intell. Res. **16**, 321–357 (2002)
21. Cho, K., Merriënboer, B., Gulcehre, C., Bougares, F., Schwenk, H., Bahdanau, D., Bengio, Y.: Learning Phrase Representations using RNN Encoder-Decoder for Statistical Machine Translation. Computer Research Repository (CoRR). arXiv: 1406.1078 (2014)
22. Chollet, F., et al.: Keras (2015). https://github.com/fchollet/keras

23. Chuang, A.: Time series analysis: univariate and multivariate methods. Technometrics **33**(1), 108–109 (1991)
24. Cohen, L.: Time-Frequency Analysis. Prentice-Hall, New York (1995). ISBN 978-0135945322
25. Covell, M.M., Richardson, J.M.: A new, efficient structure for the short-time Fourier transform, with an application in code-division sonar imaging. In: International Conference on Acoustics, Speech, and Signal Processing (ICASSP), vol. 3, pp. 2041–2044 (1991)
26. Daubechies, I.: Orthonormal bases of compactly supported wavelets. Commun. Pure Appl. Math. **41**(7), 909–996 (1988)
27. Drummond, C., Holte, R.: C4.5, class imbalance, and cost sensitivity: why undersampling beats over-sampling. In: Proceedings of the ICML'03 Workshop on Learning from Imbalanced Data Sets, Washington, DC, USA (2003)
28. Dudani, S.A.: The distance-weighted k-nearest neighbor rule. IEEE Trans. Syst. Man Cybern. **SMC-6**(4), 325–327 (1976)
29. Duhamel, P., Vetterli, M.: Fast Fourier transforms: a tutorial review and a state of the art. Signal Process. **19**(4), 259–299 (1990)
30. Fayyad, U.M., Irani, K.B.: Multi-interval discretization of continuous-valued attributes for classification learning. In: 13th International Joint Conference on Artificial Intelligence, pp. 1022–1027 (1993)
31. Ferri, C., Flach, P., Orallo, J., Lachice, N. (eds.): ECAI'2004 First Workshop on ROC Analysis in Artificial Intelligence (2004)
32. Fleuret, F.: Fast binary feature selection with conditional mutual information. J. Mach. Learn. Res. **5**, 1531–1555 (2004)
33. Freedman, D., Diaconis, P.: On the histogram as a density estimator: ℓ_2 theory. Probab. Theory Relat. Fields **57**(4), 453–476 (1981)
34. Friedman, N., Geiger, D., Goldszchmidt, M.: Bayesian network classifiers. Mach. Learn. **29**(2–3), 131–163 (1997)
35. Frigo, M., Johnson, S.G.: A modified split-radix FFT with fewer arithmetic operations. IEEE Trans. Signal Process. **55**(1), 111–119 (2007)
36. García, S., Luengo, J., Sáez, J.A., López, V., Herrera, F.: A survey of discretization techniques: taxonomy and empirical analysis in supervised learning. IEEE Trans. Knowl. Data Eng. **25**(4), 734–750 (2013)
37. Garvey, D., Wigny, R.: PHM Data Challenge 2014. PHM Society. https://www.phmsociety.org/sites/phmsociety.org/files/PHM14DataChallenge.pdf (2014)
38. Gelper, S., Schettlinger, K., Croux, C., Gather, U.: Robust online scale estimation in time series: a model-free approach. J. Stat. Plann. Inference **139**(2), 335–349 (2008)
39. Gerretzen, J., Szymańska, E., Jansen, J., Bart, J., van Manen, H.-J., van den Heuvel, E.R., Buydens, L.: Simple and effective way for data preprocessing selection based on design of experiments. Anal. Chem. **87**(24), 12096–12103 (2015)
40. Graves, A., Schmidhuber, J.: Framewise phoneme classification with bidirectional LSTM and other neural network architectures. Neural Netw. **18**(5–6), 602–610 (2005)
41. Greff, K., Srivastava, R.K., Koutník, J., Steunebrink, B.R., Schmidhuber, J.: LSTM: a search space odyssey. IEEE Trans. Neural Netw. Learn. Syst. **28**(10), 2222–2232 (2017)
42. Guha, S., Rastogi, R., Shim, K.: CURE: an efficient clustering algorithm for large databases. In: SIGMOD'98, Proceedings of the 1998 ACM SIGMOD International Conference on Management of Data, pp. 73–84 (1998)
43. Guo, L., Ma, Y., Cukic, B., Singh, H.: Robust prediction of fault-proneness by random forests. In: 15th International Symposium on Software Reliability Engineering, pp. 417–428 (2004)
44. Guyon, I., Elisseeff, A.: An introduction to variable and feature selection. J. Mach. Learn. Res. **3**(7–8), 1157–1182 (2003)
45. Guyon, I., Elisseeff, A.: An introduction to feature extraction. In: Guyon, I., Nikravesh, M., Gunn, S., Zadeh, L.A. (eds.) Feature Extraction. Studies in Fuzziness and Soft Computing, vol. 207, pp. 1–25. Springer, Berlin/Heidelberg (2006)

46. Hall, M.A.: Correlation-based feature selection for machine learning. PhD Thesis, University of Waikato, Hamilton (1999)
47. Hart, P.E.: The condensed nearest neighbor rule. IEEE Trans. Inf. Theory **14**, 515–516 (1968)
48. Hastie, T., Tibshirani, R., Friedman, J.: Pathwise coordinate optimization. Ann. Appl. Stat. **1**(2), 302–332 (2007)
49. Hastie, T., Tibshirani, R., Friedman, J.: The Elements of Statistical Learning: Data Mining, Inference, and Prediction, 2nd edn. Springer Series in Statistics. Springer, New York (2009)
50. Hastie, T., Tibshirani, R., Friedman, J.: Regularized paths for generalized linear models via coordinate descent. J. Stat. Softw. **33**(1), 1–22 (2010)
51. He, H., Bai, Y., García, E.A., Li, S.: ADASYN: adaptive synthetic sampling approach for imbalanced learning. In: IEEE International Joint Conference on Neural Networks, IEEE World Congress on Computational Intelligence, Hong Kong, pp. 1322–1328 (2008)
52. He, X., Niyogi, P.: Locality preserving projections. In: Proceedings of the 16th International Conference on Neural Information Processing Systems (NIPS'03), pp. 153–160 (2003)
53. Hinton, G., Roweis, S.: Stochastic neighbor embedding. In: Proceedings of the 15th International Conference on Neural Information Processing Systems (NIPS'02), pp. 857–864 (2002)
54. Hochreiter, S., Schmidhuber, J.: Long short-term memory. Neural Comput. **9**(8), 1735–1780 (1997)
55. Hofmann, T., Schölkopf, B., Smola, A.J.: Kernel methods in machine learning. Ann. Stat. **36**(3), 1171–1220 (2009)
56. Huang, N.E., Shen, Z., Long, S.R., Wu, M.C., Shih, H.H., Zheng, Q., Yen, N.-C., Tung, C.C., Liu, H.H.: The empirical mode decomposition and the Hilbert spectrum for nonlinear and non-stationary time series analysis. Proc. R. Soc. Lond. A Math. Phys. Eng. Sci. **454**, 903–995 (1998)
57. Hubert, M., Rousseeuw, P., Branden, K.V.: Robpca: a new approach to robust principal component analysis. Technometrics **47**, 64–79 (2005)
58. Japkowicz, N.: The Class imbalance problem: significance and strategies. In: Proceedings of the 2000 International Conference on Artificial Intelligence (IC-AI'2000): Special Track on Inductive Learning, pp. 111–117, Las Vegas, Nevada (2000)
59. Jo, T., Japkowicz, N.: Class imbalances versus small disjuncts. ACM SIGKDD Explor. Newsl. **6**(1), 40–49 (2004)
60. Johnson, N., Kotz, S., Balakrishnan, N.: Continuous Univariate Distributions. Wiley Series in Probability and Mathematical Statistics: Applied Probability and Statistics, vol. 2. Wiley, New York (1995)
61. Jolliffe, I.: Principal Components Analysis. Springer, Berlin/Heidelberg/New York (2002)
62. Jung, M., Niculita, O., Skaf, Z.: Comparison of different classification algorithms for fault detection and fault isolation in complex systems. Proc. Manuf. **19**, 111–118 (2018)
63. Kadambe, S., Boudreaux-Bartels, G.F.: A comparison of the existence of cross terms in the Wigner distribution and the squared magnitude of the wavelet transform and the short-time Fourier transform. IEEE Trans. Signal Process. **40**(10), 2498–2517 (1992)
64. Kalchbrenner, N., Danihelka, I., Graves, A.: Grid Long Short-Term Memory. Computer Research Repository (CoRR). arXiv: 1507.01526 (2015)
65. Kohavi, R., John, G.H.: Wrappers for feature subset selection. Artif. Intell. **97**(1–2), 273–324 (1997)
66. Kubat, M., Matwin, S.: Addressing the curse of imbalanced training sets: one sided selection. In: Proceedings of the Fourteenth International Conference on Machine Learning, pp. 179–186. Morgan Kaufmann, Nashville, TN (1997)
67. Kwak, N., Choi, C.: Input feature selection for classification problems. IEEE Trans. Neural Netw. **13**(1), 143–159 (2002)
68. Laurikkala, J.: Improving identification of difficult small classes by balancing class distribution. In: AIME'01, Proceedings of the 8th Conference on Artificial Intelligence in Medicine in Europe, pp. 63–66 (2001)
69. Li, M.: Fractal time series-a tutorial review. Math. Probl. Eng. **2010**, 1–26 (2010)

70. Li, D., Deogun, J., Spaulding, W., Shuart, B.: Towards missing data imputation — A study of fuzzy k-means clustering method. In: Tsumoto, S., Sowiski, R., Komorowski, J., Grzymaa-Busse, J. (eds.) Rough Sets and Current Trends in Computing (RSCTC 2004). Lecture Notes in Computer Science, vol. 3066, pp. 573–579. Springer, Berlin/Heidelberg (2004)
71. Lin, D., Tang, X.: Conditional infomax learning: an integrated framework for feature extraction and fusion. In: Leonardis A., Bischof H., Pinz A. (eds) Computer Vision – ECCV 2006. ECCV 2006. Lecture Notes in Computer Science, vol. 3951, pp. 68–82. Springer, Heidelberg (2006)
72. Ling, C., Li, C.: Data mining for direct marketing problems and solutions. In: Proceedings of the Fourth International Conference on Knowledge Discovery and Data Mining (KDD-98), pp. 73–79. AAAI Press, New York, NY (1998)
73. Loutas, T., Roulias, D., Georgoulas, G.: Remaining useful life estimation in rolling bearings utilizing data-driven probabilistic ϵ-support vectors regression. IEEE Trans. Reliab. **62**(4), 821–832 (2013)
74. Lughofer, E.: FLEXFIS: a robust incremental learning approach for evolving Takagi-Sugeno fuzzy models. IEEE Trans. Fuzzy Syst. **16**(6), 1393–1410 (2008)
75. Maaten, L., Hinton, G.: Visualizing data using t-SNE. J. Mach. Learn. Res. **9**, 2579–2605 (2008)
76. Maesschalck, R.D., Candolfi, A., Massart, D., Heuerding, S.: Decision criteria for soft independent modelling of class analogy applied to near infrared data. Chemom. Intell. Lab. Syst. **47**, 65–77 (1999)
77. Mahalanobis, P.: On the generalised distance in Statistics. Proc. Natl. Inst. Sci. India **2**(1), 49–55 (1936)
78. Mallat, S.G.: A theory for multiresolution signal decomposition: the wavelet representation. IEEE Trans. Pattern Anal. Mach. Intell. **11**(7), 674–693 (1989)
79. Maloof, M.: Learning when data sets are imbalanced and when costs are unequal and unknown. In: Proceedings of the ICML'03 Workshop on Learning from Imbalanced Data Sets, Washington, DC (2003)
80. Mann, H.B., Whitney, D.R.: On a test of whether one of two random variables is stochastically larger than the other. Ann. Math. Stat. **18**(1), 50–60 (1947)
81. Nikzad-Langerodi, R., Lughofer, E., Cernuda, C., Reischer, T., Kantner, W., Pawliczek, M., Brandstetter, M.: Calibration model maintenance in melamine resin production: integrating drift detection, smart sample selection and model adaptation. Anal. Chim. Acta **1013**, 1–12 (2018)
82. Nunkesser, R., Fried, R., Schettlinger, K., Gather U.: Online analysis of time series by the Q_n estimator. Comput. Stat. Data Anal. **53**(6), 2354–2362 (2009)
83. Oba, S., Sato, M., Takemasa, I., Monden, M., Matsubara, K., et al.: A Bayesian missing value estimation method for gene expression profile data. Bioinformatics **19**, 2088–2096 (2003)
84. Oliveira, M.A., Araujo, N.V.S., Silva, R.N., Silva, T.I., Epaarachchi, J.: Use of Savitzky-Golay filter for performances improvement of SHM systems based on neural networks and distributed PZT sensors. Sensors **18**(1), 152 (2018)
85. Pedrycz, W., Gomide, F.: Fuzzy Systems Engineering: Toward Human-Centric Computing. Wiley, Hoboken, NJ (2007)
86. Peng, H., Long, F., Ding, C.: Feature selection based on mutual information: criteria of max-dependency, max-relevance, and min-redundancy. IEEE Trans. Pattern Anal. Mach. Intell. **27**(8), 1226–1238 (2005)
87. Phua, C., Alahakoon, D.: Minority report in fraud detection: classification of skewed data. ACM SIGKDD Explor. Newsl. **6**(1), 50–59 (2004)
88. Propes, N.C., Rosca, J.: PHM Data Challenge 2016. PHM Society. https://www.phmsociety.org/sites/phmsociety.org/files/PHM16DataChallengeCFP.pdf (2016)
89. Qiu, G.: An improved recursive median filtering scheme for image processing. IEEE Trans. Image Process. **5**(4), 646–648 (1996)
90. Rezgui, W., Mouss, N.K., Mouss, L.H., Mouss, M.D., Benbouzid, M.: A regression algorithm for the smart prognosis of a reversed polarity fault in a photovoltaic generator. In: 2014 International Conference on Green Energy, pp. 134–138 (2014)

91. Roweis, S., Saul, L.: Nonlinear dimensionality reduction by locally linear embedding. Science **290**(5500), 2323–2326 (2000)
92. Rubin, D.B.: Multiple Imputation for Nonresponse in Survey, vol. 1. Wiley, New York (2008)
93. Saeys, Y., Inza, I., Larrañaga, P.: A review of feature selection techniques in bioinformatics. Bioinformatics **23**(19), 2507–2517 (2007)
94. Said, S.E., Dickey, D.A.: Testing for unit roots in autoregressive-moving average models of unknown order. Biometrika **71**(3), 599–607 (1984)
95. Sakia, R.M.: The Box-Cox transformation technique: a review. Statistician **41**(2), 169–178 (1992)
96. Savitzky, A., Golay, M.J.E.: Smoothing and differentiation of data by simplified least squares procedures. Anal. Chem. **36**(8), 1627–1639 (1964)
97. Schölkopf, B., Smola, A., Müller, K.R.: Kernel principal component analysis. In: Gerstner W., Germond A., Hasler M., Nicoud JD. (eds.) Artificial Neural Networks – ICANN'97. Lecture Notes in Computer Science, vol. 1327. Springer, Berlin/Heidelberg (1997)
98. Schölkopf, B., Smola, A.J.: Learning with Kernels - Support Vector Machines, Regularization, Optimization and Beyond. MIT Press, London (2002)
99. Serdio, F., Lughofer, E., Pichler, K., Buchegger, T., Pichler, M., Efendic, H.: Multivariate Fault Detection Using Vector Autoregressive Moving Average and Orthogonal Transformation in Residual Space. In: 2013 Annual Conference of the Prognostics and Health Management (PHM) Society, New Orleans, LA, pp. 1–8 (2013)
100. Serdio, F., Lughofer, E., Pichler, K., Buchegger, T., Efendic, H.: Residual-based fault detection using soft computing techniques for condition monitoring at rolling mills. Inf. Sci. **259**, 304–320 (2014)
101. Serdio, F., Lughofer, E., Pichler, K., Buchegger, T., Pichler, M., Efendic, H.: Fault detection in multi-sensor networks based on multivariate time-series models and orthogonal transformations. Inf. Fusion **20**, 272–291 (2014)
102. Serdio, F., Lughofer, E., Zavoianu, A.C., Pichler, K., Buchegger, T., Pichler, M., Efendic, H.: Improved fault detection employing hybrid memetic fuzzy modeling and adaptive filters. Appl. Soft Comput. **51**, 60–82 (2017)
103. Shannon, C.E.: A mathematical theory of communication. Bell Syst. Tech. J. **27**(3), 379–423 (1948)
104. Sharpley, R.C., Vatchev, V.: Analysis of the intrinsic mode functions. Constr. Approx. **24**(1), 17–47 (2006)
105. Silverman, B.W., Jones, M.C.: An important contribution to nonparametric discriminant analysis and density estimation: commentary on Fix and Hodges (1951). Int. Stat. Rev. **57**(3), 233–238 (1989)
106. Smith, M.R., Martínez, T., Giraud-Carrier, C.: An instance level analysis of data complexity. Mach. Learn. **95**(2), 225–256 (2014)
107. Smola, A.J., Schölkopf, B.: A tutorial on support vector regression. Stat. Comput. **14**, 199–222 (2004)
108. Solberg, A. H., Solberg, R.: A large-scale evaluation of features for automatic detection of oil spills in ERS SAR images. In: International Geoscience and Remote Sensing Symposium, pp. 1484–1486 (1996)
109. Tan, L., Jiang, J.: Digital Signal Processing: Fundamentals and Applications, 2nd edn. Academic/Elsevier, New York (2013)
110. Tenenbaum, J.B., Silva, V., Langford, J.C.: A global geometric framework for nonlinear dimensionality reduction. Science **290**(5500), 2319–2323 (2000)
111. Tibshirani, R.: Regression shrinkage and selection via the lasso. J. R. Stat. Soc. B **58**, 267–288 (1996)
112. Tomek, I.: Two modifications of CNN. IEEE Trans. Syst. Man Cybern. **6**, 769–772 (1976)
113. Troyanskaya, O., Cantor, M., Sherlock, G, Brown, P., Hastie, T., et al.: Missing value estimation methods for DNA microarrays. Bioinformatics. **17**, 520–525 (2001)
114. Tschumitschew, K., Klawonn, F.: Incremental quantile estimation. Evol. Syst. **1**(4), 253–264 (2010)

115. Vapnik, V: Statistical Learning Theory. Wiley, New York (1998)
116. Varmuza, K., Filzmoser, P.: Introduction to Multivariate Statistical Analysis in Chemometrics. CRC Press, Boca Raton (2009)
117. Vidal-Naquet, M., Ullman, S.: Object recognition with informative features and linear classification. In: 9th IEEE Conference on Computer Vision and Pattern Recognition, vol. 2, pp. 281–288 (2003)
118. Ville, J.: Théorie et Applications de la Notion de Signal Analytique. Câbles et Transm. **2**, 61–74 (1948)
119. Wang, C., Zhang, Y., Zhong, Z.: Fault diagnosis for diesel valve trains based on time–frequency images. Mech. Syst. Signal Process. **22**(8), 1981–1993 (2008)
120. Weaver, H.J.: Applications of Discrete and Continuous Fourier Analysis. Wiley, New York (1983)
121. Weiss, G., Provost, F.: Learning when training data are costly: the effect of class distribution on tree induction. J. Artif. Intell. Res. **19**, 315–354 (2003)
122. Welch, P.: The use of fast Fourier transform for the estimation of power spectra: a method based on time averaging over short, modified periodograms. IEEE Trans. Audio Electroacoust. **15**(2), 70–73 (1967)
123. Wilson, D.L.: Asymptotic properties of nearest neighbor rules using edited data. IEEE Trans. Syst. Man Cybern. **SMC-2**(3), 408–421 (1972)
124. Wolpert, D., Macready, W.: No free lunch theorems for optimization. IEEE Trans. Evol. Comput. **1**(1), 67–82 (1997)
125. Wu, T.Y., Chen, J., Wang, C.X.: Characterization of gear faults in variable rotating speed using Hilbert-Huang transform and instantaneous dimensionless frequency normalization. Mech. Syst. Signal Process. **30**, 103–122 (2012)
126. Wu, D., Jennings, C., Terpenny, J., Gao, R., Kumara, S.: A comparative study on machine learning algorithms for smart manufacturing: tool wear prediction using random forests. J. Manuf. Sci. Eng. **139**(7), 071018 (2017)
127. Yang, H., Moody, J.: Data visualization and feature selection: new algorithms for nongaussian data. Adv. Neural Inf. Process. Syst. **12**, 687–693 (1999)
128. Yang, B.S., Di, X., Han, T.: Random forests classifier for machine fault diagnosis. J. Mech. Sci. Technol. **22**, 1716–1725 (2008)
129. Yao, K., Cohn, T. Vylomova, K., Duh, K., Dyer, C.: Depth-Gated Long Short-Term Memory. Computer Research Repository (CoRR). arXiv: 1508.03790 (2015)
130. Zavoianu, A.C., Lughofer, E., Bramerdorfer, G., Amrhein, W., Klement, E.P.: An effective ensemble-based method for creating on-the-fly surrogate fitness functions for multi-objective evolutionary algorithms. In: International Symposium on Symbolic and Numeric Algorithms for Scientific Computing (SYNASC 2013), pp. 235–242 (2013)
131. Zhang, J., Mani, I.: kNN approach to unbalanced data distributions: a case study involving information extraction. In: Proceedings of the ICML'2003 Workshop on Learning from Imbalanced Datasets, Washington, DC, USA (2003)
132. Zhang, L., Xiong, G., Liu, H., Zou, H., Guo, W.: Bearing fault diagnosis using multi-scale entropy and adaptive neuro-fuzzy inference. Expert Syst. Appl. **37**(8), 6077–6085 (2010)
133. Zou, H. Hastie, T.: Regularization and variable selection via the elastic net. J. R. Stat. Soc. **67**(2), 301–320 (2005)

Part I
Anomaly Detection and Localization

A Context-Sensitive Framework for Mining Concept Drifting Data Streams

Chamari I. Kithulgoda and Russel Pears

1 Concept Drifting Data Streams

Data stream mining has been extensively researched over the last decade and a half. Research in this area has challenged conventional thinking and forced the research community to extend solutions that were developed for environments where the statistical properties of data remain static over time. Various researchers have catalogued the major challenges to mining open-ended streams of data. These challenges include novel management schemes to ensure that available memory does not overflow. In order to meet memory constraints, a new breed of incremental learners that scan data samples at most once was developed.

Alongside developments in novel memory management strategies and incremental learning methods, several researchers realized the critical importance of detecting changes in the underlying data distribution of samples arriving in the stream. This is the issue that ultimately distinguishes data stream mining from classical machine learning. If data is stationary in nature, there is no need to incrementally train classifiers and other types of learners, one can simply take a sample of available data and build a model on that sample using classical methods. Unfortunately, the reality is that most data streams experience flux, and this dictates that models be synchronized with changes that occur periodically in the stream. The realization that concept change detection plays a central role in data stream mining sparked off a flurry of research in this area. Many machine learning researchers turned to the statistical literature to develop change detection algorithms that would both accurately and efficiently detect changes in a data stream context. Given this central role of change detection, this chapter will present a framework that will support

C. I. Kithulgoda (✉) · R. Pears
Department of Computer Science, Auckland University of Technology, Auckland, New Zealand
e-mail: ckithulg@aut.ac.nz

© Springer Nature Switzerland AG 2019
E. Lughofer, M. Sayed-Mouchaweh (eds.), *Predictive Maintenance in Dynamic Systems*, https://doi.org/10.1007/978-3-030-05645-2_4

online learning in a non-stationary environment. In recognition of the proliferation of data streams resulting from the Internet of Things and other data sources, the emphasis will be on supporting high-speed data streams efficiently, in terms of processing and memory overheads. An implementation of this framework and the insights gained by experimenting with both synthetic and real-world data will also be presented.

Before the framework can be discussed, some important questions regarding the nature of concept drift need to be addressed. These include: In what forms do concept drift occur? How do we recognize drift? Finally, what actions do we perform once drift is detected?

1.1 Concept Drift

In essence, the change in the relationship between an outcome variable and its observed features is called a concept drift. In real-world scenarios, the reasons behind these changes are unforeseen, and neither the frequency nor the exact time of occurrence is certain. In general, three types of concept change could occur. In the first, the change is triggered by an alteration to the mapping between feature variables and the outcome. It could also be caused by a change in the joint distribution of the feature variables. Furthermore, both of these cases can happen simultaneously.

In formal terms, Gama et al. [7] define concept drift as the dissimilarity of the joint probability distribution of input features and class label at two subsequent time points t_0 and t_1. We adopt the definition given by that study:

$$\exists X : p_{t_0}(X, y) \neq p_{t_1}(X, y) \tag{1}$$

This dissimilarity of the likelihood of events X and y occurring can be caused by changes in components ([8, 13] as cited by [7]), namely the prior probabilities $p(y)$ of classes or the conditional probabilities of classes, that is $p(X|y)$. These changes result in a change in the posterior probability $p(y|X)$. This change in $p(y|X)$ over time is the reason behind accuracy fluctuations.

Accordingly, any solution should have the capability of sensing the changes in $p(y|X)$ throughout the life span of the data stream in a time-efficient manner in order to maintain classification accuracy. This is called the concept drift detection problem which has been widely studied [1, 2, 19, 21]. Another classification of concept drift is the speed with which it occurs. A sudden deviation in the $p(y|X)$ value is said to be an abrupt change whereas deviations that occur in an incremental and cumulative manner over time are said to be gradual. Abrupt changes require a fast response to the change in order to preserve accuracy. This in turn requires that an alternative model that is better suited to the new concept be deployed as soon as the drift is detected. In order to achieve this goal, a pool of learners needs to be available at any given point in time. Deployment can be done through a switch from learner

L_1 to another learner L_2 that is better suited to the new concept whenever a drift is detected. If no alternative learner is available that matches to the new concept, then accuracy will be severely compromised until one or more of the learners adjusts to the new concept.

On the other hand, gradual changes allow time for the system to adjust to the change, and individual learners may have adapted sufficiently well to cope with the new concepts. Thus, in general, their effects may not be as severe as with abrupt drift.

Yet another categorization of drift is whether the drift pattern reappears over a period of time. Such recurrences may follow a periodic pattern to a greater or lesser degree or be aperiodic and completely unpredictable in its recurrence pattern. In either case, the action that needs to be performed at detection time is a switch to a new learner, just as with the case of abrupt drift. However, unlike with the case of abrupt drift, if the recurrence pattern is strong, i.e. repeated appearances have statistical properties very similar to each other, then it could be profitable to store such concepts separately in an online repository which is separate from the pool of learners that adapt their models over time. The use of the repository will guarantee that models associated with recurring concepts are preserved in their original form in between successive recurrences. However, they may be subject to change at the next appearance, and a new version of the recurring concept may then be stored in its place in the repository. The key issue here is that its update cycle is quite different from that of concepts that change and evolve over time without recurring.

2 A Novel Framework for Online Learning in Adaptive Mode

2.1 Basic Components

In this section, we present a generic framework for online learning in non-stationary environments. The framework is generic in the sense that it is able to cope with all of the drift types that we identified in Sect. 1. Furthermore, the framework is modular in design as each component can have different implementations corresponding to different methods that have been proposed to solve a particular issue in learning with concept drift.

We start by arguing the case for each component. An incremental learner that restructures models by synchronizing changes in data patterns to models is indispensable to cope with high data arrival rates in a data stream. The synchronization of changes in data patterns is accomplished through the use of a concept drift detector. Without a drift detector, a learner will experience severe drops in accuracy from time to time, and hence it is also a mandatory component. We have seen in Sect. 1 that a repository of past concepts is useful to maintain in cases where a recurring drift pattern is present. Thus, the basic components required to support online learning in non-stationary environments are:

- an incremental classifier,
- a concept drift detector,
- an online repository of past concepts.

2.2 Optimizing for Stream Volatility and Speed

The above components are basic in the sense that they support core functionality, but other supporting elements such as memory management and support for high-speed streams are also essential. In streams that are highly volatile, many different concepts can manifest, and it may not be feasible to store all concepts in the repository even if compression were to be applied. This calls for a memory management scheme that goes beyond a simple first-in first-out strategy of populating past concepts in the repository.

At the same time, the framework should be able to take advantage of periods of low volatility to speed up processing by reusing already learnt models coupled with a minimal amount of learning that is needed to reflect changes in the recurring concepts. This would result in speeding up the learning process and would require a mechanism to sense the level of volatility in the stream. The volatility detector could then adjust the mode of learning from an intensive learning mode to a less intensive one or vice versa, as the case may be. The level of volatility could be estimated by monitoring whether the probability p_r of usage of past concepts in the repository is significantly higher than the probability p_n of usage of concepts that are evolving or new. If this is the case, then it indicates that the system is operating in a less volatile state, and learning can then be adjusted accordingly. Learning in a less volatile state can rely to a large extent on classifiers stored in the repository with minor adjustments if needed, and hence should be more efficient than learning in a high-volatile state where new concepts need to be learned. In Sect. 3, we will present the staged learning approach that will implement this key notion of sensitivity to stream volatility.

3 Implementation of a Context-Sensitive Staged Learning Framework

We refer to the framework as being context sensitive as it recognizes system behaviour and tailors the learning strategy accordingly. Thus, it is able to recognize periods of stability, stages in which concepts are in a state of change, periods of concept reoccurrence and finally, system states with different levels of volatility. We discuss the design choices needed to achieve context sensitivity, starting with choices available for each of the basic components. We then go on to discuss the staged learning approach and volatility detection.

Each of these components can be implemented in several different ways. A large number of incremental classifiers have been proposed for data stream mining including the decision tree group of classifiers, Bayesian classifiers and others. A popular class of incremental learners use ensembles of decision trees with the Hoeffding tree as the basic learning mechanism. Decision trees have proved to be a popular choice in data stream mining on account of their being efficient to learn and being able to cope with interdependence between features, unlike the Naive Bayes classifier. Some examples of incremental classifiers using ensembles of decision trees include CBDT [10], OzaBagADWIN [3], LeveragingBag [4] and Adaptive Random Forest [9]. It is clear that any of the decision tree ensembles can be used as the incremental classifier component, and we have chosen CBDT as our implementation choice for the incremental classifier component.

In terms of concept drift detectors, a large choice of drift detectors exist including EDDM [1], ADWIN [2] and SeqDrift2 [19], amongst others. All of these detectors require the same input, which is a binary stream of the truth value of classification decisions, while all of them produce an output which is a binary variable, indicating whether or not drift took place. Hence, the drift detector component is completely interchangeable amongst the drift detectors that are currently available. Our implementation choice was SeqDrift2 on account of its low false positive rate and optimized drift detection delay in relation to other drift detectors [19].

With respect to an online repository, there have been two different approaches so far proposed in the literature. The first by Ramamurthy and Bhatnagar [20] stores decision trees in their original form in the repository. The second approach used in [14, 15, 22, 23] is to compress decision trees into Fourier spectra by applying the discrete Fourier transform (as illustrated in Sect. 3.1), and then storing the resulting spectra in the repository. Classification can be performed directly on the spectra by applying the inverse Fourier transform without having to recover the original tree, thus making such a solution attractive on account of the compression achieved. The time to classify new data could also reduce due to the compact nature of the spectrum. The trade-off with better memory utilization is the transformation cost but this is a one-time cost. When concept drift is signalled by the drift detector, all classifiers (including the spectra in the repository) are polled to determine which one has the best classification accuracy on the new concept, and this particular classifier is then used on the current concept. This process and its result is independent of whether or not Fourier-based compression is used, and hence the repository component is also interchangeable. Our implementation uses Fourier spectra as classifiers in the repository. Before we elaborate further on the staged learning framework, we present the fundamentals of the Fourier transform in Sect. 3.1 as it plays a crucial role in the success of the framework.

3.1 The Use of the Discrete Fourier Transform in Classification and Concept Encoding

The use of the Discrete Fourier Transform (DFT) in data mining has been of recent origin and has been focused on deriving a Fourier spectrum from decision trees. We first present a basic overview of the derivation of the multivariate DFT from a decision tree and then go on to describe the setting in which our Fourier encoding and classification scheme is applied.

Before we present the mathematical foundations of the DFT, we map the fundamental ideas underpinning the Fourier transform to their meanings in Table 1 in order to communicate their roles in an intuitive manner.

A Fourier spectrum is derived from a Fourier basis set which consists of a set of orthogonal functions that are used to represent a discrete function. Consider the set of all d-dimensional feature vectors where the lth feature can take λ_l different discrete values, $\{0, 1, \ldots, \lambda_l - 1\}$. The Fourier basis set that spans this space consists of $\prod_{l=1}^{d} \lambda_l$ basis functions. Each Fourier basis function is defined as:

$$\psi_j^\lambda(\mathbf{x}) = \frac{1}{\sqrt[d]{\prod_{l=1}^{d} \lambda_l}} \prod_{l=1}^{d} \exp\left(\frac{2\pi l x_l j_l}{\lambda_l}\right) \tag{2}$$

where $\mathbf{j}$ and $\mathbf{x}$ are vectors of length d; and $x(l)$, $j(l)$ are the lth attribute values in $\mathbf{x}$ and $\mathbf{j}$, respectively. The vector $\mathbf{j}$ is called a partition and its order is the number of non-zero feature values it contains.

A function $f : X^d \to \mathscr{R}$ that maps a d-dimensional discrete domain to a real-valued range can be represented using the Fourier basis functions:

Table 1 Mapping of Fourier concepts to their intuitive meanings

Symbol	Meaning
x	A schema consists of a vector of feature values drawn from features that comprise the dataset. A schema is a compact way of defining a set of data instances, all of which share the same set of feature values
X	The schema set which contains the set of all possible schema for a given dataset
j	This is a partition of the feature space. Essentially, it is also a vector of feature values, just as with a schema. The only (conceptual) difference is that a schema refers to the data, whereas a partition indexes a Fourier spectrum
J	The partition set that defines the number of coefficients in the spectrum and its size
$w_\mathbf{j}$	A coefficient in the Fourier spectrum
$\psi_j^\lambda(\mathbf{x})$	This is the Fourier basis function that takes as input a feature vector and a partition vector and produces an integer for a dataset with binary-valued features or a complex number for a dataset with non-binary feature values

A Context-Sensitive Framework for Mining Concept Drifting Data Streams

$$f(\mathbf{x}) = \sum_{j \in X} \overline{\psi_j}^\lambda(\mathbf{x}) w_\mathbf{j}, \qquad (3)$$

where $w_\mathbf{j}$ is the Fourier Coefficient (FC) corresponding to the partition $\mathbf{j}$ and $\overline{\psi_j}^\lambda(\mathbf{x})$ is the complex conjugate of $\psi_j^\lambda(\mathbf{x})$. Henceforth, we shall drop the superscript λ from the ψ_j function formulation to simplify the presentation. The Fourier coefficient $w_\mathbf{j}$ can be viewed as the relative contribution of the partition $\mathbf{j}$ to the function value of $f(x)$ and is computed from:

$$w_\mathbf{j} = \prod_{i=1}^{l} \frac{1}{\lambda_i} \sum_{x \in X} \psi_\mathbf{j}(\mathbf{x}) f(\mathbf{x}) \qquad (4)$$

In a data mining context, $f(\mathbf{x})$ represents the classification outcome of a given data instance $\mathbf{x} \in X$. Each data $\mathbf{x}$ must conform to a schema and many data instances in the stream may map to the same schema. For example, in Fig. 1, many data instances for schema $(0, 0, 1)$ may occur at different points in the stream. Henceforth in the paper, we shall refer to schema instances rather than data instances as our Fourier classifier operates at the schema, rather than at the data instance level. Thus, we shall adopt the notation $\mathbf{x}$ to denote a schema instance, rather than a data instance. The set X is the set of all possible schema, and for the simple example in Fig. 1 it is of size 8.

The absolute value of $w_\mathbf{j}$ can be used as the "significance" of the corresponding partition $\mathbf{j}$. If the magnitude of some $w_\mathbf{j}$ is very small compared to other coefficients, we consider the jth partition to be insignificant and neglect its contribution. The order of a Fourier coefficient is simply the order of its corresponding partition. We will use terms like high-order or low-order coefficients to refer to a set of Fourier coefficients whose orders are relatively large or small, respectively.

The Fourier spectrum of a decision tree can be computed using the class outcomes predicted by its leaf nodes. As an example, consider the decision tree in Fig. 1 defined on a binary-valued domain consisting of three features. Its truth table derived from the predictions made by the tree and the corresponding Fourier

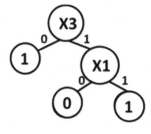

x_1	x_2	x_3	f(x)		FC	Value
0	0	0	1		w_{000}	3/4
0	0	1	0		w_{001}	1/4
0	1	0	1		w_{010}	0
0	1	1	0		w_{011}	0
1	0	0	1		w_{100}	-1/4
1	0	1	1		w_{101}	1/4
1	1	0	1		w_{110}	0
1	1	1	1		w_{111}	0

Fig. 1 A decision tree and its equivalent Fourier spectrum

spectrum that results appears in Fig. 1. Below, we illustrate the computation of jth Fourier coefficient $w_\mathbf{j}$ for a data with d binary-valued features which is given by the Boolean domain version [18] of Eq. (4):

$$w_\mathbf{j} = \frac{1}{2^d} \sum_X \psi_\mathbf{j}(\mathbf{x}) f(\mathbf{x}) \tag{5}$$

where $f(\mathbf{x})$ is the class outcome predicted by the leaf node with path vector x and $\psi_\mathbf{j}(\mathbf{x})$ the Fourier basis function given by the simplified version of Eq. (2):

$$\psi_j(\mathbf{x}) = (-1)^{(j \cdot x)} \tag{6}$$

Considering three binary-valued features X_1, X_2 and X_3 given in Fig. 1, only X_1 and X_3 are appeared in tree and hence contributed to calculation. The study of Park [18] guaranteed that coefficients for paths which are defined by attributes need to be computed since other coefficients are zero in value. Thus, coefficients $w_{010}, w_{011}, w_{110}$ and w_{111} are zero. Computation for non-zero coefficients w_{000} and w_{001} are as follows:

$$w_{000} = \frac{1}{2^3} \sum_X \psi_j(000) f(x) = \frac{1+0+1+0+1+1+1+1}{8} = 3/4$$

$$w_{001} = \frac{1}{2^3} \sum_X \psi_j(001) f(x) = \frac{1+0+1+0+1+(-1)+1+(-1)}{8} = 1/4$$

The Fourier spectrum derived from a decision tree is compact due to the two following properties:

1. The number of non-zero coefficients is polynomial in the number of features represented in the tree [11].
2. The magnitude of the coefficients $w_\mathbf{j}$ decreases exponentially with the order of the partition $\mathbf{j}$ [11, 12].

These two properties collectively make a spectrum derived from a tree very attractive. Firstly, the tree provides a natural filtering mechanism as typically only a fraction of the features have sufficient information gain to be represented in the tree. Once the tree is in place, only the set of low-order coefficients defined from partitions appearing in the tree make a significant contribution to the classification outcomes.

Kargupta and Park in [11, 12] made use of spectral energy to derive a cut-off point for coefficient order. Given a spectrum s, its energy E is defined by: $E = \sum_{j \in s} |w_\mathbf{j}^2|$. For a given energy threshold T, the subset of s (in ascending spectral order) whereby $E \geq T$ is retained; all other coefficients are deemed to be zero and removed from the array. Thus for example in the spectrum defined in Fig. 1, the first-order coefficients contain $\frac{9+0+1+1}{9+0+1+1+0+0+1+0} = 91.7\%$ of the total energy and so with a threshold of

90%, only coefficients w_{000}, w_{001}, w_{100} should be retained, thus halving the size of the spectrum that needs to be maintained.

Moreover, as discussed in [11] and implemented in [22], spectra can be aggregated with each other. Aggregation of spectra was implemented via a pair-wise algebraic summation [22] of the spectra involved as given in Eq. (7):

$$s_c(x) = \sum_i A_i \sum_i s_i(x)$$
$$= \sum_i A_i \sum_{j \in Q_i} \omega_j^{(i)} \overline{\psi}_{j(x)} \qquad (7)$$

where $s_c(x)$ denotes the ensemble spectrum produced from the individual spectra $s_i(x)$ produced at different points i in the stream; A_i is the classification accuracy of its corresponding spectrum and Q_i is the set of partitions for non-zero coefficients in spectrum s_i.

In our implementation, a pre-check is first done to determine whether or not a new spectrum should be aggregated with an already existing spectrum in the repository. If this check is passed, then aggregation proceeds with the spectrum that is most similar to the newly generated spectrum by applying Eq.(7). If the similarity is below a given threshold, the new spectrum is inserted into the repository as it is.

Aggregation of spectra brings with it two major benefits. Firstly, a reduction in space as coefficients common to spectra being aggregated need to be stored only once. Secondly, as demonstrated in our empirical study, aggregation performs a similar role to an ensemble of models and leads to better generalizability to new data arriving in the stream.

Once a Fourier spectrum is derived from a decision tree, it can fully replace the latter as classification of a newly arriving instance **x** can be computed by applying the inverse transform given in Eq. (3) over the set J that contains the reduced set of coefficients that survive the energy thresholding process.

Computing of the classification outcome for a given schema 010 through the Fourier spectrum is illustrated below:

$$f(x) = \sum_j \overline{\psi}_j(x) w_j$$

$$f(010) = \sum_j (-1)^{(j.010)} w_j$$

$$f(010) = (-1)^{(000.010)} w_{000} + (-1)^{(001.010)} w_{001} + (-1)^{(010.010)} w_{010}$$
$$+ (-1)^{(011.010)} w_{011} + (-1)^{(100.010)} w_{100}$$
$$+ (-1)^{(101.010)} w_{101} + (-1)^{(110.010)} w_{110} + (-1)^{(111.010)} w_{111}$$
$$= \frac{3}{4} + \frac{1}{4} - \frac{1}{4} + \frac{1}{4} = \frac{7}{4} = 1$$

3.2 Repository Management

As spectra in the repository may accumulate in number, it will be necessary to implement a memory management strategy to ensure that memory does not overflow in the repository. A simple strategy would be to delete the oldest spectrum when memory is not available to store a newly created spectrum. We choose instead to aggregate newly created spectra with existing spectra as a memory saving measure. This aggregation scheme is implemented by Eq. (7) as noted earlier in Sect. 3.1.

If memory is not available to store a newly created spectrum S_{new}, then its classifications on a test data segment of a certain size N are assessed against the classifications produced by spectra that exist in the repository. Suppose that C is the number of classes. Then, the winner spectrum S_{new} is aggregated with the spectrum S_a determined by Eq. (11) if the distance similarity (E) is greater than the given threshold for similarity.

$$c(S_{new}(i)) = \lceil f(S_{new}(i)) \times C \rceil \tag{8}$$

$$c(S_p(i)) = \lceil f(S_p(i)) \times C) \rceil \tag{9}$$

$$E = N - \sum_{p=1}^{P}\sum_{i=1}^{N} I(c(S_{new}(i)) \neq c(S_p(i))) \tag{10}$$

where P is the number of spectra in the repository.

$$S_a = \underset{p}{\operatorname{argmax}}(E) \tag{11}$$

In Eqs. (8) and (9), the class labels $c(S_{new}(i))$ and $c(S_p(i))$ of the ith data instance are determined for spectra S_{new} and S_p, respectively, by applying the inverse Fourier transform on the respective spectra to reconstruct a numeric approximation of the class value which is converted to a class label by multiplying by the number of classes C and then taking the ceiling of the resulting numeric value that is returned. The same operation is performed on all spectra which are already in the repository.

For example, if the number of classes $C=3$ and if the inverse Fourier value returned for data instance i with spectrum S_{new} is 0.61, then clearly instance i should be labelled with class value 2 as the class boundaries are [0.0...0.33], [0.34...0.66], [0.67...1.0]. This label of 2 is recovered by multiplying 0.61×3 and then taking the ceiling of 1.83, giving a class label of 2.

Equation (10) computes the distance similarity E between S_{new} and a spectrum S_p in the repository by counting the number of instances that return the same class labels over the test segment of size N and then subtracting the total count by N

to get the similarity score. The identity function I where $I(b) = 1$ if b is true, 0 otherwise, is used to determine if S_{new}, S_p agree or not on class outcomes.

In Eq. (11), the spectrum S_a that has the maximum similarity with S_{new} is returned. If this maximum distance similarity is smaller than the given threshold, aggregation does not take place. In the case when the repository is full and insufficient distance similarity exists to meet aggregation requirements, the least accurate spectrum is removed to make way for the newly generated spectrum.

3.3 The Staged Learning Approach

So far in our presentation, the emphasis has been on maintaining classification accuracy in the presence of concept drift and we have not yet attempted to improve performance, apart from a possible improvement in classification time resulting from more compact classifiers in the repository. As mentioned briefly in Sect. 2.2, one optimization that could result in significant improvements to system throughput would be to model a data stream as a state machine that models interactions between two states. The first state can be thought of as a "learning state" (henceforth referred to as a Stage 1) where new concepts appear, and these concepts are learned and stored as classifiers in the form of decision trees, Fourier spectra or other types of models.

The second state is a "deployment state" (henceforth referred to as Stage 2) in which concept drifts still appear but the vast majority of drifts take place between concepts already learned in the first state. If such concepts are stored online in the repository, then classifiers representing these concepts could be deployed as they were without the need for re-learning them when concepts recur. While this could yield significant gains in throughput, concepts may undergo some change when they reappear, and in practice some level of learning may also need to take place in Stage 2.

In keeping with the low-volatile nature of Stage 2, the decision tree forest is suspended and Fourier spectra are used as classifiers. When a concept drift occurs in Stage 2 processing, the spectrum S that reports the best accuracy is chosen as the classifier for the current concept. However, it is possible that some concepts may undergo change after reappearance, and at the end point of a concept's progression, as signalled by concept drift, it may happen that S may no longer be a precise representation of the concept. In order to synchronize S with the changed state of the concept, a single decision tree is used to learn any changes that might take place in the current concept. This tree is induced from spectrum S at the start of the concept and thereafter learns any changes that may take place after that point onwards. We used the tree induction algorithm proposed in [18] for this purpose. At the end of the concept, the tree is transformed into a new spectrum S_{new} which is placed into the repository if space permits, otherwise it is aggregated with its closest matching spectrum in the repository using the process described in Sect. 3.2.

An alternative strategy would have been a simple replacement of S with S_{new} but we believe that aggregation would enable historical properties of the concept to be preserved, thus offering a better generalization capability. Overall, the processing overhead in Stage 2 is much lesser than in Stage 1 as only a single tree needs to be maintained in Stage 2 in contrast to a forest of trees in Stage 1. In addition, as mentioned earlier, Fourier spectra are more compact than their decision tree counterparts and hence classification can also be expected to be more efficient. Our experimental results on throughput presented in Sect. 4 show clearly that this is the case.

The staged approach is a generalization of the "learn then deploy" paradigm used in classical machine learning on a stationary data environment. The difference here is that many cycles of learning and deployment may occur within a data stream, unlike with a stationary environment which involves just one (unless the data miner retrains a classifier periodically).

This staged approach is consistent with Kleinberg's [16] modelling of text streams as bursty collections of data in which a volatile phase containing new, previously unseen data patterns is interleaved with stable segments of text containing previously seen word patterns.

Figure 2 shows the interactions between the major components of the staged learning framework (adapted from [14]). The incremental classifier component consists of a forest of decision trees. Each tree in the decision tree forest operates under the control of a drift detector.

The system starts off in Stage 1 with the repository in an empty state. Classification is initially done in a grace period G with a randomly selected tree from the forest. This tree is designated as the "winner" classifier, meaning that it is solely

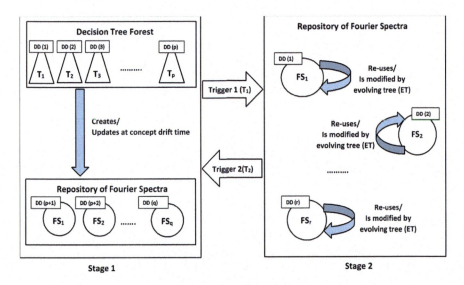

Fig. 2 Staged learning framework for context-sensitive learning

responsible for classifying data arriving in the stream until a concept drift occurs. Within the span of the grace period, the drift detection buffer of each drift detector associated with a tree is populated with its own classification decision, irrespective of whether or not it is the designated winner classifier. At the expiration of the grace period, the tree that returns the highest average accuracy is chosen as the new winner tree, and this tree is chosen to classify new data arriving in the stream beyond the grace period. This process continues until a drift signal is produced. At drift point, the classification accuracy of each tree in the forest is assessed, and a new winner is selected which will be responsible for classification until the next drift occurs. At each drift point, the winner tree is compressed by applying the DFT, and the resulting spectrum is stored in the repository. Once spectra appear in the repository, they can be used for classification, just as with trees in the forest. As spectra are classifiers in their own right, they too operate under the control of drift detectors.

In Stage 2, each Fourier spectrum is paired with its own evolving tree (ET). As described before, the tree ET is used to synchronize the current state of a concept with the spectrum that it is paired with.

The staged approach requires a mechanism for determining the stream state and for transiting between states. Transition from Stage 1 to 2 is governed by the firing of a trigger T_1 when a shift from high-stream volatility to low volatility is identified by the volatility detector. On the other hand, the reverse shift from low-stream volatility to high volatility is triggered by T_2. Details of how these triggers function appear in the subsequent section.

3.3.1 Transition Between Stages

We capture volatility shift through the application of rigorous statistical methods. We first present a formal definition of volatility and then proceed to illustrate how shifts in volatility are detected by framing the volatility shift problem in terms of a concept drift problem.

Definition 1 Volatility is defined as the rate of appearance of new concepts in the stream with respect to time. In any given stream segment of length l if n new concepts appear, then volatility is defined as the probability of appearance of a new concept and is estimated by $\frac{n}{l}$.

Note that the definition is based on the appearance of new concepts and not on the probability of concept drift taking place. Concept drift can occur as a result of concept changing over to a new, previously unseen concept or reverting to a previously seen concept. If s switches in concept take place in a stream segment of size l, then in general only $n(<s)$ of them will be new and hence the volatility rate as defined above as $\frac{n}{l}$ will be less than the rate of concept drift, $\frac{s}{l}$.

Although Definition 1 characterizes volatility, its utility in practice is limited unless a method can be found to measure the rate at which new concepts appear in a given data stream. With this in mind, we consider the role that the repository plays in classification.

With the staged transition learning framework in place, as long as a concept exists in the online repository that matches the newly emerging concept in the stream that is signalled by the drift detector, then no re-learning is required. In such cases, classification is performed with the concept stored in the repository. This suggests that the rate of re-use of objects in the repository can be taken as a proxy for the rate of appearance of new concepts. The higher the rate of re-use, the lower is the rate of appearance of new concepts in the stream, and lower is the volatility. We are now in a position to provide an operational definition for volatility.

Definition 2 At any given point in time during the operation of Stage 1 with the occurrence of s concept drifts, volatility is estimated as: $1 - \frac{\sum_{i=1}^{s} B(R)}{s}$ where $B(R)$ is a Boolean-valued function that returns "1" if the newly emerging concept i is found in the repository, otherwise it returns value "0".

Definition 2 quantifies volatility in terms of the empirical hit (success) rate of the repository in finding emerging concepts, which is given by: $h = \frac{\sum_{i=1}^{s} B(R)}{s}$. The higher the repository hit rate, the lower is the volatility. We note that with a drift detector in place that has high sensitivity and low false positive rate, the hit rate, and hence volatility can be determined. We are now in a position to determine the transition point between Stages 1 and 2.

In order to determine the transition point, a window of size w is maintained that contains samples drawn from values returned by function $B(R)$ defined above. A check for a transition detection point is made after the arrival of every s concept drifts. The window is divided into a left sub-window of size $w - s$ with the right sub-window containing the last s samples.

Definition 3 A transition from Stage 1 to Stage 2 occurs if at a concept drift point i the repository hit ratio satisfies:

- condition 1: $h_r > h_l$,
- condition 2: $h_r > \alpha$.

where h_l, h_r are the hit ratios across the left and right sub-windows, respectively, and α is a user-defined threshold on hit ratio. Definition 3 establishes that the transition point i is reached only when an upward shift in the hit ratio takes place in the window prior to the hit ratio exceeding α.

In practice, the hit ratio is a random variable, and ensuring conditions 1 and 2 requires statistical significance tests to be made. To check validity of condition 1, we formulate a one-tailed statistical hypothesis test $H_0 : h_l \geq h_r$ versus $H_1 : h_l < h_r$, where h_l, h_r represent the population means of the data across the left and right sub-windows, respectively.

Thus, it can be seen that volatility detection is essentially a second-order determination of concept drift. Amongst the set of drifts recorded in the stream if the rate of change of appearance of new concepts, as signalled by the repository hit ratio, is on a statistically significant decreasing trend, then sufficient evidence exists that the system has transited to a low-volatile state (Stage 2). The implication is that the volatility detection problem can then be framed in terms of a concept drift problem and we make use of the SeqDrift2[19] drift detector for volatility detection as well.

The SeqDrift2 detector makes use of the sample means $\hat{h}_l, \hat{h}_r$ and a threshold ϵ_1 to determine if condition 1 is satisfied. If $\hat{h}_r - \hat{h}_l > \epsilon_1$, then H_0 is rejected with probability $(1-\delta)$, else H_1 is rejected. The ϵ_1 threshold is given by: $\frac{1}{3(1-k)n_r}(p + \sqrt{p^2 + 18\sigma_s^2 n_r p})$ where $p = \ln \frac{4}{\delta}$, $k = \frac{n_r}{n_r + n_l}$, σ_s = sample variance, n_r = size of the right sub-window, n_l = size of the left sub-window and δ = drift significance level.

If H_1 is rejected, then we can conclude that no significant increase in hit ratio has occurred in the current window, and hence we proceed to update the left sub-window with samples from the right sub-window before proceeding to gather a new set of s samples in a new right sub-window for re-testing H_0 versus H_1.

If there is evidence to reject H_0 at the δ significance level, then the implication is that the stream is moving towards Stage 2 since classification is relying increasingly on the repository that contains previously captured concepts, in preference to the forest. However, as yet there is no definitive evidence to transit to Stage 2 as the recent activity may still not be high enough to justify suspending the operation of the forest of trees. Thus, a further hypothesis test for condition 2 is carried out to ascertain whether $\hat{h}_r$ is greater than some acceptable threshold value α.

To check the validity of condition 2, hypothesis H_2 tests whether $h_r \leq \alpha$ versus H_3 which corresponds to: $h_r > \alpha$. If H_2 is rejected in favour of H_3, then the stream is considered to have transited to Stage 2. If not, the stream is still in Stage 1 and at the arrival of the next s samples, condition 2 is re-evaluated. As with the test for condition 1 above, we use the sample hit ratio $\hat{h}_r$ and a threshold ϵ_2 to execute the test. If $\hat{h}_r - \alpha > \epsilon_2$, then H_2 is rejected in favour of H_3 with probability $(1-\delta)$. The threshold ϵ_2 is computed using the Hoeffding bound and is given by: $\epsilon_2 = \sqrt{\frac{\ln \frac{1}{\delta}}{2n_r}}$.

We now turn our attention to trigger T_2. The rationale behind T_2 is based on tracking how good the spectra in the repository are in classifying concepts that are evolving. To the extent that spectra return high classification accuracy, Stage 2 processing should continue. High classification accuracy can result when concepts produced by the stream are similar to those produced in the past or there is a collection of trees that is capable of reacting quickly to the arrival of several new concepts that are dissimilar to those seen in the past. Thus, when the spectra produced through growth of a regenerated tree starts to deviate sharply from those already in the repository in terms of structural similarity, then we have an indication that the concepts appearing in the stream are novel in the sense that they have not been captured previously in the stream. We are now in a position to implement T_2 once we have a definition for structural similarity in place.

Definition 4 Structural similarity sim_C between the evolving tree ET and the repository R is given by: $\text{sim}_C = \left(\frac{\max_{S \in R} \sum_i (B[S(i) = ET(i)])}{m} \right)$ where C is the current concept; B is a Boolean-valued function that returns binary "1" for data instance i if the classification outcome from spectrum S matches with the classification outcome for tree ET; if no match is produced, then binary "0" is returned in the window; m is the length of the current concept drift point—the number of data instances in the concept; and i is an index that ranges over the data instances in the current concept.

As illustrated in Definition 4, the similarity score sim_C returns the degree of agreement between the tree ET and its best matching spectrum S in the repository. It is measured at each concept drift point C. If $\text{sim}_C < \beta$, then binary "1" is written to the change detection window, otherwise "0" is recorded.

As with trigger T_1, we use SeqDrift2 as the change detector. SeqDrift2's window is split into left and right sub-windows and with the arrival of every s concept drift points, the null hypothesis $H_4 : \mu_l \geq \mu_r$ is tested against $H_5 : \mu_l < \mu_r$. If H_4 is rejected with probability $(1 - \delta)$, then we transit back to Stage 1 as the right sub-window shows a significant increase in occurrence of structural dissimilarity; otherwise, Stage 2 processing continues. The intuition behind triggering T_2 is that a state change back to Stage 1 needs to occur when the concepts stored in the form of spectra in the repository becomes sufficiently dissimilar to the currently emerging concepts in the stream.

3.4 Space and Time Complexity of Spectral Learning

The space and throughput advantages of SOL result from spectral learning. Here, we illustrate the space and time complexity for the classification based on spectra in the repository.

An upper bound for the space consumption of spectra is determined by the total number of coefficients M taken over all spectra in the repository. Assuming that there are P spectra and that Q_p is the size of the coefficient array for the pth spectrum, the space complexity can be given in the following equation:

$$M = \sum_{p=1}^{P} Q_p \tag{12}$$

The spectral size Q_p is determined by the order O_p for the given energy threshold, as described in Sect. 3.1. Hence, we have

$$Q_p = \sum_{r=0}^{O_p} {}^d C_r \tag{13}$$

where d is the number of data features and ${}^d C_r$ is the number of combinations of selecting r features from a total of d. The memory complexity M is then given by:

$$M = \sum_{p=1}^{P}\sum_{r=0}^{O_P} {}^d C_r \tag{14}$$

The Fourier classifier does not rely on a deep hierarchical tree structure but instead uses a self-indexing hashing scheme to store its schema values. This is compact due to the reasons mentioned in Sect. 3.1.

Now that an upper bound for the number of coefficients in a given spectrum has been formulated, the time complexity of the IFT operation as defined by Eq. (3) can be expressed as:

$$O(Md^2) \tag{15}$$

as d^2 multiplications are needed for the computation of the vector product between **x** and **j** for each of the coefficients in the array.

4 Empirical Study

Our empirical study consists of five basic components. Firstly, we test the effectiveness of the Staged Online Learning (SOL) approach on two key performance measures: per-stage classification accuracy and per-stage processing speed. The study compares SOL against the Ensemble Pool (EP) [22] and Recurrent Classifier (RC). The EP employs a set of stored Fourier spectra together with the CBDT[10] forest of decision tree learners. Compared to SOL, the EP approach does not employ either the staged learning approach or the reconstructed evolving tree, and hence it is one of the ideal methods to compare against. The RC is a slightly different staged learning approach that simply uses Fourier spectra generated in Stage 1 to classify data arriving in Stage 2 without any form of adaptive learning. In other words, this classifier, termed as Recurrent, assumes that concepts appearing in stage 2 are exact recurrences of those that appeared in Stage 1. This version provides a useful contrast with the SOL proposed in this study as it enables us to assess the benefits of learning through a reconstructed evolving decision tree in Stage 2.

We then go on to examine overall accuracy and throughput of SOL and compare it to several well-known classifiers. In this connection, we include two more classifiers that have been proposed for concept drifting data streams. These algorithms are state-of-the-art meta learning algorithms featured in MOA,[1] namely Adaptive Random Forest (ARF) [9] and LeveragingBag (LB) [4]. The performance of classifiers under comparison is assessed on both accuracy rank and throughput

---

[1] From http://moa.cms.waikato.ac.nz/.

rank simultaneously in the next experiment. The memory consumption evaluation follows last. In the next section, sensitivity analysis is carried out to assess the effect of SOL's parameters on its performance.

All experiments were conducted on a Windows 10 Enterprise 64-bit machine featuring Intel Core i5 processor running at 3.2 GHz and having 16 GB of RAM. The framework was implemented using C# 5.0 in .NET Framework 4.5 runtime environment.

4.1 Datasets Used for the Empirical Study

All experiments were carried out with the use of two synthetic datasets generated by MOA's stream generators (See footnote 1) and four real-world datasets.

4.1.1 Synthetic Data

We generated concepts of length 10,000 instances. Drift signals were applied at two levels on the concepts: firstly, abrupt drift was injected to produce a set of distinctive concepts, and then at the second level, recurrences of concepts generated at the first level were produced, as depicted in Fig. 3. Each concept recurred with a varying degree of change from its first appearance, depending on its cycle of repetition.

Two different types of change were introduced at level 2. Firstly, in Sect. 4.1.2, a given amount of noise was superimposed on the recurring concepts to differentiate them from their previous appearance.

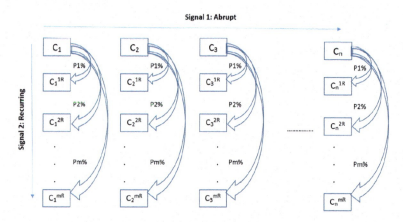

C_a^{bR} : b^{th} recurrence of ath concept, Pi: percentage of class flips

Fig. 3 Preparation of synthetic datasets with two levels of drift signals

Secondly, in addition of noise, drift patterns were injected into the data, as described in Sects. 4.1.3 and 4.1.4. In Sect. 4.1.3, a progressively increasing drift pattern was used, whereas in Sect. 4.1.4 an oscillating pattern was superimposed on the recurrence signal. To the best of our knowledge, this is the first experimental study of its kind that embeds several different drift patterns simultaneously in its data. More details can be found below.

4.1.2 Synthetic Data Recurring with Noise

For synthetic datasets, we generated several distinct concepts, each of length 10,000 instances.

In the following two synthetic datasets, we introduced a 5% noise level for each concept recurrence by inverting the binary class label of randomly selected data instances with the belief that "concepts do not repeat in exactly the same form" in reality. In this case, the probability of class flips P_i in the ith repetition remains constant at 5% throughout each of the m repetitions. This introduces random noise, not resulting in any meaningful pattern.

1. Rotating Hyperplane dataset (Noisy RH): This dataset has a total of ten attributes, and we adjusted the magnitude of change parameter in the range [0.03, 0.04, 0.05, 0.07, 0.08, 0.09] to generate six distinct concepts. The first three concepts were repeated 20 times more, with each concept being distorted by a noise level of 5% at each cycle over its base representation (i.e. its first generated state). Three previously unseen concepts were injected at the end of datasets with the intention of examining whether the staged learners would opt for adapting to the new concepts in Stage 2 or for triggering T_2 to transit back to Stage 1. The size of the dataset was 660,000 data instances.
2. RBF dataset (Noisy RBF): This ten-dimensional dataset generated concepts by changing the number of centroids. We produced 12 different concepts. The first five concepts were repeated nine more cycles with noise as per the description for RH. The remaining seven concepts were appended in order to simulate the appearance of completely new concepts in the stream. The size of this dataset was 570,000 data instances.

The objective of including several concept repetitions was to evaluate the capability of triggering T_1 which should cause our system to transit to Stage 2 when recurring concepts present themselves in the stream. Further to that, sensitivity of trigger T_2 was assessed by injecting new concepts to test whether trigger T_2 would reactivate Stage 1 operation.

4.1.3 Synthetic Data Recurring with a Progressively Increasing Pattern of Drift

With the objective of evaluating the robustness of our proposed framework to any given scenario, we created two more RH datasets by injecting two different monotonically increasing drift intensities on the first five concepts of the data stream. Whenever an instance belonging to any one of these concepts recurred with a certain attribute value, the class label for that instance was inverted to the other class with a given probability. In both of these datasets, the drifts between two consecutive concepts were abrupt while the recurrences gradually deviated from their last occurrence. Experimentation with two different drift intensities is described below:

1. RH progressive increase of 10% in flip probability over cycles (10% progressive RH): the original five concepts reappeared in ten more cycles of repetitions, each of which had 10% (where $P2 = P1 + 10\%, P3 = P2 + 10\% \ldots Pm = P(m-1) + 10\%$ as per Fig. 3) greater flip probability of its class label in comparison to the previous cycle. We challenged the process of pattern recognition by introducing 30% flips of class labels in the first repetition and thereafter 40%, 50% and so on up to 100%.
2. RH progressive increase of 20% in flip probability over cycles (20% progressive RH): the original five concepts reappeared in five more cycles of repetition, each of which had 20% (Where $P2 = P1 + 20\%, P3 = P2 + 20\% \ldots Pm = P(m-1) + 20\%$ as per Fig. 3) greater flip probability of class label than in the previous cycle. We challenged the process of pattern recognition by introducing 20% flips of class labels in the first repetition and thereafter 40%, 60% and so on up to 100%.

4.1.4 Synthetic Data Recurring with an Oscillating Drift Pattern

Further to the above, we test the learning capability of the classifiers when the pattern in between recurrences tended to oscillate, rather than being monotonic in nature:

1. RH Oscillating flips (Oscillating RH): In this dataset, we appended repetition cycles by interleaving flip probability: $P1 = 30\%, P2 = 70\%, P3 = 40\%, P4 = 80\%, P5 = 50\%, P6 = 90\%, P7 = 60\% and P8 = 100\%$ as per Fig. 3.

4.1.5 Real-World Data

1. Electricity (ELEC) dataset: NSW Electricity dataset is used in its original form.[2] There are two classes *Up* or *Down* that indicate the change of price with respect to the moving average of electricity prices in the last 24 h.
2. Flight dataset: This dataset[3] was generated by NASA's FLTz flight simulator which was designed to simulate flight conditions experienced in commercial flights. Each flight has four different concepts, corresponding to four flight scenarios: take off, climb, cruise and landing. The "Velocity" feature was discretized into two binary outcomes *Up* or *Down* depending on the directional change of the moving average in a window of size 10 data instances.
3. Covertype dataset: The original version of this dataset is available at [17]. The data was collected from Roosevelt National Forest of Northern Colorado for the task of predicting forest cover type from 54 attributes derived from 12 cartographic variables. We extracted the initial 10% of instances from the two most frequent forest types, namely "Spruce-Fir" and "Lodgepole Pine".
4. Occupancy Detection dataset: This dataset was also obtained from [17] and used by [5]. The dataset consists of measurements of temperature, humidity, light and CO_2 levels in a given room and was collected with the purpose of deciding suitability of the room for human occupancy. Ground-truth occupancy label outcomes were determined on the basis of pictures taken at intervals of 1 min.

4.2 Parameter Values

The default parameter values used in the experimentation are as follows:

Maximum number of nodes in decision tree forest: 5000, SeqDrift drift significance value $(\delta) = 0.01$, maximum number of Fourier spectra in repository: 40, repository hit ratio threshold α for T_1 is 0.5 and the similarity threshold β for T_2 is 0.7.

4.3 Effectiveness of Staged Learning Approach

In order to test whether a significant difference in performance exists between the proposed SOL classifier, EP and RC, we examined both throughput and accuracy profiles of these algorithms. Tables 2, 3, 4, 5, and 6 show average values for throughput and accuracy for five datasets. The remaining four datasets follow the

---

[2]From http://moa.cms.waikato.ac.nz/datasets/.
[3]From https://c3.nasa.gov/dashlink/resources/.

Table 2 Stage-wise throughput and accuracy profiles for the Noisy RH dataset

		Stage 1 (start–160,000)	Stage 2 (160,000–end)	Overall
SOL	Accuracy	72.2	73.8	73.4
	Throughput	7973	18,304	13,930
	Repository hit ratio	0.36	–	–
	Similarity score	–	90.0	–
EP	Accuracy	72.2	72.8	72.6
	Throughput	–	–	6403
RC	Accuracy	72.2	63.6	65.7
	Throughput	–	–	16,824

Table 3 Stage-wise throughput and accuracy profiles for the Noisy RBF dataset

		Stage 1 (start–80,000)	Stage 2 (80,000–560,000)	Stage 1 (560,000–end)	Overall
SOL	Accuracy	80.9	76.0	71.3	76.6
	Throughput	4357	4514	2976	4451
	Repository hit ratio	0.48	–	–	–
	Similarity score	–	88.8	–	–
EP	Accuracy	80.9	76.0	70.2	76.6
	Throughput	–	–	–	1435
RC	Accuracy	80.9	60.4	45.0	63.0
	Throughput	–	–	–	4777

Table 4 Stage-wise throughput and accuracy profiles for the Flight dataset

		Stage 1 (start–4000)	Stage 2 (4000–end)	Overall
SOL	Accuracy	77.1	82.7	81.8
	Throughput	1399	7201	4331
	Repository hit ratio	0.14	–	–
	Similarity score	–	97.6	–
EP	Accuracy	77.1	80.7	80.1
	Throughput	–	–	982
RC	Accuracy	77.1	52.9	56.8
	Throughput	–	–	4553

same trends and were omitted in the interest of saving space. The repository hit ratio of Stage 1 which represents the usage of stored classifiers was also presented for SOL. In addition, we tracked the average similarity score to gain insights into the extent of change in the stream during Stage 2 processing for SOL.

The objective of experimenting with the Noisy RH and Noisy RBF recurring datasets was to assess the sensitivity of trigger T_1 in detecting recurring concepts by transiting from Stage 1 to Stage 2. We also tracked the sensitivity of trigger T_2

Table 5 Stage-wise throughput and accuracy profiles for the ELEC dataset

		Stage 1 (start–8000)	Stage 2 (8000–44,000)	Stage 1 (44,000–end)	Overall
SOL	Accuracy	67.0	66.7	83.5	67.1
	Throughput	14,220	32,447	13,998	25,661
	Repository hit ratio	0.38	–	–	–
	Similarity score	–	85.7	–	–
EP	Accuracy	67.0	65.8	83.5	66.4
	Throughput	–	–	–	11,559
RC	Accuracy	67.0	64.9	83.0	65.7
	Throughput	–	–	–	34,589

Table 6 Stage-wise throughput and accuracy profiles for the Covertype dataset

		Stage 1 (start–12,000)	Stage 2 (12,000–20,000)	Stage 1 (20,000–36,000)	Stage 2 (36,000–48,000)	Stage 1 (48,000–end)	Overall
SOL	Accuracy	79.3	82.6	88.7	87.0	77.3	84.8
	Throughput	492	7314	389	2510	461	663
	Repository hit ratio	0.25	–	0.23	–	0.07	–
	Similarity score	–	93.7	–	95.2	–	–
EP	Accuracy	79.3	82.4	87.4	93.3	87.5	86.0
	Throughput			–	–	–	265
RC	Accuracy	79.3	56.8	57.8	56.8	60.0	62.7
	Throughput	–		–	–	–	1626

in transiting back to Stage 1 when newly injected concepts appeared in the stream. The results presented in Tables 2 and 3 depict the role of triggers T_1 and T_2 through necessary transitions through the stages. In case of the RH noisy dataset, we did not observe T_2 firing even though we injected three previously unseen concepts at the end of the stream. This illustrates the ability of handling new concepts without dropping accuracy while working in stage 2.

The results presented in Tables 4, 5 and 6 provide clear evidence of the presence of recurrence patterns in the real-world datasets as well (not just for RH and RBF for which they were injected) as SOL fired trigger T_1 on all of them. The fact that SOL triggered it on RH and RBF where there were known recurrences indicates that SOL is sensitive to state changes in the system. Furthermore, the results demonstrate the effectiveness of trigger T_2 in real-world scenarios as well.

In summary, we note that SOL had significant improvements in throughput over EP, ranging from 117.6% for noisy RH, 210.2% for noisy RBF, 341.0% for Flight, 122.0% for Electricity and 150.2% for Covertype dataset. As expected, the reduction in overheads caused by replacing a forest of learners with a mechanism of refining already stored concepts yielded significant gains in throughput. Obviously, the throughput gains for the RC classifier were even higher than that of SOL but they came at a heavy price in accuracy.

The competitive accuracy returned by SOL is supported by the high similarity scores registered by SOL in Stage 2 processing for all datasets. A high level of similarity between the concepts being formed and those already present in the repository implies that spectra already generated during Stage 1 are effective in classifying newly arriving data in Stage 2.

Finally, we note that the Flight dataset contains strong episodes of concept recurrence as T_2 did not fire until the end point was reached. This type of behaviour is expected to be exhibited in a number of real-world datasets. Given that Covertype does not have an explicit time dimension but rather a spatial one, yet another insight that we can gain from the Covertype results is that the staged approach is effective at capturing recurrences defined on a spatial dimension. Interestingly, we note from the relatively poor performance of RC in Stage 2 that while recurrences are present, they are not in exact form, once again underscoring the need for a limited learning capability in Stage 2 to learn small-scale changes in concepts.

Thus, we note that relying totally on spectra stored during Stage 1 is not a viable strategy. This is illustrated by the difference in classification accuracy between SOL and RC. The SOL classifier significantly outperforms RC, thus demonstrating the need for learning and refinement of the spectra in Stage 2.

4.4 Accuracy Evaluation

In this section, we compared overall accuracy of SOL against ARF, LB, EP and RC. All classifiers were run with the Hoeffding tree as the base learner. The leaf prediction method was set to majority class. The split confidence and tie confidence parameters were both set at 0.01 for all classifier ensembles in order to maintain fairness. Each meta learner was run with default settings for its internal parameters. Accuracies for each classifier was obtained in intervals of size 1000 and an overall accuracy was then computed by averaging over all intervals for a given dataset. Accuracies were stable across multiple runs for all classifiers across all datasets with a standard deviation less than 0.05 and hence were not included. The winner accuracy (with accuracies rounded to one significant place) was bolded for easy identification. The LB could not complete the classification task for algorithms symbolized by "–" due to heap space error. Ranks for each classifier on each dataset were included in parentheses. The classifier that reported the highest accuracy gets a rank of 1, rank 2 for the runner up and so forth. The last row of the table contains the average of ranks over datasets together with the overall rank of that algorithm accordingly. The discussion is continued on analysis of throughput in Sect. 4.5, and the trade-off between accuracy and throughput in Sect. 4.6.

As shown in Table 7, algorithms can be divided into two groups, comprising more accurate classifiers and less accurate classifiers. The first group consists of SOL, LB, ARF and EP, whereas the second group contains only RC. This is evident by overall average ranks of algorithms. It recapitulates the fact that RC's approach of relying totally on spectra stored during Stage 1 is not a viable strategy.

Table 7 llassification accuracy with ranking

Dataset	ARF	LB	SOL	EP	RC
RH noisy	73.9(2)	**76.5(1)**	73.4(3)	72.(4)	65.7(5)
RBF noisy	78.8(2)	**78.9(1)**	76.6(3)	76.6(3)	63.0(4)
10% progressive RH	76.8(2)	–	**77.0(1)**	76.0(3)	73.4(4)
20% progressive RH	**76.2(1)**	–	**76.2(1)**	**76.2(1)**	75.3(2)
Oscillating RH	76.5(2)	–	**76.8(1)**	74.4(3)	73.1(4)
Flight	78.7(4)	79.2(3)	**81.8(1)**	80.1(2)	56.8(5)
ELEC	65.7(3)	65.7(4)	**67.1(1)**	66.4(2)	65.7(3)
Covertype	82.8(4)	**86.3(1)**	84.8(3)	86.0(2)	62.7(5)
Occupancy	**95.9(1)**	85.2(3)	91.4(2)	91.4(2)	82.8(4)
Average rank	2.3(3)	2.2(2)	**1.8(1)**	2.4(4)	4.0(5)

The proposed algorithm SOL emerged as the winner in 5 out of 9 datasets, while it became the joint winner with ARF and EP for the 20% progressive RH dataset. The runner-up algorithm LB results in highest accuracies for RH noisy, RBF Noisy and Covertype. The Occupancy dataset reports highest accuracy with ARF. Being the winner for the majority of datasets while being the overall winner over all nine datasets verifies the robustness of the proposed stage online learning approach. More precisely, SOL is the best with three challenging synthetic datasets and two real-world datasets. This observation validates the applicability of the framework in various types of recurring and drifting scenarios including real-world datasets.

Even though we ranked algorithms based on accuracy, it is also apparent that the difference between best and the second best is negligible (less than 1%) in the majority of datasets except for RH Noisy, Flight and Occupancy. Interestingly, LB wins RH noisy by 2.6% accuracy difference compared to ARF whereas SOL is the winner for Flight by 1.7% compared to EP, while ARF outperforms SOL for Occupancy by 4.5%.

The analysis which was done above was based on the average accuracy ranks on datasets given in Table 7. The grouping that has been observed was confirmed by the non-parametric Friedman test.

This test has been recognized by the Demšar [6] as one of the best tests to use when it is necessary to compare multiple classifiers against multiple datasets. The null hypothesis H_0 used was that the average accuracy ranks across the five classifiers was the same.

Since the Friedman test statistic is greater than the critical value, null hypothesis H_0 was rejected at the 95% confidence level, thus indicating that statistically significant differences exist between the classifiers. We then subjected the classifiers to the post hoc Nemenyi test to identify exactly where those differences lay. The Nemenyi test yielded a critical difference (CD) value of 2.09 at the 95% confidence level. Figure 4 graphically illustrates that the top group consisted of SOL, LB, ARF and EP as none of the members in this group had significant differences with any

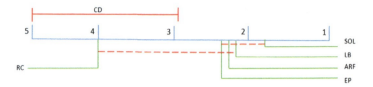

Fig. 4 Statistical comparison of algorithms by accuracy. Subsets of classifiers that are not significantly different at the 95% confidence level are connected with dashed lines

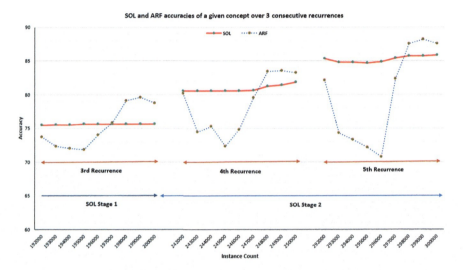

Fig. 5 Accuracy of a concept

of the other members within the group. One and only member of the other group is RC. However, the distance of RC is not significant from the subset comprising EP, ARF and LB.

4.4.1 ARF vs SOL Accuracy of a Concept

Further to the above analysis, the accuracy of one particular concept over three consecutive repetitions that manifested in the 10% progressive RH dataset was contrasted. This gives us an in-depth understanding of the rationale behind SOL's performance advantage when compared to ARF.

Figure 5 shows that SOL is significantly better at capturing recurrences of past concepts in comparison to ARF that takes a long time to re-learn the concept during which time its accuracy suffers. Once ARF learns the concept, it eventually manages to capture the concept by adjusting its forest and is able to acquire a slightly higher accuracy than SOL. However, with shorter concepts ARF would be at a disadvantage.

In addition, the accuracy of SOL is increasing with each repetition due to its property of applying modifications on previously captured patterns rather than a simplistic "use a previously stored pattern that is the best match" policy or a more expensive "re-learning the currently appearing pattern from scratch strategy".

4.5 Throughput Evaluation

In this section, we compared overall throughput of SOL against ARF, LB, EP and RC.

Throughput was measured by recording the time duration spent on processing batches of a certain size (which varied according to the dataset under consideration) and then taking an average value across the entire set of batches. The same sampling scheme was used to trace other measures such as accuracy, repository hit ratio and similarity score. The best throughput was bolded for easy identification.

From the results, it is clear that SOL is the runner-up algorithm in terms of throughput, whereas RC reported the highest throughput over all datasets. ARF and EP were at the third and fourth place, respectively, while LB was the weakest in terms of speed.

In order to gain a statistically sound indication of how these five algorithms differ from each other, the ranks presented in Table 8 were subjected to the Friedman test. As displayed in Fig. 6, there are two distinct groups. The first group consists of RC and SOL as these two classifiers have significant difference with remaining three classifiers. The other group comprises ARF, EP and LB. On the other hand, the subset of SOL, ARF and EP are not significantly different from each other in terms of speed as it formed another group (critical difference amongst these three is not significant). These insights are in accordance with the observations presented in Table 8.

Table 8 Throughput with ranking

	ARF	LB	SOL	EP	RC
RH noisy	5537(4)	659(5)	13,930(2)	6403(3)	**16,824(1)**
RBF noisy	4309(3)	221(5)	4451(2)	1435(4)	**4777(1)**
10% progressive RH	1432(3)	–	1991(2)	1046(4)	**2839(1)**
20% progressive RH	1386(3)	–	1469(2)	1176(4)	**1879(1)**
Oscillating RH	1621(3)	–	2917(2)	1150(4)	**4255(1)**
Flight	1903(3)	427(5)	4331(2)	982(4)	**4553(1)**
ELEC	5388(5)	7527(4)	25,661(2)	11,559(3)	**34,589(1)**
Covertype	1016(2)	327(4)	663(3)	265(5)	**1626(1)**
Occupancy	9057(5)	13,526(4)	19,934(2)	14,378(3)	**22,844(1)**
Average algorithm rank	3.4(3)	4.5(5)	2.1(2)	3.8(4)	1.0(1)

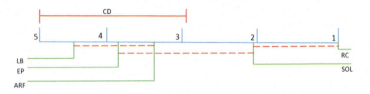

Fig. 6 Statistical comparison of algorithms by throughput. Subsets of classifiers that are not significantly different at the 95% confidence level are connected with dashed lines

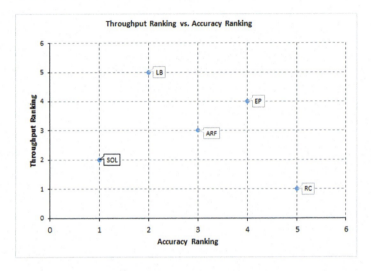

Fig. 7 Accuracy vs. throughput trade-off

4.6 Accuracy Versus Throughput Trade-Off

We studied the trade-off between accuracy and speed of each algorithm. The following Fig. 7 clearly shows that classifiers that tend to be more accurate (e.g. LB and ARF) tend to be more time consuming and vice versa. The SOL is an exception although it was not the fastest. For that reason, it is clear that SOL has provided a good balance between the two opposing characteristics of accuracy and speed. The nearest classifiers to SOL are ARF and EP as they achieved the next best compromise between accuracy and speed.

4.7 Memory Consumption Evaluation

Memory consumption was sampled at the same intervals used in collecting other performance measures that we presented in the previous section. The optimum memory consumption was bolded for easy identification.

Table 9 Average memory consumption (in kBs)

		10% prog. RH	ELEC	Flight
SOL	Stage 1 forest	1514.80	58.59	585.29
	Stage 1 repository	315.36	9.01	32.82
	Stage 1 total	1831.55	67.60	618.10
	Stage 2 total	571.79	27.16	126.55
	Weighted average total	**1243.66**	**37.27**	**342.46**
	Spectra count at the end of stream	10	9	10
EP	Total	2037.66	92.64	803.38
	Spectra count at the end of stream	10	10	10

Table 9 presents the average number of kilobytes for 10% progressive RH, ELEC and Flight datasets. Table 9 shows the gains in memory usage when SOL and EP operate with a repository that can accommodate a maximum of ten spectra. The memory gains mirror those of throughput, with improvements ranging from 38.9% for 10% progressive RH, 59.8% for Electricity and 57.4% for Flight. The memory profiles for the other datasets follow the same trends and were omitted in the interest of space.

Once again this is due to reduction of overheads in Stage 2 as SOL's memory overheads are exactly the same as that of EP in Stage 1 as they share the same learning mechanism. As SOL suspends the operation of its forest in Stage 2, its memory is released and its only memory overheads are that of the single tree that it grows and its repository, with the latter taking a small fraction of the space taken up by a forest of trees. While it is also true that EP's repository is also very compact, it suffers by maintaining its forest unnecessarily in Stage 2.

5 Sensitivity Analysis

The sensitivity analysis was done in two phases. Firstly, we investigate the effects of the threshold firing parameters α and β on all datasets and then present results for two representative datasets, namely noisy RBF and ELEC. The second phase is focused on analysing the effect of the permitted maximum for the number of spectra. The results are presented for noisy RH, 10% progressive RH, Flight and Covertype.

Table 10 shows that the noisy RBF dataset throughput is sensitive to the cut-off value used for α. As α is increased from a low value of 0.5–0.7, there was a 49.0% loss in throughput while not registering a significant difference in accuracy. This throughput loss is to be expected as it delayed the firing of T_1 by a further 40,000 instances. This exemplifies the negative effects of too high a cut-off value for α. The same effects of the α parameter were also seen for the Electricity dataset—a cut-off value of 0.7 inhibited the firing of trigger T_1, resulting in lower throughput. The effects of α on accuracy were very marginal, as illustrated by Table 10.

Table 10 Effect of alpha and beta on the noisy RBF dataset and ELEC dataset

		T_1	T_2	Throughput	Accuracy
Noisy RBF dataset					
α	0.6	Same as default			
	0.7	120,000	–	2268	76.1
β	0.5	80,000	–	4590	76.5
	0.8	80,000	520,000	4015	76.5
Default setting $\alpha = 0.5, \beta = 0.7$		80,000	560,000	4451	76.6
ELEC dataset					
α	0.6	Same as default			
	0.7	24,000	32,000	12,887	67.6
β	0.5	Same as default			
	0.9	8000	36,000	22,654	67.2
Default setting $\alpha = 0.5, \beta = 0.7$		8000	44,000	25,661	67.1

Table 10 shows that throughput was also sensitive to the value of β, albeit to a lesser extent than with the α parameter. With RBF, a β setting of 0.5 caused throughput to slightly increase by 3.1% from its default value with the 0.7 setting. The lower setting for RBF allowed it to stay in Stage 2 for a little longer (as trigger T_2 was not activated), thus resulting in better throughput. On the other hand, a 0.5 setting of β for the Electricity dataset did not cause any difference in trigger timing when compared to its default setting. This difference in behaviour is due to the difference in dynamics of the two datasets—the lower setting for Electricity had no effect on T_2 as the concept recurrence level and concept similarity were on a lesser scale than with RBF, thus enabling it to remain in Stage 2 for the same length of time as with the higher setting for β. Table 10 shows that the effects of β on accuracy were also marginal, just as with α.

As depicted in Table 11, 10% progressive and Flight datasets needed a certain amount of spectra to stay in Stage 2; with size 10, the throughput increased by 68.7% and 4.4%, respectively, over the throughput obtained with size 5. However with the Covertype dataset having five spectra was better in terms of speed while not losing significant accuracy. Having more spectra, especially in case of datasets with more attributes also results in speed disadvantages, which should be avoided. As with the other two parameters, the effects of repository size on accuracy were marginal.

6 Conclusion

This chapter has presented an in-depth examination of the staged learning paradigm that uses a two-staged approach to learning in a data stream environment. The staged learning paradigm represents a major shift in the way that data streams are mined and was motivated by the need to scale classifiers to high-speed data environments.

Table 11 Effect of repository size on the 10% progressive RH, Flight and Covertype datasets

Dataset	Repository size	T_1	T_2	Throughput	Accuracy
10% progressive RH	5	160,000 440,000	320,000	2656	76.4
	10	80,000	–	4480	74.0
	40 (Default)	160,000	–	1991	77.0
Flight	5	4000 18,000	16,000	4016	82.5
	10	4000	–	4195	81.8
	40 (Default)	4000	–	4331	81.8
Covertype	5	12,000 28,000	20,000	1081	84.4
	10	12,000 36,000	20,000	663	84.8
	40 (Default)	12,000 36,000	20,000 48,000	663	84.8

Stage duration is determined by the application of triggers which are sensitive to the rate of appearance of previously unseen concepts. Our experimental study demonstrated the effectiveness of the staged learning approach in terms of processing speed and memory without compromising on classification accuracy. The gain is significant for data streams that exhibit periods of low volatility, and hence the detection of such periods of low volatility is of critical importance to the staged approach. Classification within those low-volatility periods can be effectively dealt with through exploitation of concept recurrence.

The volatility detection strategy proved to be effective in identifying low-volatility states that enabled the computationally expensive ensemble learning component to be suspended, thus directly contributing to the performance gains that we obtained. Likewise, precise recognition of the high-volatility stage avoided potential accuracy drops by ensuring the timely reactivation of Stage 1 which is indispensable to learning a new batch of concepts unseen in the past.

Moreover, our empirical results confirmed the robustness of the staged learning platform under a variety of challenging recurrence scenarios such as patterns repeating with noise, patterns repeating with monotonically increasing drift intensity, and oscillating patterns. We also demonstrated the capability of capturing recurrences across a spatial dimension.

The sensitivity analysis conducted on three critical system parameters: α, β and repository size demonstrated the influence of these parameters on throughput.

In summary, the empirical study has shown that the staged learning paradigm of tailoring the learning strategy to the level of volatility in the stream has significant performance benefits in terms of throughput and memory savings.

7 Future Research

There are in general two different directions in which future research can proceed with the staged learning paradigm.

The first direction would be to experiment with different choices of incremental classifiers in Stage 1. The replacement of an ensemble of decision trees with other types of classifiers such as an adaptive version of neural networks could result in more accurate classification models being produced in Stage 1 and these in turn could produce higher quality spectra in Stage 2 that exhibit higher accuracy than those produced by decision trees.

The second line of research would be to dispense with the hybrid tree/spectrum learning scheme in Stage 2 and rely entirely on spectrum learning. This would entail building a Fourier classifier that would incrementally adjust its spectrum to changes in concepts that occur with time. The advantage of having such an incremental Fourier classifier is that the overhead involved in transforming spectra to trees and vice versa could be avoided at concept drift points, thus improving performance further. Some research in this direction has been reported in [15] but further research is required to examine the effectiveness of such an incremental Fourier classifier when decision models other than decision trees are used in Stage 1 as generators for Fourier spectra in Stage 2.

Finally, it is worthwhile to extend this research work to cover the area of predictive maintenance. Rather than being reactive to drifts, it is possible to start training of classifiers for the next potential concept drift in a proactive fashion. This can be done by introducing a warning period into the drift detector. The best performing classifier within the warning period can be selected as the winner classifier for the newly emerging concept in order to ensure a smooth transition, thus helping to ensure that there are no abrupt accuracy drops that would otherwise have resulted in an out-of-date classifier acting on a concept that it was not matched to.

References

1. Baena-García, M., Campo-Ávila, J., Fidalgo-Merino, R., Bifet, A., Gavald, R., Bueno, R.: Early drift detection method. In: In Fourth International Workshop on Knowledge Discovery from Data Streams, pp. 77–86 (2006)
2. Bifet, A., Gavald, R.: Learning from time-changing data with adaptive windowing. In: Proceedings of the 2007 SIAM International Conference on Data Mining, pp. 443–448 (2007). https://doi.org/10.1137/1.9781611972771.42
3. Bifet, A., Holmes, G., Pfahringer, B., Kirkby, R., Gavaldà, R.: New ensemble methods for evolving data streams. In: Proceedings of the 15th ACM SIGKDD International Conference on Knowledge Discovery and Data Mining, KDD '09, pp. 139–148. ACM, New York, NY, USA (2009). https://doi.org/10.1145/1557019.1557041
4. Bifet, A., Holmes, G., Pfahringer, B.: Leveraging Bagging for Evolving Data Streams, pp. 135–150. Springer, Berlin (2010). https://doi.org/10.1007/978-3-642-15880-3_15

5. Candanedo, L.M., Feldheim, V.: Accurate occupancy detection of an office room from light, temperature, humidity and CO2 measurements using statistical learning models. Energy Build. **112**, 28–39 (2016). https://doi.org/10.1016/j.enbuild.2015.11.071. http://www.sciencedirect.com/science/article/pii/S0378778815304357
6. Demšar, J.: Statistical comparisons of classifiers over multiple data sets. J. Mach. Learn. Res. **7**, 1–30 (2006). http://dl.acm.org/citation.cfm?id=1248547.1248548
7. Gama, J., Žliobaitė, I., Bifet, A., Pechenizkiy, M., Bouchachia, A.: A survey on concept drift adaptation. ACM Comput. Surv. **46**(4), 44:1–44:37 (2014). https://doi.org/10.1145/2523813
8. Gao, J., Fan, W., Han, J., Yu, P.S.: A General Framework for Mining Concept-Drifting Data Streams with Skewed Distributions, pp. 3–14 (2007). https://doi.org/10.1137/1.9781611972771.1
9. Gomes, H.M., Bifet, A., Read, J., Barddal, J.P., Enembreck, F., Pfharinger, B., Holmes, G., Abdessalem, T.: Adaptive random forests for evolving data stream classification. Mach. Learn. **106**(9), 1469–1495 (2017). https://doi.org/10.1007/s10994-017-5642-8
10. Hoeglinger, S., Pears, R., Koh, Y.S.: CBDT: A Concept Based Approach to Data Stream Mining, pp. 1006–1012. Springer, Berlin (2009). https://doi.org/10.1007/978-3-642-01307-2_107
11. Kargupta, H., Park, B.H.: A Fourier spectrum-based approach to represent decision trees for mining data streams in mobile environments. IEEE Trans. Knowl. Data Eng. **16**(2), 216–229 (2004). https://doi.org/10.1109/TKDE.2004.1269599
12. Kargupta, H., Park, B.H., Dutta, H.: Orthogonal decision trees. IEEE Trans. Knowl. Data Eng. **18**(8), 1028–1042 (2006). https://doi.org/10.1109/TKDE.2006.127
13. Kelly, M.G., Hand, D.J., Adams, N.M.: The impact of changing populations on classifier performance. In: Proceedings of the Fifth ACM SIGKDD International Conference on Knowledge Discovery and Data Mining, KDD '99, pp. 367–371. ACM, New York, NY, USA (1999). https://doi.org/10.1145/312129.312285
14. Kithulgoda, C.I., Pears, R.: Staged online learning: a new approach to classification in high speed data streams. In: 2016 International Joint Conference on Neural Networks (IJCNN), pp. 1–8 (2016). https://doi.org/10.1109/IJCNN.2016.7727173
15. Kithulgoda, C.I., Pears, R., Naeem, M.A.: The incremental Fourier classifier: leveraging the discrete Fourier transform for classifying high speed data streams. Expert Syst. Appl. **97**, 1–17 (2018). https://doi.org/10.1016/j.eswa.2017.12.023
16. Kleinberg, J.: Bursty and hierarchical structure in streams. Data Min. Knowl. Discov. **7**(4), 373–397 (2003). https://doi.org/10.1023/A:1024940629314.
17. Lichman, M.: UCI machine learning repository (2013). http://archive.ics.uci.edu/ml
18. Park, B.H.: Knowledge discovery from heterogeneous data streams using Fourier spectrum of decision trees. Ph.D. thesis, Washington State University, Pullman, WA, USA (2001)
19. Pears, R., Sakthithasan, S., Koh, Y.S.: Detecting concept change in dynamic data streams. Mach. Learn. **97**(3), 259–293 (2014). https://doi.org/10.1007/s10994-013-5433-9
20. Ramamurthy, S., Bhatnagar, R.: Tracking recurrent concept drift in streaming data using ensemble classifiers. In: Sixth International Conference on Machine Learning and Applications (ICMLA 2007), pp. 404–409 (2007). https://doi.org/10.1109/ICMLA.2007.80
21. Ross, G.J., Adams, N.M., Tasoulis, D.K., Hand, D.J.: Exponentially weighted moving average charts for detecting concept drift. Pattern Recogn. Lett. **33**(2), 191–198 (2012). https://doi.org/10.1016/j.patrec.2011.08.019
22. Sakthithasan, S., Pears, R., Bifet, A., Pfahringer, B.: Use of ensembles of Fourier spectra in capturing recurrent concepts in data streams. In: 2015 International Joint Conference on Neural Networks (IJCNN), pp. 1–8 (2015). https://doi.org/10.1109/IJCNN.2015.7280583
23. Sripirakas, S., Pears, R.: Mining Recurrent Concepts in Data Streams Using the Discrete Fourier Transform, pp. 439–451. Springer International Publishing, Cham (2014). https://doi.org/10.1007/978-3-319-10160-6-39

Online Time Series Changes Detection Based on Neuro-Fuzzy Approach

Yevgeniy Bodyanskiy, Artem Dolotov, Dmytro Peleshko, Yuriy Rashkevych, and Olena Vynokurova

1 Introduction

The problem of time series changes and fault detection has long been engaging the attention of researchers in many areas, and its solutions have been applied for monitoring of the manufacturing processes, control of moving objects, in medical diagnosis, bioinformatics, video stream processing, etc. Many approaches to address the problem have been proposed, among which most popular are methods based on conventional stochastic time series analysis (correlation, spectral, regression analysis, and others), mathematical models of objects generating such series (first of all, Box–Jenkins models), exponential smoothing, pattern recognition and classification, clustering and segmentation, faults detection, artificial neural networks, etc. The situation gets considerably complicated if information comes from the controllable object in real time, while the problem should be solved in online mode. And in this, two alternative types of situations may arise. First, the changes may suddenly develop in leaps and bounds. In this case, it is important to distinguish two possible options: the drastic change has happened in the controllable object (fault) or an abnormal (noisy) observation has come for processing (outlier). Clearly, those

---

Y. Bodyanskiy · A. Dolotov (✉)
Kharkiv National University of Radio Electronics, Kharkiv, Ukraine
e-mail: yevgeniy.bodyanskiy@nure.ua

D. Peleshko
IT Step University, Lviv, Ukraine

Y. Rashkevych
Ministry of Education and Science of Ukraine, Kyiv, Ukraine

O. Vynokurova
Kharkiv National University of Radio Electronics, Kharkiv, Ukraine

IT Step University, Lviv, Ukraine

© Springer Nature Switzerland AG 2019
E. Lughofer, M. Sayed-Mouchaweh (eds.), *Predictive Maintenance in Dynamic Systems*, https://doi.org/10.1007/978-3-030-05645-2_5

controlling algorithms in this way must have robust properties. Another situation occurs if changes in the controllable object develop smoothly and slowly. In this case, it is impossible to define crisp boundary between segments of signals-time series coming from the object. Certainly, in light of this, fuzzy clustering methods use is appropriate. Given that the vast majority of fuzzy clustering algorithms, both probabilistic and possibilistic, operate in batch mode, it is more feasible to use their online adaptive modifications [1], which are essentially gradient procedures of fuzzy goal function minimization, including robust ones [2].

The problem gets even more complicated when it is required to process not scalar time series but multidimensional (vector and matrix) sequences. In these situations, use of hybrid systems of computational intelligence appears to be the most effective. First of all, they are learning neuro-fuzzy systems for control and analysis of multidimensional signals, which allow linguistic interpretation of the obtained results and can detect both drastic and smooth changes of characteristics of stochastic and chaotic multidimensional time series in online mode.

2 Fuzzy Online Segmentation-Clustering

The task of clustering data arrays of multidimensional observations of different nature, the purpose of which is to find in the samples of these data the homogeneous (in accepting sense) groups (segments, clusters, and classes) of observations, is an integral part of the research area, called Data Mining [3], and its results can be used in many applications, including the task of segmentation of images of arbitrary nature [4]. The most popular today is the K-means method and its subsequent modifications thanks to the simplicity of numerical implementation, high-speed response, and the results visibility.

At the same time, there are a large number of tasks related to the processing of static images and video streams, where traditional methods of clustering-segmentation, based on the self-learning paradigm [5], appear to be ineffective because the amount of information submitted for real-time processing is too large, the data themselves are "contaminated" with different types of disturbances and interruptions, including abnormal perturbations such as "salt and pepper," and the image segments themselves, as a rule, are intersecting, forming fuzzy or "smeared" boundaries. And, if the latter problem can be solved sufficiently by using fuzzy clustering methods [6] and, above all, the classical Fuzzy C-means (FCM) method and its modifications, the processing of real-time video streams is a rather complicated problem, since known algorithms for fuzzy clustering are designed for data processing in batch mode. In addition, these algorithms do not have robust properties; that is, they are unprotected from the output data distortion.

Therefore, the development of adaptive fuzzy online algorithms of fuzzy clustering, having both filtering and tracking properties and capable of sequentially processing both static images and multidimensional time sequences generated by video sequences, is undoubtedly an interesting and useful task.

The problem of clustering multidimensional observations arriving at real-time processing is quite common in many tasks related to data mining. The traditional approach to segmentation-clustering of time series implies that each observation may belong to only one cluster [7], although more natural is the situation where the input vector of attributes with different levels of probabilities or possibilities may belong directly to several clusters or classes [6]. This situation is the subject of consideration of the fuzzy cluster analysis, which is developing intensively at this time in two main directions: a probabilistic approach and an approach based on possibilities (so-called, possibilistic approach) [6].

The problem of fuzzy cluster analysis has been widespread, and recently hybrid neuro-fuzzy systems that combine artificial neural networks and clustering methods are widely studied, and studied very actively. The results of such studies are presented, for example, in works [8, 9]. Fuzzy clustering methods have been further developed within the adaptive approach, which allows monitoring changes in the data structure and adapting the system parameters to these changes. Adaptive online methods are also capable of handling large volumes of data through sequential processing of incoming observations.

The output information for both approaches is a sample of observations generated by n-dimensional attribute vectors:

$$X = \{x(1), x(2), \ldots, x(N), \ldots\}, \quad x(k) \in X, \quad k = 1, 2, \ldots, N, \ldots \quad (1)$$

and is limited by N observations in the case of a batch approach to clustering. The result of the procedure is the output data array partitioning into m clusters with some membership levels $w_j(k)$ of the k-th attribute vector to the j-th cluster.

The processed inputs are preliminarily centered and normalized to the standard for all attributes so that all observations belong to the hypercube $[-1,1]^n$. The centering can be done either in reference to the single sample mean, calculated according to the ratio:

$$m_i(k) = m_i(k-1) + \frac{1}{k}(x_i(k) - m_i(k-1)) \quad (2)$$

or, in order to add robust properties (that is, protection against abnormal observations), in reference to the median, calculated according to the recurrent expression:

$$me_i(k) = me_i(k-1) + \eta_m \operatorname{sign}(x_i(k) - me_i(k-1)), \quad i = 1, 2, \ldots, n, \quad (3)$$

where η_m is parameter of the learning rate, which is selected in the stationary case in accordance with conditions of A. Dvoretzky, for instance in the simplest case $\eta_m = 1/k$.

Clustering methods based on goal functions [6] and designed for solving clustering problems by optimizing some preset clustering quality criterion are most correct in terms of mathematics.

2.1 Probabilistic Approach

The most popular goal function, which is used for fuzzy clustering based on probabilistic approach, is

$$E\left(w_j(k), c_j\right) = \sum_{k=1}^{N}\sum_{j=1}^{m} w_j^\beta(k) D^2\left(x(k), c_j\right) \qquad (4)$$

with constraints:

$$\sum_{j=1}^{m} w_j(k) = 1, \quad k = 1, \ldots, N, \qquad (5)$$

$$0 < \sum_{k=1}^{N} w_j(k) < N, \quad j = 1, \ldots, m, \qquad (6)$$

where $w_j(k) \in [0, 1]$ is the membership level of vector $x(k)$ to the j-th cluster; c_j is the prototype (center) of the j-th cluster; β is an integral parameter called "fuzzyfier," usually $\beta = 2$; and $D(x(k), c_j)$ is the distance between $x(k)$ and c_j according to the selected metrics.

The result of clustering is $(N \times m)$ fuzzy partitioning matrix:

$$W = \{w_j(k)\}. \qquad (7)$$

It should be noted that since the elements of the matrix W can be considered as the probability of the hypothesis of the data vectors membership to certain clusters, the procedure (4) generated at constraints (5 and 6) is called probabilistic clustering method.

As a function of the distance $D(x(k), c_j)$, the Minkowski distance in the L^p metric is usually chosen:

$$D^p\left(x(k), c_j\right) = \|x_i(k) - c_{ji}\|_{L_p}^p = \left(\sum_{i=1}^{n} |x_i(k) - c_{ji}|^p\right)^{\frac{1}{p}}, \quad p \geq 1, \qquad (8)$$

where $x_i(k)$ is the i-th component of $(n \times 1)$-vector $x(k)$, and c_{ji} is the i-th component of $(n \times 1)$-vector c_j.

Let's consider the Lagrangian function:

$$L\left(w_j(k), c_j, \lambda(k)\right) = \sum_{k=1}^{N} \sum_{j=1}^{m} w_j^\beta(k) D^2\left(x(k), c_j\right) + \sum_{k=1}^{N} \lambda(k) \left(\sum_{j=1}^{m} w_j(k) - 1\right)$$
$$= \sum_{k=1}^{N} \left(\sum_{j=1}^{m} w_j^\beta(k) D^2\left(x(k), c_j\right) + \lambda(k) \left(\sum_{j=1}^{m} w_j(k) - 1\right)\right), \quad (9)$$

where $\lambda(k)$ is an indefinite Lagrange multiplier, which ensures conditions (5 and 6). Solving the system of Kuhn–Tucker equations:

$$\begin{cases} \frac{\partial L(w_j(k), c_j, \lambda(k))}{\partial w_j(k)} = 0; \\ \frac{\partial L(w_j(k), c_j, \lambda(k))}{\partial \lambda(k)} = 0; \\ \nabla_{c_j} L\left(w_j(k), c_j, \lambda(k)\right) = \vec{0}, \end{cases} \quad (10)$$

it is easy to get the desired solution in the form:

$$w_j^{pr}(k) = \frac{\left(D^2\left(x(k), c_j\right)\right)^{\frac{1}{1-\beta}}}{\sum_{l=1}^{m} \left(D^2\left(x(k), c_l\right)\right)^{\frac{1}{1-\beta}}}, \quad (11)$$

$$\lambda(k) = -\left(\sum_{l=1}^{m} \left(\beta D^2\left(x(k), c_l\right)\right)^{\frac{1}{1-\beta}}\right)^{1-\beta}, \quad (12)$$

$$c_j^{pr} = \frac{\sum_{k=1}^{N} w_j^\beta(k) x(k)}{\sum_{k=1}^{N} w_j^\beta(k)}. \quad (13)$$

Equations (11)–(13) generate a broad class of clustering procedures. For $\beta = p = 2$, that is in the Euclidean space:

$$D^E\left(x(k), c_j\right) = \|x(k) - c_j\| = \sqrt{\left(x(k) - c_j\right)^T \left(x(k) - c_j\right)}, \quad (14)$$

we obtain a relatively simple and effective procedure for clustering fuzzy C-means of Bezdek [6]:

$$w_j^{pr}(k) = \frac{\|x(k) - c_j\|^{-2}}{\sum_{l=1}^{m} \|x(k) - c_l\|^{-2}}, \quad (15)$$

$$c_j^{\text{pr}} = \frac{\sum_{k=1}^{N} w_j^2(k)x(k)}{\sum_{k=1}^{N} w_j^2(k)}, \qquad (16)$$

$$\lambda(k) = -\sum_{l=1}^{m} \left(\frac{\|x(k) - c_l\|^{-2}}{2} \right)^{-1}. \qquad (17)$$

Process of clustering, in accordance with [6], begins with fairly arbitrary values of centroids which are further adjusted in multiepoch mode where the input data is repeatedly processed until the centroids are stabilized. Understandably, such approach is inefficient in tasks of online data processing.

Probabilistic clustering methods also include procedures of Gustafson–Kessel [10], Gath–Geva [11], and many others. In spite of the insignificant computational complexity, the procedure (15)–(16) has a disadvantage expressed in the necessity of fulfilling the condition (5), common to all probabilistic methods of fuzzy clustering.

In the simplest case of two clusters ($m = 2$), it is easy to see that the observation $x(k)$ equally belonging to both clusters, and the observations $x(p)$ not belonging to any of them can have the same membership levels $w_1^{\text{pr}}(k) = w_2^{\text{pr}}(k) = w_1^{\text{pr}}(p) = w_2^{\text{pr}}(p) = 0.5$. Obviously, this feature can significantly reduce the quality of the classification. At the same time, the possibilistic approach to fuzzy clustering [12] helps to avoid the abovementioned situation and thereby improve the quality of the classification.

2.2 Possibilistic Approach

For possibilistic approach to clustering, the minimized criterion is written as:

$$E\left(w_j(k), c_j\right) = \sum_{k=1}^{N} \sum_{j=1}^{m} w_j^\beta(k) D^2\left(x(k), c_j\right) + \sum_{j=1}^{m} \mu_j \sum_{k=1}^{N} \left(1 - w_j(k)\right)^\beta, \qquad (18)$$

where the scalar parameter $\mu_j > 0$ defines the distance at which the membership level takes the value 0.5; that is, if $D^2(x(k), c_j) = \mu_j$, then $w_j(k) = 0.5$.

Minimization of the criterion (18) with parameters $w_j(k)$, c_j, and μ_j leads to the system of equations:

$$\begin{cases} \frac{\partial E(w_j(k), c_j)}{\partial w_j(k)} = 0; \\ \frac{\partial E(w_j(k), c_j)}{\partial \mu(k)} = 0; \\ \nabla_{c_j} E(w_j(k), c_j) = \vec{0}. \end{cases} \qquad (19)$$

The solution of the first two equations gives a well-known result:

$$w_j^{pos}(k) = \left(1 + \left(\frac{D^2(x(k), c_j)}{\mu_j}\right)^{\frac{1}{\beta-1}}\right)^{-1}, \qquad (20)$$

$$\mu_j = \frac{\sum_{k=1}^{N} w_j^{\beta}(k) D^2(x(k), c_j)}{\sum_{k=1}^{N} w_j^{\beta}(k)}. \qquad (21)$$

The solution of the third equation of the system (19) for the Euclidean norm (14) is given by:

$$c_j^{pos} = \frac{\sum_{k=1}^{N} w_j^{\beta}(k) x(k)}{\sum_{k=1}^{N} w_j^{\beta}(k)}. \qquad (22)$$

We can see that possibilistic and probabilistic methods are quite similar and go through one to another by replacing the expression (20) with formula (11), and vice versa. A common disadvantage of the methods considered is the inability to work in real time when the data is received, for example, in the form of video stream.

The work of the procedure (11) and (12) begins with the task of the initial (usually random) partitioning matrix W^0. Based on its values, an initial set of prototypes c^0_j is calculated, which are then used to specify the new matrix W^1. The next step in batch mode is the calculation of c^1_j, W^2, ..., W^t, c^t_j, W^{t+1}, and so on, until the difference $\|W^{t+1} - W^t\|$ becomes less than the predefined threshold value ε. Thus, the entire sample of data is processed multiple epochs.

The solution that can be obtained using the probabilistic method is recommended to be used as the initial conditions for the possibilistic method (20)–(22), in which the initial values of the parameters of the distance μ^t_j are selected in accordance with (21) by the results of the probabilistic procedure activity.

2.3 Online Combined Approach

For information processing in online mode, instead of the Lagrange function (9), its local modification can be used:

$$L_k\left(w_j(k), c_j, \lambda(k)\right) = \sum_{j=1}^{m} w_j^\beta(k) D^2\left(x(k), c_j\right) + \lambda(k) \left(\sum_{j=1}^{m} w_j(k) - 1\right). \tag{23}$$

The optimization of the expression (23) using the Arrow–Hurwitz–Uzawa procedure leads to the procedure:

$$w_j^{\text{pr}}(k) = \frac{\left(D^2\left(x(k), c_j(k)\right)\right)^{\frac{1}{1-\beta}}}{\sum_{l=1}^{m} \left(D^2\left(x(k), c_l(k)\right)\right)^{\frac{1}{1-\beta}}}, \tag{24}$$

$$\begin{aligned}c_j^{\text{pr}}(k+1) &= c_j^{\text{pr}}(k) - \eta(k)\nabla_{c_j} L_k\left(w_j(k), c_j^{\text{pr}}(k), \lambda(k)\right) \\ &= c_j^{\text{pr}} - \eta(k) w_j^\beta(k) D\left(x(k+1), c_j^{\text{pr}}(k)\right) \nabla_{c_j} D\left(x(k+1), c_j^{\text{pr}}(k)\right),\end{aligned} \tag{25}$$

where $\eta(k)$ is the learning rate parameter which either satisfies the conventional conditions of A. Dvoretzky or is set to a rather small value ($0.01 \leq \eta(k) \leq 0.1$); though in self-learning mode, it is impossible to obtain an optimal value for $\eta(k)$ evidently; $c_j^{\text{pr}}(k)$ is the prototypes of the j-th cluster, calculated on the sample of k observations.

The procedure (24) and (25) is similar to the Chung–Lee algorithm [13], and for $\beta = p = 2$ it coincides with the gradient clustering procedure of Park–Dagher [14]:

$$w_j^{\text{pr}}(k) = \frac{\|x(k) - c_j(k)\|^{-2}}{\sum_{l=1}^{m} \|x(k) - c_l(k)\|^{-2}}, \tag{26}$$

$$c_j^{\text{pr}}(k+1) = c_j^{\text{pr}}(k) + \eta(k) w_j^2(k) \left(x(k+1) - c_j^{\text{pr}}(k)\right). \tag{27}$$

We shall notice that relationship (27) is very similar to neural gas algorithm [15], but physically, it has different sense which is based on fuzzy logic.

Within the scope of possibilistic approach, the local criterion acquires the form:

$$E_k\left(w_j(k), c_j\right) = \sum_{j=1}^{m} w_j^\beta(k) D^2\left(x(k), c_j\right) + \sum_{j=1}^{m} \mu_j \left(1 - w_j(k)\right)^\beta, \tag{28}$$

and the result of its optimization is written as:

$$E_k\left(w_j(k), c_j\right) = \sum_{j=1}^{m} w_j^{\beta}(k) D^2\left(x(k), c_j\right) + \sum_{j=1}^{m} \mu_j\left(1 - w_j(k)\right)^{\beta}, \tag{29}$$

$$\begin{aligned} c_j^{\text{pos}}(k+1) &= c_j^{\text{pos}}(k) - \eta(k) w_j^{\beta}(k) \\ &\times D\left(x(k+1), c_j^{\text{pos}}(k)\right) \nabla_{c_j} D\left(x(k+1), c_j^{\text{pos}}(k)\right), \end{aligned} \tag{30}$$

$$\mu_j(k+1) = \frac{\sum_{p=1}^{k} w_j^{\beta}(p) D^2\left(x(p), c_j(k+1)\right)}{\sum_{p=1}^{k} w_j^{\beta}(p)}. \tag{31}$$

In the quadratic case (when $\beta = 2$), the procedure (29)–(31) turns into a fairly simple structure:

$$w_j^{\text{pos}}(k) = \frac{\mu_j(k)}{\mu_j(k) + \|x(k) - c_j(k)\|^2}, \tag{32}$$

$$c_j^{\text{pos}}(k+1) = c_j^{\text{pos}}(k) + \eta(k) w_j^2(k) \left(x(k+1) - c_j^{\text{pos}}(k)\right), \tag{33}$$

$$\mu_j(k+1) = \frac{\sum_{p=1}^{k} w_j^2(p) \|x(p) - c_j(k+1)\|^2}{\sum_{p=1}^{k} w_j^2(p)}. \tag{34}$$

The parallel application of adaptive probabilistic and possibilistic algorithms leads to a combined procedure of fuzzy clustering:

$$\begin{cases} c_j^{\text{pr}}(k) = c_j^{\text{pos}}(k-1) - \eta(k) w_j^{\text{pos}\beta}(k-1) D\left(x(k), c_j^{\text{pos}}(k-1)\right) \\ \qquad \times \nabla_{c_j} D\left(x(k), c_j^{\text{pos}}(k-1)\right); \\ w_j^{\text{pr}}(k) = \left(D^2\left(x(k), c_j^{\text{pr}}(k)\right)\right)^{\frac{1}{1-\beta}} \left(\sum_{l=1}^{m}\left(D^2\left(x(k), c_j^{\text{pr}}(k)\right)\right)^{\frac{1}{1-\beta}}\right)^{-1}; \\ c_j^{\text{pos}}(k) = c_j^{\text{pr}}(k-1) - \eta(k) w_j^{\text{pr}\beta}(k) D\left(x(k), c_j^{\text{pr}}(k)\right) \nabla_{c_j} D\left(x(k), c_j^{\text{pr}}(k)\right); \\ \mu_j(k) = \left(\sum_{p=1}^{k} w_j^{\text{pr}\beta}(p) D^2\left(x(k), c_j^{\text{pos}}(k)\right)\right)\left(\sum_{p=1}^{k} w_j^{\text{pr}\beta}\right)^{-1}; \\ w_j^{\text{pos}}(k) = \left(1 + \left(\frac{D^2\left(x(k), c_j^{\text{pos}}(k)\right)}{\mu_j(k)}\right)\right)^{-1}, \quad j = 1, 2, \ldots, m. \end{cases} \tag{35}$$

The evidence of the correct finding of prototypes (and, consequently, the correct clustering), using the procedure (35), is the implementation of inequality:

$$\sum_{l=1}^{m} D^2\left(c_l^{\mathrm{pr}}(k), c_l^{\mathrm{pos}}(k)\right) \leq \varepsilon, \tag{36}$$

where ε determines the acceptable accuracy of the clustering.

For the Euclidean metric, the value of parameter $\mu_j(k)$ can be calculated according to the recurrence relationship, which is directly derived from (34):

$$\begin{cases} \beta_q(k) = \beta_q(k-1) + w_j^{\mathrm{pr}\,2}(k) s_q(x(k)), & q = 0, 1, 2; \\ \mu_j(k) = \dfrac{\beta_2(k-1) - 2c_j^{\mathrm{pos}\,T}(k)\beta_1(k-1) + \left\|c_j^{\mathrm{pos}}(k)\right\|^2 \beta_0(k-1)}{\beta_0(k-1)}, \end{cases} \tag{37}$$

where:

$$S_q(x(k)) = \begin{cases} 1, & \text{if } q = 0; \\ x(k), & \text{if } q = 1; \\ \|x(k)\|^2, & \text{if } q = 2. \end{cases} \tag{38}$$

The initial values of the parameter $\beta q(k)$ are selected as:

$$\beta_q(N) = \sum_{p=1}^{N} \left(w_j^{\mathrm{pr}}(p)\right)^2 s_q(x(p)), \quad q = 0, 1, 2. \tag{39}$$

Thus, the adaptive procedure (35) can work both in the batch mode for iterative processing of a given sample, and in a real-time mode, where the number of observations is determined by the discrete time $k = 1, 2, \ldots, N, +1, \ldots$. In the latter case, this procedure sequentially processes the observations that arrive for processing. Consequently, in the case of nonstationary data, the membership levels and cluster prototypes are rebuilt according to new data.

Most of the practical problems associated with video processing are characterized by the presence of abnormal perturbations in the data, which significantly affects the results of clustering by classical methods, which is manifested in the detection of nonsignificant clusters, the displacement of prototypes, and clusters radii.

Because of this, more and more attention is being paid to the problems of cluster analysis of data generated by distributions with slowly falling (or heavy) "tails." Various robust modifications of classical clustering procedures for processing data containing perturbations were proposed in [16].

At the same time, most of the proposed robust fuzzy clustering methods cannot be used for consistent work or real-time operation. In order to overcome this disadvantage, it is expedient to synthesize recurrent procedures for fuzzy robust clustering of time series that have adaptive properties and can be applied to the sequential processing of incoming data, and under the conditions when the properties of the system generating these data are changed with time.

2.4 Robust Approach

The estimates related to a quadratic goal function are optimal when the data belong to a class of distributions with a limited variance. The most important representative of this class is the Gaussian distribution. Changing the value of the parameter p can improve the robustness of the clustering procedure. However, it should be borne in mind that the quality assessment is determined by data distribution. For example, estimates corresponding to $p = 1$ are optimal for the Laplace distribution, but a large number of computations are required to obtain them.

The most important function for approximating the probability density close to the normal distribution is the function:

$$p(x_i, c_i) = \text{Se}(c_i, s_i) = \frac{1}{2s_i}\sinh^2\left(\frac{x_i - c_i}{s_i}\right), \tag{40}$$

where c_i, s_i—parameters that determine the center and variance of the distribution, respectively.

This function is close to a Gaussian function in vicinity of the center, but essentially differs from it by the presence of heavy "tails." The distribution (40) is related to the goal function:

$$f_i(x_i, c_i) = \beta_i \ln\left[\cosh\left(\frac{x_i - c_i}{\beta_i}\right)\right]. \tag{41}$$

Here, β_i is the parameter that determines the speed of the change of this function.

It should be noted that the function (41) is close to the quadratic in the vicinity of the center c_i, and approaches to the linear with increasing distance from the center. The derivative of this function can be written as:

$$f_{i'}(x_i) = \phi(x_i) = \tanh\left(\frac{x_i}{\beta_i}\right), \tag{42}$$

and coincides with the standard activation function of the artificial neuron.

Let's consider the function:

$$D^R(x(k), c_j) = \sum_{i=1}^{n} f_i(x_i(k), c_{ji}) = \sum_{i=1}^{n} \beta_i \ln\left[\cosh\left(\frac{x_i(k) - c_{ji}}{\beta_i}\right)\right] \tag{43}$$

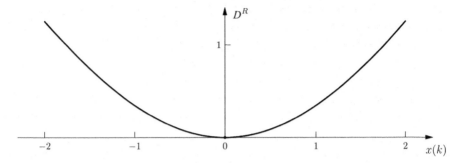

Fig. 1 Function (43) graph for $n = 1$, $\beta_i = 2$, and $c_{ij} = 0$

the graph of which for $n = 1$, $\beta_i = 2$, and $c_{ij} = 0$ is depicted in Fig. 1. As one can see, the function, in contrast to conventional quadratic one, does not amplify outliers which are far distant from its minimum.

Next, as a distance to the fuzzy clustering goal function, we will use the function:

$$D\left(x(k), c_j\right) = \left(D^R\left(x(k), c_j\right)\right)^{\frac{1}{2}}. \tag{44}$$

Let's consider the target function for robust probabilistic clustering:

$$E^R\left(w_j(k), c_j\right) = \sum_{k=1}^{N} \sum_{j=1}^{m} w_j^\beta D^R\left(x(k), c_j\right)$$
$$= \sum_{k=1}^{N} \sum_{j=1}^{m} w_j^\beta \sum_{i=1}^{n} \beta_i \ln\left[\cosh\left(\frac{x_i(k)-c_{ji}}{\beta_i}\right)\right]. \tag{45}$$

Corresponding to it, the Lagrange function is given by expression:

$$L^R\left(w_j(k), c_j, \lambda(k)\right) = \sum_{k=1}^{N} \sum_{j=1}^{m} w_j^\beta(k) \sum_{i=1}^{n} \beta_i \ln\left[\cosh\left(\frac{x_i(k)-c_{ji}}{\beta_i}\right)\right]$$
$$+ \sum_{k=1}^{N} \lambda(k) \left(\sum_{j=1}^{m} w_j(k) - 1\right). \tag{46}$$

The saddle point of the Lagrange function (46) can be found by solving the Kuhn–Tucker equation system (10) in the same way as it was done for the derivation of the procedures (11) and (12). In this case, the solution of the first and second equations of the system (10), with due account for metric (43), will coincide with (11) and (12), respectively. However, the third equation of the system:

$$\nabla_{c_j} L^R\left(w_j(k), c_j, \lambda(k)\right) = \sum_{k=1}^{N} w_j^\beta \nabla_{c_j} D^R\left(x(k), c_j\right) = \vec{0}, \tag{47}$$

obviously does not have an analytical solution.

The solution (47) can be found in the numerical form based on the local modification of the Lagrange function by means of the recurrent fuzzy clustering procedure. Finding at the same time the saddle point of the local Lagrange function (23) for the metric (43) on the basis of the Arrow–Hurwitz–Uzawa procedure, we obtain the following procedure for fuzzy robust clustering:

$$\begin{cases} w_j^{pr}(k) = \dfrac{(D^R(x(k),c_j))^{\frac{1}{1-\beta}}}{\sum_{l=1}^{m}(D^R(x(k),c_l))^{\frac{1}{1-\beta}}}; \\ c_{ji}(k+1) = c_{ji}(k) - \eta(k)\dfrac{\partial}{\partial c_{ji}}L_k^R\left(w_j(k), c_j, \lambda(k)\right) \\ \quad = c_{ji}(k) + \eta(k)w_j^\beta(k)\tanh\left(\dfrac{x_i(k)-c_{ji}(k)}{\beta_i}\right). \end{cases} \quad (48)$$

Within the scope of the possibilistic approach, the clustering criterion, with due account for the robust metric (43), is written as:

$$E^R\left(w_j(k), c_j\right) = \sum_{k=1}^{N}\sum_{j=1}^{m} w_j^\beta(k) D^R\left(x(k), c_j\right) + \sum_{j=1}^{m}\mu_j \sum_{k=1}^{N}\left(1 - w_j(k)\right)^\beta. \quad (49)$$

Solving the system of Kuhn–Tucker equations, which is similar to (19), with the use of metric (43) for the first two equations, we obtain solution in the form (20) and (21). However, the third equation of the system (19):

$$\nabla_{c_j} E^R\left(w_j(k), c_j\right) = \sum_{k=1}^{N} w_j^\beta \nabla_{c_j} D^R\left(x(k), c_j\right) = 0 \quad (50)$$

completely coincides with (47), which leads to the impossibility of its solution in an analytical form.

Let's consider the local modification of the criterion (49):

$$\begin{aligned} E_k^R\left(w_j(k), c_j\right) &= \sum_{j=1}^{m} w_j^\beta(k) D^R\left(x(k), c_j(k)\right) + \sum_{j=1}^{m}\mu_j\left(1 - w_j(k)\right)^\beta \\ &= \sum_{j=1}^{m} w_j^\beta(k)\sum_{i=1}^{n}\beta_i \ln\left[\cosh\left(\dfrac{x_i(k)-c_{ji}}{\beta_i}\right)\right] + \sum_{j=1}^{m}\mu_i\left(1 - w_j(k)\right)^\beta. \end{aligned} \quad (51)$$

Using the Arrow–Hurwitz–Uzawa procedure, we obtain the recurrent procedure of the fuzzy possibilistic clustering of the form:

$$\begin{cases} w_j^{pos}(k) = \left(1 + \left(\dfrac{D^R(x(k),c_j(k))}{\mu_j}\right)^{\frac{1}{\beta-1}}\right)^{-1}; \\ c_{ji}(k+1) = c_{ji}(k) - \eta(k)\dfrac{\partial E_k(w_j(k),c_j,\mu_j(k))}{\partial c_{ji}} \\ \quad = c_{ji}(k) + \eta(k)w_j^\beta(k)\tanh\left(\dfrac{x_i(k)-c_{ji}(k)}{\beta_i}\right). \end{cases} \quad (52)$$

The distance parameter $\mu_j(k)$ is calculated here by formula (21) for $k < N$ observations:

$$\mu_j(k+1) = \frac{\sum_{p=1}^{k} w_j^\beta(p) D^R(x(p), c_j(k+1))}{\sum_{p=1}^{k} w_j^\beta(p)}. \quad (53)$$

It should be noted that the equations for $c_{ji}(k)$ at systems (48, 52) are completely identical and are determined by the chosen metric, while other equations do not depend on the metric; that is, the choice of an arbitrary metric for the clustering procedure will only affect the setup procedures for cluster prototypes, while the equations for calculating the values of weight coefficients $w_j^{pr}(k)$ and $w_j^{pos}(k)$ will remain unchanged.

As a metric analog for a robust recurrent fuzzy clustering method, you can use the function (Fig. 2):

$$D^R(x(k), c_j) = \sum_{i=1}^{n} \left(1 - \sinh^2(x_i(k) - c_{ji})\right) (x_i(k) - c_{ji})^{\frac{2}{5}}, \quad (54)$$

which "suppresses" abnormal perturbations in observations because it stops growing as the distance from its minimum increases and, thus, far distant observations do not affect it.

This function satisfies the axioms of the metric in the vicinity of the center c_j; however, for $|x(k) - c_j| > 0.8762$, it does not satisfy the inequality of the triangle.

Using (54), we write the target function for robust clustering in the form of:

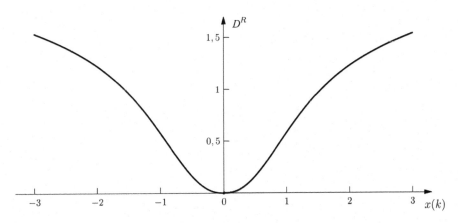

Fig. 2 Function (54) graph for $n = 1$, $c_{ji} = 0$

$$E^R\left(w_j(k), c_j\right) = \sum_{k=1}^{T} \sum_{j=1}^{m} w_j^{\beta} D^R\left(x(k), c_j\right)$$
$$= \sum_{k=1}^{T} \sum_{j=1}^{m} w_j^{\beta} \sum_{i=1}^{n} \left(1 - \sinh^2\left(x_i(k) - c_{ji}\right)\right) \left(x_i(k) - c_{ji}\right)^{\frac{2}{5}}. \tag{55}$$

and the corresponding Lagrange function in the form of:

$$L^R\left(w_j(k), c_j, \lambda(k)\right) = \sum_{k=1}^{T} \sum_{j=1}^{m} w_j^{\beta}(k) \sum_{i=1}^{n} \left(1 - \sinh^2\left(x_i(k) - c_{ji}\right)\right) \left(x_i(k) - c_{ji}\right)^{\frac{2}{5}} +$$
$$+ \sum_{k=1}^{T} \lambda(k) \left(\sum_{j=1}^{m} w_j(k) - 1\right). \tag{56}$$

Similarly to the derivation of the procedure (48), using the Arrow–Hurwitz–Uzawa method to find the saddle point of the Lagrangian (56), we obtain the following robust recurrent fuzzy clustering procedure:

$$\begin{cases} w_j(k) = \dfrac{\left(D^R(x(k), c_j)\right)^{\frac{1}{1-\beta}}}{\sum_{l=1}^{m} \left(D^R(x(k), c_l)\right)^{\frac{1}{1-\beta}}}; \\ c_{ji}(k+1) = c_{ji}(k) + \eta(k) w_j^{\beta}(k) \Big(2\sinh^2\left(x_i(k) - c_{ji}(k)\right) \\ \times \tanh\left(x_i(k) - c_{ji}(k)\right) \left|x_i(k) - c_{ji}(k)\right|^{\frac{2}{5}} + 0,4 \left(1 - \sinh^2\left(x_i(k) - c_{ji}(k)\right)\right) \\ \times \left|x_i(k) - c_{ji}(k)\right|^{-\frac{3}{5}} \text{sign}\left(x_i(k) - c_{ji}(k)\right) \Big). \end{cases} \tag{57}$$

The proposed robust recurrent fuzzy clustering method can be used both in batch mode and in one-pass version. The computational complexity of the proposed method is the same as that for other known recursive clustering procedures [13], and depends linearly on the number of observations in the data sample.

In recent years, there is an obvious increase in interest in the task of analyzing nonstationary time sequences that change their properties in a priori unknown moments of time. The problem of analyzing time sequences is inherent in tasks of speech and text processing, of "Web-mining," analysis of robot sensors, and video streams. It is important to note that these tasks should be solved in real time, provided that the new data is constantly received.

In the process of solving the above problems, the time sequence is partitioned (segmented) into internally homogeneous parts, which are subsequently presented by some more compact description for further diagnostics or processing.

In some cases, such tasks are solved using an approach based on methods for detecting changes in the properties of signals and systems. However, known methods are usually designed to detect abrupt changes and are not suitable for detecting slow variations of sequence characteristics.

In real problems, internal changes in the observed object are usually slow enough and, moreover, there are transitional states, whose characteristics belong simultaneously to several stable states.

In such cases, priority methods are fuzzy clustering-segmentation of time sequences based on known methods of fuzzy cluster analysis [6]. These methods proved to be effective in solving many problems in batch mode, but their use in real-time problems is complicated by a number of problems, some of which can be overcome using adaptive methods of recurrent cluster analysis [13]. However, these methods are effective only if the intersecting clusters are compact; that is, they do not contain sharp (abnormal) perturbations. While the real data samples usually contain up to 20% perturbations, the assumption about compactedness of clusters can be incorrect. It is precise in such situations that methods of online robust fuzzy clustering and segmentation of time series can come to the fore, when shift of controlled signal state from one segment-cluster to another is an evidence of its properties change [17–19].

For the task of segmentation of the time sequences:

$$Y = \{y(1), y(2), \ldots, y(k), \ldots, y(N)\}, \tag{58}$$

an approach based on indirect clustering of the sequence is often used. In accordance with this approach, some features are allocated for further displaying them into the transformed attribute space. Later in converted data space, the known clustering methods can be used to form clusters. As characteristics of the initial time sequence, there can be selected correlation, regression, spectral, and other characteristics, which in the case of real-time processing must be calculated using adaptive procedures [2].

To this end, estimates of the mean value, variance, and autocorrelation coefficients can be used. To provide adaptive properties, these estimates must be calculated using the exponential smoothing procedure [20]. The average value can be estimated as:

$$s(k) = \alpha y(k) + (1-\alpha) s(k-1), \quad 0 < \alpha < 1, \tag{59}$$

where $\alpha = 2/(T+1)$ is the coefficient that determines the quality of smoothing at the window of width T.

The value of the variance of the time sequence can be estimated as:

$$\sigma^2(k) = \alpha(y(k) - s(k))^2 + (1-\alpha)\sigma^2(k-1), \tag{60}$$

and coefficients of autocorrelation:

$$\rho(k, \tau) = \alpha (y(k) - s(k))(y(k-\tau) - s(k)) + (1-\alpha)\rho(k-1, \tau), \tag{61}$$

where $\tau = 1, 2, \ldots, \tau_{max}$—time delay. Use of exponential forgetting allows for excluding from consideration the outdated observations that relate to operation of the controlled object before a disorder appears in it.

So, the attribute vectors:

$$x(k) = \left(s(k), \sigma^2(k), \rho(k, 1), \rho(k, 2), \ldots, \rho(k, \tau_{max}) \right)^T \quad (62)$$

contain $n = 2 + \tau_{max}$ elements, and are calculated at each step of the discrete time k and form a set consisting N n-dimensional attribute vectors:

$$X = \{x(1), x(2), \ldots, x(N)\}, \quad (63)$$

where $x(k) \in R^n$, $k = 1, 2, \ldots N$.

The result of applying the fuzzy clustering procedure is to split-up the output data into m clusters with some membership degree $w_j(k)$ of the k-th attribute vector $x(k)$ to the j-th cluster.

Time sequences of observations, which occur consistently in time and belong to identical clusters, form the segments of the time sequence at the output. In this, along with changes of controlled process properties, clusters-segments being formed may also change because the outdated observations get forgotten in the process of learning.

In this section, the results of simulation of the developed robust methods of fuzzy real-time clustering are presented. The simulation of the presented procedures was based on the examples of solving the standard test tasks of the classification, as well as for solving the practical task of segmentation of the biological time sequences of mammalian heartbeat R–R intervals. Also in this section, the results of simulation of classical methods of clustering are presented in order to assess the quality of the solution of the problems under consideration.

To evaluate the quality of work of the proposed robust recurrent clustering procedures, a comparison of results with known methods was performed for the task of data classification on an artificially constructed sample containing three two-dimensional clusters whose observations are indicated in Fig. 3 by symbols "=," "×," and "∘." Each sample cluster is obtained randomly from the Laplace distribution, which is characterized by rather heavy "tails" and is determined by expression:

$$p(x_i) = \frac{\sigma}{1 + (x_i - c)^2}, \quad (64)$$

where σ is the mean square deviation of the distribution and c is the mathematical expectation.

The data sample contains 9000 observations (3000 in each cluster), divided into a training sample of 7200 observations and a validation sample containing 1800 observations.

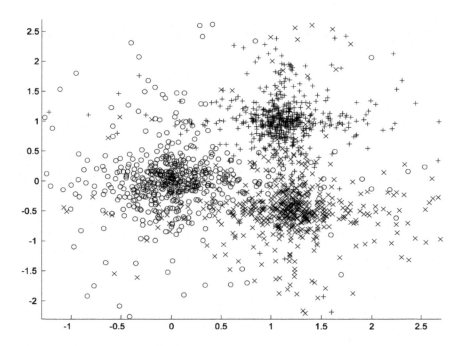

Fig. 3 Fragment of artificial data sample

The results of clustering obtained using robust probabilistic (48) and possibilistic (52) procedures were compared with the results obtained using the clustering FCM procedure of Bezdek (15 and 16). For each comparable clustering procedure, the following sequence of actions was performed:

- Observations of the training sample are fed to the input of the corresponding clustering procedure, which results in calculation of the cluster prototypes (centroids);
- Observations of the training and validation samples are classified according to the results of clustering. The membership degree of each observation is calculated according to Eqs. (15, 48), or (52) depending on the clustering procedure used;
- The cluster and the corresponding class, to which the current observation belongs, are determined by the maximum value of the membership level.

Clustering and training were carried out in real time with parameter values: $\beta = 2$, $\beta_1 = \beta_2 = \beta_3 = 1$, and $\eta(k) = 0.01$.

The results of the classification are shown in Table 1.

The average mean class error of classification (*MCE*) for training (M{MCE_{tr}}) and validation samples (M{MCE_{ts}}) are indicated here, and they were calculated as the percentage ratio of mistakenly classified observations to the size of the sample. Complete data sample was randomly divided into training and validation samples in 100 different ways. The resulting data samples are used to calculate the average values of classification errors.

Table 1 Results of classification

Clustering procedure	Classification error	
	M{MCE$_{tr}$}/observation	M{MCE$_{ts}$}/observation
Bezdek (15)–(17)	17.1%/1229	16.6%/229
Robust probabilistic (48)	15.6%/1127	15.6%/281
Robust possibilistic (52)	15.2%/1099	14.6%/263

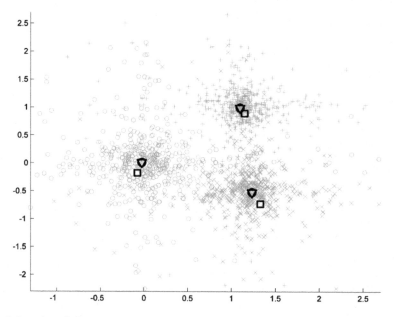

Fig. 4 Location of cluster prototypes obtained as a result of the application of various clustering procedures

The disadvantages of fuzzy clustering techniques based on quadratic target function can be visually determined by the location of cluster prototypes. In Fig. 4, we can see the shift of cluster prototypes obtained using the clustering method of Bezdek (15, 16) in comparison with the visual cluster centers, caused by density distribution of observations with heavy "tails" and, consequently, the presence of a large number of perturbations, while the proposed methods based on robust target functions (48, 52) give more accurate prototypes, which is confirmed by a lower value of classification error (Table 1).

Data sampling is a time sequence of heartbeat intervals (R–R intervals) of a hamster in the process of gradual reversal and emergence of the artificial hypometabolic state [2]. The time sequence and its segmentation are realized by an expert biologist, and the projection of the data sample on the first two principle components after the preliminary processing is shown in Fig. 5. Time sequence contains 6579 observations.

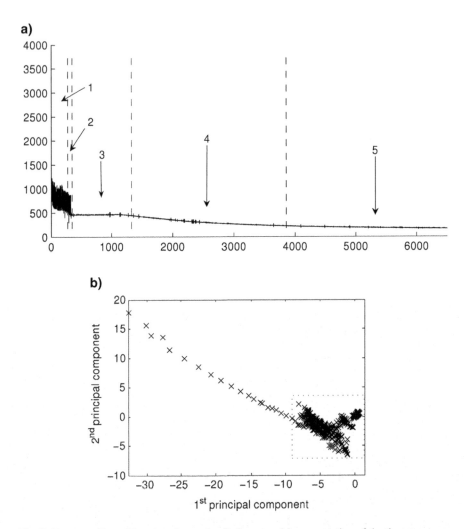

Fig. 5 Data sampling of heartbeat intervals of a hamster: (**a**) segmentation of the time sequence of heartbeat intervals, performed by an expert biologist; and (**b**) projection of preprocessed data on the first two principal components

To evaluate the quality of the result, the results of the segmentation of this time sequence, obtained by means of the proposed robust clustering procedure (57), were compared with other classical recurrent clustering methods based on various goal functions: the well-known Bezdek [6] and Gustafson–Kessel [10] clustering procedures for the fuzzy C-means.

All clustering procedures were started on the same data sample that was formed at the preprocessing stage with the following characteristics: $s(k)$, $\sigma^2(k)$, $\rho(k, 2)$, and $\rho(k, 3)$, where $\alpha = 0.095238$ (i.e., on the window of 20 observations), calculated according to (59)–(61). This subset of attributes was chosen experimentally.

The number of prototypes was assumed to be equal to $m = 5$. This value was also chosen experimentally, since smaller clusters ($m < 5$) lead to unstable clustering results that essentially depend on initial initialization, while for $m > 5$ some of the clusters obtained become indistinguishable. Consequently, assuming that there are 5 different clusters associated with changes in the conditions of animal organism functioning, their parameters were determined with the use of the developed technique.

Robust Procedure of Fuzzy Clustering Clustering with the use of procedure (57) was performed in one pass through the data sample. The parameters of the procedure (57) were taken as follows: $\beta = 2$, the training step $\eta = 0.01$. Projections of the data sample and cluster prototypes on the first two main components of the space of features are presented in Fig. 6, and the segmentation of the time sequence obtained under this procedure is shown in Fig. 7.

Bezdek Clustering Procedure The parameters of the procedure (15) and (16) were taken as follows: $\beta = 2$, the tolerance for completing the procedure $\varepsilon = 10^{-3}$. Projections of data sample and cluster prototypes on the first two principal components of the attribute space are presented in Fig. 8, and the segmentation of the time sequence carried out by this procedure is shown in Fig. 9.

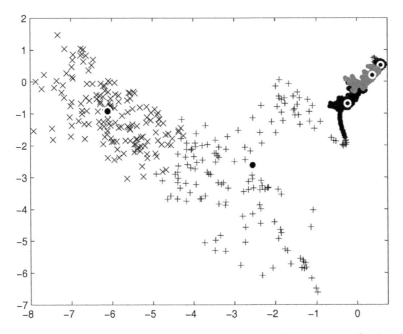

Fig. 6 Projections of data sample and cluster centers on the principal components for the robust clustering procedure

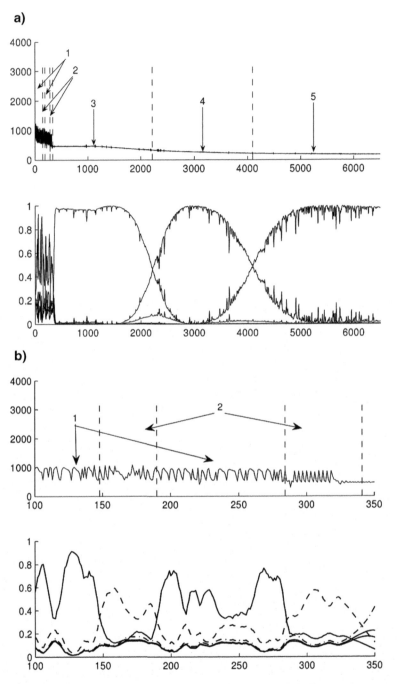

Fig. 7 Segmentation of the time sequence of heartbeat intervals for a robust fuzzy clustering procedure: (**a**) over the entire time interval, and (**b**) on a fragment of the time sequence

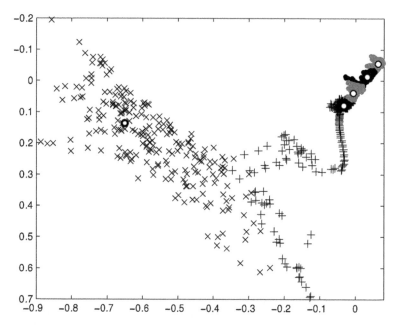

Fig. 8 Projections of sample data and cluster centers on the principal components for the Bezdek clustering procedure

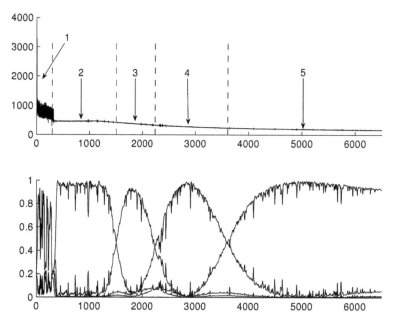

Fig. 9 Segmentation of the time series of heartbeat intervals for the Bezdek clustering procedure

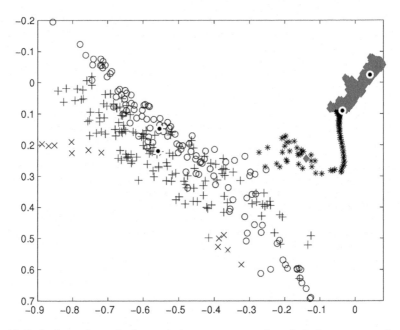

Fig. 10 Projections of sample data and cluster centers on the principal components for the Gustafson–Kessel clustering procedure

Gustafson–Kessel Clustering Procedure The parameters of the procedure were taken as follows: $\beta = 2$, the tolerance for the completion of the procedure $\varepsilon = 10^{-6}$, the predetermined value of the cluster volume $\rho_j = 1$, $j = 1, 2, ..., m$, and the maximum ratio of the maximal eigenvalue of covariance matrix to the smallest one is $\vartheta = 10^{15}$. Projections of the data sample and cluster prototypes on the first two main components of the attribute space are presented in Fig. 10, and the segmentation of the time sequence obtained by this procedure is shown in Fig. 11.

Analysis of the Results Without any other direct criteria for the determination of changes in the state of an animal's organism, such an interpretation may be accepted.

The first part of the curve, characterized by the simultaneous presence of two clusters, may reflect some degree of imbalance in the control system of the cardiac rhythm due to the initiation of an artificial hypometabolic state.

The most probable cause of such perturbations may be due to the intense slowdown in breathing frequency (up to 1 per min), which is manifested in the disturbance of the effect of modulation of breathing on the heart rate. Achieving a stable hypometabolic state with a decrease in the basic vital functions (body temperature is close to 15 °C) is presented in the second part of the curve. An appropriate cluster combines two different states, both a stable hypometabolic state, and initial stage of warming the animal, which occurs against the backdrop of gradual activation of the thermal control system. Fast activation of the latter with the greatest activation of the thermogenesis of tremor causes a cluster change. The

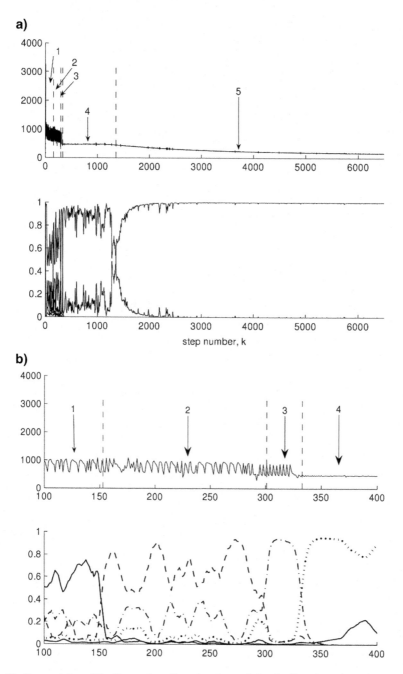

Fig. 11 Segmentation of the time series of heartbeat intervals for the Gustafson–Kessel clustering procedure: (**a**) over the entire time interval, and (**b**) on a fragment of the time sequence

normalization of the temperature homeostasis of the organism is typical for the last section of the heartbeat rhythm distribution curve and is expressed in the presence of the last cluster, whose presence may be reflected in the emergence of a stable functional state that can be assumed to be related to the completion of the most energy-intensive period of the restoration of temperature homeostasis.

Consequently, the results of solving the problem of clustering in the mode of self-training do not contradict the fact of the current state of the laboratory animal.

The proposed robust fuzzy clustering procedure gives better results of segmentation of the time sequence than other clustering methods, as can be seen from the comparison with the segmentation of the time sequence performed by the expert biologist.

Thus, the task of changes detection in data stream in self-learning mode can be considered as a task of fuzzy segmentation of time series. In this, shift of controlled signal from one segment to another is evidence of changes (smooth or sudden) in the controlled signal.

3 Robust Forecasting and Faults Detection in Nonstationary Time Series

Recently, methods of computational intelligence are increasingly used in problems of analysis and processing of nonstationary signals of arbitrary nature under uncertainty, among which hybrid neural networks can be distinguished. One of the important tasks associated with signal processing is the prediction and segmentation of nonstationary time series under uncertainty. Such tasks often arise when processing video streams.

Solving such problems is connected with exploration of local characteristics of sequences being analyzed over separate time segments. Wavelet analysis [21] can be reliably used here, which allows detecting such characteristics with high degree of accuracy.

At the intersection of wavelet analysis and artificial neural networks, the so-called hybrid wavelet-neural networks [22] arose due to their high approximating properties and their sensitivity to changes in characteristics of the analyzed processes.

To solve problems of forecasting and segmentation, a very important point is the choice of the optimization criterion for the synthesis of training algorithms for hybrid neural network systems.

Let's consider the optimization criterion in general terms:

$$E(k) = f(e(k)), \qquad (65)$$

where $e(k) = d(k) - y(k)$ is the training error, $d(k)$ is the external reference signal, $y(k)$ is the real signal of the system, and f is the target function.

In most cases, the least squares criterion (L_2—norm) is used as optimization criterion:

$$f(e(k)) = \frac{1}{2}e^2(k), \qquad (66)$$

or criterion of the smallest absolute values (L_1—norm):

$$f(e(k)) = |e(k)|. \qquad (67)$$

The experience has shown that identification methods based on the least squares criterion are extremely sensitive to the deviations of the actual data distribution law from the normal one. Under the conditions of various types of perturbations, gross errors, and non-Gaussian perturbations with "heavy tails," the methods associated with the least squares criterion lose their effectiveness. In this situation, the methods of robust estimation, which until now have been used for the study of artificial neural networks, come to the fore.

In order to reduce the influence of disturbances with the unknown distribution law in solving prediction and segmentation problems, it is proposed to use known criteria for robust statistics [23]:

– Logistic function:

$$f_L(e(k)) = \beta^2 \ln\left[\cosh\left(\frac{e(k)}{\beta}\right)\right]; \qquad (68)$$

– Huber function:

$$f_H(e(k)) = \begin{cases} \dfrac{e^2(k)}{2}, & \text{для} \quad |e(k)| \leq \beta; \\ \beta|e(k)| - \dfrac{\beta^2}{2} & \text{для} \quad |e(k)| > \beta; \end{cases} \qquad (69)$$

– Huber function with saturation (Talvar function):

$$f_T(e(k)) = \begin{cases} \dfrac{e^2(k)}{2}, & \text{для} \quad |e(k)| \leq \beta; \\ \dfrac{\beta^2}{2}, & \text{для} \quad |e(k)| > \beta; \end{cases} \qquad (70)$$

– Hampel function:

$$f_{\text{Ha}}(e(k)) = \begin{cases} \dfrac{\beta^2}{\pi}\left[1 - \cos\left(\dfrac{\pi e(k)}{\beta}\right)\right], & \text{для} \quad |e(k)| \leq \beta; \\ 2\dfrac{\beta^2}{\pi}, & \text{для} \quad |e(k)| > \beta. \end{cases} \quad (71)$$

The first derivative of such functions is also called the function of influence. We note that all four criteria are twice continuously differentiated throughout the life span.

In order to increase the convergence rate of learning algorithms and/or improve the approximating properties, it is also possible to use combined optimization criteria [24], which, in general terms, are expressed as formula:

$$E(k) = (1 - \lambda) f_1(e(k)) + \lambda f_2(e(k)), \quad (72)$$

where $f_1(e(k))$ and $f_2(e(k))$ are the corresponding estimation criteria that are convex and differentiated functions; the parameter $\lambda \in [0, 1]$ gradually decreases from 1 to 0 during the training procedure. In [24], it was suggested to use the parameter λ, which is calculated according to the rule:

$$\lambda = \lambda(E) = e^{-\frac{c}{E^2}}, \quad (73)$$

where $c > 0$ is a positive parameter and E is a generalized quality criterion.

Such a hybrid optimization criterion can combine both regular and robust local optimization criteria.

The so-called wavelet-neuron is quite attractive from the point of view of technical implementation, accuracy, and ease of training. In this case, wavelet-functions are implemented either at the level of synaptic weights or at the neuron output, and the gradient training algorithm with a constant learning rate is used for training. In order to improve the approximating properties and accelerate the training procedure, a structure called the double wavelet-neuron was introduced, as well as algorithm for its training, which has both smoothing and tracking properties.

As activating functions of the double wavelet-neuron, we can use different types of analytic wavelets. Two families—POLYWOG-wavelets and RASP-wavelets are the most interesting as of their properties.

The family of RASP-wavelets—the wavelets based on rational functions (Rational functions with Second-order Poles—RASP) that are associated with the theorem on the excesses of complex variables.

In Fig. 12, you can see two typical representatives of mother RASP-wavelets, which are described by expressions:

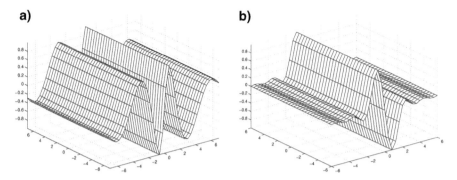

Fig. 12 Representatives of the RASP-wavelets family (**a**) shows that one achieved according to Eq. (74), (**b**) that one achieved according to Eq. (75)

$$\varphi_{ji}^1(x_i(k)) = \frac{\beta^1 \cos(x_i(k))}{x_i^2(k) + 1}, \quad \beta^1 = 2.7435, \tag{74}$$

$$\varphi_{ji}^2(x_i(k)) = \frac{\beta^2 \sin(\pi x_i(k))}{x_i^2(k) - 1}, \quad \beta^2 = 0.6111. \tag{75}$$

These wavelets are valid odd functions with a zero mean.

Another very broad wavelet family can be obtained from POLYnomials Windowed with Gaussians type of function—POLYWOG. It is interesting to note that the derivatives of these functions are also POLYWOG-wavelets and can be used as mother wavelets.

In Fig. 13, you can see some typical wavelets from the POLYWOG family, which are described by expressions:

$$\varphi_{ji}^1(x_i(k)) = \mu^1 x_i(k) \exp\left(\frac{-x_i^2(k)}{2}\right), \quad \mu^1 = \exp\left(-1/2\right), \tag{76}$$

$$\varphi_{ji}^2(x_i(k)) = \mu^2 \left(x_i^3(k) - 3x_i(k)\right) \exp\left(\frac{-x_i^2(k)}{2}\right), \quad \mu^2 = 0.7246; \tag{77}$$

$$\varphi_{ji}^3(x_i(k)) = \mu^3 \left(x_i^4(k) - 6x_i^2(k) + 3\right) \exp\left(\frac{-x_i^2(k)}{2}\right), \quad \mu^3 = 1/3; \tag{78}$$

$$\varphi_{ji}^4(x_i(k)) = \left(1 - x_i^2(k)\right) \exp\left(\frac{-x_i^2(k)}{2}\right). \tag{79}$$

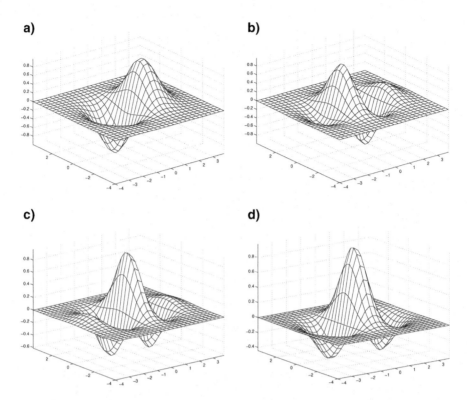

Fig. 13 Representatives of the POLYWOG-wavelets family (**a**) shows that one achieved according to Eq. (76), (**b**) that one achieved according to Eq. (77), (**c**) that one achieved according to Eq. (78), and (**d**) that one achieved according to Eq. (79)

Some wavelets of the POLYWOG family can be obtained using simple generators. Thus, in particular, the wavelets of this family can be generated with account of the properties of Hermitian nature of the derivative of the polynomial and the Gaussian function.

Let's consider the structure of the double wavelet-neuron, which is shown in Fig. 14 [25]. Apparently, the double wavelet-neuron is quite close to the structure of n—input wavelet-neuron, though it contains nonlinear wavelet-functions at the level of synaptic weights, as well as at the output of the structure.

When applying to the input of the double wavelet-neuron the vector signal $x(k) = (x_1(k), x_2(k), \ldots, x_n(k))^T$, at its output appears a scalar signal in the form of:

$$y(k) = f_0 \left(\sum_{i=1}^{n} f_i(x_i(k)) \right) = f_0(u(k))$$
$$= \sum_{l=0}^{h_2} \varphi_{l0} \left(\sum_{i=1}^{n} \sum_{j=0}^{h_1} \varphi_{ji}(x_i(k)) w_{ji}(k) \right) w_{j0} = \sum_{l=0}^{h_2} \varphi_{l0}(u(k)) w_{l0}(k), \quad (80)$$

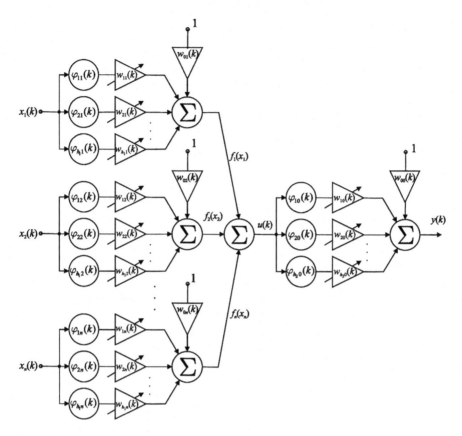

Fig. 14 Architecture of a double wavelet-neuron with nonlinear wavelet-synapses

which is determined both by configurable synaptic weights $w_{ji}(k)$ and $w_{l0}(k)$, and the values of the wavelet-functions $\varphi_{ji}(x_i(k))$ and $\varphi_{l0}(u(k))$, while it is specified that $\varphi_{00}(\bullet) = \varphi_{0i}(\bullet) \equiv 1$.

The double wavelet-neuron consists of two layers: a hidden layer, in which there are n wavelet-synapses with h_1 wavelet-functions in each one, and an output layer, consisting of a single wavelet-synapse with h_2 wavelet-functions.

In each wavelet-synapse, the wavelets are implemented that differ in their parameters of stretching (width) and shift (center).

Let's consider the robust learning algorithm for such an architecture. To study the output layer of a hybrid robust double wavelet-neuron, we introduce into consideration the error of training:

$$e(k) = d(k) - y(k), \qquad (81)$$

on the basis of which we introduce the robust criterion in the form of:

$$E(k) = \beta^2 \ln\left[\cosh\left(\frac{e(k)}{\beta}\right)\right], \qquad (82)$$

where β is a positive parameter that is chosen from empirical considerations and determines the size of the zone of insensitivity to the perturbations.

The robust algorithm for training the output layer of a double wavelet-neuron based on a gradient approach has the form of:

$$w_{j0}(k+1) = w_{j0}(k) + \eta_0(k)\beta \tanh\left(\frac{e(k)}{\beta}\right)\varphi_{j0}(u(k)), \qquad (83)$$

or in a vector form:

$$w_0(k+1) = w_0(k) + \eta_0(k)\beta \tanh\left(\frac{e(k)}{\beta}\right)\varphi_0(u(k)), \qquad (84)$$

where $w_0(k) = (w_{10}(k), w_{20}(k), \ldots, w_{h_20}(k))^T$—vector of synaptic weights, $\varphi_0(k) = (\varphi_{10}(k), \varphi_{20}(k), \ldots, \varphi_{h_20}(k))^T$—vector of wavelet-activation functions, and $\eta_0(k)$ is the learning rate to be determined.

To increase the convergence rate of the training procedure, it is necessary to move from gradient procedures to quasi-Newton algorithms, the most widely used of which is Levenberg–Marquardt algorithm.

After simple transformations, we obtain the training algorithm in the form of:

$$\begin{cases} w_0(k+1) = w_0(k) + \dfrac{\beta \tanh\left(\frac{e(k)}{\beta}\right)\varphi_0(u(k))}{\gamma_i^{w_0}(k)}; \\ \gamma_i^{w_0}(k+1) = \alpha \gamma_i^{w_0}(k) + \|\varphi_0(u(k+1))\|^2, \end{cases} \qquad (85)$$

where α is parameter of forgetting obsolete information ($0 < \alpha < 1$), which is set according to guidelines that are used in the tasks based on exponentially weighted recurrent least squares method.

The training of a hidden layer is conducted in a similar way on the basis of the error backpropagation using the same criterion, but written in the form of:

$$E(k) = \beta^2 \ln\left[\cosh\left(\frac{d(k) - f_0(u(k))}{\beta}\right)\right]$$
$$= \beta^2 \ln\left[\cosh\left(\frac{\left(d(k) - f_0\left(\sum_{i=1}^n \sum_{j=0}^{h_1} \varphi_{ji}(x_i(k))w_{ji}(k)\right)\right)}{\beta}\right)\right]. \qquad (86)$$

The robust learning algorithm for the hidden layer of double wavelet-neuron based on gradient optimization is written as:

$$w_{ji}(k+1) = w_{ji}(k) + \eta(k)\beta \tanh\left(\frac{e(k)}{\beta}\right)f_0'(u(k))\varphi_{ji}(x_i(k)), \qquad (87)$$

or in a vector form:

$$w_i(k+1) = w_i(k) + \eta(k)\beta \tanh\left(\frac{e(k)}{\beta}\right) f_0'(u(k)) \varphi_i(x_i(k)), \tag{88}$$

where $w_i(k) = (w_{1i}(k), w_{2i}(k), \ldots, w_{h_1i}(k))^T$—vector of synaptic weights, and $\varphi_i(k) = (\varphi_{1i}(k), \varphi_{2i}(k), \ldots, \varphi_{h_1i}(k))^T$—vector of wavelet-activation functions.

By analogy with (85), we can introduce a procedure:

$$\begin{cases} w_i(k+1) = w_i(k) + \dfrac{\beta \tanh\left(\frac{e(k)}{\beta}\right) f_0'(u(k)) \varphi_i(x_i(k))}{\gamma_i^{w_1}(k)}; \\ \gamma_i^{w_1}(k+1) = \alpha \gamma_i^{w_1}(k) + \|\varphi_i(x_i(k+1))\|^2. \end{cases} \tag{89}$$

The proposed approach allows signals processing under conditions of various kinds of perturbations, gross errors, and non-Gaussian perturbations with "heavy tails," and most importantly, it allows detecting changes in characteristics of such signals. It is worth noting that algorithm (89), being a procedure of gradient optimization, can get to the local minima of goal function. In such cases, it makes sense to use either set of similar procedures with different initial conditions (that is cumbersome from computational point of view) or evolutionary optimization procedures (that slows down the learning process).

An experimental study of the developed robust training algorithm for adaptive wavelet-neuron was carried out on the basis of a signal contaminated by intense perturbations. The signal was obtained using the nonlinear dynamic object of Narendra [26], whose output signal was artificially contaminated by random perturbations with a Cauchy distribution having the form of:

$$F_X^{-1}(x) = x_0 + \gamma tg[\pi(x - 0.5)], \tag{90}$$

where x_0 is the localization parameter, γ is the scale parameter ($\gamma > 0$), and x is the carrier ($x \in (-\infty; +\infty)$).

A nonlinear dynamic object was generated by the equation:

$$y(k+1) = 0.3y(k) + 0.6y(k-1) + f(u(k)), \tag{91}$$

where $f(u(k)) = 0.6\sin(u(k)) + 0.3\sin(3u(k)) + 0.1\sin(5u(k))$, $u(k) = \sin(2k/250)$.

The values of $x(k-3)$, $x(k-2)$, $x(k-1)$, and $x(k)$ were used to emulate $x(k+1)$.

Figure 15a shows the results of the emulation of a signal contaminated by interruptions. Figure 15b shows a segment of the training procedure: one can see that a high amplitude perturbation, which occurs at the beginning of the sampling, did not affect the training algorithm.

A comparison of the results of prediction based on the robust training algorithm was carried out with the results of prediction based on the gradient algorithm and the algorithm based on the recurrent method of least squares, where the structure of the network and the number of configured parameters were the same.

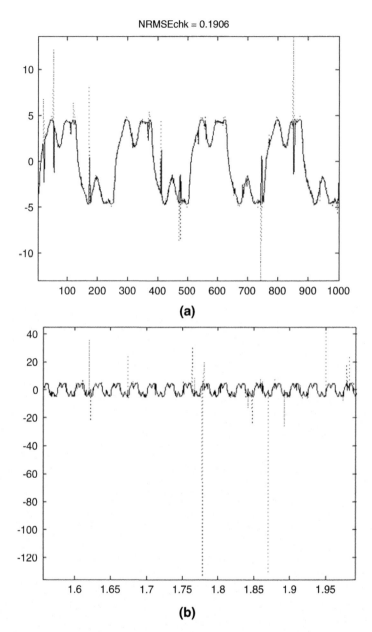

Fig. 15 (a) Result of the emulation of a signal contaminated by interruptions, (b) a segment of the training procedure

In case of training the adaptive wavelet-neuron with the use of gradient algorithm, the first perturbation at the start of sampling strongly influenced the training procedure, and as a result we have a large error of emulation. In case of training by the recurrent method of least squares, during the first perturbation a so-called explosion of parameters of the covariance matrix occurred and, as a result, the impossibility of emulating signals contaminated by abnormal perturbations.

In this way, it can be seen that the proposed robust algorithm of training the adaptive wavelet-neuron allows perform signal processing under conditions of significant contamination by perturbations. Thus, the proposed adaptive double wavelet-neuron restores the analyzed time series, eliminating abnormal observations.

After that, the signal being processed can be segmented through the use of online procedures of fuzzy clustering described above.

Thus, significant increase of forecasting error over a certain interval indicates changes in the controlled process, after which the forecasting system adjusts its parameters adaptively for the new mode. Importantly, due to the robust properties of the used learning algorithm, abnormal outliers in observations are "ignored" during the controlling process.

4 Conclusions

Approach to online detection of changes in time series properties based on hybrid systems of computational intelligence has been proposed and analyzed. The approach is guided by fuzzy sequential clustering of time series with the use of probabilistic and possibilistic procedures as well as a wavelet-neural network that is learned by robust algorithm of synaptic weights adjustment, which enables suppressing abnormal outliers present in real time series.

The proposed approach is characterized by simplicity of numerical implementation and high performance, and it can be applied in addressing practical tasks related to changes detection and fault detection in data streams.

References

1. Bodyanskiy, Y.: Computational intelligence techniques for data analysis. Lect. Notes Inf. **P-72**, 15–36 (2005)
2. Gorshkov, Y., Kokshenev, I., Bodyanskiy, Y., Kolodyazhniy, V., Shilo, O.: Robust recursive fuzzy clustering-based segmentation of biomedical time series. In: Proceedings of 2006 International Symposium on Evolving Fuzzy Systems, Lankaster, UK, 2006, pp. 101–105. (2006)
3. Han, J., Kamber, M.: Data Mining: Concepts and Techniques. Morgan Kaufmann, San Francisco (2006)
4. Bow, S.-T.: Pattern Recognition and Image Preprocessing. Marcel Dekker, Inc., New York (2002)

5. Aggarwal, C., Reddy, C.: Data Clustering: Algorithms and Applications. Chapman and Hall/CRC, Boca Raton (2014)
6. Bezdek, J.C.: Pattern Recognition with Fuzzy Objective Function Algorithms. Plenum Press, New York (1981)
7. MacQueen, J.: On convergence of k-means and partitions with minimum average variance. Ann. Math. Statist. **36**, 1084 (1965)
8. Gorshkov, Y., Kolodyazhniy, V., Bodyanskiy, Y.: New recursive learning algorithms for fuzzy Kohonen clustering network. In: Proceedings of 17th International Workshop on Nonlinear Dynamics of Electronic Systems (NDES-2009), June 21–24, 2009, Rapperswil, pp. 58–61. (2009)
9. Bodyanskiy, Y., Gorshkov, Y., Kokshenev, I., Kolodyazhniy, V.: Evolving fuzzy classification of non-stationary time series. In: Angelov, P., Filev, D.P., Kasabov, N. (eds.) Evolving Intelligent Systems Methodology and Applications, pp. 446–464. John Wiley & Sons, New York (2008)
10. Gustafson, E.E., Kessel, W.C.: Fuzzy clustering with a fuzzy covariance matrix. In: Proceedings of IEEE CDC, San Diego, California, pp. 761–766. (1979)
11. Gath, I., Geva, A.B.: Unsupervised optimal fuzzy clustering. IEEE Trans. Pattern Anal. Mach. Intell. **11**, 773–781 (1989)
12. Krishnapuram, R., Keller, J.: A possibilistic approach to clustering. IEEE Trans. Fuzzy Syst. **1**, 98–110 (1993)
13. Chung, F.L., Lee, T.: Fuzzy competitive learning. Neural Netw. **7**(3), 539–552 (1994)
14. Park, D.C, Dagher, I.: Gradient based fuzzy c-means (GBFCM) algorithm. In: Proceedings of IEEE International Conference on Neural Networks, IEEE Press, Orlando, FL, USA, pp. 1626–1631. (1994)
15. Fritzke, B.: A growing neural gas network learns topologies. Adv. Neural Inf. Process. Syst. **7**, 625–632 (1995)
16. Hoeppner, F., Klawonn, F.: Fuzzy clustering of sampled functions. In: Proceedings of 19-th International Conference North American Fuzzy Information Processing Society (NAFIPS), Atlanta, USA, pp. 251–255. (2000)
17. Khamassi, I., Sayed-Mouchaweh, M., Hammami, M., Ghedira, K.: Discussion and review on evolving data streams and concepts drift adapting. Evol. Syst. **8**(1), 1–23 (2018)
18. Lughofer, E., Weigl, E., Heidl, W., Eitzinger, C., Radauer, T.: Recognizing input space and target concept drifts with scarcely labelled and unlabeled instances. Inf. Sci. **355–356**, 127–151 (2016)
19. Lughofer, E., Pratama, M., Skrjanc, I.: Incremental rule splitting in generalized evolving fuzzy systems for autonomous drift compensation. IEEE Trans. Fuzzy Syst. **26**(4), 1854–1865 (2018)
20. Pau, L.F.: Failure diagnosis and performance monitoring. Dekker, New York (1981)
21. Chui, C.K.: An Introduction to Wavelets. Academic, New York (1992)
22. Bodyanskiy, Y., Lamonova, N., Pliss, I., Vynokurova, O.: An adaptive learning algorithm for a wavelet neural network. Expert. Syst. **22**(5), 235–240 (2005)
23. Huber, P.J.: Robust Statistics. John Wiley & Sons, New York (1981)
24. Karayiannis, N.B., Venetsanopoulos, A.N.: Fast learning algorithm for neural networks. IEEE Trans. Circuits Syst. II, Analog Digit. Signal Process. **39**, 453–474 (1992)
25. Bodyanskiy, Y., Lamonova N., Vynokurova, O.: Recurrent learning algorithm for double-wavelet neuron. In: Proceedings of XII-th International Conference "Knowledge – Dialogue – Solution", Varna, pp. 77–84. (2006)
26. Narendra, K.S., Parthasarathy, K.: Identification and control of dynamic systems using neural networks. IEEE Trans. Neural Netw. **1**, 4–26 (1990)

Early Fault Detection in Reciprocating Compressor Valves by Means of Vibration and pV Diagram Analysis

Kurt Pichler

1 Introduction

Reciprocating compressors are heavily used in modern industry, for instance in chemical industry, refinery, gas transportation, and gas storage. Economic demands of the last decades have also affected the operation of reciprocating compressors. In many cases, compressors run at full capacity without backup. Reliable performance is thus a key issue and becoming more important than ever. Customers expect reduction or even elimination of unscheduled shutdowns as well as extended maintenance intervals. These challenges are addressed by the development of advanced materials and designs. However, fatigue and wear cannot be avoided.

There is also an economic trend towards saving on labour costs by reducing the frequency of on-site inspections. Such considerations mean that modern gas storage facilities are run by remote control stations, and the compressors are monitored by automated technical systems. In this case, the system must be able to retrieve and evaluate relevant information automatically to detect faulty behaviour.

For these reasons, monitoring and diagnostics are almost an economical and technical necessity. Firstly, monitoring systems enable condition-based maintenance. Secondly, improved understanding of compressor behaviour allows evaluation and recommendations regarding efficient compressor operation.

Condition monitoring can be based on measurements of various physical states, for instance, vibrations, flow rate, power, position, temperature, and pressure. The data required for diagnostic evaluation depend mainly on the types of faults expected and observed. Broken valves, with a percentage of about 36%, are the most common reason for unscheduled shutdowns of reciprocating compressors [21], followed by

---

K. Pichler (✉)
Linz Center of Mechatronics GmbH, Linz, Austria
e-mail: kurt.pichler@lcm.at

© Springer Nature Switzerland AG 2019
E. Lughofer, M. Sayed-Mouchaweh (eds.), *Predictive Maintenance in Dynamic Systems*, https://doi.org/10.1007/978-3-030-05645-2_6

faulty pressure packings (about 18%) and piston rings (about 7%). Hence, it is obvious that especially the valves have to be monitored.

Several papers have been published about valve fault detection in reciprocating compressors. Lin et al. [27] combined time–frequency analysis of vibration data and an artificial neural network, which enabled them to differentiate between new and worn valves. Applying their approach to extended test scenarios with 15 seeded faults [28] did not lead to satisfactory validation results. However, by reducing the number of fault cases to 7, they achieved good classification results. Cyclostationary modeling of reciprocating compressors is introduced in [53] and [3]. Tiwari and Yadav [45] analysed pressure pulsation with a back-propagation neural network. The pressure pulsation (peak to peak) is modeled in relation to the leak percentage. In [37], a method using support vector machines was presented. The first four zero-lag sums of sub-band signals of the intrinsic mode functions are extracted from vibration measurements and used as input features for a support vector machine classifier. Drewes [10] described the effect of a valve fault on the pV diagram. Additionally, the effects of some other faults such as piston ring wear and damages to crank gears and pistons were discussed. Several different fault detection methods were introduced in [52]. They are divided into four main categories: time domain analysis, frequency domain analysis, orbit analysis, and trend analysis. Yang et al. [51] focus particularly on small reciprocating compressors for refrigerators at constant operation conditions. They use wavelet transform to extract features from raw noise and vibration data and classify them using neural networks and support vector machines. The changes in cylinder pressures and instantaneous angular speed for various leakage percentages were analysed visually in [13]. Based on the results, a decision table for valve faults is built. Wang et al. [48] introduce an automated evaluation of the pV diagram. They determine seven invariant moments of the pV diagram and classify them using support vector machines. In [49], the valve motion is monitored using acoustic emission signals and simulated valve motion. As the authors state in the paper, the method can distinguish between normal valve operation, valve flutter, and delayed closing, but it is not sensitive to leaks.

In this chapter, two independent methods for detecting broken reciprocating compressor valves are developed: one is based on vibration analysis, and the other one is based on analysing pV diagrams. In vibration analysis, spectrograms of accelerometer measurements are used to extract features that provide separability of faultless and faulty class in the feature space. Similarly, certain features are extracted from pV diagrams. These features provide class separability as well. Two independent approaches are developed because compressors can be equipped with different sensors, and not every compressor is equipped with the sensors required for a certain method. It depends on the operators of the compressor, what kind of sensors they want to mount on their equipment. Furthermore, upgrading the sensing system with certain sensors can be too expensive or simply not desired.

In contrast to the method proposed in this book chapter, the previously published approaches above do not mention varying load and pressure conditions. It can be assumed that those methods cannot be easily adapted to this case. Moreover, they use only one valve type each and do not extrapolate their method from one valve type to others. Another drawback of the previously published approaches is that, if an automated method is presented, training with faultless cases only is not provided.

2 Problem Statement

This section gives a brief overview of the reciprocating compressors functionality [6, 7]. Furthermore, it states the specific problems and needs of a fully automated monitoring system for compressor valves.

2.1 Reciprocating Compressor Operation

The basic elements of a reciprocating compressor are the compression cylinder and the piston. An engine, usually an electric motor, drives a crankshaft at a constant revolution speed, and the crankshaft drives the piston inside the cylinder. The piston rod is connected to the crankshaft by a connecting rod. As the piston in the cylinder is moving forwards and backwards, the volume of the cylinder changes during time. Hence, the pressure in the cylinder is increasing when the volume is decreasing and vice versa. The compression unit is called single acting if it is compressing only on one side of the piston. If it is compressing on both sides of the piston, it is called double acting. In the case of a double-acting compression cylinder, the compression chamber closer to the crank is called crank end (CE), the other compression chamber is called head end (HE). A scheme of a double-acting cylinder is shown in Fig. 1.

A reciprocating compressor is an oscillating machine, it works in repeated cycles. A typical compression cycle for the HE compression chamber is explained now. Let's start with the piston at its bottom dead centre (BDC), the discharge valve closed, the suction valve opened, and the cylinder filled with gas at suction pressure. As the piston moves towards HE, the suction valve closes immediately, and the piston movement reduces the original volume of gas with an accompanying rise in pressure, the valves remain closed. This is called the compression stroke.

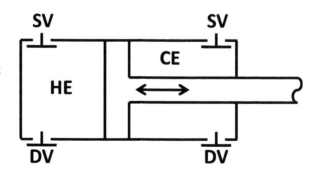

Fig. 1 Scheme of a double-acting cylinder of a reciprocating compressor with the suction valves (SV) and discharge valves (DV) at the crank end (CE) and head end (HE) compression chambers

When the pressure just exceeded the discharge pressure, the discharge valve opens. The compressed gas is flowing out through the discharge valve into the discharge chamber. This is called the delivery stroke. Just before the piston reached its top dead centre (TDC), the discharge valve is closed by its springs, leaving the clearance space filled with gas at discharge pressure. The clearance space is the volume that remains in the compression chamber when the piston is at TDC. During the following expansion stroke, both the suction and discharge valves remain closed. The gas trapped in the clearance space increases in volume, causing a reduction in pressure. This continues as the piston moves towards the CE until the cylinder pressure drops below the suction pressure. Now, the suction valve opens and gas flows into the cylinder until the end of the reverse stroke. This is called the suction stroke. When the piston passed BDC, the spring load closes the suction valve and the cycle will repeat on the next revolution of the crank. The compression cycle for the CE compression chamber is of course shifted by 180° of the crank.

The pV diagram (pressure–volume diagram) of a compression cycle plots the change in pressure in the compression chamber with respect to its volume. The pV diagram is observed for each compression cycle. As the suction and discharge pressures will usually not change significantly from one compression cycle to the next, the pV diagram has a cyclic shape, i.e. it returns to its starting pressure and volume. A key feature of the pV diagram is that the amount of energy expended or received by the system as work can be estimated. For cyclic diagrams, the net work is that enclosed by the curve. A typical (simulated) pV diagram of a reciprocating compressor together with the actual piston position is shown in Fig. 2. With the piston at TDC, the pressure is on its maximum level (the discharge pressure), while

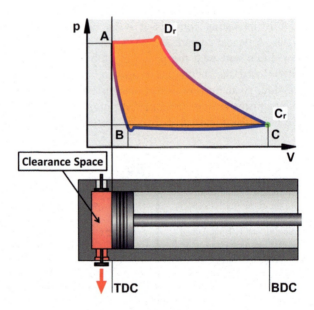

Fig. 2 The piston at TDC and the corresponding pV diagram (graphics by Hoerbiger Compression Technology)

the compression chambers' volume is at its minimum (the clearance space). The pV diagram is traversed anticlockwise.

Reciprocating compressors use automatic spring-loaded valves. The valves regulate the cycle of operation in a compressor cylinder. Automatic compressor valves are pressure activated, and their movement is controlled by the compression cycle. The valves are opened by the difference in pressure across the valve. The suction valve (SV) opens, when the pressure in the cylinder is slightly below the pressure in the suction chamber (suction pressure). The discharge valve (DV) opens, when the pressure in the cylinder is slightly above the pressure in the discharge chamber (discharge pressure). As there are so many application scenarios for reciprocating compressors, one valve type cannot cover every application. Depending on the process requirements (gas type, expected pressure and temperature conditions, crank revolution speed, lubrication, etc.), a specific valve type has to be chosen. The valve types differ in their design and material.

During normal operation, a reciprocating compressor will take in a quantity of gas from its suction line and compress the gas as required to move it through its discharge line. The compressor cannot self-regulate its capacity against a given discharge pressure; it will simply keep displacing gas until told not to. This would not be a problem if there was unlimited supply of gas to draw from and an infinite downstream to discharge into. However, in real-world refineries, chemical plants, and gas transmission lines, there are specific parameters within which to work, and the capacity is a unique quantity at any point in time. Thus, there is a real need to control the capacity of the reciprocating compressor.

The capacity (or load) of a reciprocating compressor is controlled by the reverse flow capacity control system. This system directly influences the closing time of the suction valve. An actuator keeps the suction valve open for a specified and controllable time at the beginning of the compression stroke. This allows a fraction of the gas to flow back through the suction valve into the suction chamber (Fig. 3). The discharge valve is a passive valve. It thus opens when the pressure inside the cylinder is higher than the pressure outside. When the reverse flow capacity control allows a smaller amount of gas to be compressed, the pressure for opening the discharge valve is obtained later. For that reason, the control system influences the timing of closing the suction valve as well as the timing of opening the discharge valve. In Fig. 3, the pV diagram of a compression cycle at reduced load is shown. As the suction valve is kept open, the cylinder pressure stays at suction pressure until the control system allows the suction valve to close. Then, the gas is compressed until the pressure exceeds discharge pressure and the discharge valve opens. The figure shows that the reverse flow control system affects the shape of the pV diagram only at the compression and the delivery stroke, but not at the expansion and the suction stroke.

In this chapter, the pressure conditions are defined as the pressures in the suction and discharge chamber of a compressor stage. The pressure conditions are not constant. For instance, in a two-stage compressor the suction pressure in stage 1 depends on the pressure in the reservoir from which the gas is taken from, which can of course be varying. The suction pressure in stage 2 equals the pressure built up

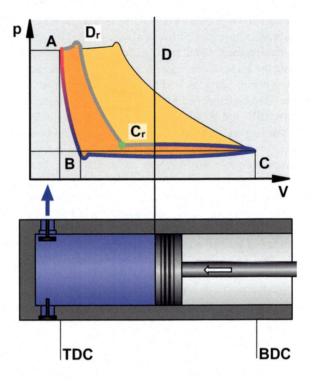

Fig. 3 The piston moving towards the head end during the compression stroke and the corresponding pV diagram (graphics by Hoerbiger Compression Technology). Due to the load control (reduced load), the suction valve is still open and the gas has not been compressed yet

in stage 1. This pressure is called the interstage pressure (IP). The discharge pressure in stage 2 depends on the pressure in the discharge vessel. By building up pressure in the discharge vessel or dumping gas from it, this pressure may be varying as well.

2.2 Problem Statement

As already mentioned in Sect. 1, broken valves are one of the main problems in the operation of reciprocating compressors. To specify, the sealing elements are most likely to break thus causing a leak. Therefore, the goal is to detect broken sealing elements of reciprocating compressor valves. In the following, the terms broken valve or faulty valve are oftentimes used for a leaking sealing element. The problem is tackled with a data-driven approach. In this context, data driven means to acquire real-world measurement data from reciprocating compressors and analyse these data. The aim of this data analysis is to quantify the differences between data acquired from faultless valves and data acquired from faulty valves (i.e. valves with a broken sealing element). This knowledge can then be used to define and quantify normal compressor operation and subsequently to detect deviations from normal compressor operation.

Of course, not only broken sealing elements can affect the acquired data. Also other components, such as pressure packings, piston rings, and many more [21], can break and influence the measured data. As these faults are already monitored with existing approaches [20], they are not considered in this chapter. When acquiring test data, it was carefully observed to measure at a compressor with all other components in faultless conditions. Hence, it is possible to concentrate on quantifying the deviation in the data originating from broken valves.

A monitoring system for reciprocating compressor valves, in fact almost every monitoring system, has to cope with changing operation conditions. For reciprocating compressors, there are several variables that might change. Among these variables are the load levels and pressure conditions described in Sect. 2.1. All these variables can affect the acquired measurement data. For obvious reasons, they should not cause over-detections.

As specified in Sect. 2.1, reciprocating compressor valves have different designs and are made of different materials. It is not desirable to have an own monitoring procedure for each valve type. Different valve types must be monitored with only one system. This should work automatically, without human operators adjusting any parameters or threshold values. Valve-type independent classification is desirable because it enables monitoring a new valve type that has not been part of the training procedure. Assume that the proposed monitoring system has been trained with a certain valve type, and it is necessary to equip the compressor with another valve type (for instance, due to changing gas characteristics or different temperature expectations). Then, it is an obvious advantage that no extensive amount of training data with, ideally, pressure and load conditions ranging from the compressors maximum to minimum is needed to train a new classifier. Furthermore, it helps to avoid mistakes from operators by selecting a wrong classifier for a certain valve type.

One final, yet very important requirement of a monitoring system concerns the training phase of the method. A data-driven method has to be trained with annotated measurement data (data with given class information) of the underlying system. As the aim is to distinguish between faultless and faulty state, it is a classification problem. Usually, classification problems are trained with data from all possible classes. Because of economic and safety considerations, it is oftentimes not possible to acquire data from broken valves. Even though test data from faultless and from leaking valves are available in this study, this is not the general case. Training the classifier with the faultless class only is thus desirable. Hence, a focus also lies on applying one-class classification (also referred to as novelty detection or data description) techniques to the feature space, which is able to operate in regular faultless data solely by extracting a model of their main characteristics. For such a one-class classification task, it is important that the training data covers a broad range of normal operating condition, i.e. load levels, pressure conditions, and so on. This will ensure that different operating conditions will not cause over-detections.

3 Vibration Analysis

Vibration analysis is a common approach in rotating machinery diagnostics, for instance for bearing [42] and gearbox monitoring [23]. It has also been applied for oscillating machines such as reciprocating compressors [27, 28]. To perform vibration analysis, data is acquired using accelerometers. When analysing vibration measurement data, mainly two basic concepts are adopted. One concept is to analyse the acquired data as a time series, some of the methods are described in [14, 40]. Oftentimes, these methods use statistical analysis procedures to evaluate the measurement data such as proposed in [15]. The other popular basic concept of vibration analysis is analysing the (time–)frequency space of the acquired data [17, 32]. Numerous concepts have been proposed, for instance observing absolute spectral density for a certain frequency or comparing whole frequency spectra.

3.1 Motivation

In earlier studies, the capability of (time–)frequency-based condition monitoring for reciprocating compressor valves has been shown. The energy content of single frequency bands was observed in [33]. An approach of comparing whole spectrograms in a metric vector space was introduced in [35] and extended to a switching model in [34]. However, all of those methods have no (or limited) valve condition monitoring capability for arbitrarily changing load conditions.

As shown in [36], valve faults or changing load conditions induce different patterns in the spectrogram. A typical time series of the accelerometer data of two compression cycles and the according spectrogram can be seen in Fig. 4. The

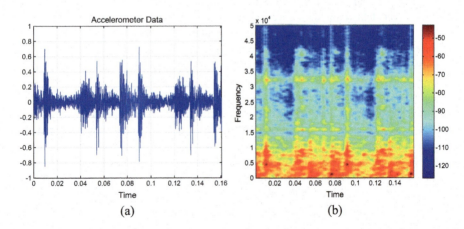

Fig. 4 (a) Raw accelerometer data and (b) spectrogram of two successive compression cycles

pointwise difference $D \in \mathbb{R}^{n_f \times n_t}$ with entries $d_{i,j}$, $i = 1, \ldots, n_f$, $j = 1, \ldots, n_t$, of a reference spectrogram (from a faultless valve) $R \in \mathbb{R}^{n_f \times n_t}$ with entries $r_{i,j}$, $i = 1, \ldots, n_f$, $j = 1, \ldots, n_t$ (with n_f and n_t denoting the number of bins in frequency and time domain), and a test spectrogram $S \in \mathbb{R}^{n_f \times n_t}$ (unknown fault state) with entries $s_{i,j}$, $i = 1, \ldots, n_f$, $j = 1, \ldots, n_t$, is defined as:

$$D = S - R \tag{1}$$

and

$$d_{i,j} = s_{i,j} - r_{i,j}, \tag{2}$$

respectively.

In the case of a test spectrogram from faultless valves at the same load level as the reference spectrogram, both spectrograms look more or less the same, and the difference D shows no significant pattern (Fig. 5).

When the test spectrogram is from faultless valves, but at a different load level, significant patterns in vertical direction reveal (Fig. 6). These patterns arise from the timing of the valve event due to the load control. In [41], it is shown that the valve events can be identified uniquely in the spectrogram. However, the experimental setup in that study was a very simple compressor geometry with only one reed valve. For the more advanced compressor and valves used in this chapter, it was not possible to identify the valve events uniquely. Nevertheless, the different time points of the valve events appear as a vertical pattern in the spectrogram difference.

Figure 7 illustrates the case when the test spectrogram is from a faulty valve at the same load level. In [33], it was empirically shown that a faulty valve is accompanied by changing amplitudes in certain frequency bands of the power spectrum. However,

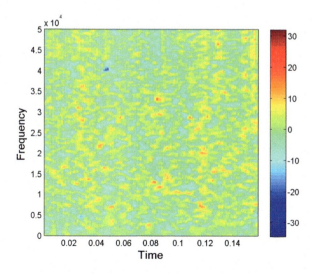

Fig. 5 Pointwise spectrogram difference of a reference spectrogram and a test spectrogram from a faultless valve at the same load level

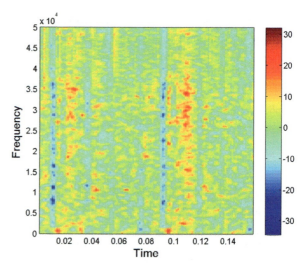

Fig. 6 Pointwise spectrogram difference of a reference spectrogram and a test spectrogram from a faultless valve at a different load level

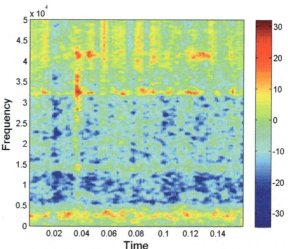

Fig. 7 Pointwise spectrogram difference of a reference spectrogram and a test spectrogram from a faulty valve at the same load level

it was not possible to determine the frequency bands of interest a priori, i.e. without measurements from a faulty valve. The effect of changed amplitudes in a certain frequency band as an indicator for a faulty valve can also be observed in the spectrogram difference. Since the frequency is plotted on the ordinate, a faulty valve corresponds to significant horizontal patterns.

It comes as no surprise that, in the case of a faulty valve at a different load level, the spectrogram difference in Fig. 8 shows significant vertical patterns (due to the load control) as well as significant horizontal patterns (due to the faulty valve).

Looking at the examples, it can be noticed that there seems to be a possibility to discriminate between the faultless and faulty case. A faulty valve shows significant horizontal (and maybe vertical) patterns in the spectrogram difference, while a

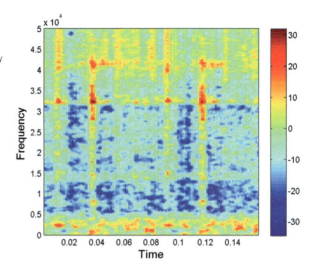

Fig. 8 Pointwise spectrogram difference of a reference spectrogram and a test spectrogram from a faulty valve at a different load level

faultless valve either shows no significant patterns or vertical patterns. The problem is that the position of these patterns within the spectrogram difference varies with the valve type and the load levels of the two spectrograms. Furthermore, the measurements are afflicted with noise. These issues make it challenging to identify the patterns automatically. By applying two-dimensional autocorrelation to the spectrogram difference, the patterns are centred and the signal-to-noise ratio is increased. Autocorrelation thus solves both tasks within one simple transformation step. This makes it easier for an automated method to recognize the patterns correctly and is therefore an important step on the way to a load independent method.

Given the pointwise spectrogram difference $D \in \mathbb{R}^{n_f \times n_t}$ (Eq. (1)) with entries $d_{i,j}$, $i = 1, \ldots, n_f$, $j = 1, \ldots, n_t$ (Eq. (2)), the two-dimensional autocorrelation $A \in \mathbb{R}^{(2n_f-1) \times (2n_t-1)}$ with entries $a_{i,j}$, $i = 1, \ldots, 2n_f - 1$, $j = 1, \ldots, 2n_t - 1$, of $D \in \mathbb{R}^{n_f \times n_t}$ is defined as [26]:

$$a_{k_f + n_f, k_t + n_t} = \sum_{i=1}^{n_f} \sum_{j=1}^{n_t} d_{i,j} \cdot d_{i-k_f, j-k_t} \quad (3)$$

for $k_f = -(n_f - 1), \ldots, n_f - 1$ and $k_t = -(n_t - 1), \ldots, n_t - 1$ and with

$$d_{i,j} := 0 \quad \text{for} \quad \begin{cases} i < 1 \\ i > n_f \\ j < 1 \\ j > n_t. \end{cases}$$

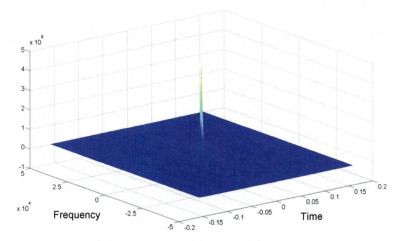

Fig. 9 Autocorrelation of the pointwise spectrogram difference (Fig. 5) of a reference spectrogram and a test spectrogram from a faultless valve at the same load level

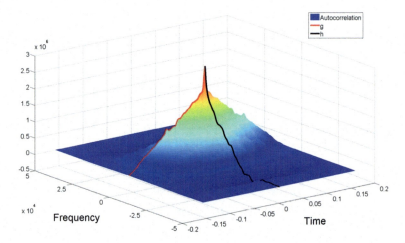

Fig. 10 Autocorrelation of the pointwise spectrogram difference (Fig. 7) of a reference spectrogram and a test spectrogram from a faulty valve at the same load level (with highlighted curves of interest)

The effect of computing autocorrelation is depicted in the following. In the case of a test spectrogram from a faultless valve at the same load level, the autocorrelation is simply a surface with a central peak and the rest approximately at level 0 (Fig. 9).

The autocorrelation surface for a test spectrogram from a faulty valve at the same load level is shown in Fig. 10. A mountain crest through the centre aligned in time dimension with a central peak can be observed.

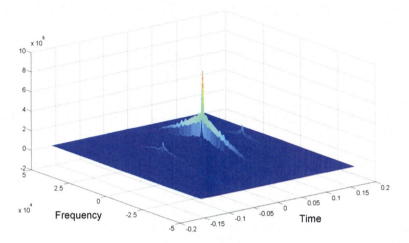

Fig. 11 Autocorrelation of the pointwise spectrogram difference (Fig. 6) of a reference spectrogram and a test spectrogram from a faultless valve at a different load level

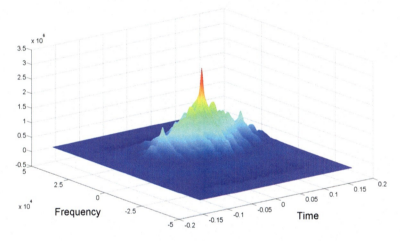

Fig. 12 Autocorrelation of the pointwise spectrogram difference (Fig. 8) of a reference spectrogram and a test spectrogram from a faulty valve at a different load level

The other two cases show consequently a mountain crest aligned in frequency dimension (faultless valve at different load level, Fig. 11) or in both dimensions (faulty valve at different load level, Fig. 12).

A characteristic for the case of a test spectrogram from a faulty valve can thus be given as a mountain crest in time dimension with a peak in its centre. Concentrating on this characteristic of the autocorrelation, it is possible to define features that enable classification of the fault state of the valve, regardless of the load levels of the reference spectrogram and the test spectrogram.

3.2 Feature Extraction

Features are extracted from the spectrogram difference $D \in \mathbb{R}^{n_f \times n_t}$ (Eq. (1)) and from its autocorrelation $A \in \mathbb{R}^{(2n_f-1) \times (2n_t-1)}$ (Eq. (3)) as well. The features are motivated by visually comparing the shapes of the plots in Sect. 3.1 and quantifying obvious differences. As in many pattern recognition scenarios, a big number of features were defined. Some of those features were motivated by the shape of the spectrograms and their autocorrelation surfaces or by expert knowledge. Some other features were defined as standard statistical features or well-known damage indicators (for instance [47]) of the data. Also, the load levels of reference and test spectrogram and the difference of their load levels were included in the original feature set. In the end, the best features were selected by feature selection and ranking criteria using filter and wrapper approaches [11, 16, 22] such as forward selection using Dy–Brodley measure or Mahalanobi distance as selection criteria. Only the four finally selected features are presented here. Selecting more than four features did not improve the classification accuracy significantly.

The first feature f_1 is given by the distance of the reference and the test spectrogram. It is computed by the matrix norm $\|.\|_M : \mathbb{R}^{n_f \times n_t} \to \mathbb{R}$ in the vector space $\mathbb{R}^{n_f \times n_t}$ that is introduced in [35] as:

$$\|D\|_M = \frac{1}{n_f \cdot n_t} \cdot \sum_{i=1}^{n_f} \sum_{j=1}^{n_t} |d_{i,j}| \qquad (4)$$

for $D \in \mathbb{R}^{n_f \times n_t}$ with matrix entries $d_{i,j} \in \mathbb{R}$. Using this norm, f_1 is defined as:

$$f_1 = \|D\|_M \qquad (5)$$

with the spectrogram difference matrix D from Eq. (1).

As shown in [35], this measure is not sufficient to obtain a load independent method. Hence, three additional features are extracted from the autocorrelation matrix $A \in \mathbb{R}^{(2n_f-1) \times (2n_t-1)}$. The significant patterns in the spectrogram difference are aligned in time and frequency dimension. They are centred by applying autocorrelation and still aligned in time and frequency dimension. For that reason, it is sufficient to observe the curves through the centre aligned in time and frequency dimension of the autocorrelation. This reduces also the amount of data to be processed and therefore saves computational time. The curves are symmetric; hence, only one half has to be observed. The curves are highlighted in Fig. 10. By defining $N_f = \left\lceil \frac{2n_f-1}{2} \right\rceil$ and $N_t = \left\lceil \frac{2n_t-1}{2} \right\rceil$, these curves $g \in \mathbb{R}^{N_t}$ and $h \in \mathbb{R}^{N_f}$ are given by the matrix entries:

$$g = a_{N_f, 1...N_t} \qquad (6)$$

and
$$h = a_{1...N_f, N_t} \tag{7}$$

of the autocorrelation matrix A (Eq. (3)).

Feature f_2, the first feature of the autocorrelation, measures the ratio of the mean values of g and h. The mean values $\bar{g}$ and $\bar{h}$ are given by:

$$\bar{g} = \frac{1}{N_t} \cdot \sum_{i=1}^{N_t} g_i \tag{8}$$

and

$$\bar{h} = \frac{1}{N_f} \cdot \sum_{i=1}^{N_f} h_i. \tag{9}$$

The feature f_2 is thus defined as:

$$f_2 = \frac{\bar{g}}{\bar{h}}. \tag{10}$$

The main advantage of using the ratio of $\bar{g}$ and $\bar{h}$ over using the two features individually is that it incorporates both dimensions (frequency and time) in only one feature value, thus reducing the dimensionality of the feature space.

The next feature, f_3, is somewhat similar to the previous one in the way that it measures the ratio of the degree of curvature of the segments g and h. Between the end points g_1 and g_{N_t}, respectively, h_1 and h_{N_f} of the curves g and h, connecting lines are linearly interpolated, that is:

$$\hat{g}_i = g_1 + \frac{g_{N_t} - g_1}{N_t - 1} \cdot (i-1) \quad \text{for} \quad i = 1, \ldots, N_t \tag{11}$$

and

$$\hat{h}_i = h_1 + \frac{h_{N_f} - h_1}{N_f - 1} \cdot (i-1) \quad \text{for} \quad i = 1, \ldots, N_f, \tag{12}$$

respectively.

Finally, feature f_3 is defined as:

$$f_3 = \frac{\sum_{i=1}^{N_f} \left(h_i - \hat{h}_i\right)^2}{\sum_{i=1}^{N_t} \left(g_i - \hat{g}_i\right)^2}. \tag{13}$$

Feature f_4 measures the ratio of deviation of the curves g and h from their linear regression lines $\tilde{g}$ and $\tilde{h}$. With $\bar{g}$ (Eq. (8)) and $\bar{h}$ (Eq. (9)) denoting the mean values of g and h, the coefficients for linear regression [9] are given by:

$$\begin{aligned}
\beta_g &= \frac{\sum_{i=1}^{N_t}\left(i-\frac{1+N_t}{2}\right)\cdot(g_i-\bar{g})}{\sum_{i=1}^{N_t}\left(i-\frac{1+N_t}{2}\right)} \\
\alpha_g &= \bar{g} - \beta_g \cdot \frac{1+N_t}{2} \\
\beta_h &= \frac{\sum_{i=1}^{N_f}\left(i-\frac{1+N_f}{2}\right)\cdot(h_i-\bar{h})}{\sum_{i=1}^{N_f}\left(i-\frac{1+N_f}{2}\right)} \\
\alpha_h &= \bar{h} - \beta_h \cdot \frac{1+N_f}{2}.
\end{aligned} \qquad (14)$$

Then, the feature:

$$f_4 = \frac{\sum_{i=1}^{N_f}(h_i - a_h - i\cdot \beta_h)^2}{\sum_{i=1}^{N_t}(g_i - a_g - i\cdot \beta_g)^2} \qquad (15)$$

is finally defined as the ratio of the sum of squares of the residuals of the curves and their linear lines of best fit.

The features f_3 and f_4 appear to be similar on first sight. However, if g or h have a distinct peak on one end and the rest is more or less constant, those two features are significantly different. The forward feature selection algorithm selected both of them as highly discriminative.

As already mentioned before, these features were not the only ones defined initially and used as an input for the feature selection algorithm. Among the other features were similar values for g and h alone (not the ratios), similar values of the twofold autocorrelation, a smoothness measure of the twofold autocorrelation, and the width of confidence intervals of second- and third-order curves fitted to g and h.

3.3 Feature Space

Finally, a first visual impression of the feature space is given. More details as well as classification results using the well-known methods logistic regression [2] and support vector machines (SVMs) [46] are provided in Sect. 6.1. Figure 13 shows the features f_1 and f_2 from some measurement runs with all valve types except for the steel valve (Sect. 5). As it is not possible to depict a four-dimensional feature space, only the projections onto a two-dimensional subspace is shown. To obtain a good

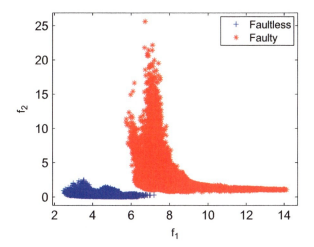

Fig. 13 Features f_1 and f_2 in vibration analysis

generalizability concerning load levels, reference spectrograms from different load levels were used. Every reference spectrogram was used to compute features for every test spectrogram of the measurement runs. The feature space thus represents observations computed for varying load levels of reference and test spectrogram. The main advantage of this approach is that once the decision boundary is trained, the reference spectrogram and the test spectrogram can have any arbitrary load level.

4 Analysis of the pV Diagram

Using the pV diagram for the detection of leaking valves has been proposed several times in the past, for instance [8, 10, 13, 31, 48]. However, none of those approaches fulfils all desired properties of a monitoring system, i.e. to be fully automated and applicable for arbitrarily varying operation conditions and different valve types.

4.1 Motivation

It is in the nature of things that a leak in a valve can only be detected when it is supposed to seal the gas stream off. The evaluation of pV diagrams can thus be restricted to the times when all valves of a compression chamber are closed, i.e. the compression and the expansion stroke. Since the load control affects the start and end time of the compression stroke, the most promising part of the pV diagram for detecting leaking valves independently of load and pulsations is the expansion stroke.

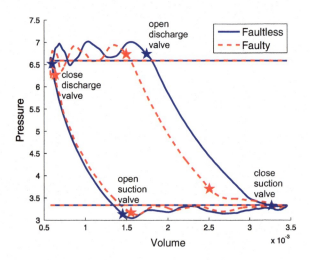

Fig. 14 pV diagrams of measurements from a faultless and a faulty discharge valve at the same pressure conditions. The horizontal lines indicate the suction (lower line) and discharge (upper line) pressure. The pV diagram is traversed anticlockwise

Figure 14 shows two pV diagrams: one from faultless valves, and the other from a leaking discharge valve. Both are measured at the same pressure conditions before (suction pressure) and after (discharge pressure) the compression cylinder. The lower horizontal lines in the figure represent the suction pressure, and the upper horizontal lines represent the discharge pressure. When looking at the expansion phase on the left-hand side of the plot (between closing the discharge valve and opening the suction valve), it is obvious that the pressure decreases slower in the faulty case. This can be explained by high-pressure gas flowing back through the leaking valve from the discharge chamber into the compression cylinder. In all test cases, the discharge valve is the leaking valve. In the case of a leaking suction valve, gas from the cylinder would flow through the valve from the cylinder into the lower pressured suction chamber. Hence, the cylinder pressure would decrease faster in this case. Moreover, the effect of load control can be seen in Fig. 14. The load of the faultless measurement is higher than the load of the faulty measurement. This is reflected on the right-hand side of the figure, where the compression stroke of the pV diagram from faultless valves starts and ends significantly later, i.e. at a lower volume. To demonstrate the effect of the pressure conditions on the pV diagram, Fig. 15 depicts two pV diagrams, both from faultless valves, at different pressure conditions. Although both measurements are from faultless valves, the shape of their expansion strokes is not equal.

According to these findings, a quantifier for the speed of pressure drop during the expansion stroke has to be found. Deviations from the normal speed can then be assumed to be caused by a leak. Of course, one has to be aware of the effect of varying pressure conditions.

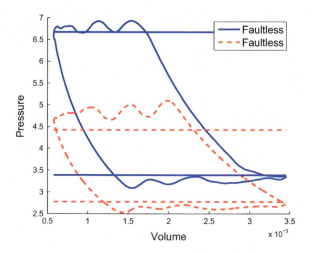

Fig. 15 Two pV diagrams of measurements from faultless valves at different pressure conditions. The horizontal lines indicate the suction (lower line) and discharge (upper line) pressure

4.2 Feature Extraction

As already illustrated in Sect. 4.1, the expansion stroke of the compression cycle can be used as an indicator for a leaking valve. The main task is to find a quantifier for the shape of the expansion stroke in the pV diagram. Focussing on the slope of the expansion stroke (for instance by integrating over the slope) has two significant drawbacks:

- The valve events cannot be determined precisely enough. The integration bounds are thus imprecise, resulting in a defective integral value.
- Different pressure conditions affect the duration of the expansion stroke as shown in Fig. 15 and therefore the integration bounds. Rescaling the axes with respect to the pressure conditions did not solve this problem.

These problems can be avoided by using the logarithmic pV diagram [30]. The compression cycle is a polytropic process. It thus follows the equation [44]:

$$p \cdot V^n = \text{const.} \qquad (16)$$

with p denoting the pressure, V denoting the volume, and n denoting the polytropic exponent. Since the compression chamber is closed during compression and expansion, the polytropic exponent is supposed to be constant during those phases [24]. According to Eq. (16), the logarithmic pV diagram delivers the equation:

$$\log p = \log c - n \cdot \log V, \qquad (17)$$

which is in fact a straight linear relationship with intercept $\log c$ and slope $-n$. Switching to logarithmic pressure and volume scales thus linearizes the compression and the expansion stroke. Figure 16 shows the same pV diagrams as Fig. 14, but

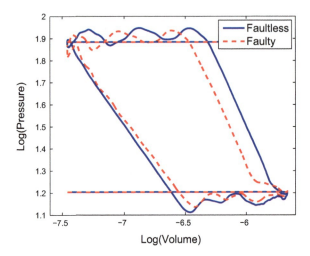

Fig. 16 The pV diagrams of Fig. 14 with logarithmic scales

with logarithmic scales. The compression and the expansion strokes are linear. Obviously, the expansion stroke in the faultless and in the faulty case have a different gradient. The leaking valve has a lower absolute gradient than the faultless valve, suggesting to use the gradient as an indicator for the leaking valve. As the pressure conditions are the same, the gradients are supposed to be equal, whereas a small difference usually already points to an (upcoming) fault. Notice that the data are from a leaking discharge valve. The lower absolute gradient can be explained by gas flowing back from the discharge chamber into the cylinder through the leaking valve leading to a slower pressure decrease in the cylinder. In the case of a leaking suction valve, gas would flow from the compression cylinder into the suction chamber. As the cylinder pressure would decrease faster in that case, this would result in a higher absolute gradient than in the faultless case.

However, the different pressure conditions still affect the gradient. This can be seen in Fig. 17, where different pressure conditions between the faultless and the faulty case appear. Even though the dashed line represents the measurement from a leaking discharge valve, its gradient in the expansion stroke appears to be very similar to the gradient of the faultless case. For that reason, both the gradient and the pressure conditions have to be considered in the feature space. Of course, the pressure conditions are not able to discriminate between the faultless and faulty case. In both cases, any pressure conditions can occur. But as pressure conditions affect the gradient of the expansion stroke, they help to discriminate whether a changed gradient is induced by a fault or by the pressure conditions. However, not only the gradient and the pressure conditions have been investigated. Just like in Sect. 3, a number of other expert-based and statistical-based features as well as previously proposed features (such as [48]) were extracted. The feature selection algorithms selected the two features mentioned before as most discriminative.

Hence, the first feature expresses the pressure difference of the compression cycle. For each compression cycle, the pressures p_S and p_D in the suction chamber

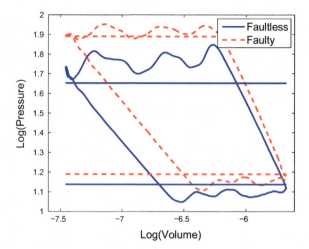

Fig. 17 Logarithmic pV diagrams of measurements from faultless and faulty valves of the same valve type with different pressure conditions. Obviously, the gradient of the expansion phase alone is not sufficient to discriminate between faultless and faulty case. Also, the pressure conditions have to be considered

and in the discharge chamber are measured. The first feature f_1 is then simply defined as the difference between those pressures, that is:

$$f_1 = p_D - p_S. \tag{18}$$

The second feature f_2 measures the gradient of the expansion stroke in the logarithmic pV diagram. This is done by a simple least squares regression [9] to fit a line to the expansion stroke. Since the method is working with measured data, the data are discrete. Let the logarithmic, discrete pressure measurement data (restricted to the expansion stroke) be denoted by $p_{\exp} \in \mathbb{R}^n$ (with $n \in \mathbb{N}$ denoting the number of samples). The noisy raw pressure signal is smoothed using Whittaker's smoother [12, 50]. Similarly, let the logarithmic, discrete volume data (restricted to the expansion stroke) be denoted by $V_{\exp} \in \mathbb{R}^n$. The volume data is obtained as a virtual signal based on the crank angle. To compute the gradient $k \in \mathbb{R}$ of the regression line of the expansion stroke, the system of equations:

$$A \cdot \begin{pmatrix} d \\ k \end{pmatrix} = p_{\exp} \tag{19}$$

has to be solved. $A \in \mathbb{R}^{n \times 2}$ is a matrix consisting of ones in the first column and $V_{\exp}$ in the second column. By multiplying both sides of the equation system in Eq. (19) with A^T and taking the inverse of the square matrix $A^T A$, the solution is obtained by:

$$\hat{\beta} = \begin{pmatrix} d \\ k \end{pmatrix} = (A^T A)^{-1} A^T \cdot p_{\exp}. \tag{20}$$

Although for small 2×2 matrices (such as $A^T A$) the likelihood is low that the matrix has a low condition and thus its inversion is numerically instable, a regularization scheme is applied following the idea of ridge regression (adding a regularization term). Then, the solution in Eq. (20) becomes

$$\hat{\beta} = \begin{pmatrix} d \\ k \end{pmatrix} = (A^T A + \alpha I)^{-1} A^T \cdot p_{\exp} \qquad (21)$$

with a regularization parameter α that is set dependent on the largest eigenvalue $\lambda_{\max}$ to:

$$\alpha = \frac{2 \cdot \lambda_{\max}}{thr} \qquad (22)$$

with $thr = 10^{15}$. For more sophisticated parameter choice techniques, see [5]; this option is used due to a good experience on data sets from various applications, see e.g. [29], and due to its low computation complexity. After solving the system of Eq. (19), the gradient k is chosen as the second feature, that is:

$$f_2 = k. \qquad (23)$$

Hence, from every compression cycle (i.e. every pV diagram) one point in a two-dimensional feature space is obtained.

4.3 Feature Space

This section gives a first visual impression of the feature space. More detailed classification results are presented in Sect. 6.2.

In Fig. 18, the features of all valve types are illustrated in one feature space. On the left-hand side, the faultless and the faulty class of all valve types are shown. The figure suggests that a classification among different valve types is not possible as the classes are overlapping. The figure on the right-hand side shows only the faultless class of six valve types. Obviously, all valve types have the same parabolic shape with a different offset in f_2-direction. Especially, the steel valve v_2 has an offset to all other (plastic) valves. The reason for that is that even faultless valves are not 100% leak tight. Compared to the synthetic valves, which have more or less a similar leak tightness, the steel valve is somewhat more leaky even though it is faultless. Nevertheless, steel valves are used for some applications as they have some other properties that plastic valves do not have, for instance, temperature resistance. One has to be aware of the offset when classifying the features of pV diagram analysis.

Finally, it has to be mentioned that the faulty class is above the faultless class in all cases shown here. This is characteristic for a leaking discharge valve. For a leaking suction valve, the faulty class would be positioned below the faultless class

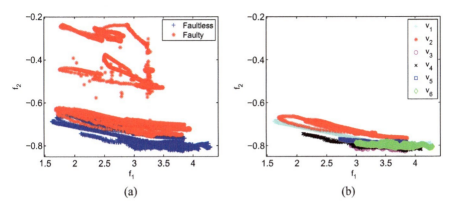

Fig. 18 Scatterplots of the feature space for different test measurement of different valve types. (**a**) Scatterplot of six valve types merged together. (**b**) Scatterplot of the faultless class of the six different valve types on the left-hand side

in the feature space. This is because of the higher absolute gradient in that case as explained in Sect. 4.2.

For obtaining a feature space that is independent of the valve type, the offset has to be removed. To remove the offset between different valve types, only the faultless class of each valve type is considered. Assume that the faultless observations in the feature space are given in the form $(x_{k,i}, y_{k,i})$ with $k = 1, \ldots, n$ indicating the valve type (n denotes the number of valve types) and $i = 1, \ldots, n_k$ indicating the n_k observations for each valve type k. The offsets are determined by minimizing a least-squares cost function on the data in the feature space. The scatterplots of the feature space (Fig. 18) show that the data from faultless valves form a parabola. Offset reduction is successful when one parabola provides the best fit for the faultless class of all valve types. Thus, an offset $d_k \in \mathbb{R}$ is added to the y-dimension yielding the observations $(x_{k,j}, y_{k,j} + d_k)$. The observations of all valve types are merged, and a parabola:

$$p(x) = ax^2 + bx + c \qquad (24)$$

is fit in a least-squares sense to the merged data. Of course, the offset of one valve type can be fixed and only the other offsets are estimated. Without loss of generality, the offset d_1 is set to 0. By defining the cost function as:

$$J(d_2, \ldots, d_n) = \sum_{k=1}^{n} \sum_{j=1}^{n_k} \left(y_{k,j} + d_k - p(x_{k,j}) \right)^2, \qquad (25)$$

it can be minimized using a gradient descent method [39] to obtain the optimal offsets d_k as:

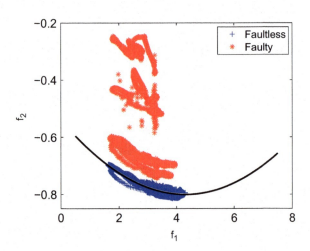

Fig. 19 Scatterplot of Fig. 18a with removed offset and the fitted parabola after the last optimization step

$$[d_2, \ldots, d_n] = \underset{d_k \in \mathbb{R},\, k=2,\ldots,n}{\arg\min} J(d_2, \ldots, d_n) \tag{26}$$

and $d_1 = 0$. The parabola is fitted after each optimization step.

By adding the optimal offset to the original data points, comparable data for all valve types are obtained. Figure 19 depicts the features of all valve types with removed offsets and the fitted parabola. It can be seen that the classes are now separable among all valve types. Please notice that the offset is only added to feature f_2.

4.4 Classification

Besides well-known classification concepts such as SVMs, a special one-class classifier has been developed for this classification problem. The faultless class has parabolic shape in the two-dimensional feature space. A leaking valve causes an excessive value of the feature f_2. Hence, the classification problem is to determine whether feature f_2 has excessive values or not (compared to the training data from faultless class). This is somehow similar to the detection step in spike sorting [25], where significant peaks of waveforms fluctuating around $y = 0$ are to be detected. Transferring this to the present classification problem, the faultless class is fluctuating around the parabola p (the regression parabola of the faultless class), and the faulty data can be interpreted as the peaks to be detected. To obtain a similar situation to spike sorting, a coordinate transformation has to be performed such that the parabola is the x-axis of the new coordinate system. The classifier training thus consists of two parts:

- Linearizing the feature space along the faultless class
- Selecting a threshold

Of course, only observations from faultless valves are considered for classifier training. Furthermore, since the offsets in the feature space are removed, there is no need to differentiate between the valve types. The data from faultless valves in the feature space can thus simply be denoted by (x_i, y_i), $i = 1, \ldots, N$, in this section.

To reduce the classification problem to a simple threshold search in one dimension, the fitted parabola is linearized. A coordinate transformation has to be found that maps the parabola onto the x-axis and the parabolas apex into the origin of a two-dimensional space. As the parabola fits the faultless class in a least squares sense, the faultless class will be represented as a scatterplot along the new x-axis. This enables fault detection by simply identifying observations with extensive distance to the new x-axis. One very natural and intuitive, but of course not the only way to achieve this goal is to choose the new y-coordinate $\hat{y}$ as the shortest orthogonal distance of a data point (x_i, y_i) to the parabola:

$$p(x) = ax^2 + bx + c \tag{27}$$

with fixed $a, b, c \in \mathbb{R}$ as computed in the last step of offset removing.

The new x-coordinate $\hat{x}$ is then the arc length from the apex of the parabola to the root point (x_i^p, y_i^p) of the orthogonal distance on the parabola. To formalize this, the parabola is parametrized as:

$$p(t) = \begin{pmatrix} t \\ at^2 + bt + c \end{pmatrix} \tag{28}$$

with $t \in \mathbb{R}$.

The root point (x_i^p, y_i^p) can be obtained by intersecting the normal straight line through (x_i, y_i) with the parabola, i.e. solving the equation:

$$\begin{pmatrix} x_i \\ y_i \end{pmatrix} + \frac{\lambda}{\sqrt{4a^2t^2 + 4abt + b^2 + 1}} \cdot \begin{pmatrix} 2at + b \\ -1 \end{pmatrix} = \begin{pmatrix} t \\ at^2 + bt + c \end{pmatrix} \tag{29}$$

for t and λ. The solution is of course not unique, picking the real solution with the shortest distance between (x_i, y_i) and (x_i^p, y_i^p), i.e. with minimal absolute value of λ, gives the final solution. Assume that this solution is denoted by t_0 and λ_0. The new coordinate $\hat{y}_i$ is given by λ, that is:

$$\hat{y}_i = \lambda_0. \tag{30}$$

The new coordinate $\hat{x}_i$ is given by the arc length of the parabola from its apex to (x_i^p, y_i^p). Hence, it can be determined by [1]:

$$\hat{x}_i = \int_0^{t_0} \sqrt{1 + 4a^2t^2 + 4abt + b^2} \cdot dt. \tag{31}$$

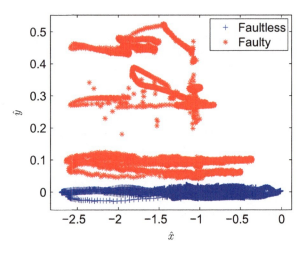

Fig. 20 Feature space of Fig. 19 linearized along the regression parabola of the faultless class according to Eqs. (30) and (31)

The new feature space is illustrated in Fig. 20. The classes are linearly separable using a simple threshold in $\hat{y}$-dimension.

For selecting the threshold in $\hat{y}$-dimension, a very simple approach inspired by spike sorting [25] is used. In the linearized feature space $(\hat{x}_i, \hat{y}_i)$, the threshold is set as proposed in [38] to:

$$\tau = 4 \cdot \underset{i=1,\ldots,N}{\mathrm{median}} \left(\frac{|\hat{y}_i|}{0.6745} \right). \tag{32}$$

The expression $\underset{i=1,\ldots,N}{\mathrm{median}} \left(\frac{|\hat{y}_i|}{0.6745} \right)$ in Eq. (32) is just an estimate for the standard deviation. However, if there are outliers in the training data, this expression is more robust than the standard deviation. Hence, the threshold τ detects 4σ outliers of the normal data.

Once the classifier is trained, it can be applied to unseen data to obtain the class information. However, there is one constraint to applying the classifier to data of a new valve: the offset to the other valve types in the feature space is unknown. Data from faultless state have thus to be provided to compute the offset $d \in \mathbb{R}$. According to reciprocating compressor operators, this is not a hindering constraint, as monitoring will start at a faultless state after deploying new valves or after a general inspection of the compressor. Data from the beginning of the compressor operation can then be used for determining the offset. From these data, the features are computed and represented by the data points (x_i, y_i), $i = 1, \ldots, n$. For determining the offset, an optimization problem similar to (25) and (26) is solved. The cost function is defined as:

$$J(d) = \sum_{i=1}^{n} (y_i + d - p(x_i))^2 \tag{33}$$

with p representing the regression parabola with fixed parameters $a, b, c \in \mathbb{R}$ as determined in the offset removal for classifier training. In this approach, the parabola parameters are fixed after classifier training. A self-adapting approach updating the parabola parameters on the fly might be a reasonable extension in future work. The cost function is minimized as:

$$d = \arg\min_{d \in \mathbb{R}} J(d) \qquad (34)$$

to obtain the offset. Once the offset is known, it is added to the feature value f_2 of each newly monitored compression cycle. Assume that such a new feature vector is denoted with (x, y). After adding the offset, the feature vector is $(x, y + d)$. Then, it is linearized along the parabola p as in Eqs. (30) and (31) resulting in the new feature vector $(\hat{x}, \hat{y})$. Finally, the value $\hat{y}$ is compared to the threshold τ.

5 Experimental Setup

The proposed methods in this chapter were developed and tested using real-world measurement data. The data were acquired by the project partner Hoerbiger Compression Technology at a reciprocating compressor test bench located at the Hoerbiger Ventilwerke in Vienna.

5.1 Compressor Test Bench

The test compressor was a two-stage Ariel JG2 reciprocating compressor. Detailed specifications of the compressor are listed in [4]. An electric motor with a constant revolution speed of 750 rpm drives the compressor—that means one cycle lasts 0.08 s. The scheme of the compressor is sketched in Fig. 21. The compressor takes in air at environmental pressure and temperature through a filter. After the first double-acting compression cylinder, the air flows to a pulsation damper and an intercooler. The next pulsation damper is followed by the second double-acting compressor stage. After cooling and damping the air again, it is discharged into a high-pressure discharge vessel. The compressed air in the discharge vessel can be dumped manually. The compressor is equipped with a reverse flow control system. The load of each stage can be controlled manually as well as automatically by a control module [19].

The test compressor was equipped with several sensors. First of all, there was a TDC sensor that measures when the piston of stage 1 is at its top dead centre. Since stage 1 and stage 2 have 180° phase shift, the piston is at its BDC in stage 2 when it is at its TDC in stage 1. The TDC information is required to split the measured signals into cycles of the compressor. As the revolution speed is constant and the

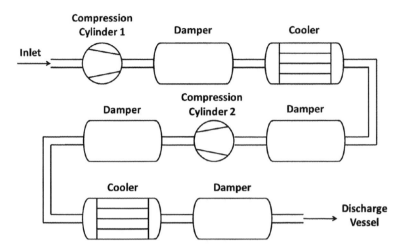

Fig. 21 A sketch of the reciprocating compressor test bench

cylinder dimensions (except for the exact clearance space) are known, it can also be used to compute the volume at each time instance. This creates a virtual volume signal that is used for plotting the pV diagrams. Furthermore, the pressure in the four compression chambers (two at each cylinder) and the suction and discharge pressure of the two stages were measured. For vibration analysis, an accelerometer was mounted at the valve cover of the DV HE2 (discharge valve at the head end of compressor stage 2). The accelerometer measures the vertical accelerations of the valve cover. All data were recorded at a sampling rate of 100 kHz. The control values can be acquired directly from the control module. This delivers one control value per cycle for each stage.

Faults were simulated by manipulating the sealing element of the valve. The manipulated (leaking) valve was positioned at DV HE2. The measurements were performed in the following way: first, measurements with a faultless valve were made. Then, this very same valve was dismounted. The sealing element was manipulated by cracking it or breaking a part out of it. Finally, the valve was mounted again, and a new measurement run with the now faulty valve was recorded. Most of the measurements were made with varying load levels and pressure conditions.

5.2 Test Runs

In multiple test runs spread over several years, test data of different valve types were acquired. The tested valve types and the according fault states are listed below. *Baseline* denotes a faultless valve, *Crack* denotes a small crack (a fissure) in the valve, and *Broken* denotes that a part of the sealing element is broken off. As a fissure in the sealing element does not cause a significant leak, it does not affect the pV diagram significantly. Hence, it cannot be expected to detect the fault state *Crack* by analysing the pV diagram. With each of the listed valve types, at least one measurement run for acquiring test data was performed. Each measurement run lasted at least 3 min.

The first part of the valve name specifies the design of the valve: R and CP are plate valves of different designs, Ring are concentric ring valves. The term in the parenthesis specifies the material of the valve. Except for the R (Steel) valve, all valves are made of synthetic material. Finally, the term in square brackets is just an abbreviation that is used for the valve (or the measurement run) in this chapter. For further details regarding the valve types, the reader is referred to [18].

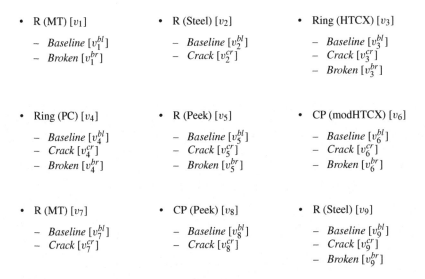

- R (MT) [v_1]
  - *Baseline* [v_1^{bl}]
  - *Broken* [v_1^{br}]

- R (Steel) [v_2]
  - *Baseline* [v_2^{bl}]
  - *Crack* [v_2^{cr}]

- Ring (HTCX) [v_3]
  - *Baseline* [v_3^{bl}]
  - *Crack* [v_3^{cr}]
  - *Broken* [v_3^{br}]

- Ring (PC) [v_4]
  - *Baseline* [v_4^{bl}]
  - *Crack* [v_4^{cr}]
  - *Broken* [v_4^{br}]

- R (Peek) [v_5]
  - *Baseline* [v_5^{bl}]
  - *Crack* [v_5^{cr}]
  - *Broken* [v_5^{br}]

- CP (modHTCX) [v_6]
  - *Baseline* [v_6^{bl}]
  - *Crack* [v_6^{cr}]
  - *Broken* [v_6^{br}]

- R (MT) [v_7]
  - *Baseline* [v_7^{bl}]
  - *Crack* [v_7^{cr}]

- CP (Peek) [v_8]
  - *Baseline* [v_8^{bl}]
  - *Crack* [v_8^{cr}]

- R (Steel) [v_9]
  - *Baseline* [v_9^{bl}]
  - *Crack* [v_9^{cr}]
  - *Broken* [v_9^{br}]

For the test measurements v_7–v_9, there are no pressure measurements available. Thus, no pV diagram analysis can be performed for these test data.

To give the reader an impression of the sealing elements and the fault types, two examples are illustrated in Fig. 22. In the left figure, a part of the sealing element of a plate valve is broken off (fault state = *Broken*). A sealing element of a plate valve with a fissure can be seen on the right side (fault state = *Crack*). Watching these figures, it is comprehensible that the fault state *Crack* does not affect the pV diagram significantly as there is no significant leak.

Fig. 22 Faulty sealing elements of a plate valve. (**a**) The sealing element of a plate valve with a part broken off (fault state = *Broken*). (**b**) The sealing element of a plate valve with a fissure (fault state = *Crack*)

6 Results

This section provides plots of the feature spaces and classification results obtained by applying classifiers to the feature spaces of vibration analysis and pV diagram analysis.

6.1 Vibration Analysis

For computing test observations of vibration analysis, a reference and a test spectrogram are compared. Test observations are generated as follows: for each valve type v_i, a reference spectrogram is computed and stored every 2.5% of the length of the measurement. The reference spectrograms are only taken from measurements from faultless valves (v_i^{bl}). Then, for all measurement runs ($v_i^{bl}, v_i^{cr}, v_i^{br}$) from this valve type, a test spectrogram is computed every 2.5% of their length. These test spectrograms are compared to every reference spectrogram of v_i, and the features are extracted. This procedure delivers a broad amount of test observations in the feature space to perform validation. During test runs, considering computational effort and detection rate, a spectrogram resolution of $n_t = 124$ and $n_f = 257$ turned out to be a good choice for a window of the length of 2 compression cycles. The number of test observations for each valve type and each fault state is provided in Table 1. As there is no desirability to discriminate between the fault states *Crack* and *Broken*, those two states are merged to one *Faulty* class for validation. The classification task is thus to discriminate between the states *Baseline/Faultless* and *Faulty*. Of course, the classifiers have to outperform at least the default classifiers, i.e. the classifiers that assign all observations the same (majority) class. With the

Early Fault Detection in Reciprocating Compressor Valves

Table 1 Number of observations of each valve type in vibration analysis

Valve type	Baseline	Crack	Broken
v_1	780	0	1600
v_2	780	1600	0
v_3	780	800	1600
v_4	780	1600	1600
v_5	780	1600	2400
v_6	1770	2400	2400
v_7	3350	5600	0
v_8	1770	4800	0
v_9	190	400	400

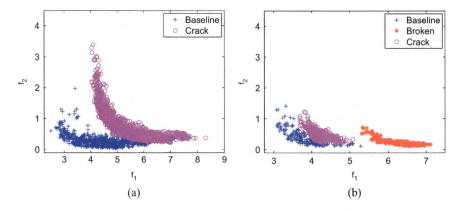

Fig. 23 Subspace f_1, f_2 of the feature space from the valves v_2 and v_9. (**a**) Steel valve v_2 with fault states *Baseline* and *Crack*. (**b**) Steel valve v_9 with fault states *Baseline*, *Crack*, and *Broken*

available test data (Table 1), the default classifier assigning every observations the state *Faulty* has an accuracy of 72.51% for plastic valves and 71.22% for steel valves.

The steel valves play a special role in vibration analysis. The difference in the spectrogram between the faultless and faulty case is of a smaller degree than for the plastic valves. This is reflected in the feature space as well. Figure 23 illustrates the subspace (f_1, f_2) of the two steel valves. The left side shows steel valve v_2 (with the fault states *Baseline* and *Crack*), and the right side shows steel valve v_9 (with the fault states *Baseline*, *Crack*, and *Broken*). The classes *Baseline* and *Crack* overlap clearly, and the feature values show less deviation in the faulty case of the steel valve than in the faulty case of the plastic valves (Fig. 13).

The subspace of the features f_1 and f_2 of all valves merged in one plot is depicted in Fig. 24. The left side shows all valves except the steel valves. Even though the valves are of different design and made of different plastic materials, actually only f_1 and f_2 provide already quite a good visual separability. On the right side, the observations from the steel valves v_2 and v_9 are added. It can be seen that the classes *Crack* and *Broken* from the steel valves are overlapping with the class *Baseline* from the plastic valves. Using other features than f_1 and f_2 shows a similar overlap. This suggests again that the features from plastic valves and steel valves cannot be classified using the same classification boundary. Furthermore, the left plot of Fig. 24 suggests that vibration analysis does not give the opportunity to distinguish between the states *Crack* and *Broken*. The two classes are overlapping in wide parts of the scatterplot. There are parts where they do not overlap; however, this may be due to the operation conditions of reference and test spectrograms and not due to the fault state. The classification task is thus to discriminate between the states *Baseline/Faultless* and *Faulty*.

Validation is performed in a leave-one-valve-out approach. This means that each valve type v_i serves for validation once. When validating the classification for this valve, the classifier is trained with all other valves. For each validation valve, the confusion matrix (CM) is computed and the CMs are summed up to obtain one final CM and therefore one accuracy value. For classification, the logistic rule (in a two-class setup) and SVMs (in two-class as well as one-class setup) are used. For one-class SVM, only the measurement runs with fault state *Baseline* are used for training the classifier. This whole validation procedure is done separately for plastic and steel valves. The obtained accuracy values are listed in Table 2. For plastic valves, all classifiers perform almost equal. Hence, even one-class classification is suited for plastic valves, reducing the amount of training data sufficiently. For the steel valve, only the two class SVM classification approach delivered satisfying accuracy.

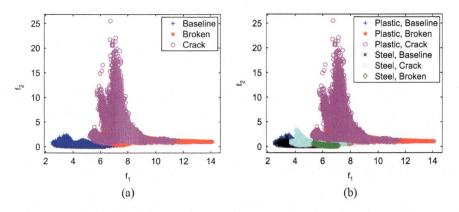

Fig. 24 Subspace f_1, f_2 of the feature space from all valves. (**a**) Valves v_1, v_3, v_4, v_5, v_6, v_7, and v_8 with fault states *Baseline*, *Crack*, and *Broken*. (**b**) Valves v_1, v_2, v_3, v_4, v_5, v_6, v_7, v_8, and v_9 with fault states *Baseline*, *Crack*, and *Broken*

Table 2 Validation accuracies [%] using logistic regression and SVM in vibration analysis

Valve material	Log. Reg.	Two-class SVM	One-class SVM
Plastic	99.49	99.97	99.37
Steel	84.96	94.66	71.60

6.2 pV Diagram Analysis

As assumed before, a small crack in the sealing element does not affect the tightness of the valve enough to have a significant effect on the pV diagram. This is illustrated by scatterplots of the feature space from valve v_3 and v_6 in Fig. 25. They show that there is actually no difference between the fault states *Baseline* and *Crack*. Similar observations can be made for all other valve types. This can be interpreted as a verification of the assumption that a very small crack does not cause a significant leak. The fault state *Crack* is thus not included in the validation scenarios for pV diagram analysis. Hence, validation is only performed for measurement runs with the fault states *Baseline* and *Broken*, that are v_1^{bl}, v_1^{br}, v_2^{bl}, v_3^{bl}, v_3^{br}, v_4^{bl}, v_4^{br}, v_5^{bl}, v_5^{br}, v_6^{bl}, and v_6^{br}. As a general information, the number of observations in the feature space, which accords to the number of compression cycles and pV diagrams, respectively, is provided here. The feature space of the measurement runs stated above consists of 48,482 observations, 27,977 from fault state *Baseline* and 20,505 from fault state *Broken*. This results in a 57.70% accuracy for the default classifier that assigns every observation the state *Baseline*. The number of observations for each single valve type is shown in Table 3.

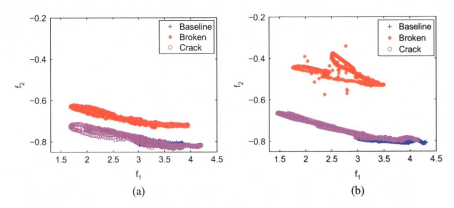

Fig. 25 Feature space of valve v_3 and v_6 containing the fault states *Baseline*, *Crack*, and *Broken*. (**a**) Feature space of valve v_3. (**b**) Feature space of valve v_6

Table 3 Number of observations of each valve type in pV diagram analysis

Valve type	Baseline	Broken
v_1	5598	3731
v_2	3728	0
v_3	3729	3731
v_4	5595	3727
v_5	3729	5589
v_6	5598	3727

First, validation in the two-class case using SVM is performed to demonstrate the good discriminative power of the proposed features. Then, the one-class classifier from Sect. 4.4 is compared with other well-known one-class classifiers.

For two-class SVM classification, two validation scenarios are considered:

1. Tenfold cross validation is applied to the feature space of each valve type separately. For all valve types, the classifier delivers an accuracy of 100%.
2. The data of all valve types are merged to one data set. Then, tenfold cross validation is applied to this feature space. The result of this validation method is an accuracy of $96.37 \pm 0.19\%$ (mean value $\pm$ standard deviation of the tenfolds) before removing the offset and 100% after removing the offset.

For evaluating the one-class classifier proposed in Sect. 4.4, it is compared to several one-class classifiers found in the literature, namely (robust) Gaussian data description (abbreviated g and rg in Table 4), mixture of Gaussians (*mog*) data description, k-means (*km*) data description, k-nearest neighbour (*knn*) data description, and *SVM* data description. Detailed information on the classifiers can be found in [43].

Each of the well-known classifiers has some free parameters to adjust, at least the rejection rate (the fraction of training observations to be rejected by the classifier). Some of the classifiers, for instance *SVMs*, have even more free parameters, and for *km* and *knn* proper k-values have to be fixed. For those classifiers, validation is performed on a reasonable grid of parameters and k-values, respectively. Finally, the values delivering the best validation accuracy are chosen. In contrast, the one-class classifier proposed in Sect. 4.4 (abbreviated *occ* in Table 4) has no free parameters. The well-known one-class classifiers in this study are not limited to a linear classification boundary (such as the classifier proposed in Sect. 4.4). Thus, they were used for classification before and after the feature space was linearized along the regression parabola. Detailed results are provided in Table 4, which shows only the accuracy for the best-case parameter values.

The validation in Table 4 is performed as follows: The classifier is trained with faultless data of five valve types. The sixth valve type is used for validation. Every valve type serves for validation once. The faultless data of the validation valve type are split into four parts of equal length. Each of the four parts serves once for computing the offset of the validation valve type. When the offset is determined, validation is performed with the other three parts of the faultless data and all faulty data of the validation valve type. Thus, each of the four parts of the faultless data

Table 4 Validation accuracies [%] for one-class classification

	occ	g	rg	mog	km	knn	SVM
Accuracy before linearizing [%]	–	97.91	89.50	94.98	98.07	98.80	99.40
Accuracy after linearizing [%]	99.96	98.88	99.85	99.76	99.24	99.74	99.54

delivers one confusion matrix. Those four matrices are added to obtain one single confusion matrix for the validation valve type. Finally, the confusion matrices of each validation valve type are added, delivering one single confusion matrix for a classifier. From this confusion matrix, the classification accuracy is determined and shown in Table 4.

As the classifier proposed in Sect. 4.4 is designed especially for the linearized data, there is no value for it in the table before linearizing. It can be seen that, with an accuracy of 99.96%, it performs best among all classifiers, even though it is not optimized with respect to any parameter value. Furthermore, all classifiers improve in classification accuracy after linearizing along the parabola. This suggests that even the linearization step alone allows an improvement.

The results of another validation scenario are presented in Table 5. Here, only one valve type was used to train the classifier. This classifier was then used to classify every other valve type. Again, the faultless data of the validation valve are split into four parts, and each part is used once to determine the offset of the validation set. The rows in the table represent the training valve type, while the columns represent the validation valve type. The two best classifiers from Table 4 are compared: the upper value shows the accuracy of the classifier proposed in Sect. 4.4. The lower value shows the accuracy of the robust Gaussian data description method. The accuracy values are worse than before due to the limited training sets: the training data do not contain the whole range of pressure conditions for all valve types, and only a couple of minutes training data are probably not representative enough to describe the feature space completely. Still, the classifier proposed in Sect. 4.4 performs better than the best of the well-known methods. Even though its accuracy is slightly below the robust Gaussian classifier in some cases, it has not as low values as the robust Gaussian classifier. This can be seen especially when the classifier is trained with v_2 and v_5.

Table 5 Validation accuracy [%] for one-class classification using only one valve type for training

Train\Valid	v_1	v_2	v_3	v_4	v_5	v_6
v_1	–	99.67	99.99	99.62	99.14	98.28
		99.64	100	100	100	100
v_2	99.22	–	99.93	99.79	99.19	97.99
	99.36		46.47	54.02	34.60	100
v_3	96.72	95.82	–	99.30	99.93	98.62
	87.05	97.38		99.96	100	99.04
v_4	99.44	99.72	100	–	100	99.80
	98.50	98.14	100		100	100
v_5	94.52	96.44	92.91	97.00	–	93.73
	82.92	74.43	89.83	94.40		90.71
v_6	96.68	95.00	99.53	98.12	99.99	–
	86.09	87.03	99.46	98.26	100	

The upper value represents the classifier proposed in Sect. 4.4, the lower value the robust Gaussian data description method

In summary, this section shows that the proposed features have a good discriminative power for the detection of leaking valves in the pV diagram. Furthermore, the proposed one-class classifier performs well even though it has no free parameter to be set.

7 Conclusions

In this chapter, two data-driven methods for detecting leaking reciprocating compressor valves are proposed. The two approaches are independent from each other, i.e. they use different measured signals to monitor the valves. One method evaluates the accelerometer data at the valve covers, and the other one evaluates pressure signals before, after, and in the compression chamber in combination with a virtual volume signal. Each of the two methods is able to cope with the challenges addressed in the problem statement (Sect. 2.2), such as varying load levels and pressure conditions, and different valve types. Both methods yield in a high validation accuracy on real-world data. However, they have some individual advantages and disadvantages.

Using vibration analysis, impending faults can be detected at an earlier stage. The experiments have shown that even a small fissure in the valve affects the vibration pattern enough to be detected by the algorithm. In contrast, pV diagram analysis detects a leak not until it is big enough that a sufficient amount of gas can flow through the closed valve, thus affecting the expansion stroke of the pV diagram significantly. Observing pV diagrams with states *Baseline* and *Crack* shows no difference at all between those two cases. Hence, it will be not possible to detect the fault *Crack* via the pV diagram only by extracting features from it.

Monitoring steel valves is a problem for the vibration analysis approach. While all of the tested plastic valves show more or less similar vibration patterns, steel valves differ significantly. Hence, separate classifiers for steel valves and plastic valves have to be used. Another problem is that the validation accuracy in the experiments for steel valves was too low for a monitoring system used in real-world applications. The pV diagram analysis is better suited for monitoring steel valves. Once the offset in the feature space is determined, steel valves are classified with the same classifier as plastic valves and with high accuracy.

The classification accuracy in vibration analysis depends strongly on the parameter choice of the classifier, for both (one-class) SVMs and logistic classification. For optimal parameter choice, a sufficient amount of annotated validation data is needed. These data were provided for developing the two approaches. But, it is doubtful whether, under real-world conditions, such a proper determination of the parameter values is possible or not. For pV diagram analysis, a parameter-free one-class classifier is proposed in Sect. 4.4. This classifier is tailor-made for the feature space of pV diagram analysis and yields in a better accuracy than several well-known (not parameter-free) one-class classifiers.

Online applicability is of course an important topic for fault detection systems. Since the feature extraction and classification steps are very quick, the basic method enables online condition monitoring of the valves. However, at the actual development state, the training phase (for instance the classifiers, the parabola, etc.) has to be performed offline with pre-recorded data. Since it is desirable for many applications to learn or update such parameters on the fly, some extensions have to be made in the future.

Acknowledgements This work has been supported by the COMET-K2 "Center for Symbiotic Mechatronics" of the Linz Center of Mechatronics (LCM) funded by the Austrian federal government and the federal state of Upper Austria.

References

1. Aichholzer, O., Jüttler, B.: Einführung in die angewandte Geometrie. Birkhäuser Basel (2014)
2. Anderson, A.: Logistic discrimination. In: Krishnaiah, P.R., Kanal, N.L. (eds.) Handbook of Statistics 2: Classification, Pattern Recognition and Reduction of Dimensionality, pp. 169–191. Gulf Publishing Company, Houston (1982)
3. Antoni, J.: Cyclostationarity by examples. Mech. Syst. Signal Process. **23**, 987–1036 (2009)
4. Ariel Corporation: Ariel JG and JGA compressors. Website (2010). https://www.arielcorp.com/JG-JGA/. Accessed 19 March 2018
5. Bauer, F., Lukas, M.: Comparing parameter choice methods for regularization of ill-posed problems. Math. Comput. Simul. **81**(9), 1795–1841 (2011)
6. Bloch, H.P.: A Practical Guide to Compressor Technology. Wiley, Hoboken (2006)
7. Bloch, H.P., Hoefner, J.J.: Reciprocating Compressors - Operation & Maintenance. Gulf Professional Publishing, Houston (1996)
8. Diab, S., Howard, B.: Reciprocating compressor management systems provide solid return on investment. In: Proceedings of the ROTATE Conference (2004)
9. Draper, N.R., Smith, H.: Applied Regression Analysis. Wiley, New York (1998)
10. Drewes, E.: Condition monitoring for reciprocating compressors. Hydrocarbon Processing (2002)
11. Dy, J.G., Brodley, C.E.: Feature selection for unsupervised learning. J. Mach. Learn. Res. **5**, 845–889 (2004)
12. Eilers, P.H.: A perfect smoother. Anal. Chem. **75**(14), 3631–3636 (2003)
13. Elhaj, M., Almrabet, M., Rgeai, M., Etiwesh, I.: A combined practical approach to condition monitoring of reciprocating compressors using IAS and dynamic pressure. World Acad. Sci. Eng. Technol. **63**, 186–192 (2010)
14. Fassios, S.D., Sakellariou, J.S.: Time-series methods for fault detection and identification in vibrating structures. Phil. Trans. R. Soc. A **365**(1851), 411–448 (2007)
15. Fugate, M.L., Sohn, H., Farrar, C.R.: Vibration-based damage detection using statistical process control. Mech. Syst. Signal Process. **15**(4), 707–721 (2001)
16. Guyon, I., Elisseeff, A.: An introduction to variable and feature selection. J. Mach. Learn. Res. **3**, 1157–1182 (2003)
17. Hlawatsch, F., Auger, F.: Time-Frequency Analysis: Concepts and Methods. Wiley, Hoboken (2008)
18. Hoerbiger Compression Technology: Compressor valves for better reliability, higher efficiency and safety. Website (2012). http://www.hoerbiger.com/upload/file/valve_overview_en.pdf. Accessed 26 March 2018

19. Hoerbiger Compression Technology: Experience real capacity control and energy savings with HydroCOM. Website (2012). http://www.hoerbiger.com/upload/file/hydrocom_en.pdf. Accessed 26 March 2018
20. Hoerbiger Compression Technology: RecipCOM - expertise, reciprocating compressor monitoring and protection tailored to your needs. Website (2012). http://www.digitalrefining.com/data/literature/file/661938396.pdf. Accessed 26 March 2018
21. Huschenbett, M., Will, G.: Thermodynamic simulation of reciprocating compressors to enable diagnostics based on measured temperatures and pressures. In: Proceedings of the 4th Conference of the European Forum of Reciprocating Compressors (2005)
22. Kohavi, R., John, G.H.: Wrappers for feature subset selection. Artif. Intell. **97**, 273–324 (1997)
23. Lebold, M., McClintic, K., Campbell, R., Byington, C., Maynard, K.: Review of vibration analysis methods for gearbox diagnostics and prognostics. In: Meeting of the Society for Machinery Failure Prevention Technology, pp. 623–634 (2000)
24. Lenz, J.R.: Polytropic exponents for common refrigerants. In: International Compressor Engineering Conference (2002)
25. Lewicki, M.S.: A review of methods for spike sorting: the detection and classification of neural action potentials. Network **9**, R53–78 (1998)
26. Lewis, J.P.: Fast template matching. In: Vision Interface, pp. 120–123. Canadian Image Processing and Pattern Recognition Society, Quebec (1995)
27. Lin, Y.H., Hu, H.S., Wu, C.Y.: Automated condition classification of a reciprocating compressor using time-frequency analysis and an artificial neural network. Inst. Phys. Publ. Smart Mater. Struct. **15**, 1576–1584 (2006)
28. Lin, Y.H., Liu, H.S., Wu, C.Y.: Automated valve condition classification of a reciprocating compressor with seeded faults: experimentation and validation of classification strategy. Inst. Phys. Publ. Smart Mater. Struct. **18**, 1576–1584 (2009)
29. Lughofer, E., Kindermann, S.: SparseFIS: data-driven learning of fuzzy systems with sparsity constraints. IEEE Trans. Fuzzy Syst. **18**(2), 396–411 (2010)
30. Machu, E.H.: Reciprocating compressor diagnostics, detecting abnormal conditions from measured indicator cards. In: International Compressor Engineering Conference, pp. 505–510 (1996)
31. Namdeo, R., Manepatil, S., Saraswat, S.: Detection of valve leakage in reciprocating compressor using artificial neural network (ANN). In: International Compressor Engineering Conference (2008)
32. Paclik, P., Duin, R.P.W.: Dissimilarity-based classification of spectra: computational issues. Phil. Trans. R. Soc. A **365**(1851), 411–448 (2007)
33. Pichler, K., Schrems, A., Huschenbett, M.: Fault detection for a reciprocating compressor by combining vibration analysis and transformation techniques. In: BINDT CM 2010 and MFPT 2010 (2010)
34. Pichler, K., Buchegger, T., Huschenbett, M.: A switching model for fault detection in reciprocating compressor valves under varying load conditions. In: IASTED International Conference on Signal and Image Processing (2011)
35. Pichler, K., Schrems, A., Buchegger, T., Huschenbett, M.: Fault detection in reciprocating compressor valves for steady-state load conditions. In: IEEE International Symposium on Signal Processing and Information Technology (2011)
36. Pichler, K., Lughofer, E., Buchegger, T., Klement, E.P., Pichler, M., Huschenbett, M.: A visual method to detect broken reciprocating compressor valves under varying load conditions. In: Mechatronics Forum (2012)
37. Ren, Q., Ma, X., Miao, G.: Application of support vector machines in reciprocating compressor valve fault diagnosis. In: Wang, L., Chen, K., Ong, Y.S. (eds.) Advances in Natural Computation, pp. 81–84. Springer, Berlin (2005)
38. Shalchyan, V., Jensen, W., Farina, D.: Spike detection and clustering with unsupervised wavelet optimization in extracellular neural recordings. IEEE Trans. Biomed. Eng. **59**, 2576–2585 (2012)

39. Snyman, J.A.: Practical Mathematical Optimization: An Introduction to Basic Optimization Theory and Classical and New Gradient-Based Algorithms. Springer, New York (2005)
40. Sohn, H., Farrar, C.R.: Damage diagnosis using time series analysis of vibration signals. Smart Mater. Struct. **10**(3), 446 (2001)
41. Spectra Quest Inc.: Vibration signatures of reciprocating compressors. Spectra Quest Tech Note (2007)
42. Tandon, N., Choudhury, A.: A review of vibration and acoustic measurement methods for the detection of defects in rolling element bearings. Tribol. Int. **32**(8), 469–480 (1999)
43. Tax, D.M.J.: One-class classification. PhD thesis, Delft University of Technology (2001)
44. Tipler, P.A., Mosca, G.: Physik: für Wissenschaftler und Ingenieure. Spektrum Akademischer Verlag, Heidelberg (2007)
45. Tiwari, A., Yadav, P.: Application of ANN in condition monitoring of a defective reciprocating air compressor. J. Instrum. Soc. India **38**(1), 13–20 (2008)
46. Vapnik, V.N.: The Nature of Statistical Learning Theory. Springer, New York (1995)
47. Wang, H.Q., Chen, P.: Fault diagnosis of centrifugal pump using symptom parameters in frequency domain. Agric. Eng. Int. CIGR Ej. **9**, 1–14 (2007)
48. Wang, F., Song, L., Zhang, L., Li, H.: Fault diagnosis for reciprocating air compressor valve using p-V indicator diagram and SVM. In: Proceedings of the 3rd International Symposium on Information Science and Engineering, pp. 255–258 (2010)
49. Wang, Y., Xue, C., Jia, X., Peng, X.: Fault diagnosis of reciprocating compressor valve with the method integrating acoustic emission signal and simulated valve motion. Mech. Syst. Signal Process. **56**, 197–212 (2015)
50. Whittaker, E.T.: On a new method of graduation. Proc. Edinb. Math. Soc. **41**(1), 63–75 (1923)
51. Yang, B.S., Hwang, W.W., Kim, D.J., Chit Tan, A.: Condition classification of small reciprocating compressor for refrigerators using artificial neural networks and support vector machines. Mech. Syst. Signal Process. **19**, 371–390 (2005)
52. Zhang, Y., Jiang, J., Flatley, M., Hill, B.: Condition monitoring and fault detection of a compressor using signal processing techniques. In: Proceedings of the American Control Conference, pp. 4460–4465 (2001)
53. Zouari, R., Antoni, J., Ille, J.L., Sidahmed, M., Willaert, M., Watremetz, M.: Cyclostationary modelling of reciprocating compressors and application to valve fault detection. Int. J. Acoust. Vib. **12**, 116–124 (2007)

A New Hilbert–Huang Transform Technique for Fault Detection in Rolling Element Bearings

Shazali Osman and Wilson Wang

1 Introduction

Rotating machinery is commonly used in various industries, such as automotive, aerospace, chemical engineering, and power generation. Such a machinery operation requires high reliability and safety but at lower costs. Therefore, accurate fault diagnosis of machine failure is vital, to the operation and maintenance of machinery to recognize an incipient machinery defect at its earliest stage so as to prevent machinery performance degradation, malfunction, and even catastrophic failures. Based on investigations, more than 50% of rotating machinery imperfections are related to defects in rolling element bearings [1, 2], and correspondingly, this work will focus on fault detection in rolling element bearings.

There are different types of rolling element bearings. According to the structure of rolling elements, for example, bearings can be classified as roller, cylindrical, taper bearings, and ball bearings. Different from gears and shafts, a rolling element bearing is, in fact, a system that is comprised of an inner ring (or race), an outer ring (or race), rolling elements, and a cage, as illustrated in Fig. 1. In general, the inner race is fastened to the shaft and rotates with the shaft; the outer ring is mounted in the bearing housing, which is usually fixed.

Bearing components are subjected to dynamic loading. Bearing defects can be classified into distributed and localized faults. The dynamic Hertzian contact loading leads to bearing fatigue damage in due course, resulting in micro-cracks and

S. Osman
Department of Electrical and Computer Engineering, Lakehead University, Thunder Bay, ON, Canada
e-mail: sosman@Lakeheadu.ca

W. Wang (✉)
Department of Mechanical Engineering, Lakehead University, Thunder Bay, ON, Canada
e-mail: wilson.wang@Lakeheadu.ca

© Springer Nature Switzerland AG 2019
E. Lughofer, M. Sayed-Mouchaweh (eds.), *Predictive Maintenance in Dynamic Systems*, https://doi.org/10.1007/978-3-030-05645-2_7

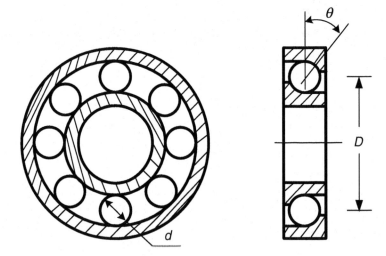

Fig. 1 The geometry of a rolling element bearing (a ball bearing in this case) and its parameters. d = diameter of the rolling element, θ = angle of contact, D = pitch diameter

localized defects that include cracks, pits, and spalls on bearing component surfaces. Distributed defects include wear, surface roughness, waviness, and misaligned races. In applications, most distributed defects originate from localized defects. Accordingly, this work focuses on the analysis of localized bearing faults.

Bearing fault detection can be undertaken based on the analysis of different information carriers, such as temperature, debris, or vibration. However, vibration-based monitoring could be the most commonly used approach due to its ease of measurement and high signal-to-noise ratio, which will be used in this work [3, 4].

Bearings generate vibration even if its components are healthy. Vibration forces can excite resonances of the surrounding structures. Although excitation is natural for rolling bearings, these forces can be greatly amplified due to imperfections or defects on the bearing components.

Many signal processing techniques have been proposed in the literature to extract representative features in the time domain, frequency domain, and time–frequency domain, for bearing fault detection. Commonly used time domain diagnostic indicators are usually determined through the analysis of probability density distribution properties, such as kurtosis, clearance factor, and impulse factor [5, 6]. Unfortunately, as the bearing damage propagates, the impulse features would take on a random pattern, and these statistical indicators may generate confusing diagnostic results.

Currently, frequency analysis could be the most common approach for bearing defect detection [4, 6]. In frequency analysis, bearing fault defection is based on the analysis of spectral information. The Fourier transform (FT) of the time signal can provide discrete information about each specific event involved. Each component of a bearing has its own characteristic frequency, as does any fault associated with that component. These frequency spectral contents can be used to examine bearing health conditions.

Consider a bearing with sound conditions and no slippage between the rolling elements and the races, as shown in Fig. 1. Assume that this bearing has a fixed outer race and a rotating inner race (the general case). For an outer race defect, the outer race defect characteristic frequency in Hz is calculated using:

$$f_{OR} = f_r \frac{z}{2}\left(1 - \frac{d}{D}\cos\theta\right) \quad (1)$$

The inner race defect characteristic frequency is calculated as:

$$f_{IR} = f_r \frac{z}{2}\left(1 + \frac{d}{D}\cos\theta\right) \quad (2)$$

If a rolling element is damaged, the defect characteristic frequency is computed by:

$$f_{BD} = f_r \frac{z}{2}\left(1 + \left(\frac{d}{D}\cos\theta\right)^2\right) \quad (3)$$

where D is pitch diameter, d is diameter of a rolling element, θ is the angle of contact, z is the number of rolling elements, and f_r is the shaft speed in Hz.

When a bearing is healthy, the related characteristic frequency is the shaft frequency f_r. As a fault occurs in a bearing component, the corresponding defect characteristic frequency and/or its harmonics will be, in theory, seen on the spectrum.

When a bearing component is damaged and the localized defect hits other bearing components, impulses are generated, which will excite support structural resonances. Envelope analysis can extract the periodic excitation of the resonance to detect the presence of a defect [4, 6]; however, this analysis requires experience in locating carrier frequencies when implementing the band-pass filter. Cepstrum analysis can be applied for detecting the periodicity of spectra corresponding to bearing fault-associated periodic impulses [6, 7]; however, it is usually difficult to detect incipient bearing damage with weak energy distribution modulation. In addition, spectrum analysis is not suitable to detect bearing defect if feature property varies in time with no baseline information.

Time–frequency domain techniques apply both time and frequency information to investigate transient feature properties. The common time–frequency analysis techniques include the short-time FT, the Wigner–Ville distribution, the wavelet transform (WT), and the Hilbert–Huang transform (HHT). In using the short-time FT, impacts that occur with different frequency spectra can be observed from the resulting map [8, 9]; however, its analysis has limited time–frequency resolutions. The Wigner–Ville distribution can be interpreted as a distribution of signal energy in the time–frequency domain with infinite resolutions [10, 11]; however, it may contain nonphysical interference (cross) terms that can deteriorate resolution. The WT uses variable size windows that can provide better resolutions and clear indication of the leading edge of impulses at higher frequencies [12, 13]; however, most WT-based analyses suffer from oscillation around singularities and shift variance, which may make it difficult to detect the individual structural resonances excited by the defect-induced impact.

The HHT-based techniques employ the empirical mode decomposition (EMD) and Hilbert transform for nonstationary signal analysis [14]. For example, a combination of the HHT, the support vector machine, and the support vector regression was suggested in [15] for bearing fault detection; however, it was difficult to track error contributions from each employed method in the proposed scheme. A marginal spectrum analysis of HHT was proposed in [16] for bearing fault detection; but, it required the fault spectrum beforehand to select intrinsic mode functions (IMFs). In general, if more IMFs are selected for HHT computation, the processing efficiency deteriorates significantly [15, 16]. To facilitate processing, classical HHT analysis usually uses the first few IMFs for bearing defect detection [17]; however, the selected IMFs may not always contain the most representative information. To solve this problem, a method was proposed in [18] to overcome the end effects and remove redundant IMFs; however, that method could not improve the stability of the HHT. The authors have also proposed a normalized HHT [19] and an enhanced HHT [20] techniques for bearing fault detection; although these methods can produce promising results in extracting bearing fault-related features under controlled operating conditions, they still require an approach to adaptively select and integrate the most representative IMFs for bearing defect detection.

On the other hand, mathematical morphology-based analysis has recently attracted more interests in signal processing and machinery fault detection [21]. For example, Huang et al. combined mathematical morphology, EMD, and power spectral density for bearing fault detection [22]; however, this method could not select appropriate structural element (SE) length for signal denoising. A flat SE was used for bearing fault detection in [23], which determined the maximum scale and length of the moving window empirically. A morphogram analysis method was suggested in [24] for bearing fault diagnosis; it used a construction index to optimize SE length, which, however, could become inaccurate when the signal is second-order pseudo-cyclostationary due to transients. In [25], a triangular SE was applied to defect bearing fault, but that method could not account for the effects of different scales and shapes of SE.

To tackle the aforementioned problems, a new enhanced Hilbert–Huang transform (eHT) technique is proposed in this work for nonstationary feature analysis and incipient bearing defect detection. The eHT technique is new in the following aspects: (1) A denoising filter is adopted to demodulate transmission impedance of the vibration signal and improve signal-to-noise ratio. (2) A novel HHT-based morphological filter technique is suggested to highlight defect-related impulse signatures for bearing fault detection. The effectiveness of the proposed eHT technique is verified experimentally using tests of different bearing conditions.

The reminder of this chapter is arranged as follows: The minimum entropy convolution denoising filter is discussed in Sect. 2. The proposed eHT technique is addressed in Sect. 3. The effectiveness of the proposed techniques is tested in Sect. 4, and some conclusion remarks are relegated to Sect. 5.

2 Minimum Entropy Deconvolution Filter

Figure 2 is the flowchart illustrating the fault detection process using the proposed eHT technique. Firstly, the measured vibration signal is denoised by the use of minimum entropy deconvolution (MED) filter. Then, the residual signal is processed using the proposed eHT technique for bearing fault detection. The denoising filter is discussed in this section.

If a bearing is damaged (e.g., a fatigue pit on the fixed ring race), impulses are generated whenever a rolling element strikes the damaged region. Due to the impedance effect of transmission path, the measured signal, using a vibration sensor, is a modulated signature of the defect-related impulses. To highlight defect-related impulses, a denoising process is taken first using the MED filter.

The MED was firstly proposed by Wiggins for deconvolving the impulsive sources on a mixture of signals [26]. The MED has shown its effectiveness to highlight the impulse excitations from a mixture of responses for machinery system condition monitoring [26, 27]. For example, the MED was combined with autoregressive models and wavelet analysis for fault detection in gear systems [28] and bearings [29]. The proposed eHT technique in this work is different from those in [26–29]; these two techniques performed fault detection using three processes: MED, autoregressive modeling, and the WT, and consequently it would be more difficult to track the processing errors contributed from each individual method whose parameters must be optimized before processing.

The implemented MED filter aims to highlight impulses by minimizing the noise (i.e., entropy) associated with signal transmission path. Entropy minimization is achieved by maximizing signal kurtosis which is sensitive to impulse-induced distortion in the tails of the distribution function. Figure 3 illustrates the MED filtering process in this signal denoising operation. The signal x represents the original form of the defect impulses. The signal N_s represents the random noise interference. The structure filter g represents the impedance effect of the transmission path from the bearing to the sensor.

Fig. 2 Flowchart of proposed eHT technique for bearing condition monitoring

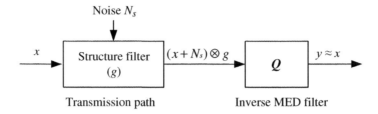

Fig. 3 The MED filtering process

The objective of the inverse MED filter Q is to find an optimal set of filter coefficients vector q to recover the original impulse signal by maximizing kurtosis or minimizing entropy. The kurtosis is determined as the fourth-order statistic measurement of a signal (an objective function) such as:

$$O_4(q(l)) = \sum_{i=1}^{N} y^4(i) \bigg/ \left[\sum_{i=1}^{N} y^2(i)\right]^2 \quad (4)$$

where y is the output signal using the inverse MED filter Q and N is the length of the signal.

The optimal filter coefficient vector q is achieved by optimizing the kurtosis of the objective function in Eq. (4), which is achieved by letting

$$\partial(O_4(q(l)))/\partial q(l) = 0 \quad (5)$$

The convolution of the inverse filter is generally given by:

$$y(j) = \sum_{j=1}^{L_m} q(l) z(j-l) \quad (6)$$

where $z = (x+N_s) \otimes g$ is the observed signal, and $\otimes$ is the convolution operator; and L_m is the length of the MED filter. Delay l is used to make the inverse filter causal [26–29].

By using $\partial y(j)/\partial q(l) = z(j-l)$ in Eq. (5) and combining Eqs. (4)–(6) yields:

$$\left[\sum_{i=1}^{N} y^2(i) \bigg/ \sum_{i=1}^{N} y^4(i)\right] \sum_{i=1}^{N} y^3(i) z(i-l) = \sum_{p=1}^{L_m} q(p) \sum_{i=1}^{N} z(i-l) z(i-p) \quad (7)$$

Equation (7) can also be represented by $B = Aq$, where B is the left-hand side of Eq. (7) and $A = \sum_{i=1}^{N} z(i-1) z(i-p)$ is the Toeplitz autocorrelation matrix of observed signal z.

The MED is conducted by the use of the following algorithm [29]:

1. Set the initial Toeplitz autocorrelation vector A as delayed impulse value, then initialize the filter coefficient $q^{(0)}$ as the delayed impulse. The autocorrelation matrix A is calculated once and is used repeatedly in the following iteration operations.
2. Determine the output signal $y(0)$ after applying the inversed MED filter, $q(0)$ and using the input signal $z(0)$ (Eq. (6)) (see Fig. 3).
3. Calculate $B^{(1)}$ from Eq. (7) and determine the new optimal filter coefficients $q^{(1)}$ by:

$$q^{(1)} = A^{-1} B^{(1)} \qquad (8)$$

4. Compute the error from the changes in filter coefficient values:

$$e = \left(q^{(1)} - \mu q^{(0)}\right) / \mu q^{(0)} \qquad (9)$$

where μ is the inner product determined by: $\mu = \left(E(q^{(0)})^2 / E(q^{(1)})^2\right)^{1/2}$ and $E(.)$ is the expectation operator.

In operation, if $E(e) > T_E$ ($T_E = 0.01$ in this case), update the filter coefficients by repeating the aforementioned algorithm starting from step 2. Otherwise, if $E(e) \leq T_E$, terminate the iteration. On the other hand, the iteration process is terminated if the number of iterations exceeds a threshold (100 in this case, selected by trial and error) or if the algorithm does not converge to the set value of $E(e)$.

In the above algorithm, MED filter length, L_m, and iteration number are determined by trial and error. As an example, consider a simulated signal consisting of three impulses and some noise as illustrated in Fig. 4a. The signal length is set to 10,000 samples. Figure 4 illustrates the processing results of three MED filters with filter lengths $L_m = 100$, 300, and 500, respectively. The respective convergence in terms of kurtosis is shown in Fig. 4. It is seen that these three distinctive impulses can be clearly highlighted if the filter length $L_m = 300$ (Fig. 4c). The filter with length $L_m = 500$ has the highest kurtosis value (Fig. 4d) than filters with lengths of 100 (Fig. 4b) and 300 (Fig. 4c); however, it cannot recognize all of the three impulses due to phase distortion.

On the other hand, Fig. 5 compares the convergence of MED filters with different lengths: 100, 300, and 500, respectively. It is seen that all three MED filters converge quickly over eight iterations only; that is, the convergence of implemented MED filter is not sensitive to filter length.

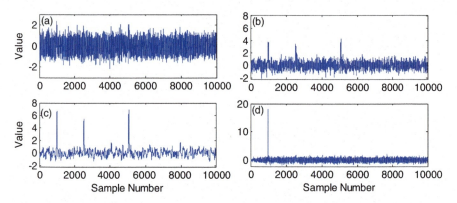

Fig. 4 Response comparison of a test signal using MED filters with different lengths: (**a**) simulated input signal, (**b**) MED response with filter length of 100, (**c**) MED response with filter length of 300, and (**d**) MED response with filter length of 500

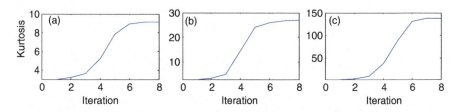

Fig. 5 Convergence comparison of MED filters with different lengths: (**a**) filter length of 100, (**b**) filter length of 300, and (**c**) filter length of 500

3 The Proposed eHT Technique for Bearing Fault Detection

The proposed eHT is applied to denoise the vibration signal from the adopted MED filter. The eHT includes two processing procedures: firstly, a novel morphological filter technique to process related signatures employing a new normality measure to select the optimal structural element (SE) length of the related IMFs. Secondly, a normalized HHT (denoted as NHHT) proposed by the authors in [19] is used to process the resulting signature for incipient bearing fault detection. In this section, the morphological filter technique is introduced first, followed by the normality measure for SE selection; the proposed eHT technique will be discussed in Sect. 3.2.

Before introducing the proposed morphological filtering technique, some related concepts of mathematical morphology will be briefly discussed first in this subsection, and more related details can be found from [30].

3.1 Brief Discussion of Mathematical Morphology Analysis

3.1.1 Structural Element (SE)

In general mathematical morphology analysis, it is assumed that all sets are closed sets, and signals are continuous. The basic morphological signal analysis is perturbed by transforming the signal through intersection with the SE. The SE is a probe that scans and modifies the input data by taking into account local information. An SE set can be either flat or non-flat, depending on applications. An important parameter in the construction of a flat SE is its length. If it is shorter, it becomes easier to extract the impulses by suppressing noise; however, it will be more difficult to demodulate the formulated signature, and vice versa [30, 31]. The literature still lacks an appropriate method to select the optimal length that can enhance both impulse extraction and demodulation. In general, the SE length is selected as approximately 0.6–0.7 times the repetition period [32, 33]; however, this may not be suitable for many dynamic system analyses such as the case of bearing fault detection, as demonstrated in Sect. 4.

3.1.2 Dilation and Erosion

Suppose $z(n)$ is the discrete-time signal over $Z = \{0, 1, \ldots, N - 1\}$, where n is the index variable and N is the length of the signal. If $g(n)$ is the discrete-time SE over the domain $G = \{0, 1, \ldots, M - 1\}$, where M is the length of the SE (usually $M \ll N$), the dilation operation, denoted by δ, is defined as:

$$[\delta_g(z, g)](n) = \vee_{m \in G} z(n + m) \qquad (10)$$

where $\vee$ represents the supremum operator and m is the length of the flat SE.

The erosion operation, ε, is defined as:

$$[\varepsilon_g(z, g)](n) = \wedge_{m \in G} z(n + m) \qquad (11)$$

where $\wedge$ denotes the infimum operator. $\vee$ and $\wedge$ can also be treated as the maximum and minimum operators of function z in a neighborhood defined by the SE (i.e., g). To simplify the mathematical morphology expression, the Minkowski algebra [30] will be used in the above equations: $\delta_g \sim z \oplus g$ and $\varepsilon_g \sim z \ominus g$. Equations (10) and (11) can be rewritten as:

$$[\delta_g(z, g)](n) = (z \oplus g)(n) \qquad (12)$$

$$[\varepsilon_g(z, g)](n) = (z \ominus g)(n) \qquad (13)$$

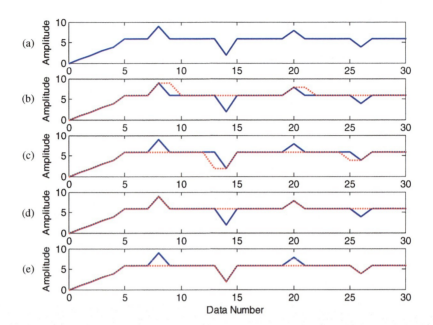

Fig. 6 Signals produced by using different morphological operators: (**a**) original signal, (**b**) after dilation (**c**), after erosion, (**d**) after closing, and (**e**) after opening. The solid blue line represents original signal and dotted red line is the modified signal

Consider a simulated signal with two scalesof impulses, which is sampled as shown in Fig. 6a. It is processed with a flat SE {0, 0, 0, 0, 0, 0, 0}, where the underlined position is the original seed point. The dilation will expand the maximum value of the signal but reduce the valleys of z as illustrated in Fig. 6b. In contrary, the erosion will reduce the peaks but increase the minima value of z (Fig. 6c). Furthermore, the dilation and erosion operators are correlated by:

$$\delta(-z) = -\varepsilon(z) \tag{14}$$

3.1.3 Closing and Opening

The closing operation, denoted by "•," and the opening operation, denoted by "o," are made by properly combining the basic dilation and the erosion, such that:

$$(z \circ g)(n) = (z \ominus g \oplus g)(n) \tag{15}$$

$$(z \bullet g)(n) = (f \oplus g \ominus g)(n) \tag{16}$$

The outputs of the closing operation and the opening operation of the simulated signal in Fig. 6a are illustrated in Fig. 6d, e, respectively.

The dilation can diminish the number of local minima, which cannot be restored by subsequent erosion. Thus, the closing will produce a simplification filtering of the signal. It can be seen from Eqs. (15) and (16) that the mirroring of the SE in the second operation (i.e., $g \ominus g$ or $g \oplus g$) can provide anti-extensivity in the closing and opening [30]. The mirroring of the SE can also make the operation independent of the SE origin. In this light, the opening (Fig. 6e) will preserve negative impulses and reduce positive impulses, but closing operates vice versa (Fig. 6d).

In general, it is difficult to obtain the prior knowledge of impulsive features from an input signal, especially when the signal contains both the positive and negative impulses such as in the case of bearing impulse signals. Our proposed approach will integrate the opening and closing operations of an appropriate SE length to filter out the noise and reconstruct the remaining objects. Correspondingly, the morphological difference filter function, $\widehat{f}$, which was also used in [30–33], will be formulated as:

$$\widehat{f} = f \bullet g - f \circ g \qquad (17)$$

The $\widehat{f}$ filter will be used to extract the impulses, and can be rewritten as:

$$\widehat{f} = f \bullet g - f \circ g = (f \bullet g - f) + (f - f \circ g) \qquad (18)$$

where $(f \bullet g - f)$ is the Black Top-Hat transform used to extract negative impulses, and $(f - f \circ g)$ is the White Top-Hat transform used to extract positive impulses [31, 32].

3.2 The Proposed Morphological Filter

The morphological difference filter function, $\widehat{f}$, as defined in Eq. (18) will be used in this work to extract the positive and negative impulsive features for bearing signal analysis. The proposed morphological filter will apply a new strategy to select an appropriate length of the SE. The selection of SE length depends on the input signal properties. Different from the methods of general autocorrelation functions [3, 4], the proposed selection method will use the Renyi entropy to process the nonlinear correlation in signals. The use of Renyi entropy is motivated by deriving a possible closed-form expression of entropy so as to avoid resorting to nonlinear measures [34]. Renyi entropy can quantify the diversity and uncertainty of random observations [34, 35], which will be defined as:

$$R_\alpha \left(\widehat{f}_q\right) = \frac{1}{1-\alpha} \log_2 \left(\sum_{n=1}^{N} p_n^\alpha\right) \qquad (19)$$

where $\widehat{f}_q$ is the morphological filter with probability p_n; $n = 1, 2, \ldots, N$, where N is the signal length; and $\alpha > 0$ is the Renyi entropy order. As R_α decreases, the function randomness decreases. The proposed morphological filter $\widehat{f}_q$ corresponds to each SE length q, and will be determined by:

$$\widehat{f}_q = (f \bullet \Gamma_q - f) + (f - f \circ \Gamma_q) \tag{20}$$

where Γ_q is the q-th selected length of SE.

In the proposed morphological filter, the indicator ϑ_q is formulated by considering kurtosis and entropy (i.e., R_α) of each filtered signal corresponding to a certain SE filter length:

$$\vartheta_q = \frac{\frac{K(\widehat{f}_q)}{R_\alpha(\widehat{f}_q)}}{\sum_{q=1}^{Q} K(\widehat{f}_q) / \sum_{q=1}^{Q} R_\alpha(\widehat{f}_q)} \tag{21}$$

where K denotes kurtosis to account for signal peakness properties, and Q is the total number of search filter lengths. ϑ_q is used to determine the optimal SE filter length to highlight impulses in the signal. Our proposed morphological filter will synthesize the outputs of Eq. (21) to select the optimal SE filter length so as to enhance the impulses and increase signal-to-noise ratio.

The higher value of ϑ_q in Eq. (21) corresponds to the sharper impulse using the filter with the chosen SE length. This is due to the fact that lower R_α and higher kurtosis K will increase ϑ_q, which corresponds to lower signal randomness and higher peakedness for the filtered signal.

As stated in Sect. 2, the most important parameter of the proposed morphological filter is the SE length. In general, without prior knowledge about the bearing health condition, the range of the filter length is selected over 10–90% of the cyclic interval [31–33]. In bearing vibration signal analysis for incipient fault detection, the generally applied filter length is 60–70% of the cyclic interval [31–33], which may deteriorate processing accuracy when the signal is nonlinear or nonstationary. If finer search spacing is used, it can result in longer filter length for optimization; however, this may lead to slower convergence in calculation.

The following example illustrates how to recognize impulses in a signal by the use of the proposed morphological filter technique. Consider a simulated signal consisting of five impulses with unity amplitude with a signal length ($N = 10{,}000$), and some random noise (with 140% of the impulse magnitude) as shown in Fig. 7a. Figure 7b–f illustrates the corresponding outputs of the proposed morphological filtering using different SE lengths $q = 0.1, 0.3, 0.5, 0.7$, and 0.9, respectively. When the filter length q is too short (e.g., $q = 0.1$ in this case), the impulsive information cannot be highlighted effectively as displayed in Fig. 7b. When the SE length is too long (e.g., $q = 0.9$), filtering distortion becomes severe as shown in Fig. 7f. If

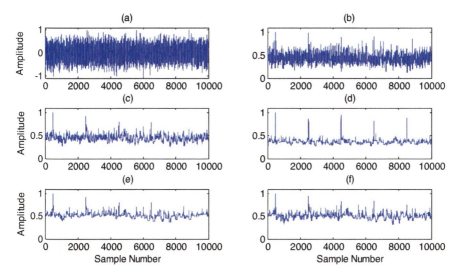

Fig. 7 Comparison of the filtering effects: (**a**) the simulated signal with noise added, (**b**) filtered signal with $q = 0.1$, (**c**) filtered signal with $q = 0.3$, (**d**) filtered signal with $q = 0.5$, (**e**) filtered signal with $q = 0.7$, and (**f**) filtered signal with $q = 0.9$

$q = 0.5$, five impulses can be clearly recognized with highest resolution as illustrated in Fig. 7d, which corresponds to the optimal SE length in this case.

Figure 8 compares the performance of different filter lengths and their corresponding kurtosis (Fig. 8a), entropy (Fig. 8b), and the proposed ϑ_q indicator (Fig. 8c), respectively. Although the filtered signal corresponding to the SE length of $q = 0.3$ has a slightly lower kurtosis values than those with SE length of 0.5, not all of the five impulses can be recognized as illustrated in the figure. Meanwhile, although filter length of 0.1 has higher entropy value than that of the filter length of 0.5, its output is distorted and not all of the five impulses are enhanced. Thus, the proposed indicator, ϑ_q, can provide the best results to select the optimal SE length to enhance impulses and improve signal-to-noise ratio.

3.3 The Proposed eHT Technique

In the proposed eHT technique, the bearing signal is firstly filtered using the optimal SE length, and then the HHT will be applied for bearing fault detection. In HHT analysis, the most distinctive IMF(s) will be selected by the use of the normality method as suggested by the authors in [19], which has been proven to be effective in processing nonstationary signals. Correspondingly, the proposed eHT technique is realized by:

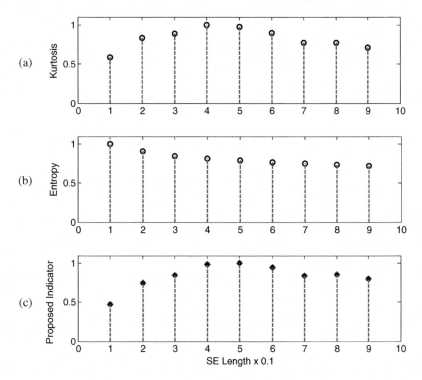

Fig. 8 Values of the simulated signal in Fig. 7 based on: (**a**) kurtosis, (**b**) entropy, and (**c**) proposed indicator

$$\text{eHT}_q = \widehat{f_q} \times \text{IMF}_w \qquad (22)$$

where IMF_w is the selected w-th IMF and $\widehat{f_q}$ is the morphological filter with the highest indicator value ϑ_q of the q-th SE length. The proposed eHT technique will be applied to process the signal in the frequency domain; its implementation and effectiveness will be evaluated in the following section.

4 Application of the Proposed eHT Technique for Bearing Fault Detection

4.1 Experimental Setup and Instrumentations

The experimental setup employed for this work is shown in Fig. 9. The system is driven by a 2-Hp induction motor, with the speed ranging from 0.3 to 70 Hz, controlled by a speed controller (VFD022B21A). A flexible coupling is utilized to damp out the high-frequency vibration generated by the motor and to accommodate

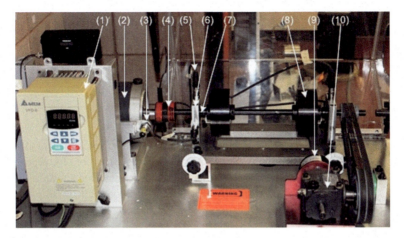

Fig. 9 Experimental setup: (1) speed control, (2) motor, (3) optical sensor, (4) flexible coupling, (5) ICP accelerometer, (6) bearing housing, (7) test bearing, (8) load disc, (9) magnetic load system, and (10) bevel gearbox

misalignment errors in assembly. Variable load is applied by a magnetic brake system through a bevel gearbox and a belt drive; the specified load torque in the following tests represents the torque value of the magnetic brake. An optical sensor is used to provide a one-pulse-per-revolution signal for shaft speed measurement. Two ball bearings (MB ER-10K) are press-fitted into the bearing housings, which have the following parameters: number of rolling elements: 8, rolling element diameter: 7.938 mm, pitch diameter: 33.503 mm, and contact angle: 0°. The bearing on the left-hand side housing is used for testing. Accelerometers (ICP-IMI, SN98697) are mounted on the housings to measure vibration signals along the vertical and horizontal directions. Considering the structure properties, the signal measured vertically is utilized for analysis in this work, whereas the signal measured from the horizontal direction is used for verification. These vibration and reference signals are fed to a computer for further processing through a data acquisition board (NI PCI-4472) which has built-in anti-aliasing filters with the cutoff frequency set at half of the sampling rate.

The sampling frequency of the verification test depends on the range of the shaft speed to collect 600–700 samples over each shaft rotation cycle. In this testing, four bearing health conditions are considered: healthy bearings, bearings with outer race defect, bearings with inner race defect, and bearings with rolling element defect. Seven different shaft speeds (i.e., 15, 20, 25, 30, 32, 35, and 40 Hz) and three break load levels (i.e., 1, 2.5, and 5 Nm) are used in this test. The outer race defect and the inner race defect each has a size of (area × depth) about 0.2 mm^2 × 0.5 mm, and the rolling element defect dimension (area × depth) is about 0.3 mm^2 × 0.5 mm. Table 1 summarizes the characteristic frequencies in terms of shaft speed orders for bearings with different health conditions (e.g., healthy bearings and bearings with

Table 1 Characteristic frequencies of the bearing in terms of shaft speed order

Bearing condition	Frequency in the order of the shaft speed
Healthy/normal	1.00
Inner race defect	3.05
Outer race defect	4.95
Rolling element defect	3.98

defects on the outer races, inner races, and rolling elements, respectively) [19, 20]. The selected techniques are implemented in MATLAB environment for processing.

In the following subsection, only some typical examples with torque load of 2.5 Nm and shaft speed of 1800 rpm ($fr = 30$ Hz) in each bearing condition will be discussed to demonstrate the effectiveness of the proposed eHT technique.

4.2 Performance Evaluation

4.2.1 Validation of Morphological-Based Filtering Technique

Firstly, a few examples are used to demonstrate how to implement the proposed morphology-based filtering technique for bearing fault detection. Equation (21) is used to determine the values of ϑ_q of each selected length of SE over [0.1, 0.9]. With the higher value of ϑ_q, the corresponding SE length is more appropriate to enhance signal feature properties (i.e., peaks and valleys). ϑ_q values are normalized and plotted in Fig. 10 corresponding to four bearing health conditions with shaft speed of 1800 rpm (or 30 Hz) and load torque of 2.5 Nm. To illustrate the effectiveness of the proposed method, only the SE length (i.e., $Q = 9$) is used for eHT analysis. The proposed eHT utilizes all the SE lengths, but it manages the contribution of each filter length according to its weights in Eq. (21) and from Eq. (22), respectively. For a healthy bearing (Fig. 10a), the most SE length is the fifth (i.e., $q = 0.5$) instead of the sixth or seventh SE length as in the classical mathematical morphology analysis [31–33]. For the bearing with an outer race defect (Fig. 10b), the most SE length is the eighth (i.e., $q = 0.8$). For the bearing with an inner race defect (Fig. 10c), the fifth SE length becomes the most significant one in this case (i.e., $q = 0.5$). For the bearing with a rolling element (ball) defect (Fig. 10d), the most significant SE length is the eighth (i.e., $q = 0.8$). On the other hand, it is seen that other order SE length, different from 0.6 or 0.7 as used in [25, 26], may also affect signal properties in this case.

4.2.2 Validation of the Normality Measure

Firstly, a few examples are used to demonstrate how to implement the proposed IMF selection technique in the eHT processing. Figure 11 shows some values of the

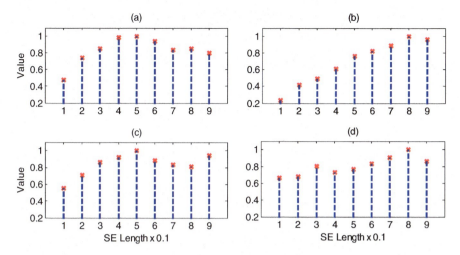

Fig. 10 Demonstration of normalized eHT indicators versus filter length corresponding to different bearing health conditions: (**a**) healthy bearing, (**b**) bearing with outer race defect, (**c**) bearing with inner race defect, and (**d**) bearing with rolling element defect

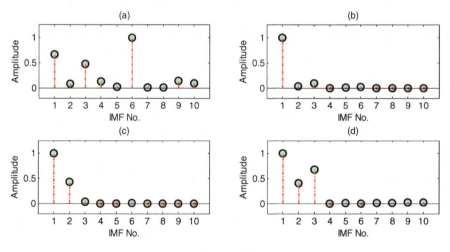

Fig. 11 Demonstration of normalized D'Agostino-Pearson normality indicators used in [19] versus IMF scales corresponding to filtered signal of different bearing health conditions: (**a**) healthy bearing, (**b**) bearing with outer race defect, (**c**) bearing with inner race defect, and (**d**) bearing with rolling element defect

D'Agostino-Pearson normality measure used in [19] to process the filtered signals corresponding to four bearing conditions with shaft speed of 30 Hz and load torque of 2.5 Nm. The D'Agostino-Pearson normality measure was used in [19] to select condition-related IMFs for bearing fault detection. The first ten IMFs will be used for eHT analysis. For a healthy bearing (Fig. 11a), the most distinguishable IMFs are the first and the sixth IMFs (instead of the first and second IMFs in the classical HHT analysis). Meanwhile, for a bearing with an outer race defect (Fig. 11b), and a bearing with an inner race defect (Fig. 11c), the most distinguishable IMFs are determined to be the first and the second IMFs, which is consistent with the classical HHT analysis. On the other hand, for a bearing with a rolling element defect (Fig. 11d), the most distinguishable functions become the first and the third IMFs. From Fig. 11, it can be noted that higher-order IMFs (higher than 2) may also contribute significantly to signal properties in this case.

Figures 12a–c show part of the collected vibration signals corresponding to different bearing conditions. Although some impulses could be recognized from these vibration signatures such as in Fig. 12b–d, it is difficult to diagnose bearing health conditions just based on these original vibration patterns. The corresponding frequency spectrums for these four bearing conditions are plotted in Fig. 12e–h, respectively. It can be seen that the bearing health condition(s) cannot be detected reliably just based on spectral analysis, especially for complex bearing systems with nonlinear and nonstationary signals.

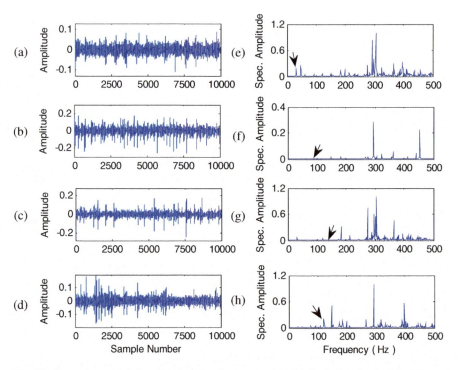

Fig. 12 Spectral maps of the vibration signals presented in (**a–d**) for bearings with different conditions using frequency analysis (**e–h**) for: a healthy bearing (**a, e**), a bearing with outer race defect (**b, f**), a bearing with inner race defect (**c, g**), and a bearing with rolling element defect (**d, h**). Arrows indicate the bearing characteristic frequency

4.3 Evaluation of the Proposed eHT Technique

The effectiveness of the proposed eHT technique will be compared to some related techniques available in the literature used for bearing fault diagnosis. Specifically, the processing results from the classical HHT method using the first two IMFs, designated as HHT, will be provided. To examine the effectiveness of the proposed morphological filter, the processing results from the HHT utilizing the morphological filtering, designated as MHT, will be compared to the HHT without using the proposed filter but applying a constant flat SE length of 0.6 (designated as CHT).

4.3.1 Condition Monitoring of a Healthy Bearing

Firstly, the bearings with healthy conditions are tested. Figure 13 shows the processing results using the related techniques. In this case, the bearing characteristic frequency is $f_r = 30$ Hz. Examining these power spectral graphs, the bearing characteristic frequency can be identified by each technique. Comparing Fig. 13a (eHT), Fig. 13b (MHT), and Fig. 11c (CHT), it can be noticed that the proposed morphological filter can denoise the signal properly and highlight characteristic frequency component ($f_r \approx 30$ Hz) and its harmonics. On the other hand, although the HHT in Fig. 13d can recognize the shaft speed f_r, its spectral maps contain more noise components due to leakage, which may reduce the reliability for bearing health condition monitoring.

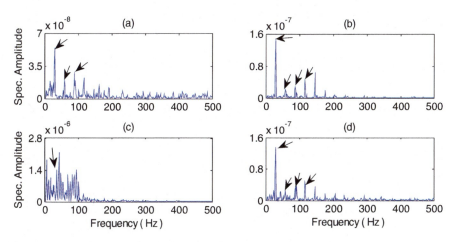

Fig. 13 Comparison of processing results for a healthy bearing using the techniques of: (**a**) eHT, (**b**) MHT, (**c**) CHT, and (**d**) HHT. Arrows indicate the characteristic frequency and its harmonics

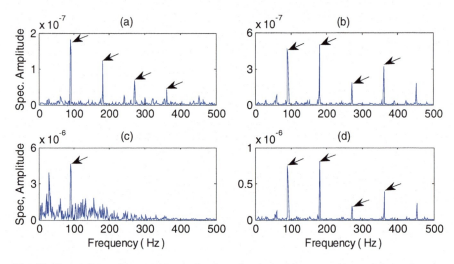

Fig. 14 Comparison of processing results for a bearing with an outer race defect using the techniques of: (**a**) eHT, (**b**) MHT, (**c**) CHT, and (**d**) HHT. Arrows indicate the characteristic frequency and its harmonics

4.3.2 Outer Race Fault Detection

When defect occurs on the fixed ring race of a bearing (the outer race in this case), its defect-induced resonance modes do not change over time. In this case, the characteristic frequency is $f_{OR} \approx 91$ Hz. Figure 14 shows the processing results using the related techniques. It is seen that the proposed eHT (Fig. 14a) outperforms not only the MHT (Fig. 14b) with higher magnitude of characteristic frequency components but also the CHT (Fig. 14c) and HHT (Fig. 14d) with the defect frequency and its harmonics dominating the spectral map. The main reason is due to its effective information processing using the suggested linearity measure and the proposed morphological filtering techniques.

4.3.3 Inner Race Fault Detection

The detection of fault on rotating elements is usually more challenging than the detection of fault on the outer race, because the resonance modes associated with the inner race impacts usually vary over time. In this case, the characteristic frequency $f_{IR} \approx 148$ Hz. Processing results using the related techniques are shown in Fig. 15. The proposed eHT (Fig. 15a) outperforms the MHT (Fig. 15b), the CHT (Fig. 10c), and the HHT (Fig. 15d) due its more efficient denoising processing. It can be seen that defect-related signatures on the maps of the CHT (Fig. 15c) and HHT (Fig. 15d) do not dominate the spectra, which can give false diagnostic results. Whereas, in the case of the MHT (Fig. 15b), the second harmonic of f_{IR} becomes lower in magnitude than that in Fig. 15a, which mitigates the redundant information for diagnosis.

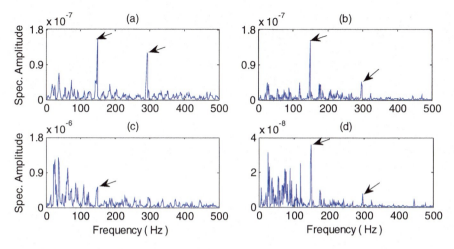

Fig. 15 Comparison of processing results for a bearing with an inner race defect using the techniques of: (**a**) eHT, (**b**) MHT, (**c**) CHT, and (**d**) HHT. Arrows indicate the characteristic frequency and its harmonics

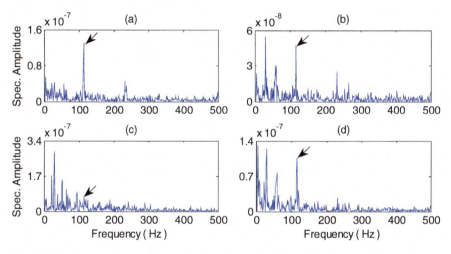

Fig. 16 Comparison of processing results for a bearing with a rolling element defect using the techniques of: (**a**) EHT, (**b**) EHTT, (**c**) CHT, and (**d**) HHT. Arrows indicate the characteristic frequency and its harmonics

4.3.4 Rolling Element Fault Detection

The detection of fault on a bearing rolling element (i.e., ball in this case) is generally considered the most challenging task in bearing fault detection. This is because a ball rolls along different directions (as well as it slides), and its resonance modes change over time. In this case, the characteristic frequency is $f_{BD} \approx 119$ Hz. Figure 16 shows the processing results using the related techniques. It is seen that the proposed eHT (Fig. 16a) outperforms other related methods for this fault detection, due to its efficient filtering and IMF integration strategies. It can mitigate disturbance and extract diagnostic information more effectively, which is important for nonstationary signal analysis. The characteristic frequency components of MHT (Fig. 16b), CHT (Fig. 16c), and HHT (Fig. 16d) do not dominate their respective spectral maps, which may lead to false or missed alarms especially in automatic online machinery health condition monitoring.

5 Conclusion

A new enhanced HHT technique, eHT in short, has been proposed in this work for incipient bearing fault detection and nonstationary signal analysis. The collected vibration signals are firstly denoised by the suggested minimum entropy deconvolution (MED) filter. Then, the signal reminder is processed by the proposed morphological filtering technique. Finally, the signal is processed by a normality indicator to select the most distinctive IMF(s) for eHT processing. The proposed MED filter is based on correlation measure and information entropy to attenuate distortion effect of the signal transmission path. The suggested morphological filtering technique is based on Renyi entropy and kurtosis analysis to reduce impedance effect of the measured vibration signal and to actively enhance the impulsive features. The effectiveness of the proposed eHT technique has been verified by experimental tests corresponding to different bearing conditions. The eHT method can effectively recognize related distinctive IMFs for nonstationary signal analysis and bearing fault detection. Test results have also shown that the proposed filter can effectively denoise the signal and highlight defect-related features. Although, testing has been conducted under controlled load and speed conditions for single incipient bearing fault detection, the proposed eHT is proven to be an effective signal processing technique and has potential for real bearing condition monitoring applications. Advanced research is undertaken for bearing fault detection in gearboxes, in which the bearing health-related features would be modulated by gear mesh operations.

Acknowledgments This research was financially supported by the Natural Sciences and Engineering Research Council of Canada (NSERC), Bare Point Water Treatment Plant, and eMech Systems Inc., in Thunder Bay, ON, Canada.

References

1. Hashemian, H.: State-of-the-art predictive maintenance techniques. IEEE Trans. Instrum. Meas. **60**(1), 226–236 (2011)
2. Donnell, P., Heising, C., Singh, C., Wells, S.: Report of large motor reliability survey of industrial and commercial installations. IEEE Trans. Ind. Appl. **23**(1), 153–158 (1987)

3. Wang, W., Lee, H.: An energy kurtosis demodulation technique for signal denoising and bearing fault detection. Meas. Sci. Technol. **24**(2), 025601 (2013)
4. Randall, R.B., Antoni, J.: Rolling element bearing diagnostics—a tutorial. Mech. Syst. Signal Process. **25**(2), 485–520 (2011)
5. Mohanty, S., Gupta, K.K., Raju, K.S.: Adaptive fault identification of bearing using empirical mode decomposition–principal component analysis-based average kurtosis technique. IET Sci. Meas. Technol. **11**(1), 30–40 (2017)
6. Borghesani, P., Pennacchi, P., Chatterton, S.: The relationship between kurtosis and envelope-based indexes for the diagnostic of rolling element bearings. Mech. Syst. Signal Process. **43**(1–2), 25–43 (2014)
7. Borghesani, P., Pennacchi, P., Randall, R.B., Sawalhi, N., Ricci, R.: Application of cepstrum pre-whitening for the diagnosis of bearing faults under variable speed conditions. Mech. Syst. Signal Process. **36**(2), 370–384 (2013)
8. Cocconcelli, M., Zimroz, R., Rubini, R., Bartelmus, W.: Kurtosis over energy distribution approach for STFT enhancement in ball bearing diagnostics. In: Condition Monitoring of Machinery in Non-Stationary Operations, pp. 51–59. Springer, Berlin, Heidelberg (2012)
9. Gao, H., Liang, L., Chen, X., Xu, G.: Feature extraction and recognition for rolling element bearing fault utilizing short-time Fourier transform and non-negative matrix factorization. Chin. J. Mech. Eng. **28**(1), 96–105 (2014)
10. Rai, A., Upadhyay, S.H.: A review on signal processing techniques utilized in the fault diagnosis of rolling element bearings. Tribol. Int. **1**(96), 289–306 (2016)
11. Li, H., Zheng, H., Tang, L.: Wigner-Ville distribution based on EMD for faults diagnosis of bearing. In International Conference on Fuzzy Systems and Knowledge Discovery, 803–812. Springer, Berlin, Heidelberg (2006)
12. Mishra, C., Samantaray, A.K., Chakraborty, G.: Rolling element bearing fault diagnosis under slow speed operation using wavelet de-noising. Measurement. **1**(103), 77–86 (2017)
13. Yan, R., Gao, R.X., Chen, X.: Wavelets for fault diagnosis of rotary machines: a review with applications. Signal Process. **1**(96), 1–5 (2014)
14. Elbouchikhi, E., Choqueuse, V., Amirat, Y., Benbouzid, M.E., Turri, S.: An efficient Hilbert–Huang transform-based bearing faults detection in induction machines. IEEE Trans. Energy Conv. **32**(2), 401–413 (2017)
15. Soualhi, A., Kamal, M., Noureddine, Z.: Bearing health monitoring based on Hilbert–Huang transform, support vector machine, and regression. IEEE Trans. Instrum. Meas. **64**(1), 52–62 (2015)
16. Li, H., Zhang, Y., Zheng, H.: Hilbert-Huang transform and marginal spectrum for detection and diagnosis of localized defects in roller bearings. J. Mech. Sci. Technol. **23**, 291–301 (2009)
17. Tsao, W.C., Li, Y.F., Le, D.D., Pan, M.C.: An insight concept to select appropriate IMFs for envelope analysis of bearing fault diagnosis. Measurement. **45**(6), 1489–1498 (2012)
18. Yan, J., Lu, L.: Improved Hilbert–Huang transform based weak signal detection methodology and its application on incipient fault diagnosis and ECG signal analysis. Signal Process. **98**, 74–87 (2014)
19. Osman, S., Wang, W.: A normalized Hilbert-Huang transform technique for bearing fault detection. J. Vib. Control. **22**(11), 2771–2787 (2016)
20. Osman, S., Wang, W.: An enhanced Hilbert-Huang transform technique for bearing condition monitoring. Meas. Sci. Technol. **24**(8), 1–13 (2013)
21. Li, Y., Zuo, M.J., Lin, J., Liu, J.: Fault detection method for railway wheel flat using an adaptive multiscale morphological filter. Mech. Syst. Signal Process. **1**(84), 642–658 (2017)
22. Huang, H., Ziwei, P.: A new bearing fault diagnosis method based on MM and EMD. In: IEEE Image Signal Process (CISP), 2010 3rd International Congress, vol. 8, pp. 3975–3979 (2010)
23. Zhang, P., Li, B., Mi, S., Zhang, Y., Liu, D.: Bearing fault detection using multi-scale fractal dimensions based on morphological covers. Shock. Vib. **19**, 1373–1383 (2011)
24. Wang, D., Tse, P.W., Tse, Y.L.: A morphogram with the optimal selection of parameters used in morphological analysis for enhancing the ability in bearing fault diagnosis. Meas. Sci. Technol. **23**(6), 1–15 (2012)

25. Chen, Q., Chen, Z., Sun, W., Yang, G., Palazoglu, A., Ren, Z.: A new structuring element for multi-scale morphology analysis and its application in rolling element bearing fault diagnosis. J. Vib. Control. **21**(4), 1–25 (2015)
26. He, D., Wang, X., Li, S., Lin, J., Zhao, M.: Identification of multiple faults in rotating machinery based on minimum entropy deconvolution combined with spectral kurtosis. Mech. Syst. Signal Process. **1**(84), 642–658 (2017)
27. Cheng, Y., Zhou, N., Zhang, W., Wang, Z.: Application of an improved minimum entropy deconvolution method for railway rolling element bearing fault diagnosis. J. Sound Vib. **7**(425), 53–69 (2018)
28. Endo, H., Randall, R.B.: Enhancement of autoregressive model based gear tooth fault detection technique by the use of minimum entropy deconvolution filter. Mech. Syst. Signal Process. **21**(2), 906–919 (2007)
29. Sawalhi, N., Randall, R.B., Endo, H.: The enhancement of fault detection and diagnosis in rolling element bearings using minimum entropy deconvolution combined with spectral kurtosis. Mech. Syst. Signal Process. **1**(84), 642–658 (2017)
30. Maragos, P., Schafer, W.: Morphological filters—part 1: their set-theoretic analysis and relations to linear shift-invariant filters. IEEE Trans. Acoust. Speech Signal Process. **35**(8), 1153–1169 (1987)
31. Chen, Q., Chen, Z., Sun, W., Yang, G., Palazoglu, A., Ren, Z.: A new structuring element for multi-scale morphology analysis and its application in rolling element bearing fault diagnosis. J. Vib. Control. **21**(4), 765–789 (2015)
32. Zhang, L., Xu, J., Yang, J., Yang, D., Wang, D.: Multiscale morphology analysis and its application to fault diagnosis. Mech. Syst. Signal Process. **22**(3), 597–610 (2008)
33. Murali, N.: Early classification of bearing faults using morphological operators and Fuzzy inference. IEEE Trans. Ind. Electron. **60**(2), 567–574 (2013)
34. Maszczyk, T., Włodzisław, D.: Comparison of Shannon, Renyi and Tsallis entropy used in decision trees. In: A.I. Soft Comp.–ICAISC 2008, vol. 5097, pp. 643–651, Springer, Berlin (2008)
35. Boškoski, P., Juričić, D.: Fault detection of mechanical drives under variable operating conditions based on wavelet packet Rényi entropy signatures. Mech. Syst. Signal Process. **31**, 369–381 (2012)

Comparison of Genetic and Incremental Learning Methods for Neural Network-Based Electrical Machine Fault Detection

Daniel Leite

1 Introduction

There is an increasing demand on reliability and safety of industrial systems subject to potential process abnormalities and component faults [1]. Electrical motors are one of the most used machines in the industry. Generally, they are critical components in automation processes. Therefore, questions related to their protection against failures have received great attention [2–6]. Condition monitoring and predictive maintenance of induction motors may lead to significant improvements of availability, quality, and productivity of production lines. Detecting faults in incipient stage is of utmost importance since functional failures may quickly occur after the initial development of a fault.

A major part of induction motor faults occurs in the stator windings [2, 7]. The inter-turns short-circuit is a primary fault that happens after insulation breakdown. Among the main reasons for insulation fail are high stator core or winding temperatures; slack core lamination, slot wedges, and joints; loose bracing for end winding; contamination due to chemical reactions, moisture, or dirt; electrical discharges due to aging of the insulating material; and leakage in cooling systems. After primary faults, the motor degradation process increases, and more serious failures, such as phase-to-phase and phase-to-ground short-circuits, appear. Usually, these types of faults result in irreversible motor damage. However, if inter-turns faults are detected at incipient stage, the faulty phase winding may, for example, be replaced, which significantly reduces financial losses and increases operational safety. Among the benefits detection systems can bring to industry are motor life extension, idling periods reduction, unnecessary disconnections avoiding,

---

D. Leite (✉)
Department of Engineering, Federal University of Lavras, Lavras, Minas Gerais, Brazil
e-mail: daniel.leite@deg.ufla.br

© Springer Nature Switzerland AG 2019
E. Lughofer, M. Sayed-Mouchaweh (eds.), *Predictive Maintenance in Dynamic Systems*, https://doi.org/10.1007/978-3-030-05645-2_8

manpower scheduling at the fault moment, repair cost minimization, human security improvement, components storage reduction, and minimization of losses.

During the last three decades, computational intelligence methods have been a promising direction for solutions of pattern recognition issues. Neural Networks and Hybrid Systems have been successfully applied to detecting different kinds of faults in electrical machines [8–12]. A difficulty of applying neural networks to condition monitoring systems, as well as to the vast majority of real-world engineering applications, concerns the selection of a suitable network structure and connection parameters. A proper selection of these is essential to lead the network to achieve a reasonable fault detection performance. Some of these parameters, viz., the number of hidden layers and the number of neurons per layer, are frequently set from a trial-and-error approach performed by a human designer. This may be an exhaustive task that can take a considerable amount of time [12]. Efforts for making neural network design more sophisticated and less human dependent is underway, especially considering information from particular application domains.

A drawback of using neural networks for fault detection, including feedforward, recurrent, convolutional, and deep networks and deep models in general [13], is the use of first- or second-order deterministic optimization algorithms for training. Since the backpropagation (BP) algorithm was discussed by Rumelhart et al. [14], researchers quite often resort to first-order learning methods and variations. Since the nature of first-order methods is to converge locally, it can be demonstrated that its solution is highly dependent on random initial weights and rarely is the global solution. Several variations of first-order optimization methods were compared to Quasi-Newton, Non-Derivative Quasi-Newton, Gauss–Newton, and Secant methods by Chen and Sheu [15]; Evolutionary Strategies and Genetic Algorithm by [16, 17]; Bayesian Regularization, Modified Levenberg–Marquardt, and Simulated Annealing in [18]; Bee and Ant Colony by [19, 20]; Adaptive Differential Evolution in [21]; and Particle Swarm by [22, 23]. All these training methods could lead a neural model to achieve better performance in terms of learning efficiency, training time, easiness-of-use, and accuracy in a class of problems.

The present study focuses on the development of architectures and weights of feedforward neural networks using different learning algorithms, viz., a properly designed genetic algorithm and an online incremental algorithm. The former produces an evolutionary neural network (EANN), while the latter generates an evolving fuzzy granular neural network (EGNN). The purpose of the neural networks is to detect and determine the number of shorted-turns in the stator windings of induction machines. The problem is formulated as a multiclass classification problem. Real dynamic environment subject to mechanical asymmetries, voltage unbalance, and measurement noise is taken into consideration.

Evolving the EANN architecture includes finding a suitable number of hidden layers and neurons per layer—being these the parameters that largely affect its generalization ability. An overly complex neural model may overfit the data and thus exhibit poor generalization, whereas a simple model may be insufficient to represent nonlinear correlations among features. GA takes into account the development of both, parameters and structure of the neural network. The main reasons for the choice of GA as learning method are: (1) GA operates on codified parameters. It results in a search for local minima independently of the continuity of error functions or the existence of derivative; (2) the search toward the best solution starts from a set of points deployed in the search space (global search–populational strategy). Thus, the probability that the solution gets stuck on local minima is minimized; (3) the search toward the best solution utilizes genetic operators, which are stochastic in nature, instead of deterministic; and (4) GA automatizes the trial-and-error approach to set up structural parameters.

EGNN encodes a set of fuzzy rules in its structure. Therefore, neural processing conforms with that of a fuzzy inference system [24]. The network is equipped with fuzzy neurons, which perform aggregation functions, and with an incremental algorithm for learning from a data stream. Fuzzy granules and rules are created gradually according to new information discovered from the data. Evolving systems from data streams is an active and promising research topic [25–35]. In particular, EGNN provides: (1) computational tractability and scalability with the number of samples and attributes; (2) improved interpretability and transparency by means of granular local models and linguistic rules; and (3) reduced cost of data processing in relation to non-evolving methods. EGNN has shown to be extremely general and able to outperform state-of-the-art evolving methods and models, including evolving classifiers [24, 36, 37].

Section 2 outlines a general framework for electrical machine fault detection assisted by the genetic neural classifier, EANN, and the incremental neurofuzzy classifier, EGNN. Section 3 addresses GA learning and describes the genetic operators for recombining, mutating, and selecting architectures and connection weights of a feedforward network. Incremental learning from data streams and development of a neurofuzzy granular network are given in Sect. 4. EANN and EGNN performance on detecting incipient faults in induction machines and discussions about genetic and incremental learning are reported in Sect. 5. True conditions of actual industrial practice, namely, different load and speed conditions, voltage unbalance, and noisy environment, are analyzed. Section 6 concludes the chapter and presents some ideas for further investigation.

2 Electrical Machine Fault Detection

A general view of the electrical machine fault detection system is shown in Fig. 1. Voltage, current, and rotor speed measurements are obtained from induction motors and properly placed sensors. The data are preprocessed and attribute vectors whose values are related to the healthy state of the machines are obtained. The *Acquisition and Data Treatment* module also disposes the machines' state variables to the *Parameters Estimation*, *Optimization*, and *Faults Simulator* modules. The latter includes faulty samples in the *Database*. Therefore, the database is composed of healthy and faulty data vectors. Faulty samples are related to different incipient fault severities, i.e., from 1 to 3 shorted-turns in the stator windings; different locations,

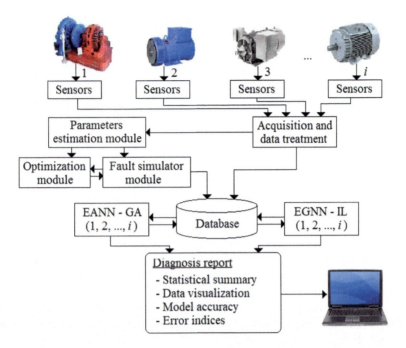

Fig. 1 General view of the electrical machine fault detection and classification system

i.e., stator phases *a*, *b*, and *c*; and different operating points and levels of noise. For EANN, a percentage of randomly selected samples are admitted for training, while the rest is used for testing. The *Genetic Algorithm* module develops the EANN architecture and its weights. It elects the best architecture and its respective best vector of weights according to a fitness function. On the other hand, a neurofuzzy EGNN structure is evolved from scratch by means of an incremental learning algorithm. In this case, training and testing are not separated procedures. In other words, EGNN provides an output—a classification for the input data sample—and then, the input–output pair is used for training. Online learning proceeds in a per-sample incremental basis. A diagnosis report is generated by both EANN and EGNN.

A description of the system modules is given below:

- *Acquisition and data treatment* → This module measures voltage, current, and rotor speed signals from induction machines. The number of motors connected to the system is limited by the number of I/O channels. The acquired data are preprocessed. Offsets are removed, and magnitude correction factors are applied. Low pass filters minimize noise and slot effect. Other calculations such as active and reactive power, power factor, rotor slip, and sequential components are carried out. The data are displayed on the interface prior to being saved in a file repository.

- *Parameters estimation module* → No load and blocked rotor tests are required to be performed to obtain fundamental machine parameters if the corresponding datasheet is not available. Additionally, online parameter estimation algorithms, namely, Recursive Least Squares and Extended Kalman Filter, operate in parallel to adapt key parameters required by the inter-turns fault simulator model. Updated parameters are important to reflect the actual condition of the motor and avoid false positives. The most significant parameters considered for adaptation over time are the mutual inductance, rotor resistance, and equivalent resistance and inductance. Refer to [7, 38] for further descriptions.
- *Optimization module* → This module provides further refinement of the parameters used by the fault simulator model. The Conditional Gradient method, also known as the Frank–Wolfe algorithm [39], is employed to optimize certain parameters, viz., the magnetizing and leakage inductances of the stator windings and the stator resistance. The objective is to allow the fault simulator to better reproduce the actual state variables. The objective function (OF) is the sum of the square error between the estimated and actual stator currents, voltage–current displacement angles, and rotor speed. Taken the derivative of the OF with respect to the states, we obtain a linearized OF for the application of the method. At each iteration, steps on the motor model parameters are given as an attempt to minimize the OF. If the OF value increases, the step on the parameters is rejected. The smaller the OF value, the more accurate the estimated states.
- *Fault simulator module* → The state-space model of induction motors is changed mainly to allow simulations of turn-to-turn short-circuit in the stator windings. Moreover, changeable loads, voltage unbalance, noise, and winding asymmetries can be simulated. For completeness of this study, the key formulas are succinctly presented below. Refer to [7, 38, 40] for detailed information.

The dynamic equations of an induction motor in state variables are

$$[\dot{I}_{abcsr}] = [L]^{-1} \left[[V_{abcsr}] - [[R] + [\dot{L}]] [I_{abcsr}] \right] \quad (1)$$

where $[L]$ and $[R]$ are 6×6 inductance and resistance matrices; $[V_{abcsr}]$ and $[I_{abcsr}]$ are 6×1 stator and rotor voltage and current matrices in the abc frame of reference. $[\dot{I}_{abcsr}]$ can be calculated by the fourth-order Runge–Kutta algorithm. The model is complemented by mechanical equations:

$$T_e = \frac{P}{2}(i_{abcs})^T \frac{\partial}{\partial \theta_r}[L'_{sr}]i'_{abcr} \quad (2)$$

$$\omega_r = \frac{P}{2} \int \frac{T_e - T_l}{J} \quad (3)$$

where T_e is the electromagnetic torque; P the number of poles; θ_r the electrical angular displacement; $[L'_{sr}]$ and i'_{abcr} the inductances and currents referred to the stator; ω_r is the rotor speed; T_l the load torque; and J the inertia.

In a condition of stator shorted-turns, inductances are calculated from:

$$L_{(1-k)} = (1-k)^2 L \tag{4}$$

$$L_k = k^2 L \tag{5}$$

$$L_{k(1-k)} = (1-k)L \tag{6}$$

where k is the percentage of turns in short-circuit; $L_{(1-k)}$ refers to the inductance of the winding fraction without fault; and L_k is the inductance of the faulty part of the winding. The latter equation refers to the mutual inductance between the part of the winding without fault and the other phases, including rotor phases. Refer to [7, 38] to comprehend how exactly the elements of the matrix L are changed due to shorted-turn faults.

A fault requires the inductance matrix to be rewritten in seven dimensions, with three lines and columns representing the fraction of the stator phases without fault, a line and column representing the fraction of the faulty stator phase, and three lines and columns representing the rotor phases. Similarly, the resistance matrix is rewritten as a seven-dimension matrix considering:

$$R_{(1-k)} = (1-k)R \tag{7}$$

$$R_k = kR \tag{8}$$

where $R_{(1-k)}$ and R_k are the resistances of the winding fraction without and with fault. Naturally, the resistance matrix is diagonal.

- *Database* → The database consists of input–output samples that are useful to train and test neural network classification models. The abc stator currents, voltage–current displacement angles per phase, and rotor speed are used in this study as input attributes to the neural classifiers. Several other attributes are available such as the dq0 and sequential components of voltages, currents, and magnetic fluxes [40]. The output variable is a class, which is associated to the number of stator shorted-turns per phase. Therefore, the neural classification models consist of a map $f : X \rightarrow Y$ so that $X \in \mathbb{R}^7$ and $Y \in \mathbb{N}$. As new data samples are available, the neural classification models can be updated if needed.
- *EANN-GA* → A feedforward neural network for each induction machine being monitored is considered. The network structure and connection weights are evolved via a specially designed genetic algorithm. EANN may have one or two hidden layers. While an inner loop deals with optimization over the parameter space, an outer loop concerns with searching for a potentially optimal solution over the structure space. After the learning process, the best network architecture and its best vector of parameters are chosen. Section 3 addresses phenotype representation; initialization of populations; recombination, mutation, and selection operators; fitness evaluation; and the stopping criteria adopted.
- *EGNN-IL* → A neurofuzzy network able to learn gradually from a data stream is developed for each induction machine. EGNN self-adapts its granular structure

and connection weights by means of a recursive procedure. Different fuzzy neurons to perform aggregation of values through the network can be chosen. Additionally, attribute weighting is an inherent characteristic of the network due to its modular structure. In general, EGNN can handle fuzzy, interval, and numerical data as well as prediction, control, and classification problems. This study focuses on numerical data processing and fault classification only. Section 4 describes how granules, rules, and weights are adapted on the fly.

- *Diagnosis report* → A diagnosis report may be displayed at any time. The report includes statistical summaries of electrical machines, graphics of specific state variables and parameters, evolution of error indices and fault patterns, and neural network classification performance.

The condition monitoring system can manage several induction machines in field applications simultaneously. As the development of inter-turn faults takes some time, constant (uninterrupted) supervision of machines is not mandatory and, therefore, a single microcomputer and a switching scheme among machines and corresponding neural classifiers is, in general, acceptable. Otherwise, distributed computing may be taken into consideration. The monitoring response time is usually short, and microcomputer availability is high since offline training of EANN classification models can be performed apart from online data processing whereas EGNN online adaptation is carried out in a matter of milliseconds, as discussed in the next sections. As a numerical example, if 10 induction motors are monitored, then switching the corresponding neural classifiers in 10-s intervals is enough.

3 Genetic Algorithm for Neural Network Learning

This section describes a genetic method for developing the architecture and setting the parameters of a feedforward neural network. Fundamentally, the genetic algorithm performs the following steps:

- Genetic representation or codification of potential solutions;
- Definition of initial parameters. These include population size, relative elitism, penalty factors, mutation rate, and training stop criteria;
- Initialization of the population with a priori knowledge about the expected behavior of the neural network;
- Application of genetic operators, i.e., mutation, recombination, and selection operators, over individuals of the population;
- Evaluation of the fitness of the individuals of a population;
- Post-processing of the fittest individual.

The GA learning procedure is shown in the flowchart of Fig. 2. In the figure, the inner loop evolves Ω generations of *weights* individuals, whereas the outer loop evolves α generations of *architectures* individuals (global search). The fittest architecture individual and its respective fittest weights are found. Post-processing

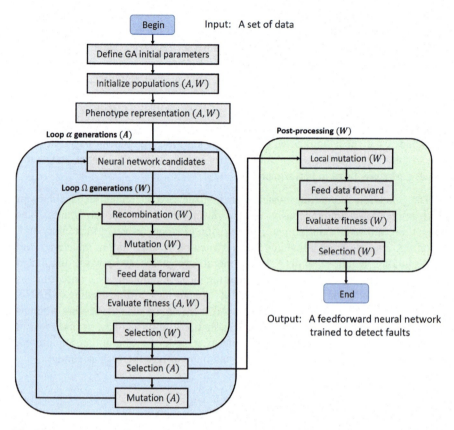

Fig. 2 General view of the genetic algorithm for developing the architecture and adapting the parameters of a feedforward neural network for fault detection

concerns searching for better solutions on the parameter space, close to the fittest vector of weights for a specific architecture (local search). In the figure, W and A are related to the weights and architectures populations, respectively. Learning is guided by an error-and-model-compactness-based fitness function.

3.1 Initialization and Parameterization

A schematic representation of the basic processing units of the evolutionary neural network is illustrated in Fig. 3. In the figure, x_j and w_j refer to the j-th input and connection weight, respectively; b is the bias; *net* is the weighted sum of the inputs, and $\varphi(.)$ is a sigmoidal logistic function, which gives the output y.

From prior knowledge about the neural network learning problem, we want the nonlinear functions $\varphi(.)$ of all neurons to be initially triggered within their

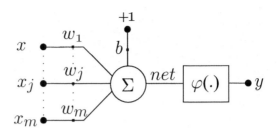

Fig. 3 Schematic representation of a neuron of the evolutionary neural network

unsaturated regions, i.e., $net \cong 0$ for sigmoid functions. Otherwise, the network would barely differentiate input data at the very beginning of the learning process, which could lead a rougher and slower adaptation. For example, if the error backpropagation algorithm is considered, a mechanism to accelerate learning convergence is to set small random initial weights. Similarly, in real genetic programming, the allele (range of possible values) of genes (elements) of chromosomes (candidate solutions) representing weight vectors can be initially adjusted to small random values around 0, e.g., $[-0.01, 0.01]$, yielding the same result.

Some remarks about the EANN learning algorithm include: (1) initial architectures and weights populations consist of 20 individuals each. We consider the Pittsburgh approach, i.e., all architecture and weight vectors compete with each other to be the fittest solution; (2) the algorithm is elitist. The fittest solution in previous generations is preserved for the next generations. The elitist approach ensures that the overall best individual remains in the population independently of the application of genetic operators; and (3) the number of individuals that compose the populations of architectures and weights is constant over generations. Although recombination operators make the populations double in size, a selection operator reduces the populations to the half.

Notice that GA global heuristic search tends to find a "good" solution by exploiting a highly dimensional parameter space whose error surface contains several hills, valleys, and plateaus. However, such solution may not be locally optimal. There are some hybrid techniques proposed in the literature for post-processing the solution through local search methods, e.g., [41, 42]. In this study, local search using a gene mutation operator is employed. Details are addressed in the subsequent sections.

3.2 Phenotype Representation

Mapping phenotypes into genotypes is crucial in the development of GA-based models, especially in constrained optimization problems. For instance, mutation and recombination operators may produce infeasible solutions. Care must be taken in both representation of individuals and definition of genetic operators. Undesirable effects such as requirement of extra manipulations of chromosomes, more complex objective functions, and premature convergence of the population may be immediate outcomes of an inadequate representation.

Encoding in GA is the form in which chromosomes and genes are expressed. There are basically two types of encoding, binary and real. The former was widely discussed, while the latter fits continuous optimization problems better and therefore is adopted in the present study. Several successful applications of real codification may be found in the literature, e.g., [17, 43].

Let P and G be phenotypic (behavioral) and genotypic (informational) spaces, respectively [44]. Phenotypes representing feedforward neural network architectures and weight vectors are encoded into genotypes by a direct mapping $M_{P \to G}$. Genotypes are assumed to be haploid chromosomes as shown in Fig. 4. Chromosomes associated to architectures are composed of a pair of genes, A and B, which refer to the number of neurons in the first and second hidden layers of EANN. In case gene A or B is zero, then the underlying EANN has a single hidden layer. The range of values, i.e., the allele, that each gene of the architecture chromosomes may assume is [0, 99], while the allele of weight chromosomes is [−1, 1].

The length of a weight chromosome:

$$L = IA + AB + BO, \tag{9}$$

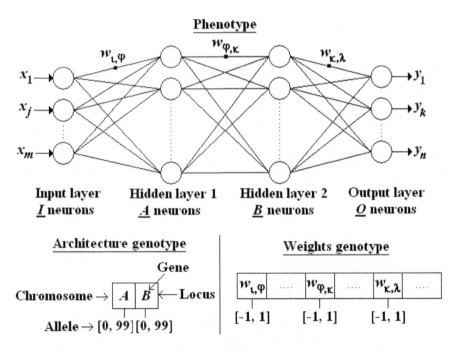

Fig. 4 Phenotype–genotype codification of architectures and weights

is variable. It depends on the values of the genes of the corresponding architecture chromosome, A and B, and on the number of inputs, I, and outputs, O, of the underlying neural network.

3.3 Recombination Operator

Defining the most appropriate recombination operator for different sorts of applications is a hard problem and an open issue. We examine common recombination operators in anomaly detection scenarios, viz., Arithmetic, Multipoint, and Local Intermediate Crossover. These operators can be either sexual or global. In the sexual form, only parents are involved on the generation of offspring. Conversely, in the global form, the whole population may contribute to generate offspring. In this study, we opt for the sexual form of recombination. In other words, 20 parents generate 20 children in each iteration of the algorithm.

3.3.1 Arithmetic Crossover

This operator is particularly suited for constrained numerical optimization problems with convex feasible region Ξ_C. Let $c_n, n = 1, \ldots, N$, represent the n-th individual of a population. As a consequence of two individuals, c_{n1} and c_{n2}, belong to Ξ_C, convex combinations of c_{n1} and c_{n2} also belong to Ξ_C. This ensures that Arithmetic Crossover produces only valid offspring.

Formally, two parent chromosomes are linearly combined to produce two offspring according to:

$$Son_1 = a\ Parent_1 + (1-a)\ Parent_2 \tag{10}$$

$$Son_2 = (1-a)\ Parent_1 + a\ Parent_2 \tag{11}$$

where a is a random value chosen before each crossover operation. As an example, consider a random list of parents, see Fig. 5. Each operation produces two children whose genes inherit a combination of the parent genes.

3.3.2 Multipoint Crossover

In multipoint crossover, children inherit sets of successive genes from two randomly selected parents. p randomly selected points along the chromosomes of the parents divide them into $p+1$ parts. Then, genes of each father are exchanged to generate offspring. An intuitive example of a p-point crossover operator is shown in Fig. 6. Similar to other crossover operators, the population doubles in size after multipoint recombination. The recombination potential, exploratory power, and learning progress of a GA using p-point crossover are discussed in [45].

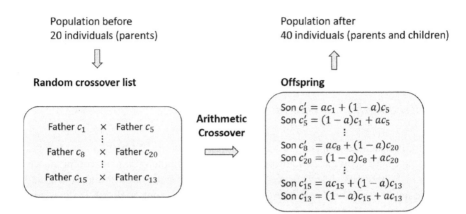

Fig. 5 Arithmetic crossover operator

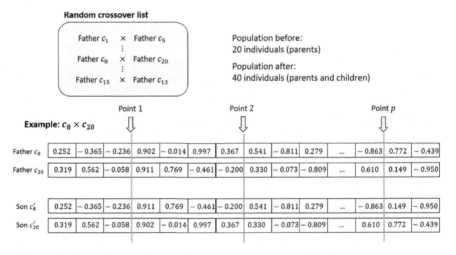

Fig. 6 Multipoint crossover operator

3.3.3 Local Intermediate Crossover

Local intermediate crossover is particularly useful when convergence to a unique solution is expected. In this operator, the average values of the genes of randomly selected parents are inherited by the single offspring, that is:

$$c'_{n1} = \frac{c_{n1} + c_{n2}}{2}. \tag{12}$$

Local intermediate crossover implies that sons receive inheritance of both parents equally. Clearly, if intermediate recombination is applied often, then the chromosomes become similar. This may lead to premature convergence of the population, especially when no other operator, such as mutation, is used to keep population diversity. Two crossover lists with 10 matches each are considered in this study to generate 20 children for the next generation.

3.4 Mutation Operator

Mutation operations involve single individuals, in contrast to recombination. This type of operator assures that there is always a probability of reaching any point within the search space. Usually, when the current solution of the problem is far from being the best according to a fitness function, a higher mutation rate can be employed as an attempt to find better solutions farther from the current ones. On the other hand, when the current solution is close from being the best, a low mutation rate can be adopted. This approach leads to search for solutions in promising regions.

Mutation commonly does not produce offspring. The mutated individuals remain in the population for later breeding. An individual c_n has its corresponding mutated value c'_n from $c'_n = m(c_n)$, where $m(.)$ is a mutation function. Gaussian and random mutation operators are considered for analysis.

3.4.1 Gaussian Mutation

Gaussian mutation is frequently applied in real-coded GA. This is mainly because it supports fine-tuning of solutions. Chromosomes c_n of an individual have their corresponding mutated values c'_n from:

$$c'_n = c_n \pm M, \qquad (13)$$

where M is a normal density function $N(mean, \Gamma)$ with $mean = 0$ and standard deviation, Γ.

Gaussian mutation is applied gene-to-gene, with a gene mutation probability rate between 1% and 10%, directly proportional to the fitness of the chromosome.

3.4.2 Random Mutation

Random mutation is a member of the class of random search optimization methods. Features that make this operator useful are the enhancement of processing speed and nonsusceptibility to local minima. Randomness is generally controlled to ensure convergence while allowing enough freedom for a complete coverage of the search space.

Random mutation creates a random solution c'_n at the vicinity of the current solution c_n using a uniform probability distribution such that all genes of the newest individual c'_n are within $[-1, 1]$ and $[0, 99]$ for weight and architecture chromosomes, respectively. The mutated genes should remain feasible with respect to these bounds. The free change of mutated genes may give rise to better solutions. Better solutions are maintained, while worse ones are rejected. Similar to the Gaussian mutation, random mutation is applied gene-to-gene with a changing probability rate from 1% to 10%. Mutated values are given as:

$$c'_n = c_n + r, \quad (14)$$

where r is a random value in $[-0.1, 0.1]$ for weight individuals; and a random value in $[-5, 5]$ rounded to an integer for the case of architecture individuals.

3.4.3 Post-Processing Based on Local Random Mutation

Post-processing neural network parameters can be done from different local search methods. Local mutation is a simple and fast approach to try to improve the solution found so far. With the EANN architecture defined, the basic idea is to change some genes of the current weight solution using random mutation in a specific way.

Mutation probability rate is restricted to 10% per gene, and $r \in [-0.1, 0.1]$, as in (14). Whenever a gene is changed, the fitness of the new solution is immediately calculated, and the new value of the gene is either accepted or ignored. New weight solutions are evaluated twice, considering $c_n + r$ and $c_n - r$. This approach promotes a local search for a better solution around the current best solution and parallel to the axes of the search space.

3.5 Fitness Function

GA mimics the principle of natural selection. A fitness measure is used to choose relatively fitter individuals in a population to evolve. The higher the fitness of an individual, the higher its survival probability [44].

To determine the fitness of weight chromosomes, we use

$$F(c_n) = \gamma (\tau_{train}^{\xi} + \tau_{test}^{\zeta}), \quad (15)$$

where $F(c_n)$ is the fitness of the chromosome c_n; ξ and ζ are parameters defined according to the emphasis on training and testing performance; τ_{train} and τ_{test} refer to the neural network learning and generalization ability:

$$\tau_{train} = \frac{C_{train}}{C_{train} + W_{train}}, \quad (16)$$

$$\tau_{test} = \frac{C_{test}}{C_{test} + W_{test}}, \quad (17)$$

where C and W are the amount of correct and wrong classifications. In addition:

$$\gamma = e^{-kL} \tag{18}$$

is a penalty factor for large network architectures; L, calculated as in Eq. (9), is the length of a weight chromosome; and $0 < k < 1$ is a constant.

The fitness of an architecture chromosome is the greatest fitness of weight chromosomes of the current generation. Naturally, F should be maximized.

3.6 Selection Operator

Selection is the operation in which individuals are chosen for later breeding. First, individuals are chosen to enter a mating pool. Operators should ensure that individuals with higher fitness have greater probability of being selected for mating, but those individuals with lower fitness still have a probability of being chosen. Having some probability of choosing worse individuals is important to assure that the search process is global and it does not simply converge to the nearest local minimum.

The original GA uses selection proportional to the fitness usually implemented with Roulette Wheel [44]. To better control the selective pressure of individuals and to avoid premature convergence, Tournament selection is considered [46].

3.6.1 Tournament Selection

Tournament selection is an alternative to fitness-proportional selection. Empirical results suggest that the tournament method can perform better and be faster than roulette selection [44, 46]. Moreover, it attenuates the selection pressure.

The operator considers the number of wins of an individual in H matches against H random opponents of the population—a one-against-one approach. The winner of a match is the individual with the best fitness compared to the direct opponent.

We use $H = 5$ in such a way that individuals winning at least three matches remain for the next generation. The procedure continues until the mating pool is full, i.e., 20 out of 40 individuals are selected. The selective pressure provided by the tournament operator is weak since a good diversity of individuals may remain in the population. Parents and children may compose the next generation.

3.6.2 Elitism

Elitism consists in maintaining the fittest individual of the population. This strategy ensures that the best solution found so far is retained. While this strategy could be applied more broadly, e.g., selecting the 2 or 3 best solutions, overuse can lead to premature convergence to a suboptimal solution. Tournament selection, as described in the previous section, is inherently an elitist approach.

3.7 Stopping Criteria

Training stopping criteria is an important issue in evolutionary modeling. Early termination may generate poor solutions, whereas late termination might cause overfitting. The proposed GA is terminated if one of the following is reached:

- Maximum number of architecture generations, α;
- Maximum number of weight generations, Ω;
- Acceptable fitness reached, $F(c_n^*) > \Xi$;
- Maximum number of weight generations without replacing the fittest c_n, δ.

In the latter case, the solution has attained a plateau such that iterations have no longer produced better results.

4 Incremental Algorithm for Neurofuzzy Network Learning

EGNN is a neurofuzzy granular network constructed incrementally from an online data stream [24, 36]. Although its learning algorithm can process mixtures of fuzzy, interval, and numerical data, this study focuses on numerical data only. Additionally, the network can play the role of a regressor, predictor, controller, or classifier [47, 48]. We emphasize EGNN for fault detection and classification. Particularly, evolving classification is a research topic under broad discussion. A number of methods have been developed with focus on typicality and eccentricity data analytics [49], robustness of Takagi–Sugeno fuzzy models [50], local strategies to smooth parameter changes [51], self-organization of fuzzy models [52], ensembles of models [53], scaffolding fuzzy type-2 models [54], semi-supervision [55], interval granular computing models [56], and on applications such as fault detection in wind turbines [57] and monitoring of waste-water treatment processes [58].

The basic processing elements of EGNN are fuzzy neurons. Its architecture encodes a set of fuzzy rules, and neural processing conforms with a fuzzy inference system. The network architecture results from a gradual construction according to new information. The consequent part of an EGNN rule may be composed of a linguistic and a functional term. Independently of the choice of fuzzy neuron, network parameters, and properties of input–output data, the linguistic term of the rule consequent produces a granular output, while the functional term gives a pointwise output. In the present study, we are interested in the pointwise output only, which is a class. Learning in EGNN means to fit new data into local granular models recursively. Granules, neurons, and connections can be added, adapted, removed, and combined. Therefore, the network captures new information from data streams and adapts itself to a new scenario.

4.1 Numerical and Fuzzy Data

Fuzzy data arise from expert knowledge, inaccurate measurements, variables that are hard to be precisely quantified, perceptions, and when preprocessing steps introduce uncertainty in numerical data. A fuzzy datum x_j has the following form:

$$x_j(z) = \begin{cases} \phi_j, & z \in [\underline{\underline{x}}_j, \underline{x}_j[\\ 1, & z \in [\underline{x}_j, \overline{x}_j] \\ \iota_j, & z \in]\overline{x}_j, \overline{\overline{x}}_j] \\ 0, & \text{otherwise} \end{cases} \qquad (19)$$

where z is a real number in X_j. If the fuzzy datum x_j is normal ($x_j(z) = 1$ for at least one $z \in \Re$) and convex ($x_j(\kappa z^1 + (1-\kappa)z^2) \geq min(x_j(z^1), x_j(z^2))$, $z^1, z^2 \in \Re$, $\kappa \in [0, 1]$), then it is a fuzzy interval [59]. In particular, if:

$$\phi_j = \frac{z - \underline{\underline{x}}_j}{\underline{x}_j - \underline{\underline{x}}_j} \text{ and} \qquad (20)$$

$$\iota_j = \frac{\overline{\overline{x}}_j - z}{\overline{\overline{x}}_j - \overline{x}_j}, \qquad (21)$$

then the fuzzy datum (19) has trapezoidal membership function and can be represented by the quadruple $(\underline{\underline{x}}_j, \underline{x}_j, \overline{x}_j, \overline{\overline{x}}_j)$. If $\underline{x}_j = \overline{x}_j$, the fuzzy datum is a fuzzy number. Numerical data arise if $\underline{\underline{x}}_j = \underline{x}_j = \overline{x}_j = \overline{\overline{x}}_j$.

4.2 Network Architecture

Let $x = (x_1, \ldots, x_n)$ be an input vector and y the output. Consider that the data stream $(x, y)^{[h]}$, $h = 1, \ldots$, is measured from an unknown function f. Inputs x_j and output y can be fuzzy data in general and numerical data in particular.

Figure 7 shows a four-layer EGNN model. The input layer receives $x^{[h]}$. The granular layer consists of a set of granules G_j^i, $j = 1, \ldots, n; i = 1, \ldots, c$, stratified from the input data. Fuzzy sets G_j^i, $i = 1, \ldots, c$, form a fuzzy partition of the

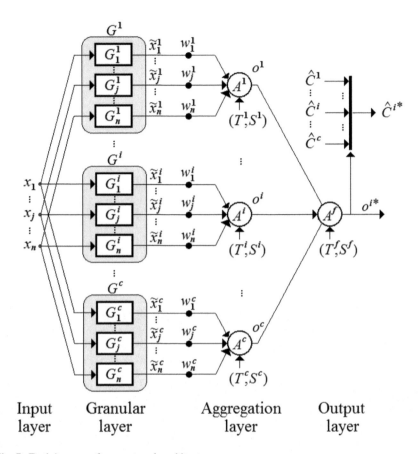

Fig. 7 Evolving neurofuzzy network architecture

j-th input domain, X_j. A granule $G^i = G_1^i \times \cdots \times G_n^i$ is a fuzzy relation, i.e., a multidimensional fuzzy set in $X_1 \times \cdots \times X_n$. Thus, G^i has membership function $G^i(x) = min\{G_1^i(x_1), \cdots, G_n^i(x_n)\}$ in $X_1 \times \cdots \times X_n$. Granule G^i may have a companion local function p^i. For classification, we use a 0-th-order function:

$$p^i(\hat{x}_1, \ldots, \hat{x}_n) = \hat{C}^i, \tag{22}$$

where $\hat{C}^i$ is the estimated class.

Define $\hat{x}_j$ as the midpoint of $x_j = (\underline{\underline{x}}_j, \underline{x}_j, \overline{x}_j, \overline{\overline{x}}_j)$. Thus:

$$\mathrm{mp}(x_j) = \hat{x}_j = \frac{\underline{x}_j + \overline{x}_j}{2}. \tag{23}$$

Naturally, if the input data are numerical, then $\hat{x}_j = x_j$.

Similarity degrees $\tilde{x}^i = (\tilde{x}_1^i, \ldots, \tilde{x}_n^i)$ are the result of matching between input $x = (x_1, \ldots, x_n)$ and fuzzy sets $G^i = (G_1^i, \ldots, G_n^i)$. In general, data and granules are trapezoidal objects. A convenient similarity measure to quantify the match between a sample and the current knowledge is

$$\tilde{x}_j^i = 1 - \frac{|\underline{g}_j^i - \underline{x}_j| + |\underline{\underline{g}}_j^i - \underline{\underline{x}}_j| + |\overline{g}_j^i - \overline{x}_j| + |\overline{\overline{g}}_j^i - \overline{\overline{x}}_j|}{4(\max(\overline{\overline{g}}_j^i, \overline{\overline{x}}_j) - \min(\underline{g}_j^i, \underline{x}_j))}. \tag{24}$$

This measure returns $\tilde{x}_j^i = 1$ for identical trapezoids and reduces linearly as any numerator term increases. Naturally, measure (24) can be applied to numerical data. In this case, $\underline{x}_j = \underline{\underline{x}}_j = \overline{x}_j = \overline{\overline{x}}_j$ [24, 48].

The aggregation layer is composed of fuzzy neurons A^i, $i = 1, \ldots, c$. A fuzzy neuron A^i combines weighted similarity degrees $(\tilde{x}_1^i w_1^i, \ldots, \tilde{x}_n^i w_n^i)$ into a single value o^i, which refers to the level of activation of rule R^i. The output layer processes $(o^1, \ldots, o^c)$ using a fuzzy neuron A^f. A^f performs the maximum S-norm in this study. The class C^{i*} associated to the most active rule R^{i*} is the network output.

Under assumption on specific weights and types of neurons, fuzzy rules extracted from the EGNN classifier, as described in this study, are of the type:

$$R^i : \text{IF}(x_1 \text{is} G_1^i) \text{AND} \ldots \text{AND}(x_n \text{is} G_n^i) \text{THEN}(\hat{y} \text{is} \hat{C}^i).$$

As: (1) fuzzy sets G_j^i, $\forall i, j$, are time varying; (2) a diversity of aggregation functions can be used in the neural body A^i; and (3) fuzzy granules overlap in the input space, thus the class separation surface provided by an EGNN model is nonstationary and can be highly nonlinear.

4.3 Fuzzy Neuron

Fuzzy neurons are neuron models based on aggregation operators. EGNN may use different types of aggregation neurons to perform information fusion. Generally, there is no guideline to choose a particular aggregation operator to construct a fuzzy neuron [60].

Aggregation operators $A : [0, 1]^n \rightarrow [0, 1]$, $n > 1$, combine input values in the unit hypercube $[0, 1]^n$ into a single value in $[0, 1]$. They must satisfy the following properties: (1) monotonicity in all arguments, i.e., given $x^1 = (x_1^1, \ldots, x_n^1)$ and $x^2 = (x_1^2, \ldots, x_n^2)$, if $x_j^1 \leq x_j^2$ $\forall j$ then $A(x^1) \leq A(x^2)$; and (2) boundary conditions: $A(0, 0, \ldots, 0) = 0$ and $A(1, 1, \ldots, 1) = 1$. The classes of aggregation operators considered in this study are summarized below. See [59, 60] for a detailed coverage.

4.3.1 Triangular Norm and Conorm

T-norms (T) are commutative, associative, and monotone operators on the unit hypercube whose boundary conditions are $T(\alpha, \alpha, \ldots, 0) = 0$ and $T(\alpha, 1, \ldots, 1) = \alpha$, $\alpha \in [0, 1]$. An example of T-norm is the minimum operator:

$$T_{min}(x) = \min_{j=1,\ldots,n} x_j, \tag{25}$$

which is the strongest T-norm because:

$$T(x) \leq T_{min}(x) \text{ for any } x \in [0, 1]^n. \tag{26}$$

The minimum is idempotent, symmetric, and Lipschitz-continuous. Further examples of T-norms include the product:

$$T_{prod}(x) = \prod_{j=1}^{n} x_j, \tag{27}$$

and the Lukasiewicz T-norm:

$$T_L(x) = \max(0, \sum_{j=1}^{n} x_j - (n-1)), \tag{28}$$

which are non-idempotent, but Lipschitz-continuous aggregation operators.

S-norms (S) are operators on the unit hypercube which are commutative, associative, and monotone. $S(\alpha, \alpha, \ldots, 1) = 1$ and $S(\alpha, 0, \ldots, 0) = \alpha$ are the boundary conditions of S-norms.

S-norms are stronger than T-norms. The maximum operator:

$$S_{max}(x) = \max_{j=1,\ldots,n} x_j, \tag{29}$$

is the weakest S-norm, that is:

$$S(x) \geq S_{max}(x) \geq T(x), \text{ for any } x \in [0, 1]^n. \tag{30}$$

The maximum is idempotent, symmetric, and Lipschitz-continuous.

4.3.2 Neuron Model

Let $\tilde{x} = (\tilde{x}_1, \ldots, \tilde{x}_n)$ be a vector of membership degrees of a sample $x = (x_1, \ldots, x_n)$ in the fuzzy sets $G = (G_1, \ldots, G_n)$. Let $w = (w_1, \ldots, w_n)$ be a weight vector such that:

$$w_j \in [0, 1], \; j = 1, \ldots, n. \tag{31}$$

Fig. 8 Fuzzy neuron model

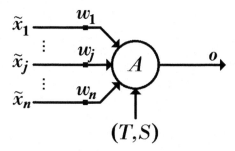

Product T-norm is used to perform synaptic processing, while an aggregation operator A is used to fuse the individual results of synaptic processing. The output of a fuzzy aggregation neuron is

$$o = A(\tilde{x}_1 w_1, \ldots, \tilde{x}_n w_n). \tag{32}$$

A fuzzy neuron produces a diversity of nonlinear mappings between neuron inputs and output depending on the choice of weights w and aggregation operator A. The fuzzy neuron model is shown in Fig. 8.

4.4 Granular Region

The support and the core of trapezoidal membership function G^i_j are

$$\mathrm{supp}(G^i_j) = [\underline{g}^i_j, \overline{g}^i_j], \tag{33}$$

$$\mathrm{core}(G^i_j) = [\underline{\underline{g}}^i_j, \overline{\overline{g}}^i_j]. \tag{34}$$

The midpoint and width of G^i_j are given by:

$$\mathrm{mp}(G^i_j) = \frac{\underline{g}^i_j + \overline{g}^i_j}{2}, \tag{35}$$

$$\mathrm{wdt}(G^i_j) = \overline{g}^i_j - \underline{g}^i_j. \tag{36}$$

The maximal allowed expandable width of fuzzy sets G^i_j is denoted by ρ, i.e., $\mathrm{wdt}(G^i_j) \leq \rho$, $j = 1, \ldots, n$; $i = 1, \ldots, c$. Let the expansion region of a fuzzy set G^i_j be

$$E^i_j = \left[\mathrm{mp}(G^i_j) - \frac{\rho}{2}, \mathrm{mp}(G^i_j) + \frac{\rho}{2} \right]. \tag{37}$$

It follows that $\mathrm{wdt}(G^i_j) \leq \mathrm{wdt}(E^i_j)$ $\forall j, i$. Values of ρ allow different representations of the same problem at different levels of detail.

4.5 Granularity Adaptation

A balance between parametric and structural adaptation is a key to capture changes of time-varying systems. The procedure described below reconciles parametric and structural changes in EGNN.

The value of ρ affects the granularity and accuracy of models. In practice, $\rho \in [0, 1]$ settles the size of expansion regions (37) and the need to either create or adapt rules to fit a new sample. EGNN starts learning with an empty rule base and with no a priori knowledge of data properties. In this case, it is worth to initialize ρ at an intermediate value, e.g., $\rho^{[0]} = 0.5$.

Let r be the number of rules created in h_r steps. If the number of rules grows faster than a rate η, i.e., $r > \eta$, then ρ is increased:

$$\rho(\text{new}) = \left(1 + \frac{r}{h_r}\right)\rho(\text{old}). \tag{38}$$

Equation (38) acts against outbursts of growth since large rule bases increase model complexity and worsen generalization. If the number of rules grows at a rate smaller than η, i.e., $r \leq \eta$, then ρ is decreased:

$$\rho(\text{new}) = \left(1 - \frac{(\eta - r)}{h_r}\right)\rho(\text{old}). \tag{39}$$

If $\rho = 1$, then EGNN is structurally stable, but unable to capture abrupt changes. Conversely, if $\rho = 0$, then EGNN overfits the data causing excessive model complexity. Adaptability is reached from intermediate values.

Reducing ρ may require a reduction of larger granules to fit them to the new requirement. In this case, the support of G^i_j is narrowed as follows:

If $\text{mp}(G^i_j) - \frac{\rho(\text{new})}{2} > \underline{g}^i_j$ then $\underline{g}^i_j(\text{new}) = \text{mp}(G^i_j) - \frac{\rho(\text{new})}{2}$
If $\text{mp}(G^i_j) + \frac{\rho(\text{new})}{2} < \overline{g}^i_j$ then $\overline{g}^i_j(\text{new}) = \text{mp}(G^i_j) + \frac{\rho(\text{new})}{2}$

Cores $[\underline{g}^i_j, \overline{g}^i_j]$ are handled similarly. Time-varying granularity is useful to avoid guesses on how fast and how often the data stream properties change.

4.6 Developing Granules

Granules are created if the support of at least one entry of $(x_1, \ldots, x_n)$ is not enclosed by expansion regions $(E^i_1, \ldots, E^i_n)$, $i = 1, \ldots, c$. This is the case that granules G^i cannot expand beyond the limit ρ to fit the sample. Otherwise, if $x^{[h]}$ is placed inside an E^i, but the class $C^{[h]} \neq \hat{C}^i$, then a new granule G^{c+1} should be created.

A new granule G^{c+1} is formed by fuzzy sets G_j^{c+1} whose parameters match the sample:

$$G_j^{c+1} = (\underline{\underline{g}}_j^{c+1}, \underline{g}_j^{c+1}, \overline{g}_j^{c+1}, \overline{\overline{g}}_j^{c+1}) = (\underline{\underline{x}}_j, \underline{x}_j, \overline{x}_j, \overline{\overline{x}}_j). \tag{40}$$

The consequent p^{c+1} is associated to a class, $\hat{C}^{c+1} = C^{[h]}$.

Adaptation of granules consists in expanding or contracting the support and the core of fuzzy sets G_j^i. Granule G^i is adapted if a sample falls within its expansion region, i.e., if $\text{supp}(x_j) \subset E_j^i$, $j = 1, \ldots, n$, and $C^{[h]}$ is the same as $\hat{C}^i$. In situations in which more than one granule encloses the sample, adapting only one of them is enough to guarantee data inclusion. In particular, we may choose G^i such that $i = arg\ max(o^1, \ldots, o^c)$, i.e., G^i has the highest activation level.

Adaptation proceeds depending on where the input datum x_j is placed in relation to the fuzzy set G_j^i:

If $\underline{\underline{x}}_j \in [\text{mp}(G_j^i) - \frac{\rho}{2}, \underline{g}_j^i]$ then $\underline{\underline{g}}_j^i(\text{new}) = \underline{\underline{x}}_j$

If $\underline{x}_j \in [\text{mp}(G_j^i) - \frac{\rho}{2}, \underline{g}_j^i]$ then $\underline{g}_j^i(\text{new}) = \underline{x}_j$

If $\underline{x}_j \in [\underline{g}_j^i, \text{mp}(G_j^i)]$ then $\underline{g}_j^i(\text{new}) = \underline{x}_j$

If $\underline{x}_j \in [\text{mp}(G_j^i), \text{mp}(G_j^i) + \frac{\rho}{2}]$ then $\underline{g}_j^i(\text{new}) = \text{mp}(G_j^i)$

If $\overline{x}_j \in [\text{mp}(G_j^i) - \frac{\rho}{2}, \text{mp}(G_j^i)]$ then $\overline{g}_j^i(\text{new}) = \text{mp}(G_j^i)$

If $\overline{x}_j \in [\text{mp}(G_j^i), \overline{g}_j^i]$ then $\overline{g}_j^i(\text{new}) = \overline{x}_j$

If $\overline{x}_j \in [\overline{g}_j^i, \text{mp}(G_j^i) + \frac{\rho}{2}]$ then $\overline{g}_j^i(\text{new}) = \overline{x}_j$

If $\overline{\overline{x}}_j \in [\overline{g}_j^i, \text{mp}(G_j^i) + \frac{\rho}{2}]$ then $\overline{\overline{g}}_j^i(\text{new}) = \overline{\overline{x}}_j$

The first and last rules perform support expansion, and the second and seventh rules execute core expansion. The remaining cases concern core contraction.

Operations on core parameters, $\underline{g}_j^i$ and $\overline{g}_j^i$, require adjustment of the midpoint of the respective fuzzy sets:

$$\text{mp}(G_j^i)(\text{new}) = \frac{\underline{g}_j^i(\text{new}) + \overline{g}_j^i(\text{new})}{2}. \tag{41}$$

As a result, support contraction may happen in two occasions:

If $\text{mp}(G_j^i)(\text{new}) - \frac{\rho}{2} > \underline{\underline{g}}_j^i$ then $\underline{\underline{g}}_j^i(\text{new}) = \text{mp}(G_j^i)(\text{new}) - \frac{\rho}{2}$

If $\text{mp}(G_j^i)(\text{new}) + \frac{\rho}{2} < \overline{\overline{g}}_j^i$ then $\overline{\overline{g}}_j^i(\text{new}) = \text{mp}(G_j^i)(\text{new}) + \frac{\rho}{2}$.

4.7 Adapting Connection Weights

Weights $w^i_j \in [0, 1]$ represent the importance of the j-th attribute of G^i_j to the neural network output. If $w^i_j = 1$, then the output is not affected. A relatively lower value of w^i_j discounts the impact of the respective attribute. The procedure described below assigns lower weight values to less helpful attributes.

If a granule G^{c+1} is created, weights are set as $w^{c+1}_j = 1, \forall j$. If it is known a priori that different attributes have different importance, then values for w^{c+1}_j can be chosen in a way to reflect that.

Weights w^i_j, $j = 1, \ldots, n$, corresponding to the most active granule G^i, $i = arg\, max(o^1, \ldots, o^c)$, are updated from:

$$w^i_j(\text{new}) = w^i_j(\text{old}) - \beta^i \tilde{x}^i_j |\epsilon|. \tag{42}$$

where $\tilde{x}^i_j$ is the similarity between x^i_j and G^i_j; β^i depends on the number of right (R^i) and wrong (W^i) classifications so far provided by G^i according to:

$$\beta^i = \frac{W^i}{R^i + W^i}; \tag{43}$$

and

$$\epsilon^{[h]} = C^{[h]} - \hat{C}^{[h]} \tag{44}$$

is the current estimation error. Equation (42) penalizes the j-th attribute of G^i in the next iterations if the estimated class is wrong.

4.8 Learning Algorithm

The learning algorithm to evolve EGNN is given below:

BEGIN
Select a type of neuron for the aggregation layer;
Set parameters $\rho^{[0]}$, h_r, η, $c = 0$;
Read input sample $x^{[h]}$, $h = 1$;
Create granule G^{c+1}, neurons A^{c+1}, A^f, and respective connections;
For $h = 2, \ldots$ do
  Read and feedforward $x^{[h]}$ through the network;
  Compute rule activation levels $(o^1, \ldots, o^c)$;
  Aggregate activation values using A^f to get an estimation $\hat{C}^{[h]}$;

// The class $C^{[h]}$ becomes available;
Compute output error $\epsilon^{[h]} = C^{[h]} - \hat{C}^{[h]}$;
If $x^{[h]}$ is not within expansion regions $E^i \forall i$ or $\epsilon^{[h]} \neq 0$
  Create granule G^{c+1}, neuron A^{c+1}, and connections;
Else
  Adapt the most active granule G^{i*}, $i* = arg\ max(o^1, \ldots, o^c)$;
  Adapt weights $w_j^{i*}\ \forall j$;
If $h = \alpha h_r, \alpha = 1, 2, \ldots$
  Adapt model granularity ρ;
END

5 Results and Discussion

Experimental results on electrical machine fault detection and classification using neural networks trained via genetic (EANN) and incremental (EGNN) algorithms are shown in this section. First, individual results for each neural classifier and discussions considering different initial parameters and operators are presented. Then, general comparisons and statistical analyses are performed. We look forward to concise models and high accuracy and processing speed.

5.1 Preliminaries

The dataset was generated from the validated mathematical model described in Sect. 2. An induction motor properly designed for insertion of stator shorted-turns and the experimental setup for model validation are shown in Fig. 9. Stator phase windings were fractionated such that short-circuits on a number of turns could be imposed externally (see white wires on the top of the motor).

The characteristics of the underlying induction motor are: power, 5 Hp; voltage (Y), 127 V; poles, 4; stator turns per phase, 84; inertia, 0.00995 J m^2; rated torque, 2.1 kgf m; rated speed, 1715 RPM; stator resistance, 0.730 Ω; rotor resistance, 0.360 Ω; stator and rotor leakage inductance, 0.006 H; and mutual inductance 0.027 H. Table 1 shows the conditions of shorted-turns in the stator windings and a summary of the 10-class balanced classification problem. The dataset contains 350 7-attribute samples. The attributes are the abc stator currents, voltage–current displacement angles, and the rotor speed. Voltage unbalance in the 3-phase system (127 $\pm$ 10 V), current measurement noise ($\pm$0.1 A), and variable load ([0, 6] Nm) were considered to generate 35 samples that represent each of the 10 classes.

Fig. 9 Instrumental setup and motor external connections

Table 1 Classes of the 10-class balanced classification problem

Class	k_a (turns)	k_b (turns)	k_c (turns)	No. of samples
1	0	0	0	35
2	1	0	0	35
3	0	1	0	35
4	0	0	1	35
5	2	0	0	35
6	0	2	0	35
7	0	0	2	35
8	3	0	0	35
9	0	3	0	35
10	0	0	3	35

Offline learning methods, such as EANN, use 200 random samples for training and 150 samples for testing. Data stream learning methods, such as EGNN, employ a sample-per-sample testing-before-training approach.

5.2 Genetic EANN for Fault Detection

Genetic mutation and recombination operators are compared in this section within the framework of electrical machine fault detection. Additionally, the overall performance of the detection system using the GA-based neural network is given.

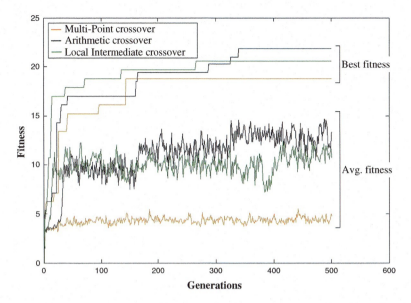

Fig. 10 Comparison between crossover operators

First, the effect of applying arithmetic, multipoint, and local intermediate recombination operators is evaluated assuming the other GA operators fixed. Figure 10 illustrates the evolution of the average and best fitness of the population over the generations using the different recombination operators. For 500 generations of weight individuals, arithmetic crossover provided the overall fittest individual and the highest average fitness.

A second experiment concerns the evaluation of Gaussian and random mutation operators with all other GA operators fixed. Figure 11a shows the detection system performance under different Gaussian mutation rates. The fittest individual was reached under an 8% mutation rate, while the best average fitness was obtained under a 5% rate. Figure 11b shows the result for random mutation under distinct mutation rates. Random mutation under a 5% rate generated the fittest individual and the highest average fitness.

Genetic operators that generated the fittest individual, i.e., arithmetic crossover and random mutation under a 5% rate, were chosen for subsequent experiments. In addition, $k = 0.9$ in Eq. (18). We evaluated neural network architectures considering recombination, mutation, tournament selection with elitism, and postprocessing based on local random mutation. Figure 12a presents the development of various EANN architectures over the generations. In a small amount of architecture generations—30 generations, an 88.77% accuracy on fault detections was reached by the architecture [7; 21; 5; 1]. This notation indicates the number of neurons in the input, first and second hidden, and output layers, respectively.

In a further experiment, the fault detection system using the trained EANN with 21 and 5 neurons in the hidden layers was subject to different sets of test data.

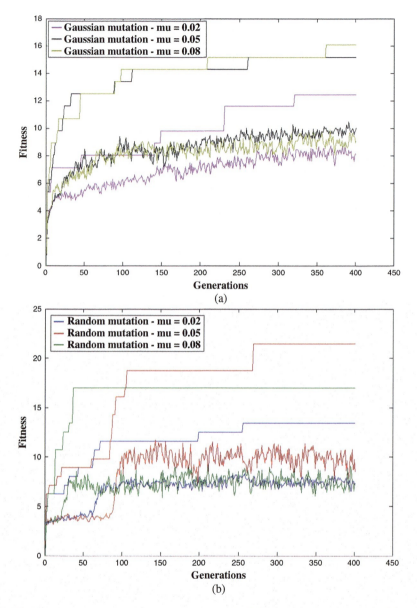

Fig. 11 Comparison between mutation operators under different probability rates. (**a**) Gaussian mutation. (**b**) Random mutation

Datasets were built considering different maximum levels of voltage unbalance and measurement noise (zero-mean white noise) on current waveforms. Each dataset contains 150 samples, which are used to test the EANN—being 15 samples of each class. Figure 12b shows the performance of EANN and that of a Multilayer

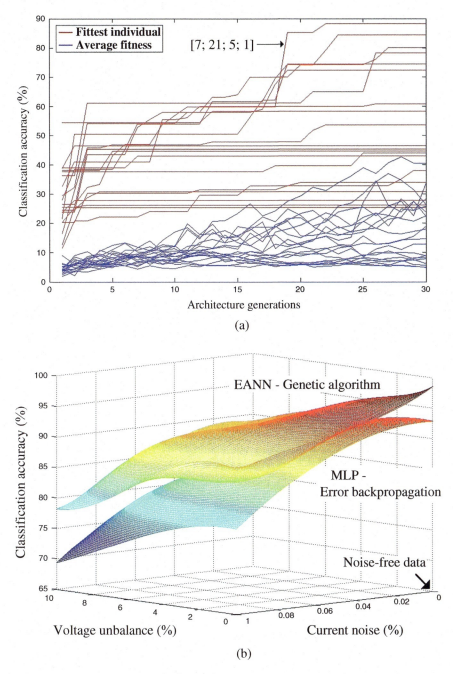

Fig. 12 General results for inter-turns fault detection using EANN. (**a**) Evolution of the fitness/accuracy of EANN architectures over time. (**b**) Performance comparison between a deterministic error-backpropagation-based MLP neural network and EANN, which carries out a global search for parameters prior to local search using genetic operators

Perceptron MLP neural network for detecting stator inter-turns short-circuit. The MLP neural network has similar architecture as that of EANN, but was trained via backward propagation of errors—a gradient descent optimization method. Conversely, GA carried out a global search for parameters prior to the typical local search, which supports deterministic methods such as the backpropagation algorithm.

Notice from Fig. 12b that EANN outperformed MLP in all situations. For example, in a less noisy environment with balanced voltages, close to the right upper corner of the graph, the detection system using EANN achieved 94.67% of correct classifications against 91.33% of the MLP neural network. The total training time for ten thousand epochs of the MLP backpropagation algorithm was about 19 min, while the time to evolve 30 architecture generations with 20 individuals each, 50 weight generations with 20 individuals each, and post-processing the fittest solution was about 55 min. In general, EANN provided greater robustness to voltage unbalance, variable loads, and measurement noise as shown by the 16% gap between the accuracy surfaces in the left lower corner of Fig. 12b.

5.3 Incremental EGNN for Fault Detection

A neurofuzzy EGNN classifier was evolved based on a data stream from the induction machine. The network uses seven input attributes, viz., abc stator currents, voltage–current displacement angles, and rotor speed. 350 samples, one at a time, became available for testing and training. No data is available before EGNN learning starts, and no data is stored during the learning process.

The initial parameters for the learning algorithm are $\rho^{[0]} = 0.7$, $h_r = 30$, and $\eta = 1$. First, different types of aggregation neurons A^i $\forall i$, viz., minimum, product, and Lukasiewicz were evaluated. The output neuron A^f performs S_{max} aggregation. A summary of the results from 10 runs obtained using the different aggregation neurons A^i is shown in Table 2. The average number of rules during the iterations and the total processing time are also shown.

Notice that the EGNN construction that uses product T_{prod} and maximum S_{max} fuzzy neurons in the aggregation and output layers, respectively, performed better than the other configurations using approximately from 12 to 13 granules. EGNN learning algorithm alone, disregarding any other acquisition and data processing procedure, can handle 2906 samples per second in the worst case.

Table 2 Evaluation of different types of EGNN fuzzy aggregation neurons

Aggregation	Avg no. of rules	Avg $Acc(\%)$	Best $Acc(\%)$	Avg time (ms)
T_{min}	12.8 ± 2.4	91.51 ± 1.76	94.57	109.2
T_{prod}	12.3 ± 2.4	93.62 ± 1.36	96.28	97.1
T_L	12.2 ± 2.1	91.22 ± 0.92	92.57	104.4

The need to calculate effective values of the voltage and current, and phase angle based on a 60-Hz power system imposes a limit on the provision of input samples to the network. If the effective values are calculated using half-wave cycle, then a maximum of 120 input data samples per second are available for EGNN processing. Therefore, the bottleneck of the fault detection system in terms of processing capacity is certainly not imposed by the classifier so that parallel programming environments are needless.

For gradual and small changes of attribute values, the learning algorithm adapts the parameters of granules and connections. EGNN is able to handle new classes and abrupt changes on the data stream, e.g., due to the development of a fault, load change, or voltage unbalance. In these cases, the algorithm creates additional granules, connections, and neurons to maintain its accuracy.

Figure 13 depicts a typical behavior of the $T_{prod} - S_{max}$ EGNN model using $\rho^{[0]} = 0.7$, $h_r = 30$, and $\eta = 3$. The figure shows that the accuracy of the classifier increases quickly along with an increase in the number of rules during the first iterations. EGNN makes use of approximately 12.7 rules with a maximum of 16

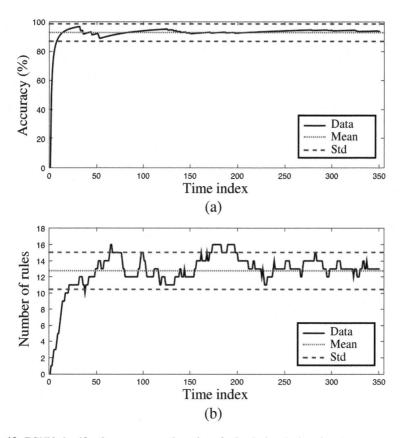

Fig. 13 EGNN classification accuracy and number of rules during the learning steps

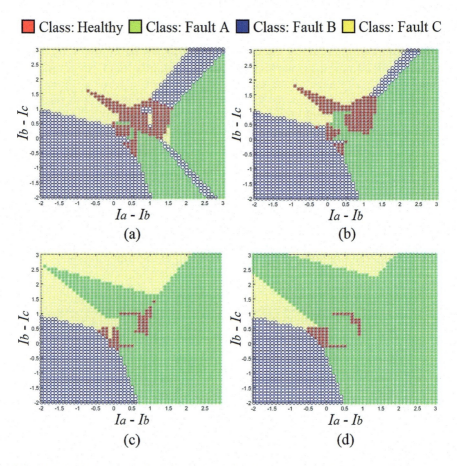

Fig. 14 EGNN classification boundaries during the development of a fault in stator phase A: (**a**) healthy induction machine. Boundaries after (**b**) 1, (**c**) 2, and (**d**) 3 shorted-turns

rules to support a 94.5% classification accuracy on this simulation. After about 70 time steps, the performance of the EGNN classifier achieved a quasi-steady state in spite of recurrent structural and parametric updates.

Nonstationary decision boundaries drawn by EGNN granules during the progress of inter-turns short-circuit in the stator phases A, B, and C are illustrated in Figs. 14, 15, and 16, respectively. Granular regions representing a healthy and faulty induction machine are visible in the first quadrant of the figures since in previous iterations the neural network was exposed to samples of all classes provided by the Fault Simulator model. After the occurrence of a shorted-turn in one of the stator windings, the neural network changed structurally and parametrically to fit the new samples, which reflect the fault occurrence. Therefore, as the number of shorted-turns evolves from 1 to 3 turns, the granular regions related to the underlying faulty winding expand toward the other regions. Samples bringing information about the faulty class predominate in the data stream.

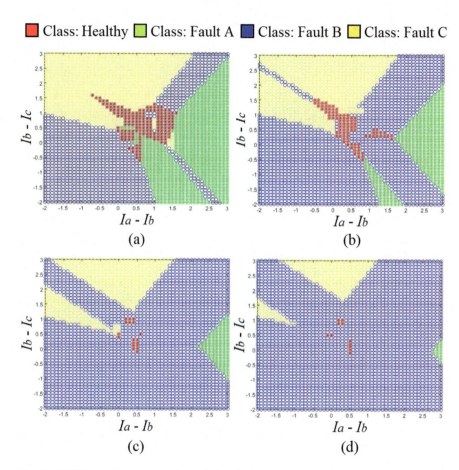

Fig. 15 EGNN classification boundaries during the development of a fault in stator phase B: (**a**) healthy induction machine. Boundaries after (**b**) 1, (**c**) 2, and (**d**) 3 shorted-turns

5.4 Comparative Analyses and Discussion

The evolutionary EANN and evolving fuzzy granular EGNN neural networks were able to detect shorted-turns in the stator windings of an induction machine with a reasonable degree of success. While EGNN achieved a 96.28% best accuracy using product T_{prod} and maximum S_{max} fuzzy neurons in the aggregation and output layers, EANN reached a 94.67% best correct classification rate in a similar scenario using arithmetic crossover, random mutation, tournament selection, and subsequent local search based on local random mutation.

Both learning methods, genetic and incremental, addressed issues such as convergence to a local optimum, dependence on initial parameters and trial-and-error approach on choosing an architecture for the neural classifier. To shed light

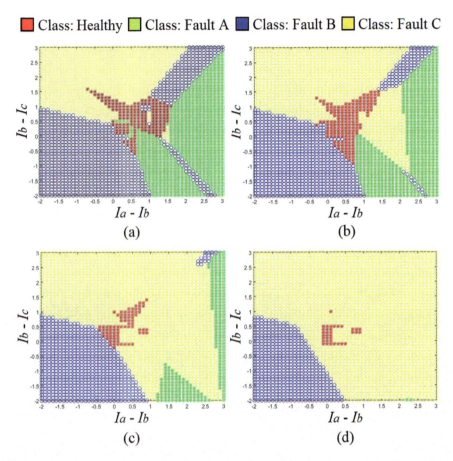

Fig. 16 EGNN classification boundaries during the development of a fault in stator phase C: (**a**) healthy induction machine. Boundaries after (**b**) 1, (**c**) 2, and (**d**) 3 shorted-turns

on these issues, a Multilayer Perceptron MLP neural network with the same structure as that of the fittest EANN, but trained with a gradient algorithm, was used for comparison. The MLP classifier achieved a 91.33% success rate on fault classification.

While from one side the accuracy of the evolutionary EANN, evolving EGNN, and non-evolving MLP classifiers are relatively close to each other (within a 5% range) for the underlying problem, from the other side the time spent by the learning algorithms to provide such performance was about 55 min, 19 min, and 0.1 s, respectively, for the offline genetic, error backpropagation, and online incremental algorithm. Notice that MLP and EANN would certainly benefit from parallel processing, whereas EGNN supports the volume of data in question.

The main point on time and space complexity is that the parameter space of neural networks is usually very large and in principle of undefined dimension.

Online incremental learning is based on a bottom-up approach that starts from scratch and considers new neurons and connections (dimensions in the parameter space) only if necessary. The size of the neural network model is controlled by the granularity adaptation mechanism. Moreover, recursive formulas do not require accumulation of samples and multiple passes over the same data, but a data stream and a single scan of the samples. This explains the enormous difference on computational complexity of the algorithms and supports incremental learning from sequential data as a mainstream of research to deal with complex and big data applications.

The effectiveness of the mathematical model to simulate shorted-turns in the stator windings of electrical machines, real genetic programming, and of the evolving fuzzy granular approach for condition monitoring and pattern classification was verified in this study. The fault simulation model, together with the identification and optimization algorithms, is an important alternative tool for destructive tests. Moreover, fault simulation allowed a variety of practical situations to be incorporated and considered as entries of the dataset. Therefore, the neural classifiers could be trained to be immune to voltage unbalance and load change situations.

6 Conclusion

Early detection of incipient fault conditions in induction machines, such as the turn-to-turn short-circuit fault, is of utmost importance because functional failures may occur minutes or hours after the initial development. The faulty winding should be restored or replaced to prevent complete loss of the motor. Operational hazards as well as significant financial losses can be avoided if the machine is stopped for maintenance.

In this study, learning methods are proposed to evolve architectures and weights of a feedforward neural network aiming at fault detection and classification. In particular, genetic and incremental learning methods are addressed producing, respectively, evolutionary EANN and evolving EGNN models. Both neural models were able to detect shorted-turns successfully. EGNN achieved a 96.28% accuracy using product T_{prod} and maximum S_{max} fuzzy neurons, whereas EANN reached a 94.67% accuracy on correct classification using arithmetic crossover, global and local random mutation, and tournament selection. Issues such as convergence to local optima, dependence on initial parameters, and choice of a neural architecture by trial and error were overcome. Moreover, a Multilayer Perceptron MLP model trained with a gradient algorithm was considered to highlight the advantages of performing global search for parameters prior to local search. The MLP classifier achieved a 91.33% accuracy. The striking difference concerns the time spent to produce such results, which was about 55 min, 19 min, and 0.1 s, respectively, for the genetic, gradient, and online incremental algorithms. Bottom-up incremental learning from scratch takes into account new neurons and connections (dimensions in the parameter space) only if necessary for a better classification accuracy.

The effectiveness of a mathematical model to simulate shorted-turns in the stator windings was verified (validated) in this and related studies. The fault simulation model, together with the identification and optimization algorithms briefly presented, is an important alternative tool for destructive tests. Moreover, fault simulation granted a diversity of practical situations to be incorporated into datasets. Therefore, the neural classifiers could be trained to be immune to voltage unbalance and load change situations.

Further work will address methods to control the specificity of information granules in evolving granular neurofuzzy networks; linguistic data approximation, and missing data imputation due to sensor malfunction or saturation. In addition to sensor faults, airgap eccentricity and multi-fault models based on dq0 and sequential components will be approached.

References

1. Gao, Z., Cecati, C., Ding, S.: A survey of fault diagnosis and fault-tolerant techniques - part I: fault diagnosis with model-based and signal-based approaches. IEEE Trans. Ind. Electron. **62**(6), 3757–3767 (2015)
2. Nandi, S., Toliyat, A.: Condition monitoring and fault diagnosis of electrical motors - a review. IEEE Trans. Energy Convers. **20**(4), 719–729 (2005)
3. Bessa, I., Palhares, R., D'Angelo, M.F., Filho, J.E.: Data-driven fault detection and isolation scheme for a wind turbine benchmark. Renew Energy **87**(1), 634–645 (2016)
4. D'Angelo, M.F., Palhares, R. Cosme, L., Aguiar, L., Fonseca, F., Caminhas, W.: Fault detection in dynamic systems by a Fuzzy/Bayesian network formulation. Appl. Soft Comput. **21**, 647–653 (2014)
5. Frosini, L., Harlişca, C., Szabó, L.: Induction machine bearing fault detection by means of statistical processing of the stray flux measurement. IEEE Trans. Ind. Electron. **62**(3), 1846–1854 (2015)
6. Chang, H.-C., Lin, S.-C., Kuo, C.-C., Hsieh, C.-F.: Induction motor diagnostic system based on electrical detection method and fuzzy algorithm. Int. J. Fuzzy Syst. **18**(5), 732–740 (2016)
7. Leite, D., Hell, M., Costa Jr., P., Gomide, F.: Real-time fault diagnosis of nonlinear systems. Nonlinear Anal. Theory Methods Appl. **71**(12), 2665–2673 (2009)
8. Ghate, V., Dudul, S.: Cascade neural-network-based fault classifier for three-phase induction motor. IEEE Trans. Ind. Electron. **58**(5), 1555–1563 (2011)
9. Fuente, M., Moya, E., Alvarez, C., Sainz, G.: Fault detection and isolation based on hybrid modelling in an AC motor. IEEE Int. Conf. Neural Netw. **3**, 1869–1874 (2004)
10. Gandhi, A., Corrigan, T., Parsa, L.: Recent advances in modeling and online detection of stator interturn faults in electrical motors. IEEE Trans. Ind. Electron. **58**(5), 1564–1575 (2011)
11. Sun, W., Shao, S., Zhao, R., Yan, R., Zhang, X., Chen, X.: A sparse auto-encoder-based deep neural network approach for induction motor faults classification. Measurement **89**, 171–178 (2016)
12. Chow, M.-Y.: Methodologies of Using Neural Network and Fuzzy Logic Technologies for Motor Incipient Fault Detection. World Scientific Publishing Co. Pte. Ltd., Singapore (1998)
13. Goodfellow, I., Bengio, Y., Courville, A.: Deep Learning. Adaptive Computation and Machine Learning. MIT Press, Cambridge (2017)
14. Rumelhart, D., Hinton, G., Willians, R.: Learning internal representations by error propagation. In: Parallel Distributed Processing: Explorations in the Microstructure of Cognition, vol. 1, pp. 318–362. MIT Press, Cambridge (1986)

15. Chen, O., Sheu, B.: Optimization schemes for neural network training. IEEE Int. Conf. Neural Netw. **2**, 817–822 (1994)
16. Yao, X., Liu, Y.: Towards designing artificial neural networks by evolution. Appl. Math. Comput. **91**(1), 83–90 (1998)
17. Sexton, R., Gupta, J.: Comparative evaluation of genetic algorithm and backpropagation for training neural networks. Inf. Sci. **129**(1–4), 45–59 (2000)
18. Huang, H.-X., Li, J.-C., Xiao, C.-L.: A proposed iteration optimization approach integrating backpropagation neural network with genetic algorithm. Expert Syst. Appl. **42**, 146–155 (2015)
19. Chen, X., Chau, K., Busari, A.: A comparative study of population-based optimization algorithms for downstream river flow forecasting by a hybrid neural network model. Eng. Appl. Artif. Intell. **46**(A), 258–268 (2015)
20. Blum, C., Socha, K.: Training feed-forward neural networks with ant colony optimization: an application to pattern classification. In: International Conference on Hybrid Intelligent Systems, 6p (2005)
21. Wang, L., Zeng, Y., Chen, T.: Back propagation neural network with adaptive differential evolution algorithm for time series forecasting. Expert Syst. Appl. **42**(2), 855–863 (2015)
22. Taormina, R., Chau, K.-W.: Neural network river forecasting with multi-objective fully informed particle swarm optimization. J. Hydroinformatics **17**(1), 99–113 (2015)
23. Ren, C., An, N., Wang, J., Li, L., Hu, B., Shang, D.: Optimal parameters selection for BP neural network based on particle swarm optimization: a case study of wind speed forecasting. Knowl.-Based Syst. **56**, 226–239 (2014)
24. Leite, D., Costa, P., Gomide, F.: Evolving granular neural networks from fuzzy data streams. Neural Netw. **38**, 1–16 (2013)
25. Lughofer, E., Pratama, M.: Online active learning in data stream regression using uncertainty sampling based on evolving generalized fuzzy models. IEEE Trans. Fuzzy Syst. **26**(1), 292–309 (2018)
26. Rubio, J.J.: USNFIS: uniform stable neuro fuzzy inference system. Neurocomputing **262**(1), 57–66 (2017)
27. Silva, A., Caminhas, W., Lemos, A., Gomide, F.: A fast learning algorithm for evolving neo-fuzzy neuron. Appl. Soft Comput. **14**(B), 194–209 (2014)
28. Mohamad, S., Moamar, S.-M., Bouchachia, A.: Active learning for classifying data streams with unknown number of classes. Neural Netw. **98**, 1–15 (2018)
29. Leite, D., Ballini, R., Costa, P., Gomide, F.: Evolving fuzzy granular modeling from nonstationary fuzzy data streams. Evol. Syst. **3**(2), 65–79 (2012)
30. Mirzamomen, Z., Kangavari, M.: Evolving fuzzy min-max neural network based decision trees for data stream classification. Neural Process. Lett. **45**(1), 341–363 (2017)
31. Soares, E., Costa, P., Costa, B., Leite, D.: Ensemble of evolving data clouds and fuzzy models for weather time series prediction. Appl. Soft Comput. **64**, 445–453 (2018)
32. Andonovski, G., Music, G., Blazic, S., Skrjanc, I.: Evolving model identification for process monitoring and prediction of non-linear systems. Eng. Appl. Artif. Intell. **68**, 214–221 (2018)
33. Lopes, P.A., Camargo, H.A.: FuzzStream: fuzzy data stream clustering based on the online-offline framework. In: IEEE International Conference on Fuzzy Systems (2017)
34. Sayed-Mouchaweh, M., Lughofer, E.: Learning in Non-Stationary Environments: Methods and Applications. Springer, New York (2012)
35. Bezerra, C., Costa, B., Guedes, L., Angelov, P.: An evolving approach to unsupervised and real-time fault detection in industrial processes. Expert Syst. Appl. **63**(30), 134–144 (2016)
36. Leite, D., Costa, P., Gomide, F.: Evolving granular neural network for semi-supervised data stream classification. In: International Joint Conference on Neural Networks, 8p (2010)
37. Silva, S., Costa, P., Gouvea, M., Lacerda, A., Alves, F., Leite, D.: High impedance fault detection in power distribution systems using wavelet transform and evolving neural network. Electr. Power Syst. Res. **154**, 474–483 (2018)
38. Leite, D., Hell, M., Diez, P., Gariglio, B., Nascimento L., Costa P.: Real-time model-based fault detection and diagnosis for alternators and induction motors. In: IEEE International Electric Machines & Drives Conference, 6p. (2007)

39. Bertsekas, D.: Nonlinear Programming. Athena Scientific, Belmont (1999)
40. Krause, P., Wasynczuk, O., Sudhoff, S.: Analysis of Electric Machinery. IEEE Press, New York (1995)
41. Chen, S., Wu, Y., Luk, L.: Combined genetic algorithm optimization and regularized orthogonal least squares learning for radial basis function networks. IEEE Trans. Neural Netw. **10**(5), 1239–1243 (1999)
42. Brown, A., Card, H.: Cooperative coevolution of neural representations. Int. J. Neural Syst. **10**(4), 311–320 (2000)
43. Ahmadizar, F., Soltanian, K., Tab, F., Tsoulos, I.: Artificial neural network development by means of a novel combination of grammatical evolution and genetic algorithm. Eng. Appl. Artif. Intell. **39**, 1–13 (2015)
44. Fogel, D.: Evolutionary Computation: Toward a New Philosophy of Machine Intelligence, 3rd edn. Wiley-Blackwell, Hoboken (2006)
45. Kaya, M.: The effects of two new crossover operators on genetic algorithm performance. Appl. Soft Comput. **11**(1), 881–890 (2011)
46. Miller, B., Goldberg, D.: Genetic algorithms, tournament selection, and the effects of noise. Complex Syst. **9**, 193–212 (1995)
47. Leite, D., Santana, M., Borges, A., Gomide, F.: Fuzzy granular neural network for incremental modeling of nonlinear chaotic systems. In: IEEE International Conference on Fuzzy Systems, pp. 64–71 (2016)
48. Leite, D., Palhares, R., Campos, V., Gomide, F.: Evolving granular fuzzy model-based control of nonlinear dynamic systems. IEEE Trans. Fuzzy Syst. **23**(4), 923–938 (2015)
49. Kangin, D., Angelov, P., Iglesias, J.A.: Autonomously evolving classifier TEDAClass. Inf. Sci. **366**, 1–11 (2016)
50. Lughofer, E.: FLEXFIS: a robust incremental learning approach for evolving Takagi-Sugeno fuzzy models. IEEE Trans. Fuzzy Syst. **16**(6), 1393–1410 (2008)
51. Shaker, A., Lughofer, E.: Self-adaptive and local strategies for a smooth treatment of drifts in data streams. Evol. Syst. **5**(4), 239–257 (2014)
52. Gu, X., Angelov, P.: Self-organising fuzzy logic classifier. Inf. Sci. **446**, 36–51 (2018)
53. Mirza, B., Lin, Z., Liu, N.: Ensemble of subset online sequential extreme learning machine for class imbalance and concept drift. Neurocomputing **149**, 315–329 (2015)
54. Pratama, M., Lu, J., Lughofer, E., Zhang, G., Anavatti, S.: Scaffolding type-2 classifier for incremental learning under concept drifts. Neurocomputing **191**, 304–329 (2016)
55. Kim, Y., Park, C.H.: An efficient concept drift detection method for streaming data under limited labeling. IEEE Trans. Inf. Syst. **E100**(10), 2537–2546 (2017)
56. Leite, D., Costa, P., Gomide, F.: Granular approach for evolving system modeling. In: International Conference on Information Processing and Management of Uncertainty in Knowledge-Based Systems, pp. 340–349. Springer, Berlin (2010)
57. Toubakh, H., Sayed-Mouchaweh, M.: Hybrid dynamic data-driven approach for drift-like fault detection in wind turbines. Evol. Syst. **6**(2), 115–129 (2015)
58. Dovzan, D., Logar, V., Skrjanc, I.: Implementation of an evolving fuzzy model (eFuMo) in a monitoring system for a waste-water treatment process. IEEE Trans. Fuzzy Syst. **23**(5), 1761–1776 (2015)
59. Pedrycz, W., Gomide, F.: Fuzzy Systems Engineering: Toward Human-Centric Computing. Wiley/IEEE Press, Hoboken (2007)
60. Beliakov, G., Pradera, A., Calvo, T.: Aggregation Functions: A Guide for Practitioners. Springer - Studies in Fuzziness and Soft Computing Series, vol. 221 (2007)

Evolving Fuzzy Model for Fault Detection and Fault Identification of Dynamic Processes

Goran Andonovski, Sašo Blažič, and Igor Škrjanc

1 Introduction

A process fault is defined as an unpermitted deviation of at least one characteristic property or variable of the system from acceptable/usual/standard behavior [28]. Further, the fault identification defines the kind, location, and time of detection of a fault and usually follows the step of fault detection.

In general, many engineering systems are typical dynamic processes with complex structure and frequently operating under changing environmental conditions. To ensure a high production quality and to match the economic requirements, industrial processes are becoming increasingly complicated in both their structure and the degree of automation. Numerous sensors and actuators are part of such processes, and all the data should be processed in real time. On the other hand, there is an increasing demand on safety, reliability, and protection of the industrial systems subjected to potential process faults, failures, and abnormalities. For successful and optimal operation of any process, it is important to detect and to predict undesired events as early as possible. Real-time and data-driven monitoring techniques could play a crucial rule in solving such tasks. **Answer No. 3.** Therefore, the evolving systems [2, 5, 20] because of their data-driven and adaptive nature appear to be an appropriate approach when dealing with complex and nonlinear processes. Moreover, the evolving models are evolved on-the-fly based on streaming data, and the model structure is extended (evolved) on demand (based on the characteristics of the new data). The evolving principles are often used in combination with fuzzy rule-based (FRB), neuro-fuzzy (NF), or neural-network (NN) models.

G. Andonovski (✉) · S. Blažič · I. Škrjanc
Faculty of Electrical Engineering, University of Ljubljana, Ljubljana, Slovenia
e-mail: goran.andonovski@fe.uni-lj.si; saso.blazic@fe.uni-lj.si; igor.skrjanc@fe.uni-lj.si

As we said above, a fault is defined as an unpermitted deviation of one component or parameter of the system. The faults can be caused by different elements that are part of the system, e.g., loss of a sensor value, physical blocking of an actuator, or disconnection of a system component. As explained in [12], the faults can be classified as sensor faults, actuator faults, and plant faults. All these types of faults can change the dynamic properties of the system, influence the input/output relation through changed control actions, and in the worst case the fault can damage or collapse the whole system.

We can find different fault detection methods in the literature that can be categorized into four groups: model-based, signal-based, knowledge-based, and hybrid-based (combination of last two) methods [12, 13]. Model-based fault detection methods are developed to monitor the consistency between the measured and the model outputs. The used models are obtained by system identification techniques or by physical principles of the process. Signal-based methods utilize the measured signals and the extracted features for diagnostic decision. These methods assume that the faults and its symptoms are reflected in the measured signals. Knowledge-based methods apply a variety of artificial intelligent techniques to extract the process variables' dependencies (could be either statistical or nonstatistical methods).

Data-based statistical methods for system monitoring mainly concentrate on the input/output relations of the data collected from the processes. Some of them are principle component analysis (PCA) [9, 14, 19] and partial least squares (PLS) [9, 18, 30] which are one of the basic techniques. More recently, independent component analysis (ICA) [27, 29] has received a lot of attention and has seen great success in practice.

The applicability of statistical data-based methods can be improved by considering the system dynamics. For example, in [16] improved version of PCA (DPCA) is presented to deal with the problem of system dynamics. Also, the subspace aided approach (SAP) presented in [11] shows sufficient results when dealing with such systems. An overview on signal-based fault detection methods, that combine the data-driven techniques with statistical approaches, has been presented in [22]. As dynamic processes are tackled, the evolving-based methods play an important role. In this case, the process model is not known a priori but is developed recursively from the data stream. Furthermore, the acquired knowledge is used for the prediction of the future system behavior.

Answer No. 4. The goal of this chapter is to propose an evolving cloud-based fuzzy method for fault detection of complex dynamic processes. The method combines the statistical knowledge in a recursive manner with the evolving mechanisms to cope with dynamic and nonstationary conditions. Furthermore, this method does not make any assumption about the data distribution.

At this point, it is necessary to mention some existing work about evolving-based data-driven methods for fault detection purposes. In [24], a residual-based approach is presented, where the original signals are transformed into a model space by identifying the multidimensional relationships. The model architecture consists of pure linear models, generic Box–Cox models for weak and Takagi–

Sugeno fuzzy model for more complex nonlinearities. Adaptive fault detection and diagnosis method based on participatory evolving clustering was proposed in [17]. The method uses an incremental unsupervised clustering algorithm, and the new detected operation modes are labeled manually by the operator of the process. In [6], autonomous fault detection procedure is presented. The method is based on TEDA (Typicality and Eccentricity Data Analytics) [3] and uses the Chebyshev inequality as a measure for anomaly detection.

The proposed fault detection algorithm is based on the simplified fuzzy rule-based (FRB) system AnYa [4]. AnYa system does not require any assumptions of the data distribution and it is based on the concept of data clouds [4]. The membership functions are calculated using recursive density estimation (RDE). The previous work proposed in [10] calculates the mean RDE using the time thresholds for detecting faults and normal conditions. This can be very dependent on the process's time constant. The concept presented in this paper differs from the previous ones in the way how the data density is calculated.

To evaluate the proposed method, a model of HVAC (Heating Ventilation Air Conditioning) system was used. The parameters of the model are tuned/adjusted using real data acquired from the local facility. The model copes with the statistical and the dynamic properties of the real process [25].

The chapter is organized as follows. Section 2 describes the evolving fuzzy model and how the problem space is partitioned to normal process operation and faults. Next, in Sect. 3 the fault detection procedure is described, where the stages of learning, detection, and identification are explained. In Sect. 4, the HVAC model is described and further, the main focus of this section is on the possible faults that can occur on the process. Finally, the experimental results in Sect. 5 are presented, and the conclusions are justified in Sect. 6.

2 Evolving Fuzzy Model

Fuzzy systems are general approximation tools for the modeling of nonlinear dynamic processes. In this paper, we use a fuzzy rule-based (FRB) system with a nonparametric antecedent part presented by Angelov and Yager [4]. The main difference comparing to the classical Takagi–Sugeno [26] and Mamdani [21] fuzzy systems is the simplified antecedent part that relies on the relative data density.

2.1 Fuzzy Cloud-Based Model Structure

The rule-based form in this chapter is used for classification purposes. The antecedent part of the fuzzy model contains the evolving mechanism, while the consequent part is used as classifier. The fuzzy rule-base of the ith rule is defined as follows:

$$\mathscr{R}^i : \text{IF} \quad (x_k \sim X^i) \quad \text{THEN} \quad x_k \in Class^i \qquad (1)$$

where $x_k = [x_k(1), x_k(2), \ldots x_k(m)]$ is m-dimensional input vector. The operator $\sim$ is linguistically expressed as "*is associated with*," which means that the current data x_k is related to one of the existing clouds X^i according to the membership degree (the normalized relative density of the data):

$$\lambda_k^i = \frac{\gamma_k^i}{\sum_{j=1}^{c} \gamma_k^j} \quad i = 1, \ldots, c \qquad (2)$$

where γ_k^i is the local density of the current data x_k with the ith cloud. The local density is defined using the Cauchy kernel as follows [4]:

$$\gamma_k^i = \frac{1}{1 + \frac{\sum_{j=1}^{M^i}(x_k - x_j^i)^T A^i (x_k - x_j^i)}{M^i}}, \quad i = 1, \ldots, c \qquad (3)$$

where M^i is the number of data points x_j^i that belong to the i-th cloud. Equation (3) could be rewritten in the recursive form for easier implementation, as follows:

$$\gamma_k^i = \frac{1 + T^i}{1 + (x_k - \mu_{M^i}^i)^T A^i (x_k - \mu_{M^i}^i) + T^i}, \quad i = 1, \ldots, c \qquad (4)$$

where $\mu_{M^i}^i$ is vector of mean value (center) of the data that are part of the ith cloud. The recursive form of $\mu_{M^i}^i$ is calculated as follows:

$$\mu_{M^i}^i = \frac{M^i - 1}{M^i} \mu_{M^i-1}^i + \frac{1}{M^i} x_k, \quad \mu_0^i = x_k \qquad (5)$$

In (4), the matrix $A^i \in \mathbb{R}^{m \times m}$ is an identity matrix or an inverse of the covariance matrix in case of Euclidean or Mahalanobis distance, respectively. In (4), the scalar T^i is calculated as follows:

$$T^i = \frac{M^i - 1}{M^i} trace(A^i \Sigma_{M^i}^i) \qquad (6)$$

where $\Sigma_{M^i}^i \in \mathbb{R}^{m \times m}$ is the covariance matrix of the i-th cloud and it is calculated as follows:

$$\Sigma_{M^i}^i = \frac{1}{M^i - 1} \sum_{k=1}^{M^i} (x_k^i - \mu_{M^i}^i)(x_k^i - \mu_{M^i}^i)^T \qquad (7)$$

The recursive way of calculating the covariance matrix (7) is explained in the following. Firstly, the un-normalized covariance matrix is computed as:

$$S^i_{M^i} = S^i_{M^i-1} + (x_k - \mu^i_{M^i-1})(x_k - \mu^i_{M^i})^T \quad (8)$$

and the covariance matrix is then obtained as:

$$\Sigma^i_{M^i} = \frac{1}{M^i - 1} S^i_{M^i} \quad (9)$$

The final algorithm for recursive calculating of the covariance matrix (9) can be summarized using the following instructions [7]:

$$M^i \leftarrow M^i + 1 \quad (10)$$

$$\mu^i_{M^i} \leftarrow \frac{M^i - 1}{M^i} \mu^i_{M^i-1} + \frac{1}{M^i} x_k \quad (11)$$

$$S^i_{M^i} \leftarrow S^i_{M^i-1} + (x_k - \mu^i_{M^i-1})(x_k - \mu^i_{M^i})^T \quad (12)$$

$$\Sigma^i_{M^i} \leftarrow \frac{1}{M^i - 1} S^i_{M^i} \quad (13)$$

where the states are initialized with $M^i = 0$, $\mu^i_0 = x_1$, and $S^i_0 = 0$.

2.2 Evolving Mechanism

The evolving nature of the proposed method means that new clouds in the rule-base system (1) could be added if some criteria are satisfied. We should note that at the beginning of the experiment there are no predefined clouds. The first cloud with the properties $X^1_1 \in \{\mu^1_1 = x_1, S^1_1 = 0\}$ is defined with the first data point x_1 received.

At each time stamp k, all the partial densities $\gamma^i_{\delta k}$ between the current data x_k and the existing clouds X^i are calculated ($i = 1, \ldots, c$). The current active cloud is the one with the maximal partial density $\max_i \gamma^i_{\delta k}$ ("winner takes all"). If this value is lower than a predefined threshold γ_{max} ($\max_i \gamma^i_{\delta k} < \gamma_{max}$), then a new cloud is added. The value of density threshold was chosen $\gamma_{max} = 0.85$. Further, to protect adding new clouds based on outliers, the evolving mechanism is frozen for $n_{add} = m + 1$ samples after a new cloud is added [1], where m is the dimensionality of the input vector.

We have to note here that only the properties μ^i_k, and S^i_k of the currently active cloud (with maximal partial density) are updated, and the properties of all the other clouds are kept constant.

3 Fault Detection and Identification

In this section, the fault detection procedure will be described. For this purpose, three different data sets are necessary. The first data set describes the normal process operation (without faulty states, $F = 0$), while second data set contains faults ($F = 1$). The third data set is used for testing purposes and this data set contains areas of normal process operation as well as areas where there are faults. Therefore, the proposed method is trained on the first two data sets and its effectiveness is tested on the third data set. In the following subsection, this is explained in detail.

3.1 Learning/Training Phase

As we said above, in the learning/training phase we use two types of data. The first one does not contain any faults and the second one contains faulty states. Using the first data set, we obtain a quantitative model that describes normal process operation. This model contains c_0 data clouds which are labeled as $X^i \in \{\mu_k^i, A_k^i, F = 0\}$ (shortened notation $X_{F=0}^i$), where $i = 1, \ldots, c_0$. Next, for each type of fault, that we want to detect, we use a separate data set to discover the diagnostic model for that particular fault. Again, each of the obtained models contains different number of data clouds labeled as $X^i \in \{\mu_k^i, A_k^i, F = 1\}$ (shortly notation $X_{F=1}^i$). The total number of data clouds for each model are noted as $c_1, , c_2, \ldots,$ for each type of fault $F1, F2, \ldots,$ respectively.

To summarize, at the end of the learning phase we have obtained one model for normal process operation and other models for each type of fault, respectively. A synthetic example is presented in Fig. 1 where three models are shown. First model, for normal process operation, contains three clouds ($X_{F=0}^1$, $X_{F=0}^2$, $X_{F=0}^3$), second model for fault has two clouds ($X_{F=1}^4$, $X_{F=1}^5$), and the third one three clouds ($X_{F=1}^6$, $X_{F=1}^7$, $X_{F=1}^8$).

3.2 Fault Detection Phase

In the fault detection phase, we use the models acquired in the learning phase to detect the faulty states. Therefore, for each data point received we have to check to which model it is more associated with. For that purpose, we have to calculate all the local densities (4) of the current data x_k to all the detected clouds from the previous phase (see Fig. 1). Next, for each model we calculate the global density for that model as follows:

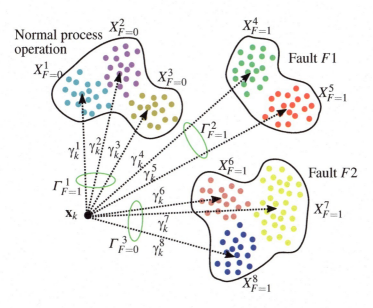

Fig. 1 Example of classifying the current sample x_k as fault or normal state

$$\Gamma_k^i = \frac{\sum_{j=1}^{c^i} \gamma_k^j}{c^i}, \qquad i = 1, 2, \ldots \qquad (14)$$

where c^i is the number of data clouds of the i-th model.

A fault is detected if the following criterion is satisfied:

$$\max_i \Gamma_k^i(X_{F=0}^i) < \max_i \Gamma_k^i(X_{F=1}^i) \qquad (15)$$

where we compare the maximal local density of the model for normal process operation $\max_i \Gamma_k^i(X_{F=0}^i)$ with maximal local density over all the data clouds for faults $\max_i \Gamma_k^i(X_{F=1}^i)$.

3.3 Fault Identification Phase

If the criteria (15) is not satisfied, then the data point x_k is classified as normal process operation. If the faulty state is detected using (15), then for the fault identification purposes the global densities (14) are compared. The kind of the fault is determined (isolated) according to the maximal global density $\max_i \Gamma_k^i(X_{F=1}^i)$ of the faulty models. The detected fault is associated with the model with the maximum global density.

4 Description of the HVAC Process Model

In this section, the model of Heating, Ventilation, and Air Condition (HVAC) process will be described. The general purpose of the HVAC system is to provide thermal comfort in the rooms and to achieve the required indoor air quality parameters. The HVAC system controls the temperature, the humidity, and the pressure (or the air flow through air channel) for each sector (room) that the system is responsible for. Such systems present an important segment of each modern and energy-efficient building. They have been widely used in the residential and commercial buildings. According to [8], the contribution of buildings to overall energy consumption in developed countries is between 20% and 40% of which 40% belongs to HVAC systems. Therefore, an efficient management and monitoring of these systems can significantly reduce the energy consumption.

The HVAC system used in this chapter is presented in Fig. 2 and consists of the following components. First, the blinds for input and output air, which are open to 100% when the system is turned on. There are also different types of air filters (G3, F5, and F7) that remove the solid particulates such as dust and pollen. The air flow is divided into four separate zones: outdoor, supply, return, and exhaust air which are shown (see Fig. 2) in green, blue, yellow, and orange color, respectively. The direction of the air flow is shown by the arrows in Fig. 2. We can see that the return air is not mixed with the fresh outdoor air due to the required indoor air quality standards. The main (electrically controlled) parts of the system are: recuperator, heater, humidifier, cooler, and supply and return fan (see Fig. 2).

Answer No. 6. The main control algorithm makes decisions according to the outdoor air condition and the indoor air quality requirements. Due to the fact that the control algorithm is a product and property of an automation company, we will explain at this point only the basic principles and not all the details. In general, we can divide the control algorithm into three parts: temperature, humidity, and air flow control. These three parts are intertwined between them considering the Mollier diagram principles and some internal logic. Two PI (Proportional–Integral) controllers take care of sufficient (required) flow of the supply and the return air. The temperature and humidity control parts are connected through a sequencer, which decides which control mode is active: heating, cooling, or dehumidifying. This part consists of four PI controllers and directly influences on the heater, the cooler, and the recuperator. The humidification control part consists of two PI controllers which control the humidifier's valve.

In [25], a model of the presented HVAC system was built for testing and developing new fault detection methods and for optimizing the control algorithm. The parameters of the model were tuned using the real data acquired from the real HVAC system. Therefore, as shown in [25] the model sufficiently represents the static and dynamic properties of the real process.

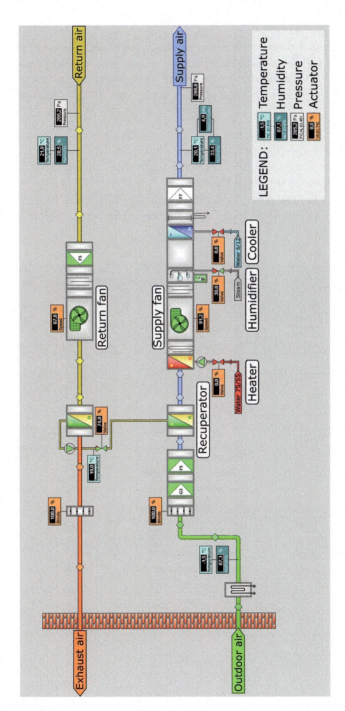

Fig. 2 HVAC system

4.1 Possible Faults on HVAC System

The list of signals that are available on the described HVAC system is presented in Table 1. The signals are separated into two groups: actuators and sensors. The possible faults that can appear on the HVAC system can be influenced by the actuators and sensors as well. Some of the possible types of faults (also see Fig. 3) are:

1. *Communication error*—The real value of the signal is not the same as the value that the control algorithm sees (first subplot in Fig. 3).
2. *Signal offset*—The real value of the sensor/actuator is higher/lower for some offset value (second subplot in Fig. 3).
3. *Signal drift*—The real signal starts drifting (third subplot in Fig. 3).
4. *Step error*—The real signal is on its maximum or minimum value (fourth subplot in Fig. 3).

Table 1 List of signals on HVAC system

Description	Symbol	Range
Actuators		
Valve: recuperator	V_r	0–100%
Valve: heater	V_h	0–100%
Valve: humidifier	V_{hm}	0–100%
Valve: cooler	V_c	0–100%
Fan: supply air	F_{sa}	0–100%
Fan: return air	F_{ra}	0–100%
Blinds: outdoor air	B_{oa}	0–100%
Blinds: exhaust air	B_{ea}	0–100%
Sensors		
Temperature: outdoor air	T_{oa}	−40–80 °C
Temperature: supply air	T_{sa}	0–50 °C
Temperature: return air	T_{ra}	0–50 °C
Temperature: recuperation water	T_{rw}	−40–80 °C
Temperature: hot water for heater	T_{hw}	0–100 °C
Temperature: cold water for cooler	T_{cw}	0–50 °C
Humidity: outdoor air	RH_{oa}	0–100%
Humidity: supply air	RH_{sa}	0–100%
Humidity: return air	RH_{ra}	0–100%
Pressure: supply air	P_{sa}	0–300 Pa
Pressure: return air	P_{ra}	0–300 Pa

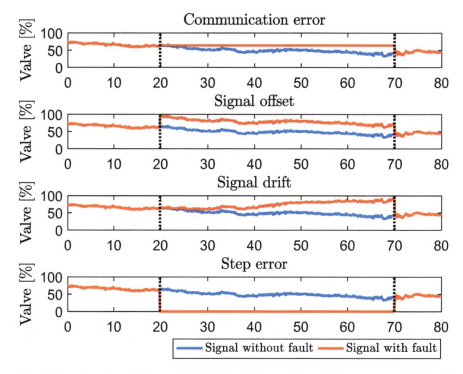

Fig. 3 Examples of possible types of faults that can appear on the actuators or sensors. The dotted black lines denote the start and the end of the faults

5 Experimental Results

In this section, we will present the experimental results of the proposed cloud-based method. As we already mentioned in the previous sections, the proposed cloud-based data-driven fault detection method was tested on model of HVAC system. The results were compared to established fault detection method DPCA [15, 23, 25]. For the statistical comparing of the methods, we used the True Positive Rate (TPR), the False Positive Rate (FPR), and the overall Accuracy (ACC) measure:

$$TPR = \frac{TP}{samples\ (F=1)} \times 100\,[\%] \quad (16)$$

$$FPR = \frac{TN}{samples\ (F=0)} \times 100\,[\%] \quad (17)$$

$$ACC = \frac{TP+TN}{total\ samples} \times 100\,[\%] \quad (18)$$

Table 2 List of the tested faults of HVAC system

Fault no.	Signal description	Symbol	Fault type
$F1$	Temperature: cold water	T_{cw}	Offset ($-40\,°C$)
$F2$	Temperature: cold water	T_{cw}	Offset ($+40\,°C$)
$F3$	Valve: heater	V_h	Step (100%)
$F4$	Valve: recuperator	V_r	Step (100%)

Table 3 Settings of the proposed method for each fault

Fault no.	Signal description	Regressors	Parameters
$F1$	Temperature: cold water	$T_{sa} - T_{oa}$	$\gamma_{max} = 0.85\ c_{max} = 100\ n_{add} = m+1$
		V_c	
		T_{cw}	
$F2$	Temperature: cold water	$T_{sa} - T_{oa}$	
		$SP_{T_{sa}}$	
		V_c	
		T_{cw}	
		$RH_{sa} - RH_{oa}$	
$F3$	Valve: heater	$T_{sa} - T_{oa}$	
		V_h	
		V_r	
$F4$	Valve: recuperator	$T_{sa} - T_{oa}$	
		V_h	
		V_r	

where TP (true positives) and TN (true negatives) are the number of correctly detected faults and normal states, respectively.

In this experiment, four different faults were considered as presented in Table 2. For each of these faults, three data sets (two training and one testing) were acquired from the HVAC model presented in [25] (Table 3).

In Figs. 4, 5, 6, and 7, the fault detection results of the proposed method for the faults $F1$, $F2$, $F3$, and $F4$ are presented, respectively. The upper plots show the maximal density $\max_i \Gamma^i_{\delta k}(X^i_{F=0})$ for normal process operation (blue line) and fault $\max_i \Gamma^i_{\delta k}(X^i_{F=1})$ (red line). The light blue shaded areas indicate the actual period where the particular fault appears. Analogously to the upper plot, the bottom plot shows the performance analysis. The green line shows the detected fault by the proposed evolving algorithm. From all figures, we can notice that majority of wrongly detected samples are in the area where fault is not present, and therefore in this case we have just false alarms. On the other hand, there are just few wrongly detected (actually undetected) faults. One of the reasons for wrong detections we see is the poorly designed control strategy which causes parameter oscillations. The TPR, the FPR, and the overall accuracy for each fault are shown in Tables 4, 5, and 6, respectively. Moreover, the results of the proposed method are compared to DPCA method presented in [25]. From the results in the tables, we can notice that using the proposed evolving method we improve the overall accuracy comparing to the DPCA technique.

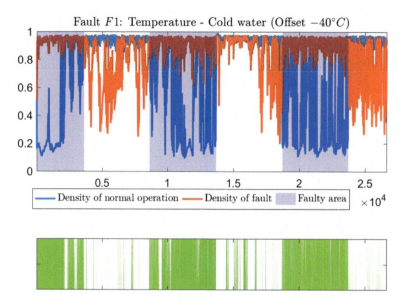

Fig. 4 Fault detection results for the fault $F1$

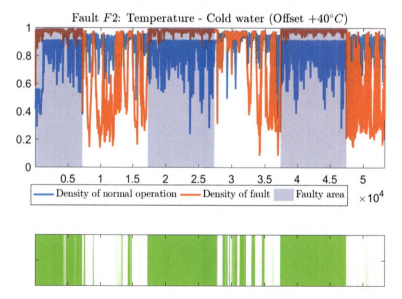

Fig. 5 Fault detection results for the fault $F2$

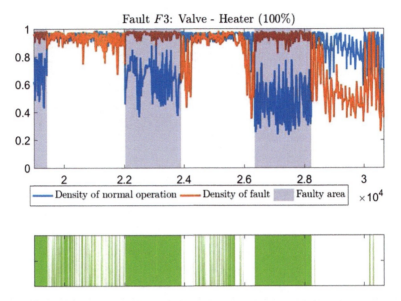

Fig. 6 Fault detection results for the fault $F3$

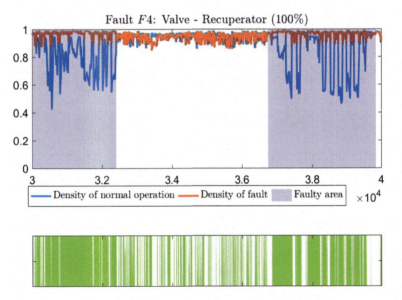

Fig. 7 Fault detection results for the fault $F4$

Table 4 True positive rate ($TPR[\%]$) comparison between DPCA and proposed method

	F1	F2	F3	F4
$DPCA$ [25]	81.30	81.30	65.20	71.60
$CB_{optimized}$	**90.59**	**99.23**	**77.73**	**72.42**

The highest TPR among both methods for each fault type is highlighted in bold font

Table 5 False positive rate ($FPR[\%]$) comparison between DPCA and proposed method

	F1	F2	F3	F4
$DPCA$ [25]	**92.10**	**92.10**	**79.00**	66.20
$CB_{optimized}$	89.78	78.26	74.40	**67.53**

The highest FPR among both methods for each fault type is highlighted in bold font

Table 6 Accuracy ($ACC[\%]$) comparison between DPCA and proposed method

	F1	F2	F3	F4
$DPCA$ [25]	86.59	86.59	73.35	67.90
$CB_{optimized}$	**90.20**	**89.03**	**75.76**	**69.04**

The highest accuracy among both methods for each fault type is highlighted in bold font

6 Conclusion

In this paper, we introduced an evolving cloud-based method for fault detection purposes on dynamic processes. The proposed approach is based on simplified fuzzy model AnYa and uses an evolving mechanism for partitioning the problem space according to the local data density measure. The method obtains separate models (data clouds), one for normal process operation and other for faults. According to the affiliation of the current data to these models, we classify the data as fault or normal state. The proposed procedure was tested on data acquired from HVAC model, and the results were compared to the well-established fault detection method DPCA (Dynamic Principle Component Analysis). Four different faults were investigated, and the obtained results show that the cloud-based method could be competitive to the DPCA method.

References

1. Andonovski, G., Mušič, G., Blažič, S., Škrjanc, I.: Evolving model identification for process monitoring and prediction of non-linear systems. Eng. Appl. Artif. Intell. **68**(October), 214–221 (2018)
2. Angelov, P.: Evolving Rule-Based Models: A Tool for Design of Flexible Adaptive Systems, 1st edn. Springer, London (2002)
3. Angelov, P.: Anomaly detection based on eccentricity analysis. In: IEEE Symposium on Evolving and Autonomous Learning Systems (EALS), Orlando, pp. 1–8 (2014)

4. Angelov, P., Yager, R.: Simplified fuzzy rule-based systems using non-parametric antecedents and relative data density. In: Symposium Series on Computational Intelligence (IEEE SSCI 2011) - IEEE Workshop on Evolving and Adaptive Intelligent Systems (EAIS 2011), pp. 62–69 (2011)
5. Angelov, P., Filev, D., Kasabov, N.: Evolving Intelligent Systems: Methodology and Applications. Wiley, New Jersey (2010)
6. Bezerra, C.G., Costa, B.S.J., Guedes, L.A., Angelov, P.P.: An evolving approach to unsupervised and Real-Time fault detection in industrial processes. Expert Syst. Appl. **63**, 134–144 (2016)
7. Blažič, S., Angelov, P., Škrjanc, I.: Comparison of approaches for identification of all-data cloud-based evolving systems. In: 2nd IFAC Conference on Embedded Systems, Computer Intelligence and Telematics CESCIT 2015, Maribor, pp. 129–134 (2015)
8. Bruton, K., Coakley, D., Raftery, P., Cusack, D.O., Keane, M.M., O'Sullivan, D.T.: Comparative analysis of the AHU InFO fault detection and diagnostic expert tool for AHUs with APAR. Energy Effic. **8**(2), 299–322 (2015)
9. Chen, A., Zhou, H., An, Y., Sun, W.: PCA and PLS monitoring approaches for fault detection of wastewater treatment process. In: 2016 IEEE 25th International Symposium on Industrial Electronics (ISIE), pp. 1022–1027 (2016)
10. Costa, B.S.J., Angelov, P.P., Guedes, L.A.: Fully unsupervised fault detection and identification based on recursive density estimation and self-evolving cloud-based classifier. Neurocomputing **150**(Part A), 289–303 (2015)
11. Ding, S.X., Zhang, P., Naik, A., Ding, E.L., Huang, B.: Subspace method aided data-driven design of fault detection and isolation systems. J. Process Control **19**(9), 1496–1510 (2009)
12. Gao, Z., Cecati, C., Ding, S.X.: A survey of fault diagnosis and fault-tolerant techniques part I: fault diagnosis. IEEE Trans. Ind. Electron. **62**(6), 3768–3774 (2015). https://doi.org/10.1109/TIE.2015.2417501
13. Gao, Z., Cecati, C., Ding, S.X.: A survey of fault diagnosis and fault-tolerant techniques-part II: fault diagnosis with knowledge-based and hybrid/active approaches. IEEE Trans. Ind. Electron. **62**(6), 3768–3774 (2015)
14. Gertler, J., Cao, J.: PCA-based fault diagnosis in the presence of control and dynamics. AIChE J. Process Syst. Eng. **50**(2), 388–402 (2004)
15. Ketelaere, B.D., Hubert, M., Schmitt, E.: Overview of PCA-based statistical process-monitoring methods for time-dependent, high-dimensional data. J. Qual. Technol. **47**(4), 318–335 (2015)
16. Ku, W., Storer, R.H., Georgakis, C.: Disturbance detection and isolation by dynamic principal component analysis. Chemom. Intell. Lab. Syst. **30**(1), 179–196 (1995)
17. Lemos, A., Caminhas, W., Gomide, F.: Adaptive fault detection and diagnosis using an evolving fuzzy classifier. Inf. Sci. **220**, 64–85 (2013)
18. Li, G., Qin, S.J., Zhou, D.: Geometric properties of partial least squares for process monitoring. Automatica **46**(1), 204–210 (2010)
19. Li, W., Yue, H., Valle-Cervantes, S., Qin, S.J.: Recursive PCA for adaptive process monitoring. J. Process Control **10**, 471–486 (2000)
20. Lughofer, E.: Evolving Fuzzy Systems Methodologies, Advanced Concepts and Applications, 1st edn. Springer, Berlin (2011)
21. Mamdani, E., Assilian, S.: An experiment in linguistic synthesis with a fuzzy logic controller. Int. J. Man Mach. Stud. **7**(1), 1–13 (1975)
22. Precup, R.E., Angelov, P., Costa, B.S.J., Sayed-Mouchaweh, M.: An overview on fault diagnosis and nature-inspired optimal control of industrial process applications. Comput. Ind. **74**, 75–94 (2015)
23. Rato, T.J., Reis, M.S.: Defining the structure of DPCA models and its impact on process monitoring and prediction activities. Chemom. Intell. Lab. Syst. **125**, 74–86 (2013)
24. Serdio, F., Lughofer, E., Pichler, K., Buchegger, T., Efendic, H.: Residual-based fault detection using soft computing techniques for condition monitoring at rolling mills. Inf. Sci. **259**, 304–320 (2013)

25. Stržinar, Ž.: Modeliranje in zaznavanje napak v klimatskih sistemih. Master thesis, University of Ljubljana (2017)
26. Takagi, T., Sugeno, M.: Fuzzy identification of systems and its applications to modeling and control. IEEE Trans. Syst. Man Cybern. **SMC-15**(1), 116–132 (1985)
27. Tsai, D.M., Wu, S.C., Chiu, W.Y.: Defect detection in solar modules using ICA basis images. IEEE Trans. Ind. Inf. **9**(1), 122–131 (2013)
28. van Schrick, D.: Remarks on Terminology in the Field of Supervision, Fault Detection and Diagnosis. IFAC Proc. Vol. (IFAC-PapersOnline) **30**(18), 959–964 (1997)
29. Zhang, Y., Qin, S.J.: Improved nonlinear fault detection technique and statistical analysis. AIChE J. Process Syst. Eng. **54**(12), 3207–3220 (2008)
30. Zhang, Y., Zhou, H., Qin, S.J., Chai, T.: Decentralized fault diagnosis of large-scale processes using multiblock kernel partial least squares. IEEE Trans. Ind. Inf. **6**(1), 3–10 (2010)

An Online RFID Localization in the Manufacturing Shopfloor

Andri Ashfahani, Mahardhika Pratama, Edwin Lughofer, Qing Cai, and Huang Sheng

1 Introduction

Radio Frequency Identification (RFID) technology has been used to manage object's location in the manufacturing shopfloor. It is more popular than similar technologies for object localization, such as Wireless Sensor Networks (WSN) and WiFi, due to the affordable price and the easy deployment [12, 25].

In the Maintenance, Repair, and Overhaul (MRO) industry, for example, locating the equipments and trolleys manually over the large manufacturing shopfloor area results in time-consuming activities and increases operator workload. Embracing RFID technology for localization will help companies improve productivity and efficiency in the industry 4.0. Instead of manually locating the tool-trolleys, RFID localization is utilized to monitor the location real time. Despite much work and progress in RFID localization technology, it still has challenging problems. The key disadvantage of RFID is that it has low-quality signal, which is primarily altered by the complexity and severe noises in the manufacturing shopfloors [3].

Generally, an RFID localization system comprises of three components, i.e., RFID tag, RFID reader, and the data processing subsystem. The reader aims to identify the tag ID and obtain the received signal strength (RSS) information from

---

A. Ashfahani (✉) · M. Pratama · Q. Cai
Nanyang Technological University, Singapore, Singapore
e-mail: andriash001@e.ntu.edu.sg; mpratama@ntu.edu.sg

E. Lughofer
Fuzzy Logic Laboratorium Linz-Hagenberg, Department of Knowledge-Based Mathematical Systems, Johannes Kepler University Linz, Linz, Austria
e-mail: edwin.lughofer@jku.at

Huang Sheng
Singapore Institute of Manufacturing Technology, Singapore, Singapore
e-mail: shuang@SIMTech.a-star.edu.sg

© Springer Nature Switzerland AG 2019
E. Lughofer, M. Sayed-Mouchaweh (eds.), *Predictive Maintenance in Dynamic Systems*, https://doi.org/10.1007/978-3-030-05645-2_10

tags. An object's location can be estimated by observing the RSS. However, the RSS quality in the real world is very poor; moreover, it keeps changing over time. As an illustration, although RFID tag is used in the static environment, the RSS keeps changing over time. Moreover, a minor change in the surrounding area can greatly fluctuate the RSS. The major factors causing the phenomena are multipath effect and interference. Therefore, obtaining the accurate location relying on RSS information is a hard task. In several research programs, these problems are tackled by employing computational techniques, and thus the object's location can be estimated accurately [12].

There are several techniques which can be utilized to estimate the object's location. First of all, the distance from an RFID tag to an RFID reader can be calculated via the two-way radar equation for a monostatic transmitter. It can be obtained easily by solving the equation. Another approach, LANDMARC, is proposed for the indoor RFID localization [11]. It makes use of K reference tags, and then it evaluates the RSS similarity between reference tags and object tags. A higher weight will be assigned to the reference tags which possess the similar RSS information to the object tags. In the realm of machine learning, support vector regression (SVR) is implemented for the indoor RFID localization [3]. It is a one-dimensional method and is designed for stationary objects in the small area. Another way to obtain better accuracy is by employing Kalman filter (KF), and it has been demonstrated to deal with wavelength ambiguity of the phase measurements [23].

The strategy to estimate the object location via the so-called radar equation is easy to execute. However, the chance to obtain acceptable accuracy is practically impossible due to the severe noises. LANDMARC manages to improve the localization accuracy. Nonetheless, it is difficult to select the reference tag properly in the industrial environments, where interference and multipath effect occurred. Improper selection of reference tags can alter the localization accuracy. Similarly, SVR is also designed to address object localization problem in the small area, and it even encounters an over-fitting problem. Meanwhile, integrating KF in some works can improve localization accuracy and it also has low computational cost. KF has better robustness and good statistical properties. However, the requirement to calculate the correlation matrix burdens the computation. In addition, the use of KF is also limited by the nonlinearity and nonstationary condition of the real world [3, 13]. The data generated from a nonstationary environment can be regarded as the data stream [14].

Evolving intelligent system (EIS) is an innovation in the field of computational intelligence to deal with data stream [1, 9]. EIS has an open structure, and it implies that it can start the learning processes from scratch or zero rule base. Its rules are automatically formed according to the data stream information. EIS adapts online learning scenario, and it conducts the training process in a single-pass mode [20]. EIS can either adjust the network parameters or generate fuzzy rule without retraining process. Hence, it is capable to deal with the severe noises and the systems dynamics [10, 16, 20]. In several works, the gradient descent (GD) is utilized to adjust the EIS parameters. However, it is vulnerable to noises due to its sensitivity [13]. In the premise part, the Gaussian membership function (GMF) is usually employed to capture the input features of EIS. The drawback of GMF is its inadequacy to detect the overlaps between classes. Several works have employed quantum membership function (QMF) to tackle the problem [4, 8, 21]. Nevertheless, it is type-1 QMF which lacks robustness to deal with uncertainties in real-world data streams [20].

This research proposes an EIS, namely evolving Type-2 Quantum Fuzzy Neural Network (eT2QFNN). The eT2QFNN adopts online learning mechanism, and it processes the incoming data one-by-one and the data is discarded after being learned. Thus, eT2QFNN has high efficiency in terms of computational and memory cost. The eT2QFNN is encompassed by two learning policies, i.e., the rule growing mechanism and parameters adjustment. The first mechanism enables eT2QFNN to start the learning process from zero rule base. It can automatically add the rule on demand. Before a new rule is added to the network, it is evaluated by a proposed formulation, namely modified Generalized Type-2 Datum Significance (mGT2DS). The second mechanism performs parameters adjustment whenever a new rule is not formed. It aims to keep the network adapted to the current data stream. This mechanism is accomplished by decoupled extended Kalman filter (DEKF). It is worth noting that the DEKF algorithm performs localized parameter adjustment, i.e., each rule can be adjusted independently [13]. In this research, the adjustment is only undertaken on a winning rule, i.e., a rule with the highest contribution. Therefore, DEKF is more efficient than extended Kalman filter (EKF), and yet it still preserves the EKF performance. On the premise part, the interval type-2 QMF (IT2QMF) is proposed to approximate the desired output. It worth noting that it is a universal function approximator which has been demonstrated by several researchers[4, 8, 21]. Moreover, it is proficient to form a graded class partition, such that the overlaps between classes can be identified [4].

The major contributions of this research are summarized as follows: (1) This research proposes the IT2QMF with uncertain jump position. It is the extended version of QMF. That is, IT2QMF has the interval type-2 capability in terms of incorporating data stream uncertainties. (2) The eT2QFNN is equipped with rule growing mechanism. It can generate its rule automatically in the single-pass learning mode, if a condition is satisfied. The proposed mGT2DS method is employed as the evaluation criterion. (3) The network parameter adjustment relies on DEKF. The mathematical formulation is derived specifically for eT2QFNN architecture. (4) The effectiveness of eT2QFNN has been experimentally validated using real-world RFID localization data.

The remainder of this book chapter is organized as follows. Section 2 presents the RFID localization system. The proposed type-2 quantum fuzzy membership function and the eT2QFNN network architecture are presented in Sect. 3. Section 4 presents the learning policies of eT2QFNN, and the DEKF for parameter adjustment is discussed. Section 5 provides empirical studies and comparisons to state-of-the-art algorithms to evaluate the efficacy of eT2QFNN. Finally, Sect. 6 concludes this book chapter.

2 RFID Localization System

RFID localization technology has three major components, i.e., RFID tags, RFID readers, and data processing subsystem. There are two types of RFID tags, i.e., the active and passive tag. The active RFID tag is battery powered and has its own transmitter. It is capable of sending out beacon message, i.e., tag ID and RSS information, actively at specified time window. The transmitted signal can be read up to 300-m radius. In contrast, passive tag does not have independent power source. It exploits the reader power signal, and thus it cannot actively send the beacon message in a fixed period of time. The signal can only be read around 1-m radius from the reader, which is very small compared to the manufacturing shopfloor. After that, the transmitted signal is read by the RFID reader. And then, it is propagated to the data processing subsystem, where the localization algorithm is executed [12]. The configuration of RFID localization is illustrated in Fig. 1.

The RSS information can be utilized to estimate the object location. It can be achieved by solving the two-way radar equation for a monostatic transmitter as

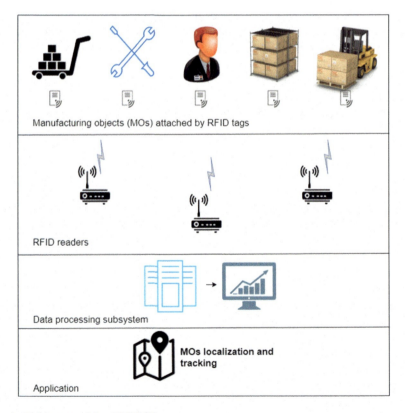

Fig. 1 Architecture of the eT2QFNN

per (1). The variables are explained as follows. P_T, G_T, λ, σ, and R are the reader signal power, the antenna gain, carrier wavelength, tag radar cross-section, and the distance between reader and tag, respectively. However, due to the occurrence of multipath effect and interference, the RSS information becomes unreliable and keeps changing over time. Consequently, the satisfying result cannot be obtained [3, 12]. Instead of employing (1) to locate the RFID, this research utilizes eT2QFNN to process the RSS information to obtain the precise object location:

$$P_R = \frac{G_T^2 \lambda^2 \sigma}{(4\pi)^2 R^4} \text{ (watts)} \quad (1)$$

One of many objectives of this research is to demonstrate the eT2QFNN to deal with the RFID localization problem. The RSS information of reference tags are utilized to train the network. The reference tags are tags placed at several known and static positions. The eT2QFNN can estimate the new observed tags location according to its RSS. It worth noting that eT2QFNN learning processes are achieved in the evolving mode, and it keeps the network parameters and structure adapted to the current data stream. This benefits the network to deal with multipath effect and interference occurred in the manufacturing shopfloor.

Now, suppose that there are I reference tags which are deployed in M locations, then the RSS measurement vector at nth time-step can be expressed as $X_n = [x_1 \ldots x_i \ldots x_I]^T$. Afterward, the network outputs for RFID localization problem can be formulated into a multiclass classification problem. As an illustration, if there exist $M = 4$ reference tags deployed in the shopfloor, it will indicate that the number of classes is equal to 4.

3 eT2QFNN Architecture

This section presents the network architecture of eT2QFNN. The network architecture, as illustrated in the Fig. 2, consists of a five-layer, multi-input-single-output (MISO) network structure. It is systematized into I input features, M outputs nodes, and K-term nodes for each input feature. The rule premise is compiled of IT2QMF and is expressed as follows:

$$\mathbf{R}_j : \textbf{IF } x_1 \text{ is close to } \widetilde{Q}_{1j} \text{ and} \ldots \text{ and } x_I \text{ is close to } \widetilde{Q}_{IK}, \textbf{ THEN } y_j^o = X_e \widetilde{\Omega}_j \quad (2)$$

where x_i and y_j^o are the ith input feature and the regression output of the oth class in the jth rule, respectively. $\widetilde{Q}_{ij} = [\overline{Q}_{ij}, \underline{Q}_{ij}]$ denotes the set of upper and lower linguistic term of IT2QMF, X_e is the extended input, and $\widetilde{\Omega}_j = [\overline{\Omega}_j, \underline{\Omega}_j]$ is the set of upper and lower consequent weight parameters which are defined as $\widetilde{\Omega}_j \in \Re^{M \times 2(I+1)}$.

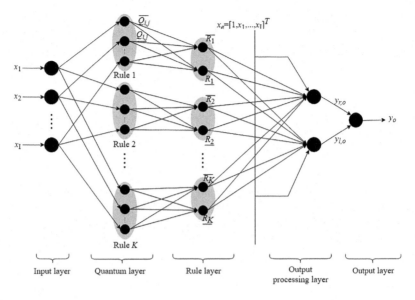

Fig. 2 The RFID localization system

The membership function applied in this study is different from the typical QMF and GMF. The QMF concept is extended into interval type-2 membership function with uncertain jump position. Thus, the network can identify overlaps between classes and capable to deal with the data stream uncertainties. The IT2QMF output of the jth rule for the ith input feature is given in (3):

$$\widetilde{Q}_{ij}(x_{ij}, \beta, m_{ij}, \widetilde{\theta}_{ij}) = \frac{1}{n_s} \sum_{r=1}^{n_s} \left[\left(\frac{1}{1 + \exp(-\beta x_i - m_{ij} + |\widetilde{\theta}_{ij}^r|)} \right) U(x_i; -\infty, m_{ij}) \right.$$
$$\left. + \left(\frac{\exp(-\beta(x_i - m_{ij} - |\widetilde{\theta}_{ij}^r|))}{1 + \exp(-\beta(x_i - m_{ij} - |\widetilde{\theta}_{ij}^r|))} \right) U(x_i; m_{ij}, \infty) \right] \quad (3)$$

where m_{ij}, β, and n_s are mean of ith input feature in jth rule, slope factor, and number of grades, respectively. $\widetilde{\theta}_{ij} = [\overline{\theta}_{ij}, \underline{\theta}_{ij}]$ is the set of uncertain jump position, and it is defined as $\widetilde{\theta}_{ij} \in \mathfrak{R}^{2 \times I \times n_s \times K}$. The upper and lower jump position is expressed as $\overline{\theta}_{ij} = [\overline{\theta}_{1j}^1 \ldots \overline{\theta}_{1j}^{n_s}; \ldots; \overline{\theta}_{Ij}^1 \ldots \overline{\theta}_{Ij}^{n_s}]$ and $\underline{\theta}_{ij}^r = [\underline{\theta}_{1j}^1 \ldots \underline{\theta}_{1j}^{n_s}; \ldots; \underline{\theta}_{Ij}^1 \ldots \underline{\theta}_{Ij}^{n_s}]$. It is defined that $\overline{\theta}_{ij}^r > \underline{\theta}_{ij}^r$, thus the execution of (3) leads to interval type-2 inference scheme which produces a footprint of uncertainties [20], and it can be clearly seen in Fig. 3. The eT2QFNN operation in each layer is presented in the following passages.

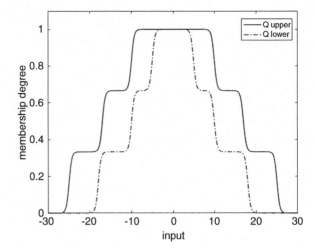

Fig. 3 Interval type-2 quantum membership function with $n_s = 3$

3.1 Input Layer

This layer performs no computation. The data stream is directly propagated to the next layer. The input at nth observation is defined by $X_n \in \Re^{1 \times I}$. And, the output of the ith node is given as follows:

$$u_i = x_i \tag{4}$$

3.2 Quantum Layer

This layer performs fuzzification step. IT2QMF is utilized to calculate the membership degrees of X_n in each rule. The number of rules is denoted as K. The quantum layer outputs mathematically can be obtained via (5) and (6):

$$\overline{Q}_{ij} = \widetilde{Q}_{ij}(x_i, \beta, m_{ij}, \overline{\theta}_{ij}^r) \tag{5}$$

$$\underline{Q}_{ij} = \widetilde{Q}_{ij}(x_i, \beta, m_{ij}, \underline{\theta}_{ij}^r) \tag{6}$$

3.3 Rule Layer

This layer functions to combine the membership degree of jth rule denoted as $\widetilde{R}_j = [\overline{R}_j, \underline{R}_j]$, which is known as spatial firing strength. This can be achieved by employing product T-norm of IT2QMF, as per (7) and (8). The set of upper and lower firing strengths are expressed as $\overline{R} = [\overline{R}_1 \ldots \overline{R}_K]$ and $\underline{R} = [\underline{R}_1 \ldots \underline{R}_K]$, respectively.

$$\overline{R}_j = \prod_{i=1}^{I} \overline{Q}_{ij} \qquad (7)$$

$$\underline{R}_j = \prod_{i=1}^{I} \underline{Q}_{ij} \qquad (8)$$

3.4 Output Processing Layer

The calculation of the two endpoints output, i.e., $y_{l,o}$ and $y_{r,o}$, is conducted here. These variables represent the lower and upper crisp output of the oth class, respectively. The design factor $[q_l, q_r]$ are employed to convert the interval type-2 variable to type-1 variable, and this is known as the type reduction procedure. This requires less iterative steps compared to the Karnik–Mendel (KM)-type reduction procedure [19]. The design factor will govern the proportion of upper and lower IT2QMF, and it is defined such that $q_l < q_r$. The design factor is adjusted using DEKF, and thus the proportion of upper and lower outputs $[y_l, y_r]$ keeps adapting to the data stream's uncertainties. The lower and upper outputs are given as:

$$y_{l,o} = \frac{(1-q_{l,o})\sum_{j=1}^{K} \underline{R}_j \, \underline{\Omega}_{jo} x_e^T + q_{l,o}\sum_{j=1}^{K} \overline{R}_j \, \underline{\Omega}_{jo} x_e^T}{\sum_{j=1}^{K}(\overline{R}_j + \underline{R}_j)} \qquad (9)$$

$$y_{r,o} = \frac{(1-q_{r,o})\sum_{j=1}^{K} \underline{R}_j \, \overline{\Omega}_{jo} x_e^T + q_{r,o}\sum_{j=1}^{K} \overline{R}_j \, \overline{\Omega}_{jo} x_e^T}{\sum_{j=1}^{K}(\overline{R}_j + \underline{R}_j)} \qquad (10)$$

where $q_l = [q_{l,1}, \ldots, q_{l,M}]$ and $q_r = [q_{r,1}, \ldots, q_{r,M}]$ are the design factors of all classes, while $\overline{\Omega}_{jo} = [\overline{w}_{ij}^o, \ldots, \overline{w}_{(I+1)j}^o]$ and $\underline{\Omega}_{jo} = [\underline{w}_{ij}^o, \ldots, \underline{w}_{(I+1)j}^o]$ express the upper and lower consequent weight parameters of the jth rule for the oth class. In addition, $x_e \in \Re^{(I+1)\times 1}$ is the extended input vector. For example, X_n has I input features $[x_1, \ldots, x_I]$, then the extended input vector is $X_e = [1, x_1, \ldots, x_I]$. The entry 1 is included to incorporate the intercept of the rule consequent and to prevent the untypical gradient [20].

3.5 Output Layer

The crisp network output of the oth class is the sum of $y_{l,o}$ and $y_{r,o}$ as per (11). Furthermore, if the network structure of eT2QFNN is utilized to deal with multiclass classification, the multi-model (MM) classifier can be employed to obtain the final classification decision. The MM classifier splits the multiclass classification problem into K binary subproblems, then K MISO eT2QFNN is built accordingly. The final class decision is the index number o of the highest output, as per (12):

$$y_o = y_{l,o} + y_{r,o} \tag{11}$$

$$y = \underset{o=1,\dots,M}{\arg\max}\, y_o \tag{12}$$

4 eT2QFNN Learning Policy

The online learning mechanism of eT2QFNN consists of two scenarios, i.e., the rule growing and the parameter adjustment which is executed in every iteration. The eT2QFNN starts its learning process with an empty rule base and keeps updating its parameters and network structure as the observation data comes in. The proposed learning scenario is presented in Algorithm 1, while Sects. 4.1 and 4.2 further explain the learning scenarios.

Algorithm 1 Learning policy of eT2QFNN

Define: input–output pair $X_n = [x_1, \dots, x_I]^T$, $T_n = [t_1, \dots, t_M]^T$, n_s, and η
\\Phase 1: **Rule Growing Mechanism**\\
If $K = 0$ **then**
Initiate the first rule via (26), (29), and (30)
else
Approximate the existing IT2QMF via (18)
Initiate a hypothetical rule R_{K+1} via (26), (27), and (31)
for $j = 1$ **to** $K + 1$
Calculate the statistical contribution E_j via (20)
end for
If $E_{K+1} \geq \rho \sum_{j=1}^{K} E_j$ **then**
$K = K + 1$
end if
end if
\\Phase 2: **Parameter Adjustment using DEKF**\\
If $K(n) = K(n-1)$ **then**
Calculate the spatial firing strength via (7) and (8)
Determine the winning rule j_w via (34)
Do DEKF adjustment mechanism on rule $\mathbf{R}_{j_w}$ via (37) and (39)
Update covariance matrix of the winning rule via (38)
else
Initialize the new rule consequents weight $\widetilde{\Omega}_{K+1}$ and covariance matrix and as (31) and (32)
for $j = 1$ **to** $K - 1$ **do**
$P_j(n) = P_j(n-1)\left(\frac{K^2+1}{K^2}\right)$
end for
end if

4.1 Rule Growing Mechanism

The eT2QFNN is capable of automatically evolving its fuzzy rule on demand using the proposed mGT2DQ method. First of all, it is achieved by forming a hypothetical rule from a newly seen sample. The initialization of hypothetical rule parameters is presented in the Sect. 4.2.1. Before it is added to the network, it is required to evaluate its significance. The significance of the jth rule is defined as an L^2-$norm$ of $E_{sig}(j)$ weighted by the input density function $p(x)$ as follows [7]:

$$E_{sig}(j) = ||\omega_j|| \left(\int_{R^I} \exp(-2||X - m_j||^2/\sigma_j^2) p(X) dX \right)^{1/2} \quad (13)$$

From the (13), it is obvious that the input density $p(X)$ greatly contributes to $E_{sig}(j)$. In practical, it is hard to be calculated a priori, because the data distribution is unknown. Huang et al. [5] and [6] calculated (13) analytically with the assumption of $p(X)$ being uniformly distributed. However, Zhang et al. [26] demonstrated that it leads to performance degradation for complex $p(X)$. To overcome this problem, Bortman and Aladjem [2] proposed Gaussian mixture model (GMM) to approximate the complicated data stream density. The mathematical formulation of GMM is given as:

$$\hat{p}(X) = \sum_{h=1}^{H} \alpha_h \mathcal{N}(X; v_h, \Sigma_h) \quad (14)$$

$$\mathcal{N}(X; v_h, \Sigma_h) = \exp(-(X - v_h)^T \Sigma_h^{-1}(X - v_h)) \quad (15)$$

where $\mathcal{N}(X; v_h, \Sigma_H)$ is the Gaussian function of variable X as per (15), with the mean vector $v_h \in \Re^I$, variance matrix $\Sigma_h \in^{I \times I}$, H denotes the number of mixture model, and α_h represents the mixing coefficients ($\sum_{h=1}^{H} \alpha_h = 1; \alpha_h > 0 \forall h$).

In the next step, the estimated significance of jth rule $\hat{E}_{sig}(j)$ is calculated. Vuković and Miljković [24] derived the mathematical formulation to obtain $\hat{E}_{sig}(j)$ by substituting (14) to (13) and then solving the closed-form analytical solution, it yields to the following result:

$$\hat{E}_{sig}(j) = ||\omega_j||(\pi^{I/2} \det(\Sigma_j)^{1/2} N_j A^T)^{1/2} \quad (16)$$

where $A = [\alpha_1, \ldots, \alpha_H]$ is the vector of GMM mixing coefficients, Σ_j denotes the positive definite weighting matrix which is expressed as $\Sigma_j = \text{diag}(\sigma_{1,j}^2, \ldots, \sigma_{I,j}^2)$, and N_j is given as:

$$N_j = [\mathcal{N}(m_j - v_1; 0, \Sigma_j/2 + \Sigma_1), \mathcal{N}(m_j - v_2; 0, \Sigma_j/2 + \Sigma_2), \ldots$$
$$\ldots, \mathcal{N}(m_j - v_H; 0, \Sigma_j/2 + \Sigma_H)] \quad (17)$$

where m_j is the mean vector of jth rule defined as $m_j = [m_{1,j}, \ldots, m_{I,j}]$. And, the GMM parameters v_h, Σ_h, and A can be calculated by exploiting $N_{history}$ prerecorded data. This technique is feasible and easy to implement, because the prerecorded input data is most likely to be stored especially in the era of data stream. The number of prerecorded data is somewhat smaller than the training data, and it is denoted as $N_{history} \ll N$. It is not problem-specific, and it can be set to a fixed value [14]. In this research, it is set as 50 for simplicity.

The method (16) could not, however, be applied directly to estimate the eT2QFNN rule significance, because eT2QFNN utilizes IT2QMF instead of Gaussian membership function (GMF). The key idea to overcome this problem is by approximating IT2QMF using interval type-2 Gaussian Membership Function (IT2GMF). The mathematical formulation of this approach can be written as follows:

$$\tilde{Q}_{i,j}(x_i, \beta, m_{i,j}, \tilde{\theta}_{ij}) \approx \tilde{\mu}_{i,j} = \exp\left(-\frac{(x_i - m_{i,j})^2}{\tilde{\sigma}_{i,j}}\right) \quad (18)$$

$$\tilde{\sigma}_{i,j} = [\underline{\sigma}_{i,j}, \overline{\sigma}_{i,j}]$$

$$\underline{\sigma}_{i,j} = \min \underline{\theta}_{i,j}; \quad \overline{\sigma}_{i,j} = \min \overline{\theta}_{i,j} \quad (19)$$

The mean of IT2GMF is defined to equal the mean of IT2QMF, i.e., m_{ij}. And, the width of upper and lower IT2GMF are obtained by taking the minimum value of $\tilde{\theta}_{ij}$ as per (19). By selecting these criteria, the whole area of IT2GMF will be located inside the area of IT2QMF. As illustrated in Fig. 4, both upper and lower area of IT2GMF covers the appropriate area of IT2QMF. Therefore, this approach can provide a good approximation of IT2QMF.

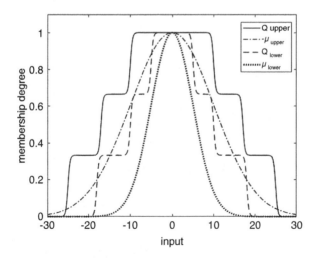

Fig. 4 The comparison result of IT2QMF and IT2GMF

Afterward, the proposed method to estimate eT2QFNN rule significance can be derived by executing (16) with the design factor as per (20). In (21) and (22), $\overline{\Omega}_j = [\overline{\Omega}_{j,1}, \ldots, \overline{\Omega}_{j,M}]^T$ and $\underline{\Omega}_j = [\underline{\Omega}_{j,1}, \ldots, \underline{\Omega}_{j,M}]^T$ are denoted as the upper and lower consequent weight parameters of all classes, respectively. The variance matrices $\overline{\Sigma}_j$ and $\underline{\Sigma}_j$ are formed of $\overline{\sigma}_{i,j}$ and $\underline{\sigma}_{i,j}$ as per (24), while $\overline{N}_j$ and $\underline{N}_j$ are given in (25). The hypothetical rule will be added to the network as a new rule $\mathbf{R}_{K+1}$ if it possesses statistical contribution over existing rules. The mathematical formulation of rule growing criterion is given in (23), where the constant $\rho \in (0, 1]$ is defined as the vigilance parameter and in this research it is fixed at 0.65 for simplicity.

$$\hat{E}_j = |\hat{E}_{j,l}| + |\hat{E}_{j,r}| \tag{20}$$

$$\hat{E}_{j,l} = \|q_l\| \cdot \|\overline{\Omega}_j\| \cdot (\pi^{1/2} \det(\overline{\Sigma}_j)^{1/2} \overline{N}_j A^T)^{1/2}$$
$$+ (1 - \|q_l\|) \cdot \|\underline{\Omega}_j\| \cdot (\pi^{1/2} \det(\underline{\Sigma}_j)^{1/2} \underline{N}_j A^T)^{1/2} \tag{21}$$

$$\hat{E}_{j,r} = \|q_r\| \cdot \|\overline{\Omega}_j\| \cdot (\pi^{1/2} \det(\overline{\Sigma}_j)^{1/2} \overline{N}_j A^T)^{1/2}$$
$$+ (1 - \|q_r\|) \cdot \|\underline{\Omega}_j\| \cdot (\pi^{1/2} \det(\underline{\Sigma}_j)^{1/2} \underline{N}_j A^T)^{1/2} \tag{22}$$

$$E_{K+1} \geq \rho \sum_{j=1}^{K} E_j \tag{23}$$

$$\overline{\Sigma}_j = \mathrm{diag}(\overline{\sigma}_{1,j}^2, \ldots, \overline{\sigma}_{I,j}^2), \quad \underline{\Sigma}_j = \mathrm{diag}(\underline{\sigma}_{1,j}^2, \ldots, \underline{\sigma}_{I,j}^2) \tag{24}$$

$$\overline{N}_j = [\mathcal{N}(m_j - v_1; 0, \overline{\Sigma}_j/2 + \Sigma_1), \mathcal{N}(m_j - v_2; 0, \overline{\Sigma}_j/2 + \Sigma_2), \ldots$$
$$\ldots, \mathcal{N}(m_j - v_H; 0, \overline{\Sigma}_j/2 + \Sigma_H)],$$

$$\underline{N}_j = [\mathcal{N}(m_j - v_1; 0, \underline{\Sigma}_j/2 + \Sigma_1), \mathcal{N}(m_j - v_2; 0, \underline{\Sigma}_j/2 + \Sigma_2), \ldots$$
$$\ldots, \mathcal{N}(m_j - v_H; 0, \underline{\Sigma}_j/2 + \Sigma_H)] \tag{25}$$

4.2 Parameter Adjustment

This phase comprises of two alternative strategies. The first strategy is carried out to form a hypothetical rule. It will be added into the network structure if the condition in (23) is satisfied. This strategy is called the fuzzy rule initialization. And, the second mechanism is executed whenever (23) is not satisfied. It is aimed to adjust the network parameters according to the current data stream. This is called the winning rule update. These strategies are elaborated in the following sub-subsections.

4.2.1 Fuzzy Rule Initialization

The rule growing mechanism first of all is conducted by forming a hypothetical rule according to the current data stream. The data at nth time-step X_n is assigned as the new mean of IT2QMF, as per (26). And then, the new jump position is achieved via distance-based formulation inspired by Lin and Chen [8], as per (27). In this research, however, it is modified such that the new distance $\tilde{\sigma}_{i,K+1}$ is obtained utilizing the mixed mean of GMM $\hat{v}$, as per (28). Thanks to the GMM features which is able to approximate the mean and variance of very complex input. For this reason, instead of using $\tilde{\sigma}_{i,K+1}$ to calculate the jump position of the first rule in (29), the eT2QFNN utilizes the diagonal entries of the mixed variance matrix, as per (30). The constant δ_1 is introduced to create the footprint of uncertainty. In this study, it is set $\delta_1 = 0.7$ for simplicity.

In the next stage, the consequent weight parameters of hypothetical rule are determined. It is equal to the consequent weight of the winning rule as per (31). The key idea behind this strategy is to acquire the knowledge of winning rule in terms of representing the current data stream [13]. The way to select the winning rule is presented in Sect. 4.2.2. Finally, if the hypothetical rule passes the evaluation criterion in (23), it is added as the new rule ($\mathbf{R}_{K+1}$), and its covariance matrix is initialized via (32).

In contrast, a consideration is required to adjust the covariance matrices of other rules, because the new rule formation corrupts those matrices. This phenomenon has been investigated in SPLAFIS [13]; the research revealed that those matrices need to be readjusted. The proper readjustment technique is achieved by multiplication of those matrices and $\left(\frac{K^2+1}{K^2}\right)$ as per (33). This strategy is signified to take into account the contribution that a new rule would have if it existed from the first iteration. It, therefore, will decrease the corruption effect [13].

$$m_{K+1} = X_n \qquad (26)$$

$$\overline{\theta}^r_{i,K+1} = \frac{1}{((n_s+1)/2)} \cdot r \cdot \overline{\sigma}_{i,K+1},$$

$$\underline{\theta}^r_{i,K+1} = \frac{1}{((n_s+1)/2)} \cdot r \cdot \underline{\sigma}_{i,K+1} \qquad (27)$$

$$\overline{\sigma}_{i,K+1} = |X_n - \hat{v}|, \quad \underline{\sigma}_{i,K+1} = \delta_1 \cdot \overline{\sigma}_{i,K+1} \qquad (28)$$

$$\hat{v} = \sum_{h=1}^{H} \alpha_h \cdot v_h$$

$$\overline{\theta}^r_{i,1} = \frac{1}{((n_s+1)/2)} \cdot r \cdot \overline{\sigma}_{i,1},$$

$$\underline{\theta}^r_{i,1} = \frac{1}{((n_s + 1)/2)} \cdot r \cdot \underline{\sigma}_{i,1} \tag{29}$$

$$\overline{\sigma}_{i,1} = \hat{\sigma}_i, \quad \underline{\sigma}_{i,1} = \delta_1 \cdot \overline{\sigma}_{i,1} \tag{30}$$

$$\hat{\Sigma} = \sum_{h=1}^{H} \Sigma_h \cdot v_h, \quad \hat{\Sigma} = \mathrm{diag}(\hat{\sigma}_1^2, \ldots, \hat{\sigma}_I^2)$$

$$\widetilde{\Omega}_{K+1} = \widetilde{\Omega}_{j_w} \tag{31}$$

$$P_{K+1}(n) = I_{Z \times Z} \tag{32}$$

$$P_j(n) = \left(\frac{K^2 + 1}{K^2}\right) P_j(n-1) \tag{33}$$

4.2.2 Winning Rule Update

The hypothetical rule would not be added to the network structure if it failed the evaluation in (23). To maintain the eT2QFNN performance, the network parameters are required to be adjusted according to the information provided by the current data stream. In this research, the adjustment is only undertaken on the winning rule which is defined as a rule having the highest average of the spatial firing strength. The mathematical formulation is given in (34). It worth noting that spatial firing strength represents the degree to which the rule antecedent part is satisfied. The rule having higher firing strength possesses higher correlation to the current data stream [16], and therefore it deserves to be adjusted.

$$j_w - \arg\max_j \widetilde{R}_j \tag{34}$$

$$\widetilde{R}_j = \frac{\overline{R}_j + \underline{R}_j}{2} \tag{35}$$

Previously, DEKF is employed to adjust the winning rule parameters of type-1 fuzzy neural network. It is capable of maintaining local learning property of EIS, because it can adjust parameters locally [13]. In this research, DEKF is utilized to update the winning rule parameters of eT2QFNN. The local parameters are classified by rule, i.e., the parameters in the same rule are grouped together. This leads to the formation of block-diagonal covariance matrix $\breve{P}(n)$ as per (36). There is only one block covariance matrix updated in each time-step, i.e., $P_{j_w}(n)$. The localized adjustment property of DEKF enhances the algorithm efficiency in terms of computational complexity and memory requirements; moreover, it still maintains the same robustness as the EKF [22].

$$\tilde{P}(n) = \begin{bmatrix} P_1(n) & \cdots & 0 & \cdots & 0 \\ \vdots & \ddots & & & \vdots \\ 0 & & P_j(n) & & 0 \\ \vdots & & & \ddots & \vdots \\ 0 & \cdots & 0 & \cdots & P_K(n) \end{bmatrix} \quad (36)$$

The mathematical formulations of DEKF algorithm are given in (37)–(39). The designation of each parameter in the equation is as follows. $G_{j_w}(n)$ and $P_{j_w}(n)$ are the Kalman gain matrix and covariance matrix, respectively. The covariance matrix represents the interaction between each pair of the parameters in the network. $\vec{\theta}_{j_w}(n)$ is the parameter vector of $\mathbf{R}_{j_w}$ at nth iteration, and it consists of all the network parameters which are about to be adjusted. It is expressed as $\vec{\theta}_{j_w}(n) = [\underline{\Omega}_{j_w}^T, \overline{\Omega}_{j_w}^T, q_l^T, q_r^T, m_{j_w}^T, \underline{\theta}_{j_w}^T, \overline{\theta}_{j_w}^T]^T$, which is respectively given in (40)–(46). The length of $\vec{\theta}_{j_w}(n)$ is equal to $Z = 2 \times M \times (2 + I) + I \times (2 \times n_s + 1)$. The Jacobian matrix $H_{k_w}(n)$, presented in (47), contains the output gradient with respect to the network parameters, and it is arranged into Z-by-M matrix. The gradient vectors are specified in (48) and is calculated using (49)–(52). The output and target vectors are defined as $y(n) = [y_1(n) \ldots y_M(n)]$ and $t(n) = [t_1(n) \ldots t_M(n)]$. It is utilized to calculate the error vector in (39). The last parameter, η, is a learning rate parameter [22]. This completes the second strategy to maintain the network adapted to the current data stream.

$$G_{j_w}(n) = P_{j_w}(n-1) H_{j_w}(n) [\eta I_{M \times M} + H_{k_w}^T(n) P_{j_w}(n-1) H_{j_w}(n)]^{-1} \quad (37)$$

$$P_{j_w}(n) = [I_{Z \times Z} - G_{j_w}(n) H_{j_w}^T(n)] P_{j_w}(n-1) \quad (38)$$

$$\vec{\theta}_{j_w}(n) = \vec{\theta}_{j_w}(n-1) + G_{j_w}(n)[t(n) - y(n)] \quad (39)$$

$$\underline{\Omega}_{j_w} = [\underline{\Omega}_{1,j_w}^1, \ldots, \underline{\Omega}_{I+1,j_w}^1, \ldots, \underline{\Omega}_{1,j_w}^M, \ldots, \underline{\Omega}_{I+1,j_w}^M]^T \quad (40)$$

$$\overline{\Omega}_{j_w} = [\overline{\Omega}_{1,j_w}^1, \ldots, \overline{\Omega}_{I+1,j_w}^1, \ldots, \overline{\Omega}_{1,j_w}^M, \ldots, \overline{\Omega}_{I+1,j_w}^M]^T \quad (41)$$

$$q_l = [q_{l,1}, \ldots, q_{l,M}]^T \quad (42)$$

$$q_r = [q_{r,1}, \ldots, q_{r,M}]^T \quad (43)$$

$$m_{j_w} = [m_{1,j_w}, \ldots, m_{I,j_w}]^T \quad (44)$$

$$\underline{\theta}_{j_w} = [\underline{\theta}_{1,j_w}^1, \ldots, \underline{\theta}_{I,j_w}^1, \ldots, \underline{\theta}_{1,j_w}^{n_s}, \ldots, \underline{\theta}_{I,j_w}^{n_s}]^T \quad (45)$$

$$\overline{\theta}_{j_w} = [\overline{\theta}_{1,j_w}^1, \ldots, \overline{\theta}_{I,j_w}^1, \ldots, \overline{\theta}_{1,j_w}^{n_s}, \ldots, \overline{\theta}_{I,j_w}^{n_s}]^T \quad (46)$$

$$H_{j_w}(n) = \begin{bmatrix} \frac{\partial y_1}{\partial \underline{\Omega}_{j_w,1}} & \cdots & 0 & \cdots & 0 \\ 0 & \cdots & \frac{\partial y_o}{\partial \underline{\Omega}_{j_w,o}} & \cdots & 0 \\ 0 & \cdots & 0 & \cdots & \frac{\partial y_M}{\partial \underline{\Omega}_{j_w,M}} \\ \frac{\partial y_1}{\partial \overline{\Omega}_{j_w,1}} & \cdots & 0 & \cdots & 0 \\ 0 & \cdots & \frac{\partial y_o}{\partial \overline{\Omega}_{j_w,o}} & \cdots & 0 \\ 0 & \cdots & 0 & \cdots & \frac{\partial y_M}{\partial \overline{\Omega}_{j_w,M}} \\ \frac{\partial y_1}{\partial q_{l,1}} & \cdots & 0 & \cdots & 0 \\ 0 & \cdots & \frac{\partial y_o}{\partial q_{l,o}} & \cdots & 0 \\ 0 & \cdots & 0 & \cdots & \frac{\partial y_M}{\partial q_{l,M}} \\ \frac{\partial y_1}{\partial q_{r,1}} & \cdots & 0 & \cdots & 0 \\ 0 & \cdots & \frac{\partial y_o}{\partial q_{r,o}} & \cdots & 0 \\ 0 & \cdots & 0 & \cdots & \frac{\partial y_M}{\partial q_{r,M}} \\ \frac{\partial y_1}{\partial m_{j_w,1}} & \cdots & \frac{\partial y_o}{\partial m_{j_w,o}} & \cdots & \frac{\partial y_M}{\partial m_{j_w,M}} \\ \frac{\partial y_1}{\partial \underline{\theta}_{j_w,1}} & \cdots & \frac{\partial y_o}{\partial \underline{\theta}_{j_w,o}} & \cdots & \frac{\partial y_M}{\partial \underline{\theta}_{j_w,M}} \\ \frac{\partial y_1}{\partial \overline{\theta}_{j_w,1}} & \cdots & \frac{\partial y_o}{\partial \overline{\theta}_{j_w,o}} & \cdots & \frac{\partial y_M}{\partial \overline{\theta}_{j_w,M}} \end{bmatrix} \quad (47)$$

$$\frac{\partial y_o}{\partial \underline{\Omega}_{j_w,o}} = \left[\frac{\partial y_o}{\partial \underline{w}^o_{1,j_w}}, \ldots, \frac{\partial y_o}{\partial \underline{w}^o_{I+1,j_w}} \right]^T, \quad \frac{\partial y_o}{\partial \overline{\Omega}_{j_w,o}} = \left[\frac{\partial y_o}{\partial \overline{w}^o_{1,j_w}}, \ldots, \frac{\partial y_o}{\partial \overline{w}^o_{I+1,j_w}} \right]^T,$$

$$\frac{\partial y_o}{\partial m_{j_w,o}} = \left[\frac{\partial y_m}{\partial m^o_{1,j_w}}, \ldots, \frac{\partial y_o}{\partial m^o_{I,j_w}} \right]^T,$$

$$\frac{\partial y_o}{\partial \underline{\theta}_{j_w,o}} = \left[\frac{\partial y_o}{\partial \underline{\theta}^1_{1,j_w,o}}, \ldots, \frac{\partial y_o}{\partial \underline{\theta}^1_{I,j_w,o}}, \ldots, \frac{\partial y_o}{\partial \underline{\theta}^{n_s}_{1,j_w,o}}, \ldots, \frac{\partial y_o}{\partial \underline{\theta}^{n_s}_{I,j_w,o}} \right]^T,$$

$$\frac{\partial y_o}{\partial \overline{\theta}_{j_w,o}} = \left[\frac{\partial y_o}{\partial \overline{\theta}^1_{1,j_w,o}}, \ldots, \frac{\partial y_o}{\partial \overline{\theta}^1_{I,j_w,o}}, \ldots, \frac{\partial y_o}{\partial \overline{\theta}^{n_s}_{1,j_w,o}}, \ldots, \frac{\partial y_o}{\partial \overline{\theta}^{n_s}_{I,j_w,o}} \right]^T \quad (48)$$

$$\frac{\partial y_o}{\partial \underline{w}^o_{1,j_w}} = \left[\frac{(1-q_{l,o})\underline{R}_{j_w} + q_{l,o}\overline{R}_{j_w}}{\sum_{j=1}^{K}(\underline{R}_{j_w} + \overline{R}_{j_w})} \right] x_{e,i}(n),$$

$$\frac{y_o}{\partial \overline{w}^o_{1,j_w}} = \left[\frac{(1-q_{r,o})\underline{R}_{j_w} + q_{r,o}\overline{R}_{j_w}}{\sum_{j=1}^{K}(\underline{R}_{j_w} + \overline{R}_{j_w})} \right] x_{e,i}(n) \quad (49)$$

$$\frac{\partial y_o}{\partial q_{l,o}} = \left[\frac{-\sum_{j=1}^{K}\underline{R}_j\underline{\Omega}_o + \sum_{j=1}^{K}\overline{R}_j\underline{\Omega}_o}{\sum_{j=1}^{K}(\underline{R}_{jw} + \overline{R}_{jw})}\right]X_e(n),$$

$$\frac{\partial y_o}{\partial q_{r,o}} = \left[\frac{-\sum_{j=1}^{K}\underline{R\Omega}_o + \sum_{j=1}^{K}\overline{R\Omega}_o}{\sum_{j=1}^{K}(\underline{R}_{jw} + \overline{R}_{jw})}\right]X_e(n) \quad (50)$$

$$\frac{\partial y_o}{\partial m_{i,jw}} = \frac{\partial y_o}{\partial y_{l,o}}\left[\frac{\partial y_{l,o}}{\partial \underline{R}_{jw}}\frac{\partial \underline{R}_{jw}}{\partial m_{i,jw}} + \frac{\partial y_{l,o}}{\partial \overline{R}_{jw}}\frac{\partial \overline{R}_{jw}}{\partial m_{i,jw}}\right]$$
$$+ \frac{\partial y_o}{\partial y_{r,o}}\left[\frac{\partial y_{r,o}}{\partial \underline{R}_{jw}}\frac{\partial \underline{R}_{jw}}{\partial m_{i,jw}} + \frac{\partial y_{r,o}}{\partial \overline{R}_{jw}}\frac{\partial \overline{R}_{jw}}{\partial m_{i,jw}}\right] \quad (51)$$

$$\frac{\partial y_o}{\partial \underline{\theta}^r_{i,jw}} = \frac{\partial y_o}{\partial y_{l,o}}\frac{\partial y_{l,o}}{\partial \underline{R}_{jw}}\frac{\partial \underline{R}_{jw}}{\partial \underline{\theta}^r_{i,jw}} + \frac{\partial y_o}{\partial y_{r,o}}\frac{\partial y_{r,o}}{\partial \underline{R}_{jw}}\frac{\partial \underline{R}_{jw}}{\partial \underline{\theta}^r_{i,jw}},$$

$$\frac{\partial y_o}{\partial \overline{\theta}^r_{i,jw}} = \frac{\partial y_o}{\partial y_{l,o}}\frac{\partial y_{l,o}}{\partial \overline{R}_{jw}}\frac{\partial \overline{R}_{jw}}{\partial \overline{\theta}^r_{i,jw}} + \frac{\partial y_o}{\partial y_{r,o}}\frac{\partial y_{r,o}}{\partial \overline{R}_{jw}}\frac{\partial \overline{R}_{jw}}{\partial \overline{\theta}^r_{i,jw}} \quad (52)$$

$$\frac{\partial y_o}{\partial y_{l,o}} = \frac{\partial y_o}{\partial y_{l,o}} = 1$$

$$\frac{\partial y_{l,o}}{\partial \underline{R}_{jw}} = \frac{(1-q_{l,o})\underline{\Omega}_{jw,o}X_e(n)}{\sum_{j=1}^{K}(\underline{R}_j + \overline{R}_j)} - \frac{(1-q_{l,o})\underline{R\Omega}_m X_e(n) + q_{l,o}\overline{R\Omega}_m X_e(n)}{(\sum_{j=1}^{K}(\underline{R}_j + \overline{R}_j))^2}$$

$$\frac{\partial y_{l,o}}{\partial \overline{R}_{jw}} = \frac{q_{l,o}\overline{\Omega}_{jw,o}X_e(n)}{\sum_{j=1}^{K}(\underline{R}_j + \overline{R}_j)} - \frac{(1-q_{l,o})\underline{R\Omega}_m X_e(n) + q_{l,o}\overline{R\Omega}_m X_e(n)}{(\sum_{j=1}^{K}(\underline{R}_j + \overline{R}_j))^2}$$

$$\frac{\partial y_{r,o}}{\partial \underline{R}_{jw}} = \frac{(1-q_{r,o})\underline{\Omega}_{jw,o}X_e(n)}{\sum_{j=1}^{K}(\underline{R}_j + \overline{R}_j)} - \frac{(1-q_{r,o})\underline{R\Omega}_m X_e(n) + q_{r,o}\overline{R\Omega}_m X_e(n)}{(\sum_{j=1}^{K}(\underline{R}_j + \overline{R}_j))^2}$$

$$\frac{\partial y_{r,o}}{\partial \overline{R}_{jw}} = \frac{q_{r,o}\overline{\Omega}_{jw,o}X_e(n)}{\sum_{j=1}^{K}(\underline{R}_j + \overline{R}_j)} - \frac{(1-q_{r,o})\underline{R\Omega}_m X_e(n) + q_{r,o}\overline{R\Omega}_m X_e(n)}{(\sum_{j=1}^{K}(\underline{R}_j + \overline{R}_j))^2}$$

$$\frac{\underline{R}_{jw}}{\partial m_{i,jw}} = \prod_{i'=1, i'\neq i}^{I}\underline{Q}_{i',jw}(m_{i',jw}) \cdot \frac{1}{n_s}\sum_{r=1}^{n_s}\widetilde{\Psi}_r(\underline{\theta}^r_{i'jw})$$

$$\frac{\overline{R}_{jw}}{\partial m_{i,jw}} = \prod_{i'=1, i'\neq i}^{I}\overline{Q}_{i',jw}(m_{i',jw}) \cdot \frac{1}{n_s}\sum_{r=1}^{n_s}\widetilde{\Psi}_r(\overline{\theta}^r_{i'jw})$$

$$\frac{\partial \underline{R}_{j_w}}{\partial \underline{\theta}^r_{i,j_w}} = \prod_{i'=1, i' \neq i}^{I} \underline{Q}_{i',j_w}(\underline{\theta}^r_{i,j_w}) \cdot \frac{1}{n_s} \tilde{\Phi}_r(\underline{\theta}^r_{i,j_w})$$

$$\frac{\partial \overline{R}_{j_w}}{\partial \overline{\theta}^r_{i,j_w}} = \prod_{i'=1, i' \neq i}^{I} \overline{Q}_{i',j_w}(\overline{\theta}^r_{i,j_w}) \cdot \frac{1}{n_s} \tilde{\Phi}_r(\overline{\theta}^r_{i,j_w})$$

$$\tilde{\Psi}_r(\theta) = \begin{cases} -\frac{\beta \exp(-\beta(x_{i'} - m_{i',j_w} + |\theta|))}{(1 + \exp(-\beta(x_{i'} - m_{i',j_w} + |\theta|)))^2}, & -\infty < x_{i'} < m_{i'j_w} \\ \frac{\beta \exp(-\beta(x_{i'} - m_{i',j_w} + |\theta|))}{(1 + \exp(-\beta(x_{i'} - m_{i',j_w} + |\theta|)))^2}, & m_{i'j} \leq x_{i'} < \infty \end{cases}$$

$$\tilde{\Phi}_r(\theta) = \begin{cases} \frac{\beta \exp(-\beta(x_{i'} - m_{i',j_w} + \theta))}{(1 + \exp(-\beta(x_{i'} - m_{i',j_w} + \theta)))^2}, & -\infty < x_{i'} < m_{i'j_w}, \theta \geq 0 \\ -\frac{\beta \exp(-\beta(x_{i'} - m_{i',j_w} + \theta))}{(1 + \exp(-\beta(x_{i'} - m_{i',j_w} + \theta)))^2}, & m_{i'j} \leq x_{i'} < \infty, \theta \geq 0 \\ \frac{\beta \exp(-\beta(x_{i'} - m_{i',j_w} - \theta))}{(1 + \exp(-\beta(x_{i'} - m_{i',j_w} - \theta)))^2}, & -\infty < x_{i'} < m_{i'j_w}, \theta < 0 \\ -\frac{\beta \exp(-\beta(x_{i'} - m_{i',j_w} - \theta))}{(1 + \exp(-\beta(x_{i'} - m_{i',j_w} - \theta)))^2}, & m_{i'j} \leq x_{i'} < \infty, \theta < 0 \end{cases}$$

5 Experiments and Data Analysis

In this section, the application of eT2QFNN for RFID localization is discussed. Several experiments are conducted in the real-world environment to evaluate the efficacy of eT2QFNN embracing the MM classifier. The results are compared against four state-of-the-art algorithms: gClass [15], pClass [17], eT2Class [19], and eT2ELM [18]. Five performance metrics are used, and those are classification rate, the number of fuzzy rules, and the time for execution, training, and testing processes (execution time). The experiments are conducted under cross-validation and periodic hold-out scenario. The technical details of this experiments are elaborated in Sect. 5.1, while Sect. 5.2 presents the consolidated results.

5.1 Experiment Setup

The experiments were conducted at the SIMTech Laboratory, Singapore. The environment is arranged to resemble the RFID smart rack system. The system utilizes RFID technology to improve the workflow efficiency by providing the static location of tools and materials for production purposes. As illustrated in Fig. 5, this system consists of one RFID reader, four passive RFID tags as references which are fixed in four locations, and a data processing subsystem. The dimension of rack is 1510 mm × 600 mm × 2020 mm. The rack has five shelves, and each shelf can

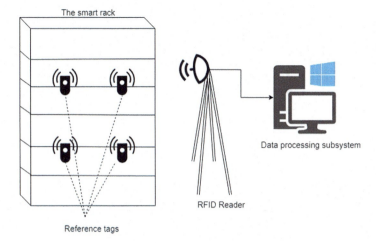

Fig. 5 The illustration of RFID smart rack

load up to six test objects. The number of reference tags indicates that there are four class label considered in this experiments. The RFID reader is placed at 1000-mm distance in front of the rack. The antenna is at 2200-mm height above the ground. The reader is then connected to an RFID receiver which functions to transmit the signal into a data processing subsystem. Ethernet links are utilized to accommodate the signal transmission. Notably, one may install more reference tags and RFID reader for larger smart rack system to increase the localization accuracy [3].

The data processing subsystem has two main components, i.e., data acquisition and the algorithm execution component. The Microsoft Visual C++-based PC application is developed to acquire the RSS information data from all tags, while the localization algorithm is executed on the MATLAB 2018a online environment. The Reader is configured to report the RSS information every 1 s. The experiment had been conducted for 20 h. There are 283,100 observations obtained via the experiment, each reference tag sent up to 70,775 observations. It is obvious that these data obtained from the same real-time experiment, and therefore all of them pose the same distribution. Finally, the observation data can be processed to identify the object location by executing the localization algorithm.

5.2 Comparison with Existing Results

To further investigate the performance of eT2QFNN, it is compared to the existing classification method, i.e., gClass, pClass, eT2Class, and eT2ELM. The comparison is conducted in the same computational environments, i.e., MATLAB Online R2018a. The gClass and pClass utilize generalized type-1 fuzzy rule, while the eT2Class and eT2ELM are built upon generalized type-2 fuzzy rule. These methods are able to grow and prune its network structure according to the information provided by the current data stream. All of them except pClass are also capable to merge similar rules. Further, the eT2ELM is encompassed with active learning and feature selection scenario which helps to discard the unnecessary training data. The eT2QFNN utilizes MM classifier, while others make use of MIMO classifier. This classifier is very dependent on rule consequents, because it establishes a first-order polynomial for each class. Another characteristic of this classifier is a transformation of true class label to either 0 or 1. As an illustration, if there are four class labels and the target class is two, then it will be converted into [0, 1, 0, 0] [17].

There are two experiments conducted to test the algorithm, i.e., 10-fold cross-validation and direct partition experiments. The first experiment is aimed to test the algorithms consistency while delivering the result. The experiment is started by dividing the data into 10 folds, and nine-fold data is for training, while one-fold data is for validation. The performance metrics are achieved by averaging the results of 10-fold cross-validation. In the second experiment, the periodic hold-out evaluation scenario is conducted. The algorithms take 50,000 data for training and 233,100 for validation. The classification rate for the experiments is measured only in the validation phase. In contrast, the execution time is taken into account since the beginning of training phase. In this experiment, we vary $n_s = [0, 10]$ and set $\eta = 0.001$. The results are presented in Tables 1 and 2.

It can be seen from Table 1 that the eT2QFNN delivers most reliable classification rates. Although it employs four sub-models to obtain this result which burden the computation, eT2QFNN has the fastest execution time second to pClass. In terms of network complexity, eT2QFNN generates a comparable number of fuzzy rules. It worth noting that eT2QFNN is not encompassed with the rule merging and pruning scenario. Further, the number of eT2QFNN rules is less than eT2ELM

Table 1 Results of the cross-validation experiment compared to the benchmarked algorithms

Algorithms	Results	
MM-eT2QFNN	Classification rate	**0.99 ± 0.05**
	Rule	6.23 ± 0.68
	Execution time	**618.63 ± 31.64**
gClass	Classification rate	0.97 ± 0.006
	Rule	2.4 ± 1.2
	Execution time	1004.36 ± 97.78
pClass	Classification rate	0.97
	Rule	2
	Execution time	**369.28 ± 9.99**
eT2Class	Classification rate	0.97 ± 0.008
	Rule	2
	Execution time	447.36 ± 11.60
eT2ELM	Classification rate	0.95 ± 0.018
	Rule	37.2 ± 5.95
	Execution time	1324.1 ± 109.47

Table 2 Results of the direct partition experiment compared to the benchmarked algorithms

Algorithms	Results	
MM-eT2QFNN	Classification rate	0.97
	Rule	**4.75**
	Execution time	**131.5**
gClass	Classification rate	0.99
	Rule	4
	Execution time	290.83
pClass	Classification rate	0.98
	Rule	2
	Execution time	225.8
eT2Class	Classification rate	0.98
	Rule	2
	Execution time	330.71
eT2ELM	Classification rate	0.98
	Rule	5
	Execution time	**41.73**

which has rule pruning and merging scenario. Table 2 confirms the consistency of eT2QFNN while delivering good result. It worth noting that the second experiment utilizes less data for training; however, eT2QFNN maintains the classification rate around 97% which is still comparable to other methods. The execution time is lower than other methods except eT2ELM. It is obvious because eT2ELM has the online active scenario which can reduce the training sample.

6 Conclusions

This paper presents an evolving model based on the EIS, namely eT2QFNN. The fuzzification layer relies on IT2QMF, which has a graded membership degree and footprints of uncertainties. The IT2QMF is the extended version of QMF which are able to both capture the input uncertainties and to identify overlaps between input classes. The eT2QFNN works fully in the evolving mode; that is, the network parameters and the number of rules are adjusted and generated on the fly. The parameter adjustment scenario is achieved via DEKF. Meanwhile, the rule growing mechanism is conducted by measuring the statistical contribution of the hypothetical rule. The new rule is formed when its statistical contribution is higher than the sum of others multiplied by vigilance parameter. The proposed method is utilized to predict the class label of an object according to the RSS information provided by the reference tags. The conducted experiments simulate the RFID smart rack system which is constructed by four reference tags, one RFID reader, and a data processing subsystem to execute the algorithm. The experimental results show that eT2QFNN is capable of delivering comparable accuracy benchmarked to state-of-the-art algorithms while maintaining low execution time and compact network.

Acknowledgements This project is fully supported by NTU start-up grant and Ministry of Education Tier 1 Research Grant. We also would like to thank Singapore Institute of Manufacturing Technology which provided the RFID data that greatly assisted the research. The third author acknowledges the support by the LCM–K2 Center within the framework of the Austrian COMET-K2 program.

References

1. Angelov, P., Zhou, X.: Evolving fuzzy systems from data streams in real-time. In: 2006 International Symposium on Evolving Fuzzy Systems, pp. 29–35. IEEE, Piscataway (2006)
2. Bortman, M., Aladjem, M.: A growing and pruning method for radial basis function networks. IEEE Trans. Neural Netw. **20**(6), 1039–1045 (2009)
3. Chai, J., Wu, C., Zhao, C., Chi, H.L., Wang, X., Ling, B.W.K., Teo, K.L.: Reference tag supported RFID tracking using robust support vector regression and Kalman filter. Adv. Eng. Inform. **32**, 1–10 (2017)
4. Chen, C.H., Lin, C.J., Lin, C.T.: An efficient quantum neuro-fuzzy classifier based on fuzzy entropy and compensatory operation. Soft Comput. **12**(6), 567–583 (2008)
5. Huang, G.B., Saratchandran, P., Sundararajan, N.: A recursive growing and pruning RBF (GAP-RBF) algorithm for function approximations. In: Proceedings of the 4th International Conference on Control and Automation, 2003. ICCA'03, pp. 491–495. IEEE, Piscataway (2003)
6. Huang, G.B., Saratchandran, P., Sundararajan, N.: An efficient sequential learning algorithm for growing and pruning RBF (GAP-RBF) networks. IEEE Trans. Syst. Man Cybern. Part B (Cybern.) **34**(6), 2284–2292 (2004)
7. Huang, G.B., Saratchandran, P., Sundararajan, N.: A generalized growing and pruning RBF (GGAP-RBF) neural network for function approximation. IEEE Trans. Neural Netw. **16**(1), 57–67 (2005)
8. Lin, C.J., Chen, C.H.: A self-organizing quantum neural fuzzy network and its applications. Cybern. Syst. Int. J. **37**(8), 839–859 (2006)
9. Lughofer, E.: FLEXFIS: a robust incremental learning approach for evolving Takagi–Sugeno fuzzy models. IEEE Trans. Fuzzy Syst. **16**(6), 1393–1410 (2008)
10. Lughofer, E., Cernuda, C., Kindermann, S., Pratama, M.: Generalized smart evolving fuzzy systems. Evol. Syst. **6**(4), 269–292 (2015)
11. Ni, L.M., Liu, Y., Lau, Y.C., Patil, A.P.: LANDMARC: indoor location sensing using active RFID. Wirel. Netw. **10**(6), 701–710 (2004)
12. Ni, L.M., Zhang, D., Souryal, M.R.: RFID-based localization and tracking technologies. IEEE Wirel. Commun. **18**(2) (2011)
13. Oentaryo, R.J., Er, M.J., Linn, S., Li, X.: Online probabilistic learning for fuzzy inference system. Expert Syst. Appl. **41**(11), 5082–5096 (2014)
14. Pratama, M.: PANFIS++: a generalized approach to evolving learning. arXiv:170502476 (2017, preprint)
15. Pratama, M., Lu, J., Anavatti, S., Lughofer, E., Lim, C.: An incremental meta-cognitive-based scaffolding fuzzy neural network. Neurocomputing **171**, (2015) 10.1016/j.neucom.2015.06.022
16. Pratama, M., Anavatti, S.G., Angelov, P.P., Lughofer, E.: PANFIS: a novel incremental learning machine. IEEE Trans. Neural Netw. Learn. Syst. **25**(1), 55–68 (2014)
17. Pratama, M., Anavatti, S.G., Joo, M., Lughofer, E.: pClass: an effective classifier for streaming examples. IEEE Trans. Fuzzy Syst. **23**(2), 369–386 (2015)
18. Pratama, M., Lu, J., Zhang, G.: A novel meta-cognitive extreme learning machine to learning from data streams. In: 2015 IEEE International Conference on Systems, Man, and Cybernetics (SMC), pp. 2792–2797. IEEE, Piscataway (2015)

19. Pratama, M., Lu, J., Zhang, G.: Evolving type-2 fuzzy classifier. IEEE Trans. Fuzzy Syst. **24**(3), 574–589 (2016)
20. Pratama, M., Lughofer, E., Er, M.J., Rahayu, W., Dillon, T.: Evolving type-2 recurrent fuzzy neural network. In: 2016 International Joint Conference on Neural Networks (IJCNN), pp. 1841–1848. IEEE, Piscataway (2016)
21. Purushothaman, G., Karayiannis, N.B.: Quantum neural networks (QNNs): inherently fuzzy feedforward neural networks. IEEE Trans. Neural Netw. **8**(3), 679–693 (1997)
22. Puskorius, G.V., Feldkamp, L.A.: Neurocontrol of nonlinear dynamical systems with Kalman filter trained recurrent networks. IEEE Trans. Neural Netw. **5**(2), 279–297 (1994)
23. Soltani, M.M., Motamedi, A., Hammad, A.: Enhancing cluster-based RFID tag localization using artificial neural networks and virtual reference tags. Autom. Constr. **54**, 93–105 (2015)
24. Vuković, N., Miljković, Z.: A growing and pruning sequential learning algorithm of hyper basis function neural network for function approximation. Neural Netw. **46**, 210–226 (2013)
25. Yang, Z., Zhang, P., Chen, L.: RFID-enabled indoor positioning method for a real-time manufacturing execution system using OS-ELM. Neurocomputing **174**, 121–133 (2016)
26. Zhang, R., Sundararajan, N., Huang, G.B., Saratchandran, P.: An efficient sequential RBF network for bio-medical classification problems. In: 2004 IEEE International Joint Conference on Neural Networks, 2004. Proceedings, vol. 3, pp. 2477–2482. IEEE, Piscataway (2004)

Part II
Prognostics and Forecasting

Physical Model-Based Prognostics and Health Monitoring to Enable Predictive Maintenance

Tiedo Tinga and Richard Loendersloot

1 Introduction

Nowadays, only a limited number of systems are operated in completely stable conditions. Most of the systems, like ships, wind turbines, military vehicles and infrastructures are facing largely variable operating conditions and environments. At the same time, failures in any of the associated subsystems or components may have large consequences, e.g. high costs (loss of revenues, high logistics costs due to remote locations) or large safety and environmental impacts. To control the number of failures, typically preventive maintenance at predetermined intervals is performed. By replacing the components in time, failures can be prevented, but this is a rather expensive policy when the operational profile is largely varying. The preventive maintenance intervals must be set to very conservative values to assure that also severely loaded subsystems do not fail. This is a costly process, but it also limits the availability of the system, as it must be available for maintenance tasks quite often.

To improve this process, reduce the costs and at the same time increase the system availability, a better prediction of failures for systems operated under specific (and mostly dynamic) conditions is required. Only when such a prediction is available, maintenance can be performed in a just-in-time manner. This is the promise that predictive maintenance as the ultimate maintenance policy is giving. However, although a lot of research has already been done on this topic in the past decade, still a generically applicable concept is not yet available. This chapter will discuss

T. Tinga (✉) · R. Loendersloot
Dynamics Based Maintenance, University of Twente, Enschede, The Netherlands
e-mail: t.tinga@utwente.nl; r.loendersloot@utwente.nl

the challenges encountered in developing predictive maintenance concepts, and will provide insights and decision support tools that can assist in further improving the existing methods.

The chapter is organized as follows. In the next section, the main challenges in this field will be introduced. Then Sect. 3 covers the topic of structural health and condition monitoring. Section 4 will address prognostics, comparing the approaches of data analytics and physical model-based prognostics. Then Sect. 5 will discuss a number of decision support tools that can assist an asset owner in applying the appropriate predictive maintenance concept. In Sect. 6, a number of cases are presented, showing how the concepts and tools introduced in the previous sections can be applied in practice. Finally, Sect. 7 will forward some conclusions.

2 Challenges in the Field of Predictive Maintenance

The main premise of predictive maintenance is that decisions are taken based on an accurate assessment of the present condition of the system, and on a detailed prognosis of the remaining useful life. The aspects treated in the following subsections make clear that this is not a trivial task. In the subsequent sections of this chapter solutions will be proposed for these challenges.

2.1 Combining Diagnosis and Prognosis

In predictive maintenance, the challenge is to detect or foresee an upcoming failure in time, such that repair or replacement can be done before the system actually fails. This is often called condition-based maintenance (CbM), since maintenance is only performed when the system condition actually requires it. The basic way to achieve this is to monitor the system (either by continuous monitoring or by periodic inspections) to obtain a timely diagnosis of a degrading system.

As this is in most cases achieved by comparing a measured condition (e.g. vibration level) to a predetermined threshold, reaching the threshold means that action is required almost immediately. Although a failure can be prevented effectively and the actual condition of the system is always known, this so-called diagnostic approach is not optimal from a planning perspective. The vast majority of the diagnostic methods only provides a warning or alarm, and does not provide any information on the remaining lifetime of the system.

At the same time, a lot of effort has been put in the development of prognostic methods, which could accurately predict when a system is expected to fail. As will be discussed later, several approaches are possible, primarily divided into data-driven and model-based approaches—the former focus on finding patterns and anomalies in large sets of collected data, whereas the latter include knowledge on the physics of failure to assess the remaining useful life.

One of the big challenges here is to validate the prognostic methods, which requires comparison of predicted failure times with actually observed failures. The performance of a model can only be tested at one point in time, i.e. at the failure point. Before that moment, no information from the system is available to compare with the model prediction. However, for many critical applications preventive maintenance policies are in place, which largely reduce the number of actually observed failures. This significantly complicates the validation of prognostic methods.

By comparing the diagnostic and prognostic methods, it can be concluded that the drawback of a diagnostic method (no future prediction) is the strong point of prognostics. The other way around, by using a diagnostic method (e.g. condition monitoring (CM) technique) a lot of additional information on the level of degradation already becomes available before the actual failure, which largely increases the possibilities for validation of the prognostic method. But again, presently applied prognostic methods, either data-driven or model-based, do typically not use any diagnostic information, but purely focus on the prognosis.

The observation that diagnosis and prognosis actually are complementary is the prime motivation to discuss both health and condition monitoring concepts (i.e. diagnostic methods, see Sect. 3) and prognostic approaches (Sect. 4) in this chapter, as the authors are convinced that only a combination of these two approaches will lead to accurate and successful predictive maintenance concepts.

2.2 System Versus Component Level

The second big challenge in the field of predictive maintenance is the gap between system and component level. Typical assets like ships, trains, aircraft and infrastructures are complex systems containing large numbers of subsystems and components [1]. A system diagram of a typical (naval) ship is shown in Fig. 1. The ship is subdivided into five main functions (e.g. platform functionality), which are each again subdivided into one up to four subfunctions (e.g. mobility/propulsion). Finally, each of the subfunctions is realized with 1 up to 11 installations, like a diesel engine, sewage system or navigation radar. However, it should be realized that this is not the lowest level, as each of the installations consists of numerous components. For the diesel engine, these are, e.g. bearings, liners, pistons, etc.

The problem now is that prognostic methods, especially physics of failure-based methods, are typically developed at this lowest (component) level. But asset owners and operators are interested in the functioning and maintenance optimization on the highest (system/ship) level.

The challenge is thus to connect the system level maintenance optimization to the component level prognostic methods. For an effective preventive maintenance concept, ideally prognostic models for all individual components would be available. This would enable the prediction of any failure occurring in the ship, giving the operator the opportunity to take appropriate action before the actual failure occurs.

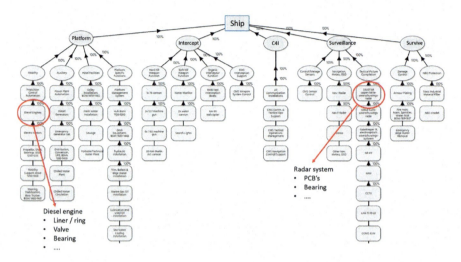

Fig. 1 Naval ship system diagram, showing the complexity and indicating some typical components

However, due to the large numbers of components and the effort required to develop prognostic methods, this full coverage of all components is not feasible in the practice of complex maritime assets. The consequence is that a suitable selection method is required to select those components for which developing prognostic methods is useful, i.e. that are dominant in the system failure behaviour. This issue, denoted as the critical part selection, will be discussed in Sect. 5.2, as one of the decision support tools for predictive maintenance. And at the same time, solutions must be found to cover all the other components and subsystems with simple and quickly available methods, as in certain situations also these non-critical components could lead to failures and associated costs and downtime.

2.3 Monitoring of Usage, Loads, Condition or Health

To be able to diagnose a system, structural health monitoring (SHM) or condition monitoring (CM) systems can be deployed. Although the purpose of both SHM and CM is to assess the condition of a system, their origin and approach are slightly different, as is discussed in [2]. SHM methods are typically focusing on measuring and interpreting the dynamic response of a system, aiming to detect, localize and quantify damage, as will be explained in detail in Sect. 3.2. Condition monitoring covers a wide range of techniques, measuring various quantities that can indicate an upcoming failure in the monitored system. For rotating equipment, vibration monitoring is a well-known CM technique, but also lubrication oil analysis and corrosion monitoring (see, e.g. [3–5]) can be considered as CM.

One of the challenges associated with monitoring is to decide for each specific application what the most suitable monitoring strategy is (usage, load, condition), what quantity should be measured at which location and what sensor type is preferred. These issues will be addressed in Sects. 3 (SHM), 4 (load/usage monitoring) and 5.1 (selection of most suitable CM technique).

2.4 Interpretation of Monitoring Data

In the present era sensors are everywhere, all systems are connected (internet of things, IoT) and data storage is not an issue anymore. In practice many original equipment manufacturers (OEMs) apply a lot of sensors to the assets and systems they produce, enabling the owners and operators to collect a lot of data on their systems, with the promise that the system can be maintained condition-based.

However, just applying a number of sensors to a system does not mean that the condition of the system is assessed. Translating the collected raw data into useful information on asset condition is in many cases challenging. Sometimes just observing a trend in a monitored parameter, or comparing the measured value with a predefined threshold provides the required insights. But in most cases this is not sufficient, and a thorough understanding of the normal (dynamic) system behaviour, as well as the system failure behaviour is required.

Vibration monitoring of bearings is, for example, a field that is so well-developed (after being in use for many decades), that understanding the details of bearing (failure) behaviour is not needed to properly diagnose a faulty bearing. But interpreting the vibration behaviour of a more complex system like a bridge or wind turbine rotor blade, aiming to detect damage, is much more challenging. Section 3 will discuss this challenge, and demonstrate how knowledge on the (dynamic) behaviour assists in diagnosing such a system.

2.5 Data-Driven or Model-Based Prognostics

Prognostics can be based on either data-driven or model-based approaches. The data-driven approaches use large amounts of data, preferably from various sources, and apply data analytics techniques like machine learning and artificial neural networks to discover patterns and relations in the data sets. This means that in principle no knowledge on the system characteristics or failure behaviour is required, which makes the approach popular and widely accessible. However, the lack of system knowledge can also lead to the discovery of trivial or accidental (non-casual) relations. For example, a high correlation between fuel flow and temperature in an engine could be discovered from a data set, but that relation is trivial from an engineering point of view. Further, the artificial intelligence (AI) methods used in this approach must be trained with data to enable to learn the patterns. This means that in principle only patterns (i.e. failures) can be predicted that have been observed before (and were included in the training set).

The alternative approach for prognostics is the use of physical failure models. In this approach the failure mechanism, like fatigue, wear or corrosion, is captured in a mathematical model, relating the usage or loading of a system or component to a degradation rate or lifetime prediction. Monitoring of the usage or loads on an individual system then enables to predict the (remaining) time to failure. Although the development of these types of models is rather time-consuming, they solve some of the limitations of the data-driven approaches: the models do not need a large set of failure data (which is typically not available for critical systems), and they also work for situations not previously encountered. The main challenge on this topic is to decide which approach is most suitable in a specific situation. The pros and cons of the approaches will be further discussed in Sect. 4, while the selection of the most suitable approach will be treated in Sect. 5.1. Some cases on mainly the model-based approach will be shown in Sect. 6.

2.6 Selection of Most Suitable Approach and Technique

Knowing that many different approaches for diagnosing the system and predicting the remaining useful life (RUL) exist, a typical asset owner has quite some difficulty in selecting the most suitable approach for the specific situation. The choice between data-driven and model-based prognostics has already been addressed in the previous subsection, but also on the diagnostic side there are many options.

Health and condition monitoring techniques can be adopted for diagnosing a system. The interpretation of monitoring data has already been discussed in Sect. 2.4, and will be treated in more detail in Sect. 3. But before a measurement can be done, a user has to decide which condition monitoring technique is most suitable for the ambition level and (technical and financial) possibilities in a specific situation. As many condition monitoring techniques have been developed, and are commercially available, selecting the most suitable technique is not trivial. This challenge will be treated in Sect. 5.1.

But for the diagnostic part, also other approaches than health and condition monitoring are available. In the absence of (monitoring) data, experience-based methods, using the knowledge and experience of experts, can be adopted to improve the maintenance process for a certain asset. Typically qualitative tools like failure mode, effect and criticality analysis (FMECA), fault tree analysis (FTA) or root cause analysis (RCA) are applied then.

Yet another approach would be the more mathematical approach of reliability engineering, where historic (failure) data are used to assess the typical behaviour of a fleet of assets, for example by determining the mean time between failures (MTBF). Although less specific than condition monitoring, this approach can be valuable for systems that are operated in a relatively constant manner (e.g. production machines in a factory).

The main challenges here are to get an overview of all approaches available, get insight in their strong and weak points, and select the most suitable approach for a specific situation. These challenges will be addressed in Sect. 5.1.

2.7 Data Quality

The final challenge in the field of predictive maintenance is the availability, accessibility and quality of data. Except for the most simple experience-based approaches, all approaches discussed so far require the input of data. In most cases, the accuracy of the diagnosis and prognosis largely depends on the quality of the data. But for many practical applications, there are quite some challenges in getting the required data.

The first problem encountered with many companies is that the required data is not available. This might be due to the fact that the required parameter is not monitored or (manually) registered, or that the data are not accessible due to security, formatting or other reasons.

The second problem is the size of the data set. This can be caused by the sample rate of the data, which is especially for manual registrations often too low. For example, for a fleet of vehicles, the status (up or down) of each vehicle might be reported only once a month. Changes in the fleet average availability will then only be visible after a couple of months, resulting in a very low response rate. If the registration would be done daily, reactions to changes in availability can be arranged much faster. Further, the size of the data set can also be limited by the history of measurements that is stored. In some applications data storage is mainly intended for troubleshooting in case of failures. Typically data are overwritten after a relatively short period, e.g. some weeks or months. This means that a long history of data will never be available, thus limiting the possibilities for reliability engineering or data analytics.

The third problem related to data is the quality of the data. In practice it occurs quite often that gaps appear in a time history of certain parameters, especially for sensor data. This can be due to a failing sensor, or due to problems in data transmission or storage. For manually registered quantities, e.g. failure reporting in the computerized maintenance management system (CMMS), the human factor plays an important role. When the system has been designed properly, engineers are forced to select from a short list of predefined options, which reduces the amount of ambiguity in the data. However, if there is a possibility to use a free text field, analysis of the data is already much more difficult, as every person has his own way of formulating a certain failure. Moreover, inexperienced engineers or operators, who are typically responsible for registering the failures, often do not have the knowledge to determine the cause of the failure, and thus register the failure incorrectly. The latter problem will be addressed in Sect. 5.2, demonstrating an expert system (ES) assisting in this process.

3 Structural Health and Condition Monitoring

Following the reasoning introduced in the previous section, key to the successful implementation of predictive maintenance methods are data collection and processing. Terminologies such as 'smart systems', 'smart industries' and other combinations with the word 'smart' all refer to the usage of sensors in a product, system or installation. These sensors provide the necessary data to assess the current loading and/or performance of the structure being monitored. However, this data collection does not yet make the system to a smart system, as was discussed in Sect. 2.4. The smartness is embedded in the processing of the data—converting it to information—and the subsequent decision process.

It is clear though, that decisions cannot be made in the absence of data: data are a prerequisite. The question is whether it is possible to define upfront what data is needed, which sensors are best suited for this task and which signal processing techniques are to be applied, a topic addressed in Sect. 5.1. This is not a sequential design process, but more a parallel and iterative process: on the one hand, the selection of signal processing techniques influences the choice of sensors and hence determines which data can be made available. On the other hand, the system being monitored and the conditions in which the monitoring should take place set constraints on the sensors and thus dictate the choice of sensors. These inherently delimit the choice of processing techniques from a different angle. This calls for an integrated design approach, such as proposed by Sanchez Ramirez et al. [6]. An important observation in this work is that a distinction must be made between the monitored and the monitoring system.

This section discusses the most common used sensors for monitoring systems relying on the dynamic response of the system. Other forms of monitoring, for example monitoring of lubricants, are not addressed.

3.1 Sensors

The first step after having established the importance of the acquisition of data is creating a categorization of suitable sensor technologies and data acquisition systems. The discussion in this section is focused on sensors and systems suitable for dynamic or vibration measurements.

One of the most frequently selected sensors is the strain gauge. Its ease of application and broad experience of application are the most common motivations for selecting this type of sensors. A strain gauge is a passive sensor in the sense it can only sense. Strain gauges can measure both static and dynamic strains and thus provide data on the local strain field, hence the use of strain gauges aligns well with the concept of load monitoring. Typical applications are fatigue dominated structures, in which a link between dynamic loads and the consumed fatigue life is established based on, e.g. the rain flow counting method [7], or the dynamic amplification factor (DAF) [8]. The use of load monitoring in prognostics will be further discussed in Sect. 2.3.

Accelerometers are widely used to capture the dynamic response of structures. Typically, a relatively low number of accelerometers is used, as their price is significantly higher than that of strain gauges. Accelerometers are also passive sensors and in most cases need an external power supply to function. Accelerometers operate in a specific frequency range. Frequencies outside this range, bound by a higher as well as a lower limit, are not captured accurately. The lower the lower bound of the frequency range, the more bulky the accelerometers gets: the most common principle used in accelerometers is based on a references mass and a position measurement of this mass.

MEMS-based accelerometers, such as those found in mobile phones, have pushed the use of accelerometers, bearing in mind that their size limits the lowest frequency that can be measured. However, the level of integration that can be reached with these devices and their significantly more favourable energy consumption are strong advantages. The lack of signal quality can (partly) be compensated by following a crowd sensing big data approach [9–12].

Optical fibres have received a lot of attention over the past years. Several reasons drive this interest: firstly, optical fibres can relatively easily be integrated in the structure; secondly, a single fibre can have multiple sensors (multiplexing) and thirdly, a very high accuracy can be reached, be it primarily in static mode. The downside of optical fibres is the still expensive and hard to integrate interrogator. Optical fibres can be embedded in composites, but distortion of the fibres of the composite has a negative effect on the mechanical properties, including fatigue. Moreover, the optical fibre, more precisely the cladding around it, will be deformed during production, leading to distortions in the signal. The reader is referred to Ref. [13] for further details on optical fibres.

Piezo-electric transducers (PZT) have excellent options for integration into the structure [14–16] at a relatively low cost. Moreover, PZTs can be used both in sensor and actuator mode, which makes them very flexible in use. The frequency range in which PZTs can be used is also very broad, be it that excitation at lower frequencies typically requires more power than the (average) PZT can produce. PZTs are applied to measure the structural dynamic response [17] ($\mathscr{O}$(kHz)) or the nonlinear response [18, 19] ($\mathscr{O}$(10 kHz)) of, e.g. composite structures, as well as to generate and measure propagating waves, such as guided waves in composite materials [20–23] ($\mathscr{O}$(100kHz)) or ultrasonic waves in, for example plastics and cementitious materials as typically used for drinking water mains [24, 25] ($\mathscr{O}$(MHz)).

In sensing mode, piezo-electric transducers are passive sensors: no power needs to be supplied. The mechanical motion of the structure causes a current to flow as a direct result of the piezo-electric effect. This opens the door for another application of PZTs: energy harvesting [26, 27]. Energy harvesting is a key element for smart and autonomous sensor nodes, as they either rely on batteries—having a finite, relatively short endurance—or local energy generation.

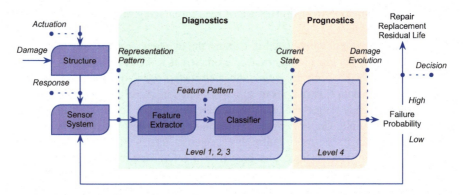

Fig. 2 Schematic overview of the vibration-based monitoring concept

3.2 Vibration and Vibration-Based Monitoring

Vibrations are omnipresent: structures in operation are in a lot of cases subjected to vibrations. The vibrations can be introduced by a driving system, such as an engine, a drive train (e.g. wind turbine) or traffic, or by environmental conditions, such as wind and waves. Generally, a distinction is made between *vibration monitoring* and *vibration-based monitoring*. The general concept of vibration (based) monitoring, shown in Fig. 2, is that the condition or state of the structure affects the dynamic response. This implies that *feature extraction* is a main step in the vibration (based) monitoring process. A *feature* is defined as a characteristic or set of characteristics in the dynamic response that can be used as an indicator for the condition or health assessment. Depending on the complexity of the case and the required level of monitoring, the signal processing involved can be relative simple to highly advanced.

So far, vibration monitoring and vibration-based monitoring are described using the same terms. The distinction between both is that the first focuses on the system response as acquired by individual sensors, e.g. observing vibration levels, natural frequencies of the system or more generic statistical parameters such as root mean square (RMS), skewness and kurtosis of the vibration signal [28–30]. The latter, on the other hand, typically combines sensor readings to construct, e.g. mode shapes and their derivatives [31–34] which are then used to identify a specific damage. Vibration-based (health) monitoring thus requires more advanced analysis methods to be applied to the raw vibration data.

The following items can be recognized in the schematic overview of vibration-based monitoring in Fig. 2:

Actuation and response Actively applied or operational (dynamic) loading on the structure and the resulting signal received from the sensors;
Representation pattern Transformation of the measured signal (raw data) to a pattern representing the response;

Feature pattern Transformation of the representation pattern to a pattern of damage or condition sensitive parameters;
Current state Quantitative measure representing the condition in which the structure currently is;
Damage evolution Model describing the evolution of the damage or condition (degradation process) in time;
Decision Risk-based decision regarding action to be taken.

An important observation is that the feature pattern is compared to the pattern of some other state. In many cases a reference measurement is used, which can either be a measurement in 'new state', executed at the start of the operational life of the structure, or a previous measurement. Some will argue that monitoring is also done in a so-called *baseline-free* manner, not relying on any reference state at all. However, as Worden et al. [35] argue, there is always an underlying assumption used in those cases that serves as a reference. The state analysis leads to a *diagnosis* of the current state. A *prognosis* of the future state is based on an extrapolation of the evolution of the states measured over time. Most vibration-based monitoring applications are limited to the diagnostic phase, as the presence of damage and its location are in those cases sufficient to determine the necessity of a maintenance action, for example delamination detection in composite structures [31].

Most vibration-based monitoring applications focus on the current state estimation, e.g. the presence, location and size of a delamination in a composite structure [17–22] or the level of degradation of a plastic or cementitious material [24, 25]. In some cases, this is sufficient, as the user is interested in a limit value, such as set by the *barely visible impact damage* (BVID) criterion [36] for composite materials. Reaching or passing this limit implies immediate action is required and further use of the system is not safe. This approach is often followed in case the damage is induced by a single, isolated event, such as an impact. It cannot be predicted when an impact will occur, but if it occurs, it must be detected if the damage inflicted has a certain, predefined critical size. The drawback of a diagnosis requiring immediate action, and the wish to combine it with a prognostic method, was already discussed as one of the challenges in the field (Sect. 2.1).

A PF-curve can be used to explain the position of diagnostic and prognostic monitoring systems, see Fig. 3. In time, several points can be recognized: Firstly, the point of onset of the deterioration (damage formation or material degradation). It should be noted that this point can in fact be at the very beginning of the operational life of a system, hence can also be interpreted as the moment the failure starts to grow. Secondly, the point at which the failure becomes observable (P). This can be interpreted in absolute sense, e.g. in terms of a minimum crack size, or in a relative sense, e.g. relative to the capabilities of monitoring technique. Finally, the point of functional failure (F), i.e. the moment the system cannot perform its intended function with preset limits. The point when the damage or deterioration becomes observable is of interest. Using the relative interpretation of this point implies that the location of this point is determined by the sensitivity and accuracy

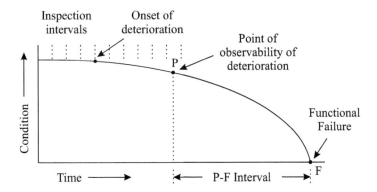

Fig. 3 PF-curve, indicating the point when a failure is initiated, when it becomes observable and when a function failure occurs

of the monitoring technique selected. In other words, the development of monitoring techniques encompasses in general terms moving point P to an earlier time in the operational life of the system i.e. to the left in Fig. 3.

The second phase, prognostics, includes the future and is looking beyond point P in the PF-curve. It encompasses the prediction of the curve beyond the point P. The uncertainty in these estimations is reduced by using as much information as possible from the present and past—hence from the diagnostics, possibly complemented with physical models. These damage evolution models enhance the reliability as more limited historic data are needed and deterioration of the system in an unknown fashion can still be predicted based on the physics-based damage evolution models, as was argued in Sect. 2.5 and will be elaborated more in Sect. 4.

It is evident that prognostics involves estimations. Less clear may be that the current state is also an estimated state. In reality it is impossible to deterministically exclude all disturbances affecting the signal. Either environmental disturbances or noise from the sensors, the connecting wires or the electronics inside data acquisition system itself will distort the signal and making an exact determination of the condition impossible.

Finally, the difference between *inspection* and *monitoring* is relevant: the first is an isolated action, resulting in an immediate advice (green or red flag), while the second is a more continuous process, as measured states are compared to each other, allowing to following the evolution of the degradation. This evolution can then be used to estimate the remaining useful life. Typical examples are systems suffering from fatigue dominated damage (due to cyclic loading) or from time dependent degradation (e.g. physical ageing of plastics).

As will be further elaborated in Sect. 4, load monitoring concepts combined with rain flow counting, Palmgren-Miner rule [7] or dynamic amplification factor methods [8] do provide a means to estimate the remaining useful life in these fatigue dominated cases, yet with a high uncertainty. More precisely, Derriso et al. [37] point out the uncertainty in the onset of damage growth is the main cause of the uncertainty of the life prediction. Predicting this point is very hard, but once it is known, models seem to predict the remaining lifetime fairly accurately. This stresses the need for reliable diagnostic methods.

4 Physical Model-Based Prognostics

While the previous section focused on health and condition monitoring concepts that aim to diagnose a system, i.e. assess the present condition, the present section will discuss prognostics, which aim to predict the (remaining) lifetime of a system or component.

The motivation to develop prognostic methods can be found in the conservatism that is present in many traditional maintenance policies. This means that in these preventive policies—applied to many critical systems, like aircraft, nuclear plants, oil and gas installations and infrastructure—components are typically replaced far before they actually reach the end of their lifetime.

It can be demonstrated [38] that this conservatism is largely due to uncertainty, as was mentioned at the end of Sect. 3, in the initiation, but also in the damage evolution in the components. In other words, since it is unknown how much damage is present in an individual component, its replacement interval will be chosen on the (very) safe side, for example based on the most extreme load case that can possibly be encountered by the system. In other cases, the uncertainty in the future usage is covered by safety factors. These types of maintenance policies are far from just-in-time, and thus lead to high maintenance costs and a waste of component lifetime.

However, if the relation between actual usage of individual components and their degradation can be quantified, the uncertainty can be reduced. In other words, if an accurate prognostic method is available, the optimal moment of replacement can be obtained for any component, based on the (monitored) usage or loading of that part [39]. In that case, just-in-time maintenance becomes feasible, and a considerable cost saving can be achieved.

As was discussed already in Sect. 2.5, basically two approaches are available for prognostics: data-driven and physical model-based. Although a lot of papers recently appeared on the data-driven approach [40, 41], the authors strongly believe that for systems operating in a variable environment (e.g. military systems, offshore wind turbines, maritime systems), the model-based approach is more suitable. Also others [42–44] have indicated the potential of this latter approach. This section will therefore focus on the model-based approach, showing the various steps in the approach and demonstrating its benefits. In the final subsection, this approach will be compared to the data-driven approach.

4.1 Relation Between Usage, Loads and Degradation Rate

One of the key ingredients of physical model-based prognostic methods is the ability to quantitatively relate the usage or loading of a system to its degradation rate. This relation is schematically shown in Fig. 4, and will be discussed next.

Normally the usage of a system, in terms of operating hours, power settings, number of starts, etc. is known to the operator or can be monitored rather easily. However, the remaining life of the system determines when maintenance actions must be performed, whereas the relation between the usage and the remaining life is in many cases unclear. Insight in this relation can be obtained by zooming in to the level of the material point, since that is the level at which the physical failure mechanisms are active. This requires translation of the usage (on the global level) to the local loads (e.g. stress, strain, temperature, electrical current, etc.) on the material level (Fig. 4).

The loads are then related to the capacity of the material by some failure model (e.g. fracture, fatigue, creep, arc flash), which yields the damage accumulation, degradation rate or life consumption rate at the present load. Finally, assuming that the usage and/or loads can be estimated, a prognosis can be given for the remaining life of the system.

Two important relations in Fig. 4 are the usage-to-load relation and the load-to-life relation, denoted by the numbers 1 and 2, respectively. These relations can be assessed in a quantitative sense only when the physical background of the loading and the failure mechanism is understood. If accurate models are available for these relations, any usage history of the system can be translated into the associated damage accumulation or life consumption.

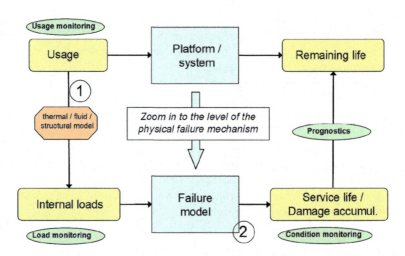

Fig. 4 Schematic representation of the relation between usage, loads, condition and life consumption. The most important relations are (1) the usage-to-load and (2) the load-to-life relations

Note that the failure mechanism is modelled on the material point level in a component, whereas a system may contain numerous components with several failure mechanisms each. This was mentioned as one of the challenges in the field in Sect. 2.2. Therefore, before the method illustrated in Fig. 4 can be applied, a critical part selection must be performed to determine which mechanism(s) in which component is/are critical to the service life of the complete system. This selection process will be discussed in Sect. 5.2.

Monitoring of the usage or loading of the system is essential in a model-based approach. Figure 4 shows that monitoring can be performed at different levels. The lowest level of monitoring is usage monitoring, which implies registration of quantities like operating hours, rotational speeds or number of starts. Although usage monitoring is in most cases rather easy to perform, relating the data to degradation rates is generally not straightforward and requires quite some models and calculations. Load monitoring is one level higher, since it directly assesses the internal loading of components. This can be realized by applying sensors like thermocouples (to measure the temperature) or strain gauges (deformation). Monitoring at this level is generally somewhat more complex than usage monitoring, but the obtained information is related more directly to the component condition. The highest level of monitoring is condition monitoring, where the actual condition (i.e. the amount of degradation) is assessed directly and no calculations are required as was discussed in Sect. 3. However, this level of monitoring generally requires rather sophisticated sensors and is not always feasible, either technically (accessibility of the component) or economically.

4.2 Developing a Prognostic Method

To develop and apply a prognostic method for a specific application, the following steps will have to be taken:

1. Select the most critical part in a system;
2. Determine the physical mechanism responsible for the failure of the critical part and define a (physical) model for this mechanism;
3. Assess the loads that govern this failure mechanism, and determine how these are related to the operational use of the system;
4. Collect data on the variation of the usage or load;
5. Predict the time to failure, given a certain (measured or assumed) usage profile;
6. Validate the model by comparing the prediction to actual failure data.

Each of these steps will now be discussed in some more detail, using a case study of a helicopter part [45] as example.

Step 1: Critical Part Selection As was discussed in Sect. 2.2, selecting the critical parts in a system is a crucial first step in many analysis. Large systems typically contain hundreds or thousands of parts, so analysing each of them is not feasible.

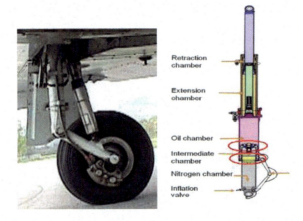

Fig. 5 Landing gear shock absorber on helicopter (left) and its schematic representation (right) showing the seals (in red circle)

Especially the development of a model-based prognostics method will typically be too time-consuming to repeat for many parts. A structured approach for selecting critical parts will be discussed in Sect. 5.2.

By assessing the historic data of the helicopter, the cost drivers (failures that lead to high costs) and availability killers (failures leading to long downtimes) could be determined. From this analysis, the landing gear shock absorber (see Fig. 5) was selected as a critical part, for which a prognostic method will be developed. This shock absorber contains an oil chamber, which contains two sets of polymer seals. After a certain period of time, these seals start to leak oil, triggering a replacement (and off-line repair) of the complete shock absorber.

Step 2: Determine the Failure Mechanism and Associated Physical Model The next step is to determine the physical mechanism that leads to this failure, i.e. the failure mechanism and a suitable physical model. The former requires to execute a root cause analysis (RCA), since it is not sufficient to just determine the failure mode (as is often done in practice). A failure mode describes the functional failure of system, and can be defined at many different hierarchical levels. For example, a failure mode for the shock absorber is 'leakage of the seal', but also 'non-functioning shock absorber'. However, both of these failure modes do not specify the actual failure mechanism at the material level, which is strictly required for prognostics (and can be found with the RCA). Once the mechanism has been determined, a suitable physical model has to be selected. Many models for a range of failure mechanisms are available in literature [46], so only in exceptional cases new models need to be developed.

In the case of the shock absorber, the failure mechanism is wear, as two parts (cylinder and seal) are sliding along each other, leading to loss of material (in the seal). This ultimately leads to the observed oil leakage. A physical model typically used for wear is the Archard's wear law [47], relating the lost volume V to the normal force in the contact (F) and the sliding distance (s) through a proportionality constant (the specific wear constant k):

$$V = kFs \tag{1}$$

Knowing the value of k, the volume loss due to wear can thus be calculated for any combination of normal force and sliding distance.

Step 3: Assess Governing Loads and Relation to Operational Use The objective of a prognostic methods is to predict the failure based on the monitored usage or loads. Therefore, the next important step is to determine which loads are governing the failure and how these loads are related to the operational usage of the system. Loads can in a general sense be classified in various types, like mechanical, thermal, electric, but in the end need to be specified as either an external load applied by the operator (e.g. force, moment, voltage) or an internal load acting on the material level (e.g. stress, strain, temperature, electric field) [46]. Moreover, it is important to understand how these loads are affected by the operational usage of the system. In many cases this is quite trivial. For example, if the weight on a bridge is increased, the bending loads (and stresses) will increase proportionally. But in other cases the relation between usage and loads is more complex.

In the case of the shock absorber, the loads that govern the wear mechanism are the normal force in the contact between the cylinder and seal and the sliding distance between the two parts. For this specific application, the sliding distance is directly related to the operational usage of the helicopter. It is obvious that the shock absorber will only be compressed when the helicopter is landing, and in that case the weight of the helicopter determines how far the cylinder is compressed. The sliding distance can thus be obtained by just accumulating the stroke of the cylinder during each landing.

Step 4: Collect Data on the Variation of the Usage or Load Once the physical model is in place, and it is clear which loads or usage parameters govern the identified failure, data collection can be started. In some cases the data can be directly obtained from a monitoring system or a control system. For example, in many production machines, but also in wind turbines, a SCADA (supervisory control and data acquisition) system is present collecting a lot of details on the operational use of the system. When these monitoring systems are not present, manual registrations of operating hours, number of start/stops, etc. can also be used as input for the predictive model.

In the case of the helicopter shock absorber, the data could be obtained from the health and usage monitoring system (HUMS). This system collects a lot of data on the operational use of the helicopter (altitude, speed, temperature), and specifically for this case the registration of number of landings and the weight of the helicopter during each landing proved to be very useful. This means that for each individual helicopter in the fleet, the HUMS enabled to reconstruct the operational usage in terms of landings.

Step 5: Predict the Time to Failure, Given a Usage Profile By combining the monitored data (step 4) with the physical model (step 2), a prediction of the time to failure can now be made. For the shock absorber, the load sequence (sliding distance) for an individual seal could be retrieved from the HUMS, enabling to predict the expected moment of failure (in terms of flight hours). The results will be shown in the next step.

Step 6: Validate Model by Comparing the Prediction and Actual Failure Data
The final step in the procedure of developing a prognostic method is the validation of the method. This a crucial step, as it will demonstrate the correctness and accuracy of the developed method. However, in practice it is also a very difficult step, as this can only be done when (1) a certain number of failures is present and (2) for each failure a detailed registration of the load/usage history is available. The first criterion, as was already discussed in Sect. 2.1, is compromised by the preventive maintenance policies of critical parts that limit the number of observed failures. But even when a number of failures have occurred, the lack of a detailed registration of the loads makes it still very hard to validate a prognostic method. For example, in [1] a prognostic method for printed circuit boards (PCB) in radar systems is discussed. Although quite some failures in these components have been registered, a usage profile (operating hours, amount of switching on/off) for individual PCBs could not be retrieved, making the detailed method validation unfeasible.

For the shock absorber case the validation was possible, since a number of failures occurred, and the HUMS provided a detailed load history of each individual shock absorber in the fleet. To demonstrate the need for a prognostic method, first the 11 failures were plotted versus the number of flight hours at failure (which is the traditional way of maintaining aircraft systems), see the left-hand side plot in Fig. 6. This clearly shows that flight hours (FH) is a bad predictor for this type of failure, as some oil leakages already occurred after 40 FH, while other seals only failed after 200 FH. Then the developed prognostic method was applied, and the predicted amount of wear for the same 11 failures is shown in the right-hand side plot in Fig. 6. Although the first two failures deviate somewhat, six other failures show a very similar amount of wear at failure (as indicated by the solid red line). Apparently 30 mm^3 wear marks the failure threshold for this part (then oil starts leaking). Three other failures (see dashed red line) clearly show a higher amount of

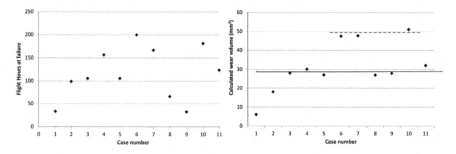

Fig. 6 Number of flight hours at failure for 11 events (left) and predicted amount of wear for the same events (right)

wear at failure. These three seals appeared to be modified seals, as the OEM reacted to the failing seals problem. As a single value for the specific wear parameter was used, the increased amount of wear at failure in fact means that this modified seal can accumulate more landings before leakage occurs. The rather accurate prediction of the amount of wear at failure for different helicopters validates the developed method.

The six steps in the development of a physical model-based prognostic method have been discussed, and should give guidance in developing such methods for other applications. Some other examples will be shown in Sect. 6.

4.3 Comparison to Data-Driven Approaches

In this section the model-based approach in prognostics has been discussed extensively. The alternative approach is the data-driven approach, where large data sets are analysed to achieve anomaly detection or predictions of remaining useful life. In recent years a large number of machine learning and deep learning algorithms have been proposed and applied in predictive maintenance cases. The interested reader is referred to Lee et al. [40], who have reviewed a large number of these techniques. Continuing the initial discussion in Sect. 2.5, the main differences between the two approaches, and thus their advantages and disadvantages will be discussed here.

The first difference between the two approaches is the data requirement. Data-driven methods fully rely on a sufficiently large set of data, but as was mentioned before, the availability of especially failure data is often limited due to the criticality of the systems. In such cases, a certain amount of system and domain knowledge may make up for the lack of data, as physical laws and first principles can be used.

Secondly, many data-driven approaches use artificial intelligence to recognize patterns and do predictions. These methods must be trained first, where various learning strategies can be followed. The two basic strategies are supervised and unsupervised learning. In a supervised learning scenario, the algorithm is trained with known failures. In that way that algorithm can learn which features in the data must be associated with specific (upcoming) failures. In an unsupervised learning scenario, only unlabeled raw data is available. In that case, the algorithm can only search for deviations from normal behaviour (anomaly detection) not be told which failure. This learning process associated with data-driven prognostics implies that typically only situations/failures that are present in the training sets will be recognized by the methods. This is not a problem for assets that are operated in a rather constant manner. However, it may not work properly for systems operated in a very variable manner or in changing environments, like military systems or off-shore wind turbines. Failures not previously encountered, or usage profiles/environments that are very different from previous experience will typically not be recognized. One solution for this problem could be the application of recently developed self-adaptive data-driven approaches [48]. But also physical models, on the other hand, do include the full system (failure) behaviour. When different usage profiles or scenarios are entered into these models, they will still provide a realistic prediction.

Finally a drawback of the physical model-based approach should be mentioned. The development of the methods is generally quite time-consuming. This is on the one hand due to the high level of detailed knowledge (on both system and failure behaviour) that is required, and on the other hand since these models are applied on the component level (see system vs component level challenge described in Sect. 2.2).

This discussion shows that both approaches have their pros and cons. However, it is expected that combining both approaches in a hybrid approach leads to very well-performing algorithms. As was mentioned above, physical models could be used to fill the gaps in data sets, and data analytics techniques could be applied to speed up the physical model development.

5 Decision Support Tools

As was discussed in the previous sections, many methods, approaches and techniques are available for predictive maintenance. For a potential user of these techniques, it is very difficult to determine which approach and what technique are the most suitable for the specific situation and application. Therefore, various decision support tools have been developed recently to assist in this process. In Sect. 5.1 some tools to select the best preventive maintenance approach and the most appropriate condition monitoring technique are described. The tools discussed in Sect. 5.2 provide support in selecting the most critical parts in a system and assess the right failure mechanism for a specific failure. A final decision that always has to be taken in developing a predictive maintenance approach is whether the business case is positive: do the benefits of the new approach out-weigh the investments that are needed in the development, implementation and operation of the method? A method for this challenge is discussed in [46] and will not be detailed any further.

5.1 Guidelines for Selecting Suitable Approach

This subsection will focus on guiding users in selecting the most suitable option for (1) the predictive maintenance approach and (2) a condition monitoring technique. The former selection is quite generic, as it aims to distinguish between a wide range of approaches, ranging from the rather simple experience-based approaches to the more sophisticated condition-based maintenance approaches. The latter selection is much specific: in case a user already has decided to apply condition-based maintenance, how can the most suitable condition monitoring technique be selected?

For the selection of the most suitable predictive maintenance approach, a selection framework has recently be developed [49, 50]. All approaches have first been classified in the following categories:

I. Experience-based predictions of failure times, based on knowledge and previous experience outside (e.g. OEM) or within the company.
II. Reliability statistics prediction techniques are based on historical (failure) records of comparable equipment without considering component specific (usage) differences.
III. Stressor-based predictions are based on historical records supplemented with stressor data, e.g. temperature, humidity or speed, to include environmental and operational variances.
IV. Degradation-based predictions are based on the extrapolation of a general path of a degradation measure to a failure threshold. By measuring symptoms of incipient failure the system can be diagnosed.
V. Model-based predictions give the expected remaining lifetime of a specific system under specified conditions. Two types of model-based approaches can be followed:
   a. Physical model-based
   b. Data model-based

It was discovered that a selected approach is suitable for a specific company when there is a good match between the ambition level of the company and the availability of data. The ambition level has been defined in five classes, as is shown in Fig. 7. It firstly specifies whether individual systems are to be addressed, or that a fleet-wide average is also sufficient. And the ambition level secondly depends on whether changing operational conditions are to be incorporated in the maintenance policy.

Once the ambition level has been determined, the type and level of detail of the data and information available at the company determines whether that ambition level is achievable. This is indicated in Fig. 8, showing the relation between potential

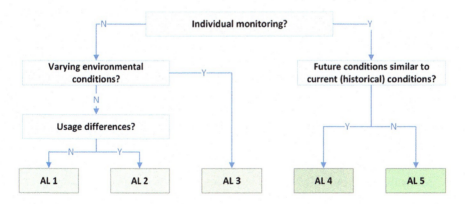

Fig. 7 Guideline for the selection of the ambition level

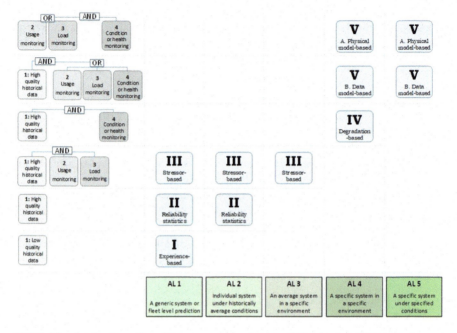

Fig. 8 Mapping the preventive maintenance approaches to ambition level and data types

approaches versus (1) ambition level and (2) data availability. The ambition level now specifies the required predictive maintenance approach, and Fig. 8 then clearly shows the data requirements. If these do not match, there are two options: either organize the collection of the proper type of data or reduce the ambition level. As companies in practice do not use this kind of reasoning, they often fall into a lengthy trial-and-error process that only in some cases results in a satisfactory solution.

Another selection challenge is in the field of condition monitoring. Many companies decide to adopt a condition-based maintenance policy for their assets, but then have to select the most suitable technique to monitor their asset. Also for this process a decision support tool has been developed recently [51].

Figure 9 shows the different steps in this process. Similar to the process of developing a prognostics method, as was discussed in Sect. 4, it is also essential in this case to first understand the failure behaviour of the asset or component (steps 2 and 3 in Fig. 9). Once the failure mechanism has been established (e.g. corrosion or wear), steps 4, 5 and 6 will guide the user in selecting a suitable CM technique. This is done by filtering a long list of potential CM techniques on a number of attributes. These attributes have been defined in three classes:

1. Attributes of the monitored system (accessibility, material type, motion);
2. Attributes of the (ambitioned) CM technique (diagnose/prognose, response time, severe environment);
3. Attributes related to the failure mechanism (local/uniform, cyclic loads).

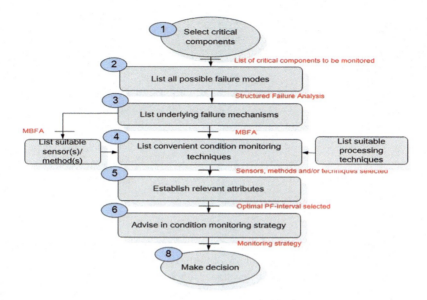

Fig. 9 Procedure for selecting the most suitable condition monitoring technique

This process has been implemented in an expert system (ES). By answering a certain amount of questions, the attributes in the three categories are set, and the ES is able to eliminate all CM techniques that do meet the requirements. In the end, the ES advises a limited number of CM techniques.

5.2 Critical Part Selection

Another selection challenge in the field of predictive maintenance is the selection of critical parts. As predictive maintenance cannot be applied to any component in a complex system, see also the discussion in Sect. 2.2, the most critical components have to be selected. An extension to this problem is the determination of the failure mechanism that is responsible for a failure. This is in fact a root cause analysis challenge, which is a rather crucial step in developing a prognostic method or a condition monitoring technique, as well as in the proper registration of failures in the computerized maintenance management system (CMMS). The latter is essential when the data has to be used later on for reliability engineering analyses. Incorrect or incomplete registrations will then lead to inaccurate results.

The traditional way of selecting critical parts in a system is to assess cost drivers or performance killers, e.g. using Pareto analysis, fault tree analysis (FTA) or failure mode, effect and criticality analysis (FMECA). Upon applying these traditional methods in several cases, the authors discovered that these do not always

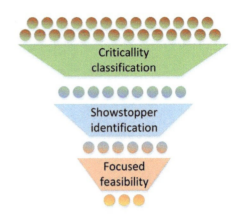

Fig. 10 Procedure for selecting the critical parts

lead to satisfactory results. Therefore, a more extensive method has recently been proposed [52], as is schematically shown in Fig. 10.

The methodology consists of three stages, each acting as a filter. The first filter is the criticality classification using the traditional methods (Pareto, FMECA). The added value of the present method are the two additional filters: the identification of show-stoppers and a focused feasibility study.

In these steps, firstly a differentiation is made between three ambitioned results of prognosis: Detection, diagnosis or prognosis. Determining the ambitioned outcome by differentiating between these three levels helps to firstly describe the requirements of the prognostic system and secondly explore the possibilities and impossibilities by recognizing the potential show-stoppers. The potential show-stoppers for predictive maintenance (PdM) to be applied on the considered component are then categorized in four groups:

Clustering: Can PdM extend the interval of this component to the next planed cluster of tasks, is the component driving the cluster of tasks?

Technical feasibility: Can a failure of this component be detected/diagnosed/predicted with current or future technology?

Economic feasibility: Is PdM for this component affordable? Sufficient failures to earn back the investment?

Organizational feasibility: Is sufficient domain knowledge for this component present? Is there sufficient trust in monitoring technique?

After identifying the show-stoppers, only a limited set of potentially critical parts will be present. For these parts, a focused feasibility study can be executed as the third step in the procedure. Such a study will typically address the show-stoppers that could not be fully checked in the previous step, e.g. a detailed business case analysis or a more thorough technical feasibility study.

Another set of decision support tools focuses on the assessment of the failure mechanism (step 2 in the prognostic method development in Sect. 4.2). The first method is the mechanism-based failure analysis (MBFA) as depicted in Fig. 11 [53]. The MBFA starts with selecting the most important failure modes using FTA and

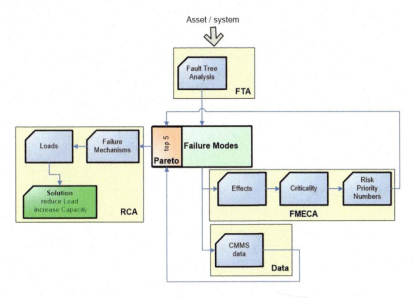

Fig. 11 Procedure for a mechanism-based failure analysis (MBFA)

Pareto analysis. Failure frequencies used as input for the Pareto are either obtained from a CMMS database or from the risk priority numbers (RPN) in a FMECA. For the most important failure modes, a root cause analysis (RCA) is performed to assess the failure mechanisms and governing loads. Preventing the failure to reoccur is then simply achieved by either increasing the capacity of the part or reducing the loads.

Finally, determining the failure mechanism for a specific failure can be quite difficult, especially for the non-experts that are typically expected to register the failures in the CMMS. Also for this process a decision support tool, implemented in an expert system (ES), has been developed: FAME-X [54]. Similar to the ES advising in the condition monitoring technique selection, this ES also asks questions to the user. These questions are clustered in two phases: the first phase does a rough estimate of the basic failure mechanism (e.g. fatigue, creep, wear, corrosion), whereas the second phase provides a more specific identification (e.g. crevice corrosion, low cycle fatigue, etc.). In each cluster, the questions are related to different sets of attributes: service life conditions (loads, environment), age, post-mortem characteristics (striations, corrosion products) and material (metal, polymer, ceramic). By checking a sufficient number of attributes, the ES is able to discriminate between potential failure mechanisms quite efficiently.

6 Case Studies

In this final section, a number of case studies in different sectors of industry are described, demonstrating how the concepts, methods and tools introduced in the previous sections can be applied in practice. The first case study on marine diesel engines and the second case study on rail infrastructure demonstrate the model-based prognostic approach. The third case study on wind turbines also touches this approach, but after that mainly focuses on vibration-based health monitoring.

6.1 Maritime Systems

Maritime systems, like ships and all their subsystems, are typically operated in a harsh and largely variable environment. At the same time, failures in any of the subsystems or components may have large consequences, e.g. high costs (loss of revenues, high logistics costs due to remote locations) or environmental impacts. To improve the (preventive) maintenance process, a better prediction of failures for systems operated under specific conditions is required. In this subsection, a prognostic method for a (mechanical) diesel engine component will be shown. Details on this analysis can be found in [1], where also a similar method for electronic parts in a radar is presented.

In a diesel engine, the liner covers the inside of the cylinder, in which the piston is reciprocating, see Fig. 12. The piston contains a number of rings, which are in lubricated contact with the liner. The reciprocating motion maintains the lubricant oil film, but at the top and bottom reversal points, the film thickness is less, and mechanical wear can occur due to the relative motion of the two metal parts.

Fig. 12 Position of cylinder liner and piston rings in a diesel engine

A model has been developed [55] that describes the physical degradation processes of sliding wear occurring at the interface between liner and ring. The model applied is the Archard's wear model [47], Eq. (1) introduced in Sect. 4.2.

This model requires a number of inputs that can be related to either the properties or the operating conditions of the specific engine:

- The normal force F is directly related to the pressure in the cylinder and therefore depends on the engine operating condition (e.g. power).
- The sliding distance s per cycle that the liner experiences is equal to two times the accumulated width of the cylinder rings that pass a certain location on the liner two times each cycle (up and down). Further, the speed of the engine (rpm) determines the number of passings per time unit.
- The wear parameter k is the proportionality constant, which can be estimated for the specific combination of materials (liner and rings) and wear mechanism. If, for example, the lubricant is contaminated with particles, the mechanism will switch from adhesive to abrasive wear, which will change the value of the wear parameter.
- Finally, in a lubricated contact the amount of wear will be negligible, since there is no metal-to-metal contact. This means that only wear occurs when the lubricant film thickness of the cylinder is lower than the critical film thickness.

The first step in the analysis is then to calculate the oil film thickness along the liner surface. This is shown in Fig. 13 for a certain engine and operating condition, in this case as a function of crack angle (a four stroke diesel engine rotates over 720° each cycle, so crank angles of 90, 270, 450 and 630° represent the same location on the liner). Also the critical film thicknesses for adhesive and abrasive wear are indicated. Note that the absolute values of the film thickness depend on several model parameters, which have to be determined from experiments. In this stage of

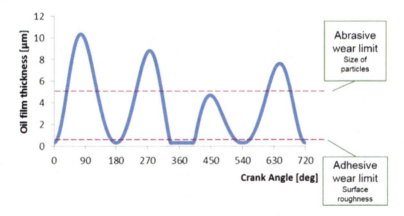

Fig. 13 Film thickness variation with crank angle, including wear limits for adhesive and abrasive wear

Table 1 Calculated wear depth and remaining life after 15,000 h for three different operating scenarios

Scenario	Max wear depth [mm]	Remaining life [hrs]
A (high speed/high load)	0.271	1.603
B (high speed/var load)	0.247	3.224
C (var speed/var load)	0.236	4.103

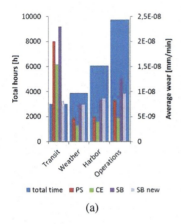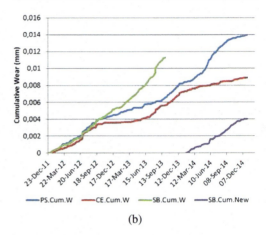

(a) (b)

Fig. 14 Distribution of operating hours over three engines and four operating profiles (**a**) and calculated wear depth evolution over time for the three engines (**b**)

the modelling process, typical values are used. This graph clearly shows that without oil contamination, wear will only occur in some small regions around the reversal points of the piston (180, 360, etc.).

Using this variation of oil film thickness, and together with the calculated normal force (as obtained from the cylinder pressure), the wear rate distribution along the liner can be calculated for three different artificial operating scenarios. These scenarios all represent 15,000 h of operation, but differ in their division over high/medium/low speed, respectively, high/medium/low load, see Table 1. The table also shows the calculated remaining service life of the liners, assuming a replacement at 0.3 mm wear depth.

The final step in the predictive method is to relate the wear of the liner to variations in operating conditions of real engines. This has been done for a ship with three diesel engines (SB: starboard, CE: centre and PS: portside) operating in four different profiles. The distribution of operating hours over these four profiles is shown in Fig. 14a, which is based on the monitoring data obtained from the asset owner. During the simulated period the SB engine has been replaced by a new one, explaining the double bar for SB in the diagram. Figure 14b shows the calculated evolution of the wear depth for the three engines over time. The plot clearly shows variations in wear rate, which can be associated with changes in operating profile. Further, it can be concluded that the three engines are operated differently, and thus also show a considerable difference in wear behaviour.

6.2 Railway Infrastructure

Maintenance of railway infrastructures is a critical factor in the performance of a network. Insufficient maintenance may lead to catastrophic failures of the network, with which high costs for repair and network unavailability are associated—apart from potential human casualties, which are to be avoided at any cost. Too much maintenance makes transport by train uncompetitive and results in a waste of materials and resources. Infra managers therefore have a high interest in optimizing their maintenance policies, by basing it on predictive maintenance. Predictive maintenance relies on monitoring and, as argued in this chapter, on physical models, supporting the interpretation of the data and guiding which parameters to monitor.

Monitoring of railway infrastructure has many elements in which large amounts of data are involved. Data are collected either in rapid pace or over large stretches of track. Moreover, various stake-holders are involved, requiring different (types of) information, resulting in a variety of monitoring systems. This case study focuses on the wear of the track.

Railway tracks inevitably wear, implying that the ability to predict wear is crucial for an optimized maintenance planning and cost reduction. Rail wear is inseparably connected to rail-wheel interaction. Moreover, wear prediction tool is heavily relying on numerical simulation tools, primarily multi-body dynamics (MBD) based. These models are used to determine the contact points, the amount of spin and slip in the contact area [56]. Combining Kalker's contact model [57] with Archard's wear model [47] (see also Eq. (1)) results in an amount of material wear of the cross-section of the track or the wheel.

As indicated by Meghoe et al. [58], most of the research focused on wear of wheels rather than on wear of the track. Moreover, the process of estimating the wear is cumbersome, as it requires a nearly continuous update of the wheel or track profile, as the combination of profiles affects the train dynamics and hence the contact points, pressure and finally the wear rate. For this reason, it is difficult for the contractors maintaining the track to plan their maintenance activities efficiently. The current process of rail wear prediction [59] is shown in Fig. 15.

The approach followed by Meghoe et al. [58] is based on the use of measured profiles of both new and worn track and wheels, rather than on new wheels that are updated, together with the track, during the simulation. Subsequently, a sensitivity analysis is ran with a multi-body dynamics simulation, accounting for variables such as speed and weight of the train, material hardness, track curvature and geometry irregularities (step 3 in Sect. 4.2). The resulting meta-function can then be used to estimate the wear, given a set of input parameters, available for the infra managers and maintenance contractors (step 5). The method is validated (step 6) using measured track profiles on a Dutch line (Weesp–Almere). The profiles were measured six times in the period of 2014–2016. Preliminary results show that this type of modelling can predict the wear of tracks and even indicate which parameters most strongly influence the wear. The predicted wear is shown in Fig. 16, in which the measured wear is also plot. The measured wear is a lower bound, due

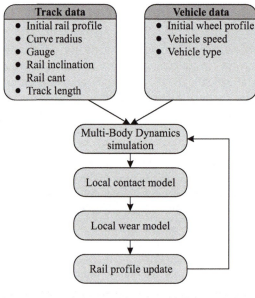

Fig. 15 Process of rail wear prediction [59]

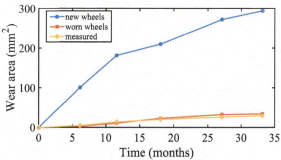

Fig. 16 Outer rail wear area versus time for a track curve of $R = 1800$ m as predicted by simulations with new and worn wheels compared to measured values

to the method of measuring the rail profiles [58]. Clearly, this method contributes significantly to the maintenance policies and strategies that can be exploited.

6.3 Wind Turbines

Over the past years, the wind energy sector has grown significantly. The need for green energy to meet climate targets has pushed the development of off-shore wind farms in particular. The industry has focused on the development of larger wind turbines. According to Wilkinson et al. [60], the operational and maintenance costs of off-shore wind turbines are five times as high as those of their onshore counterparts. An important factor in these costs are the higher failure rates that are reported [61, 62]. Following the physics-based methodology, Breteler et al. [63]

implemented a basic qualitative model for the prediction of the remaining useful lifetime of a wind turbine gearbox using SCADA data and shaft misalignment information to feed a physics-based model. The base of this model is formed by the design models as specified in ISO standards, complemented with numerical models. Adding the real loading to these models and assessing the affect of misalignments provides an indication of the actual remaining useful life.

The method discussed above is suitable for the drive train of the wind turbine, yet is less applicable for monitoring the blades. Blades, mostly made of glass fibre reinforced plastic, are predominantly fatigue loaded. Composites under fatigue loading develop micro-cracks, which are transverse cracks. These cracks initiate stochastically in time and space and grow as the loading continues. Micro-cracks join or increase their grow rate until they formed a delamination. This results in a significant drop in structural integrity and a functional failure of the component, possibly without any visual indication. Clearly, damage accumulation is an important parameter to monitor to allow the prediction of the remaining useful life.

Monitoring of damage accumulation in a specific part of a wind turbine blade is studied [64]. In this project, the vibration-based health monitoring approach, as was introduced in Sect. 3.2, is followed.

An initial investigation into the failure mechanisms of wind turbine blades [65] revealed that, among a few others, fatigue in the spar cap is a common failure mechanism with a significant effect on the structural integrity, hence the functionality of the blade (step 1 and step 2 in Sect. 4.2). Current practice in diagnosing a structural sample is to execute, for example a three-point bending fatigue test. The force is measured for a given displacement amplitude during the measurement. Typically, the force–displacement relation is not affected by the damage accumulation up to close to the moment the structure fails: the global bending stiffness is not affected by the small cracks growing in the interior of the material.

The method adopted by the authors is based on the use of piezo sensors (PZT) on plate-like (thin) composite structures [66–68] to identify a delamination (separation of internal layers of the composite material). Here, however, the objective is to identify an accumulation of small cracks in the material, effectively achieved by frequently monitoring the dynamic response of the structure (see also Sect. 3.2).

The PZTs, in this area of research often referred to as piezo-electric wafer active sensor (PWAS), are bonded on the structure and are activated with short burst signals in the low ultrasonic frequency range ($\mathcal{O}(10)$ kHz–$\mathcal{O}(100)$ kHz). A network of PWAS is formed (Fig. 17), where each transducer is sequentially appointed as actuator, while the others act as sensor. A set of signals from PWAS i to PWAS j is thus acquired. Using the reconstruction algorithm for probabilistic inspection of damage (RAPID) [69], the region in which the damage is accumulating can be estimated.

The RAPID algorithm is based on the comparison of the signals of each of the actuator–sensor paths in pristine and post-damage state. A difference between these two signal does not indicate a location yet, as the difference is condensed to a single number—the damage index ρ. Typically, the correlation coefficient between the two

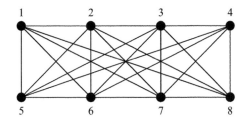

Fig. 17 Network of piezo-electric wafer active sensors (PWAS). Each PWAS is sequentially assigned as actuator, while the others act as receivers

signals S is used as damage index [70, 71], but a range of alternative methods is available that can be used to calculate the damage index [21, 72]. The choice of the method depends on the application. Venterink et al. [23] concluded the signal amplitude peak squared (SAPS) percentage difference algorithm provided the best results for this particular case. The algorithm is given by:

$$\rho_{SAPS,k} = 1 - \left(\frac{\max\left(S_{H,k}\right) - \max\left(S_{D,k}^{\tau}\right)}{\max\left(S_{H,k}\right)} \right)^2 \qquad (2)$$

where the subscript H refers to the (healthy) reference state, D to the current state and k to the actuator–sensor path number and with the small time span (here two oscillation cycles of the actuation frequency) defined as:

$$S_{D,k}^{\tau} = S_D \left(t_{\max}^{H} - \Delta t : t_{\max}^{H} + \Delta t \right) \qquad (3)$$

A probability function is subsequently used, indicating the probability that an anomaly at location (x, y) causes a change in the signal sent by transducer i and received by transducer j. It uses a geometrical function $R(x, y)$, specifying the distance from point (x, y) to the direct line between the two transducers i and j, ceiled by the threshold value β:

$$R(x, y) = \begin{cases} \dfrac{\sum\limits_{k=i,j} \sqrt{(\Delta x_k + \alpha \Delta x_{kn})^2 + (\Delta y_k + \alpha \Delta y_{kn})^2}}{(1 - 2\alpha)\sqrt{\Delta x_{ij}^2 + \Delta y_{ij}^2}} & \text{for } R(x, y) < \beta \\ \beta & \text{for } R(x, y) \geq \beta \end{cases} \qquad (4)$$

$$\Delta x_k = (x - x_k), \quad \Delta y_k = (y - y_k) \quad \text{with } k = i, j; \quad \Delta x_{ij} = x_i - x_j$$

$$n = \begin{cases} i & \text{if } k = j \\ j & \text{if } k = i \end{cases}$$

where (x_i, y_i) and (x_j, y_j) indicate the locations of transducer i and j, respectively. The values of α and β can be optimized based on minimization of blind zones, deviation in probability distribution values and the kurtosis [73].

Subsequently overlaying all path results will give a probability intensity map of the possible damage. The damage intensity probability I at an arbitrary position (x, y) is given by:

$$I(x, y) = \sum_{k=1}^{N_p} \left((1 - \rho_k) \left(\frac{\beta - R(x, y)}{\beta - 1} \right) \right) \qquad (5)$$

with ρ_k being the damage indicator of the kth actuator–sensor path, N_p the number of paths.

The experiment executed by the knowledge centre for wind turbine materials and constructions (WMC) is a three-point bending fatigue test of a thick composite beam. The uni-directional, 96 layer non-crimp glass fibre fabric reinforced plastic (Hexion RIM 135) beam was manufactured by WMC, yet instrumented by the University of Twente. The dimension of the beam is $l \times b \times h = 900 \times 60 \times 56 \, \text{mm}^3$. Eight transducers, four on top and four on the bottom, were bonded on the structure, centred around the mid-point of the beam, as shown in Fig. 18.

The LabVIEW program controlling the acousto-ultrasonic measurements was configured to communicate with the WMC system controlling the fatigue test. The fatigue test is paused at predefined intervals, shortening with increasing number of total cycles, to allow the acousto-ultrasonic measurements to be executed. Once these are finished, the fatigue test continues. The fatigue test was paused every 2000

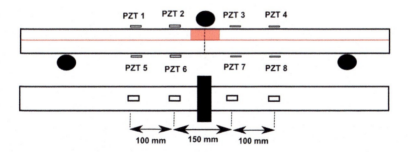

Fig. 18 Schematic representation of the three-point bending test on the PZT instrumented glass fibre reinforced beam. The transducer locations are marked along with their number. The marked red areas indicate the expected locations of the fatigue damage. The three black circles in the side view are the plunger (top one) and the two supports of the three-point bending setup

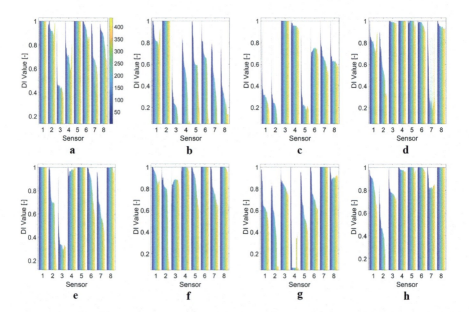

Fig. 19 The DI values of using SAPS2 and an actuation frequency of 200 kHz. The colours refer to the measurement number. The first dark blue one is the reference measurement and the last yellow one is the last measurement prior to failure (approximately after 2.65 million cycles). (**a**) Actuator PWAS 1. (**b**) Actuator PWAS 2. (**c**) Actuator PWAS 3. (**d**) Actuator PWAS 4. (**e**) Actuator PWAS 5. (**f**) Actuator PWAS 6. (**g**) Actuator PWAS 7. (**h**) Actuator PWAS 8

cycles until a total of 950,000 cycles was reached, after which the fatigue test was paused every 1000 cycles. The beam finally failed after nearly 2.7 million cycles.

The exact nature of the waves generated by the actuation signal will not be studied here. It is well known that lamb waves propagate in thin plate like structures, but the ultrasonic waves in a thick (steel) structure are more complex [66, 67]. The usage of composite materials in the present work will further increase the complexity. It will however be shown that a detailed understanding of the wave forms is not necessary.

The time signals are expected to change once damage starts to accumulate. Hence, the signals of all subsequent measurements are compared to the reference state, using Eq. (2). The damage index values, based on a 200 kHz actuation signal, are depicted in Fig. 19. The colour refers to the measurement number, where the darkest blue corresponds to the first, reference, measurement and most yellow to the last measurement before failure (approximately after 2.65 million cycles).

The damage index evaluation as a function of the number of fatigue cycles shows clear drops for some of the actuator–sensor pairs, implying the waveform is gradually changing. The damage index increases for higher cycle numbers in some cases, e.g. on the path from PWAS 1 to PWAS 4 (Fig. 19a), which is attributed to the complexity of the waveforms and the interaction with damage. Physically, an increase is impossible as it would indicate healing.

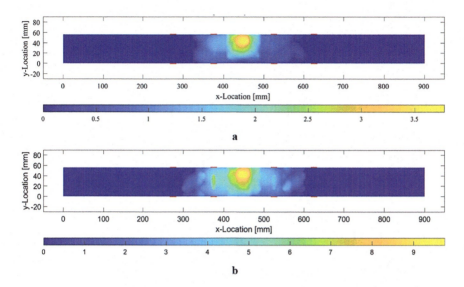

Fig. 20 The damage probability (RAPID) maps, based on the SAPS algorithm and an actuation frequency of 2000 kHz, after (**a**) 87,000 cycles; and (**b**) 2,652,000 cycles. The red lines indicate the location of the transducers

The damage probability (RAPID) maps are constructed to estimate the region of damage accumulation, using Eq. (5). The result, again for an actuation frequency of 200 kHz, is shown in Fig. 20. The figure shows the damage probability map after 87,000 cycles and after 2,652,000 cycles—just before failure. The red lines indicate the location of the transducers.

Clearly, the intensity of the damage is growing with increasing number of fatigue cycles. Note that the colour scale is different for the two images in Fig. 20. As expected, the damage starts to accumulate directly underneath the centre punch. To follow the damage accumulation over time, the probability maps need to be converted to a single number, representing the intensity of the damage accumulation. Initially, the maximum damage probability value is taken, leading to the graph shown in Fig. 21.

The maximum of the damage probability value shows a sharp increase during the first 100,000 cycles. This is attributed to damage directly inflicted by the punch. A region with a relative constant slope then follows. This indicates a steady growth of damage inside the beam. Based on Fig. 20, this damage is formed in the centre of the beam, just underneath the punch. The variation in slope of the maximum damage probability value between 100,000 and 2,300,000 cycles is attributed to both the stochastic nature of crack formation and the data processing method. Analysis of the data revealed that variations in the signal strengths can cause small variations. These variations also explain the negative slope of the maximum damage probability value between 2,300,000 and 2,500,000 cycles. A correction for this is proposed in [23].

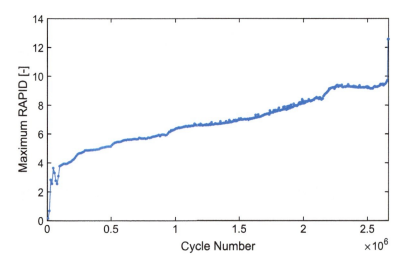

Fig. 21 Maximum damage probability value against the cycle number of the fatigue test for an actuation frequency of 200 kHz

Although the slope of the curve is fairly constant, it appears to show some small jumps around 500,000 and 1,000,000 cycles, followed by a larger jump around 2,300,000 cycles. This last jump seems to be preceded by a small increase in the slope. Sound evidence is missing, by the lack of ultrasonic inspection of the beam after each acousto-ultrasonic measurement. However, a plausible explanation is that the jump represents the formation of a delamination. Photos of the beam, taken at different moments in time, reveal a delamination is formed, just underneath the top surface and roughly running from PWAS 3 to PWAS 2, as shown in Fig. 22a. Finally the beam failed due to a larger delamination in this area (Fig. 22b).

It is demonstrated with this experiment that acousto-ultrasonics can be used for diagnosis of the damage accumulation in a thick glass fibre reinforce beam. The gradual increase of the diagnostic value (here the maximum of the RAPID plot) indicates the potential for prognostics, hence combining diagnosis and prognosis. Further research is necessary to explore the predictive capabilities. This will, for example, allow the use of the gradient of the line rather than a threshold value; differences in the loading and failure pattern may require a re-calibration of the threshold, which obviously is not practical and compromises the robustness of the method.

7 Conclusions

This chapter started with introducing a number of challenges in the field of predictive maintenance. It was argued that these challenges are the main reasons that predictive maintenance is promising, but is yet lacking the maturity for broad

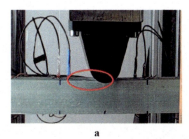
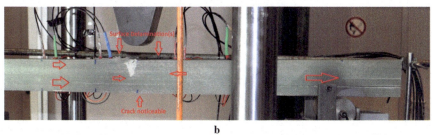

Fig. 22 Photo of the state of the beam after (**a**) approximately 300,000 cycles; (**b**) failure, approximately 2,700,000. A first delamination starts to form in the area indicated by the red ellipse (between PWAS 2 and PWAS 3), which is the location where the final delamination failure occurs

application in industrial practice. Then a number of methods and tools have been presented, which addressed most of these challenges, and provided possible ways to increase the maturity of predictive maintenance:

- Vibration-based structural health monitoring techniques can assist in detecting damage, diagnosing a system and (ultimately) predicting the remaining lifetime. It has been shown that a thorough understanding of the system dynamics is required to develop well-performing algorithms;
- An approach to develop physical model-based prognostic methods has been presented, stressing the need to understand the physics of failure when predicting the lifetime of systems operated in highly variable operational conditions;
- Decision support tools have been presented, assisting users in
  - the selection of the most suitable predictive maintenance approach;
  - the selection of an appropriate condition monitoring technique;
  - the selection of the critical parts in a larger system;
  - the determination of the failure mechanism responsible for a certain failure;

Finally, a number of cases were presented, demonstrating how the developed methods ad tools are implemented in a range of different industries.

References

1. Tinga, T., Tiddens, W.W., Amoiralis, F., Politis, M.: Predictive maintenance of maritime systems. In: Cepin, M., Bris, R. (eds.) Proceedings of the 27th European Safety and Reliability Conference (ESREL). Safety & Reliability - Theory and Applications, pp. 421–429. Taylor & Francis, Abingdon (2017)
2. Tinga, T., Loendersloot, R.: Aligning PHM, SHM and CBM by understanding the system failure behaviour. In: Bregon, A., Daigle, M.J. (eds.) Proceedings of the European Conference of the Prognostics and Health Management Society, pp. 162–171. PHM society, Nantes, France (2014)
3. Homborg, A.M., Leon Morales, C.F., Tinga, T., de Wit, J.H.: Detection of microbiologically influenced corrosion by electrochemical noise transients. Electrochim. Acta **136**, 223–232 (2014)
4. Homborg, A.M., Tinga, T., van Westing, E., Zhang, X.: A critical appraisal of the interpretation of electrochemical noise for corrosion studies. Corrosion **70**(10), 971–987 (2014)
5. Homborg, A.M., van Westing, E., Tinga, T., Zhang, X.: Novel time-frequency characterization of electrochemical noise data in corrosion studies using Hilbert spectra. Corros. Sci. **66**, 97–110 (2013)
6. Sanchez Ramirez, A., Loendersloot, R., Jauregui Becker, J.M., Tinga, T.: Design framework for vibration monitoring systems for helicopter rotor blade monitoring using wireless sensor networks. In: Chang, F.K. (ed.) Proceedings of the 9th International Workshop on Structural Health Monitoring, Stanford, CA, U.S.A., pp. 1023–1030. DEStech publishing, Lancaster (2013)
7. Boller, C.: Structural health monitoring – its association and use. In: New Trends in Structural Health Monitoring. Springer, Vienna (2013). ISBN 978-3-7091-1389-9 (Print) 978-3-7091-1390-5 (online)
8. Cantero, D., O'Brien, E.J., Karoumi, R.: Extending the assessment dynamic ratio to railway bridges. In: Pombo, J. (ed.) Proceedings of the Second International Conference on Railway Technology: Research, Development and Maintenance, 13pp. Civil-Comp Press, Stirlingshire (2014)
9. Seraj, F., van der Zwaag, B.J., Dilo, A., Luarasi, T., Havinga, P.J.M.: RoADS: a road pavement monitoring system for anomaly detection using smart phones. In: Big Data Analytics in the Social and Ubiquitous Context, pp. 128–146. Lecture Notes in Computer Science. Springer, Cham (2016)
10. Seraj, F., Meratnia, N., Havinga, P.J.M.: Rovi: continuous transport infrastructure monitoring framework for preventive maintenance. In: 2017 IEEE International Conference on Pervasive Computing and Communications, PerCom 2017, pp. 217–226. IEEE, Piscataway (2017)
11. Seraj, F.: Rolling vibes: continuous transport infrastructure monitoring. Ph.D. thesis, University of Twente (2017)
12. Blake, B.M., Kallol, D., Zand, P., Havinga, P.J.M.: Industrial wireless monitoring with energy-harvesting devices. IEEE Internet Comput. **21**(1), 12–20 (2017)
13. Güemes, J.A., Sierra-Pérez, J.: Fiber optics sensors. In: New Trends in Structural Health Monitoring. Springer, Vienna (2013). ISBN 978-3-7091-1389-9 (Print) 978-3-7091-1390-5 (online)
14. Ooijevaar, T.H., Warnet, L., Loendersloot, R., Akkerman, R., de Boer, A.: Vibration based damage identification in a composite T-beam utilising low cost integrated actuators and sensors. In: Boller, C. (ed.) Proceedings of the Sixth European Workshop on Structural Health Monitoring, pp. 232–239. DGZfP e.V., Dresden (2012)
15. Schmidt, D., Kolbe, A., Kaps, R., Wierach, P., Linke, S., Steeger, S., von Dungern, F., Tauchner, J., Breu, C., Newman, B.: Development of a door surrounding structure with integrated structural health monitoring system. In: Smart Intelligent Aircraft Structures (SARISTU): Proceedings of the Final Project Conference, pp. 935–945. Springer, Cham, 2015

16. Hwang, J.S., Loendersloot, R., Tinga, T.: Experimental evaluation of vibration-based damage identification methods on a composite aircraft structure with internally-mounted piezodiaphragm sensors. In: Chang, F.K., Guemes, A. (eds.) Proceedings of International Workshop on Structural Health Monitoring, p. 8. DEStech Publications, Lancaster (2015)
17. Ooijevaar, T.H., Warnet, L., Loendersloot, R., Akkerman, R., Tinga, T.: Impact damage identification in composite skin-stiffener structures based on modal curvatures. Struct. Control. Health Monit. **23**(2), 198–217 (2016)
18. Ooijevaar, T.H., Rogge, M.D., Loendersloot, R., Warnet, L., Akkerman, R., Tinga, T.: Nonlinear dynamic behavior of an impact damaged composite skin-stiffener structure. J. Sound Vib. **353**, 243–258 (2015)
19. Ooijevaar, T.H., Rogge, M.D., Loendersloot, R., Warnet, L., Akkerman, R., Tinga, T.: Vibro-acoustic modulation–based damage identification in a composite skin–stiffener structure. Struct. Health Monit. **15**(4), 458–472 (2016)
20. Moix Bonet, M., Wierach, P., Loendersloot, R., Bach, M.: Damage assessment in composite structures based on acousto-ultrasonics - evaluation of performance. In: Smart Intelligent Aircraft Structures (SARISTU): Proceedings of the Final Project Conference, pp. 617–629. Springer, Cham (2015)
21. Loendersloot, R., Buethe, I., Michaelides, P., Moix Bonet, M., Lampeas, G.: Damage identification in composite panels - methodology and visualisation. In: Smart Intelligent Aircraft Structures (SARISTU): Proceedings of the Final Project Conference, pp. 579–604. Springer, Cham (2015)
22. Loendersloot, R., Moix Bonet, M.: Damage identification in composite panels using guided waves. In: Proceedings of the 5th CEAS Air & Space Conference, p. 14 (2015)
23. Venterink, M., Loendersloot, R., Tinga, T.: The detection of fatigue damage accumulation in a thick composite beam using acousto ultrasonics. In: Proceedings of the First HEAMES Conference, London, UK, p. 10 (2018)
24. Demcenko, A., Akkerman, R., Nagy, P.B., Loendersloot, R.: Non-collinear wave mixing for non-linear ultrasonic detection of physical ageing in PVC. NDT & E Int. **49**, 34–39 (2012)
25. Hernandez Delgadillo, H., Loendersloot, R., Akkerman, R., Yntema, D.R.: Development of an inline water mains inspection technology: detection of acidic deterioration in cement-based water pipes with ultrasonic pulse-echo technique. In: Proceedings of 2016 IEEE International Ultrasonics Symposium (IUS), p. 4. IEEE International, Piscataway (2016)
26. Da Silva Souza, F., Oki, N., Filho, J.V., Loendersloot, R., Berkhoff, A.P.: Accuracy and multi domain piezoelectric power harvesting model using VHDL-AMS and SPICE. In: Proceedings of IEEE Sensors, p. 3. IEEE International, Piscataway (2016)
27. Gomez Casseres Espinosa, A.F., Sanchez Ramirez, A., Combita Alfonso, L.F., Loendersloot, R., Berkhoff, A.P.: Development of a piezoelectric based energy harvesting system for autonomous wireless sensor nodes. In: Le Cam, V., Mevel, L., Schoefs, F. (eds.) 7th European Workshop on Structural Health Monitoring, pp. 205–212 (2014)
28. Thobiani, F.A., Tran, V.T., Tinga, T.: An approach to fault diagnosis of rotating machinery using the second-order statistical features of thermal images and simplified fuzzy ARTMAP. Engineering **9**(6), 524–539 (2017)
29. Tran, V.T., Thobiani, F.A., Tinga, T., Ball, A.D., Niu, G.: Single and combined fault diagnosis of reciprocating compressor valves using a hybrid deep belief network. Proc. IME C. J. Mech. Eng. Sci. (2017). https://doi.org/10.1177/0954406217740929
30. Fassois, S.D., Kopsaftopoulos, F.P.: Statistical time series methods for vibration based structural health monitoring. In: New Trends in Structural Health Monitoring. Springer, Vienna (2013). ISBN 978-3-7091-1389-9 (Print) 978-3-7091-1390-5 (online)
31. Ooijevaar, T.H.: Vibration based structural health monitoring of composite skin-stiffener structures. Ph.D. thesis, University of Twente (2014)
32. Fan, W., Qiao, P.: Vibration-based damage identification methods: a review and comparative study. Struct. Health Monit. **10**, 83–111 (2011)
33. Montalvao, D., Maia, N.M.M., Ribeiro, A.M.R.: A review of vibration-based structural health monitoring with special emphasis on composite materials. Shock Vib. Digest **38**, 295–324 (2006)

34. Carden, E.P., Fanning, P.: Vibration based condition monitoring: a review. Struct. Health Monit. **3**(4), 355–377 (2004)
35. Worden, K., Farrar, C.R., Manson, G., Park, G.: The fundamental axioms of structural health monitoring. Proc. R. Soc. **437**, 1639–1664 (2007)
36. Baaran, J.: Visual inspection of composite structures. Technical report, Institute of Composite Structures and Adaptive Systems, DLR Braunschweig (2009)
37. Derriso, M.M., DeSimio, M.P., McCurry, C.D., Schubert Kabban, C.M., Olson, S.O.: Industrial age non-destructive evaluation to information age structural health monitoring. Struct. Health Monit. **13**(6), 591–600 (2014)
38. Tinga, T.: Application of physical failure models to enable usage and load based maintenance. Reliab. Eng. Syst. Saf. **95**(10), 1061–1075 (2010)
39. Tinga, T.: Physical model based component prognostics. In: Maintenance Modelling and Applications, pp. 166–184. DNV Hovik (2011)
40. Lee, J., Wu, F., Zhao, W., Ghaffari, M., Liao, L., Siegel, D.: Prognostics and health management design for rotary machinery systems – reviews, methodology and applications. Mech. Syst. Signal Process. **42**(1), 314–334 (2014)
41. Khan, S., Yairi, T.: A review on the application of deep learning in system health management. Mech. Syst. Signal Process. **107**, 241–265 (2018)
42. Jardine, A.K.S., Lin, D., Banjevic, D.: A review on machinery diagnostics and prognostics implementing condition-based maintenance. Mech. Syst. Signal Process. **20**, 1483–1510 (2006)
43. Farrar, C.R., Lieven, N.A.J.: Damage prognosis: the future of structural health monitoring. Phil. Trans. R. Soc. A **365**, 623–632 (2006)
44. Zio, E.: Reliability engineering: old problems and new challenges. Reliab. Eng. Syst. Saf. **94**, 125–141 (2009)
45. Tinga, T.: Predictive maintenance of military systems based on physical failure models. Chem. Eng. Trans. **33**, 295–300 (2013)
46. Tinga, T.: Principles of Loads and Failure Mechanisms: Applications in Maintenance, Reliability and Design. Springer Series in Reliability Engineering. Springer, London (2013)
47. Archard, J.F.: Contact and rubbing of flat surfaces. J. Appl. Phys. **24**(8), 981–988 (1953)
48. Sayed-Mouchaweh, M.: Learning from Data Streams in Dynamic Environments. SpringerBriefs in Applied Sciences and Technology. Springer, New York (2016)
49. Tiddens, W.W., Braaksma, A.J.J., Tinga, T.: Towards informed maintenance decision making: guiding the application of advanced maintenance analyses. In: Optimum Decision Making in Asset Management. IGI Global, Hershey (2017). https://doi.org/10.4018/978-1-5225-0651-5.ch013
50. Tiddens, W.W., Braaksma, A.J.J., Tinga, T.: Framework for the selection of the optimal preventive maintenance approach. (to be submitted, 2018)
51. Mouatamir, A.: Decision support system for condition monitoring technologies. PDEng. thesis, University of Twente (2018)
52. Tiddens, W.W., Braaksma, A.J.J., Tinga, T.: Selecting suitable candidates for predictive maintenance. Int. J. Prognostics Health Manag. **9**(1), 020, 1–14 (2018)
53. Tinga, T.: Mechanism Based Failure Analysis. Improving Maintenance by Understanding the Failure Mechanisms. Netherlands Defence Academy, Den Helder (2012)
54. Karampelas, D.: FAME-X : failure mechanism identification expert system, in engineering technology. PDEng. thesis, University of Twente, Enschede, The Netherlands (2018)
55. Duplex, P.: Design of a life prediction tool for high-speed diesel engines, PDEng Thesis, University of Twente, Enschede (2018)
56. Sichani, M.S.: On efficient modelling of wheel-rail contact in vehicle dynamics simulation. Ph.D. thesis, KTH Royal Institute of Technology (2016)
57. Kalker, J.J.: Three-Dimensional Elastic Bodies in Rolling Contact. Solid Mechanics and Its Applications, vol. 2, 1st edn. Springer, Dordrecht (1990)

58. Meghoe, A.A., Loendersloot, R., Bosman, R., Tinga, T.: Rail wear estimation for predictive maintenance: a strategic approach. In: Proceedings of the Prognostic Health Management Conference, p. 11. PHM Society, Utrecht, Netherlands (2018)
59. Jendel, T.: Prediction of wheel profile wear-comparisons with field measurements. Wear **253**(1–2), 89–99 (2002)
60. Wilkinson, M., Spinato, F., Knowles, M., Tavner, P.: Towards the zero maintenance wind turbine. In: Proceedings of Power Engineering Conference Newcastle, pp. 74–78 (2006)
61. Perez, J.M.P., Marquez, F.P.G., Tobias, A., Papaelias, M.: Wind turbine reliability analysis. Renew. Sust. Energ. Rev. **23**, 463–472 (2013)
62. Crabtree, C.: Operational and reliability analysis of offshore wind farms. Ph.D. thesis, School of Engineering and Computing Sciences (2012)
63. Breteler, D., Kaidis, C., Tinga, T., Loendersloot, R.: Physics based methodology for wind turbine failure detection, diagnostics & prognostics. In: Rosmi, A. (ed.) EWEA, p. 9 (2015)
64. Loendersloot, R., Venterink, M., Kruse, A., Lahuerta, F.: Acousto-ultrasonic damage monitoring in a thick composite beam for wind turbine applications. In Proceedings of the European Workshop on Structural Health Monitoring, pp. 1–12 , Manchester, UK (2018)
65. Lahuerta, F.: Identification of typical failures in composite rotor blades and structural health monitoring. Technical report TKI SLOWIND, Knowledge center WMC, Wieringerwerf, The Netherlands (2016)
66. Greve, D.W., Oppenheim, I.J., Zheng, P.: Lamb waves and nearly-longitudinal waves in thick plates. In: Proceedings of SPIE - The International Society for Optical Engineering (2008)
67. Giurgiutiu, V.: Structural Health Monitoring with Piezoelectric Waver Active Sensors. Elsevier, Columbia (2008)
68. Liu, X.L., Jiang, Z.W., Ji, L.: Investigation on the design of piezoelectric actuator/sensor for damage detection in beam with lamb waves. Exp. Mech. **53**(3), 485–492 (2013)
69. Zhao, X., Gao, H., Zhang, G., Ayhan, B., Yan, F., Kwan, C., Rose, J.L.: Active health monitoring of an aircraft wing with embedded piezoelectric sensor/actuator network: I. defect detection, localization and growth monitoring. Smart Mater. Struct. **16**(4), 1208–1217 (2007)
70. Su, Z., Ye, L., Lu, Y.: Guided lamb waves for identification of damage in composite structures: a review. J. Sound Vib. **295**(3–5), 753–780 (2006)
71. Su, Z., Ye, L.: Identification of Damage Using Lamb Waves – From Fundamentals to Applications. Lecture Notes in Applied and Computational Mechanics, vol. 48. Springer, London (2009)
72. Wu, Z., Liu, K., Wang, Y., Zheng, Y.: Validation and evaluation of damage identification using probability-based diagnostic imaging on a stiffened composite panel. J. Intell. Mater. Syst. Struct. **26**(16), 2181–2195 (2014)
73. Moix Bonet, M., Eckstein, B., Loendersloot, R., Wierach, P.: Identification of barely visible impact damages on a stiffened composite panel with a probability-based approach. In: Chang, F.K., Guemes, A. (eds.) Proceedings of International Workshop on Structural Health Monitoring, p. 8. DEStech Publications, Lancaster (2015)

On Prognostic Algorithm Design and Fundamental Precision Limits in Long-Term Prediction

Marcos E. Orchard and David E. Acuña

1 Introduction

Failure prognostic algorithms typically generate long-term predictions to characterize the evolution in time of a condition indicator. These long-term predictions are then used to estimate the Time-of-Failure (ToF) of a faulty system. This procedure can only be effective if the algorithm is able to adequately incorporate information about the underlying degradation processes and future operating profiles, as well as an effective characterization of all associated uncertainty sources.

Probability-based methods, and particularly Bayesian approaches [1], stand out as a class of algorithms that allow to easily characterize uncertainty sources in failure prognostic problems [2]. The mathematical and theoretical framework associated with probability-based algorithms allows to implement filtering, smoothing, and prediction approaches in a straightforward manner, even in the case of nonlinear, non-Gaussian, dynamic processes [3]. For this reason, many Bayesian state estimation methods have been applied in the past to determine initial conditions for the generation of system long-term predictions [4], while others have been used to characterize future loading (or stress) profiles [5].

In failure prognostics, an effective characterization of future uncertainty sources is important because it allows to avoid catastrophic events and take preventive measures [4, 6]. Although this problem can be solved if we assume that both the actual system condition and degradation model are known, by performing Monte Carlo (MC) simulation [7], the computational cost associated with this method is significant and nearly impossible to handle for real-time decision-making processes. The Prognostic and Health Management (PHM) community has chosen

---

M. E. Orchard (✉) · D. E. Acuña
Department of Electrical Engineering, Faculty of Mathematical and Physical Sciences, University of Chile, Santiago, Chile
e-mail: morchard@ing.uchile.cl; davacuna@ing.uchile.cl

© Springer Nature Switzerland AG 2019
E. Lughofer, M. Sayed-Mouchaweh (eds.), *Predictive Maintenance in Dynamic Systems*, https://doi.org/10.1007/978-3-030-05645-2_12

particle-filtering-based algorithms [4] as the de facto alternative to MC [8], since particle-filters (PFs) offer an interesting balance between efficiency and efficacy in state estimation problems. However, it is still not clear how to measure the efficacy of particle-filtering-based prognostic methods in terms of the generated results, because the PHM community has not yet established adequate performance metrics. This lack of standards that affects implementation of prognostic algorithms is in part due to varied end-user requirements: forecasting is a topic of interest for a number of different domains, including aerospace, automotive, electronics, finance, medicine, nuclear power, and weather.

The general agreement is that better algorithms will exhibit better accuracy (related to estimates biases) and precision (related to variance of an estimate). This idea sounds natural and intuitive. However, it is easy to artificially "improve" the precision exhibited by an algorithm by modifying hyper-parameters of the model that defines the evolution of the state over time (state transition model). It is only natural to wonder which is the fundamental limit for these "improvements." This book chapter aims at describing an interesting approach to define prognostic performance metrics based on the concept of Bayesian Cramér–Rao Lower Bounds (BCRLBs) for the predicted state mean square error (MSE), conditional to measurement data and model dynamics [9]. Furthermore, a step-by-step design methodology to tune prognostic algorithm hyper-parameters is discussed, allowing to guarantee that obtained results do not violate fundamental precision bounds. As an illustrative example, this design methodology is applied to the problem of End-of-Discharge (EoD) time prognostics in lithium-ion batteries.

The structure of the book chapter is as follows. BCRLBs are first introduced in Sect. 2. Section 3 focuses on the definition of a prognostic performance metric based on BCRLBs, as well as on a step-by-step methodology for prognostic algorithm design based on that metric. Section 4 shows the application of the proposed metrics and design methodology to the EoD problem in lithium-ion batteries in two different case scenarios: (1) when the future operating profile is assumed to be known (Sect. 4.4), or (2) when the future operating profile is unknown and statistically characterized as a new uncertainty source (Sect. 4.5). Section 5 presents main conclusions.

2 Cramér–Rao Lower Bounds

The Cramér–Rao Lower Bound (CRLB) [10, 11] is a fundamental limit that establishes a lower bound for the mean square error (MSE) of any estimator. The most conventional version of this bound was developed for the assessment of the performance of unbiased estimators for unknown deterministic parameters. Later, Van Trees developed an analogous bound applicable to the case of random parameters, where the assumption of unbiasedness is no longer required: the Bayesian Cramér–Rao Lower Bound (BCRLB) [12].

2.1 Bayesian Cramér–Rao Lower Bounds

Let $x \in \mathbb{R}^{n_x}$ be a vector of random parameters to be estimated and $y \in \mathbb{R}^{n_y}$ a random vector of observations. Let $\hat{x}(y)$ be an estimator of x obtained as a function of the observations y. The Bayesian Cramér–Rao inequality [12] establishes that

$$\mathbb{E}_{p(x,y)}\{[\hat{x}(y) - x][\hat{x}(y) - x]^T\} \geq J^{-1} \qquad (1)$$

where $p(x, y)$ is a joint probability density function and J is the Bayesian Information Matrix (BIM) (called Fisher Information Matrix in the conventional setting of deterministic parameter estimation), defined as

$$J = \mathbb{E}_{p(x,y)}\{-\Delta_x^x \log p(x, y)\} \qquad (2)$$

The operator Δ denotes the second-order derivative

$$\Delta_x^y = \nabla_x \nabla_y^T, \qquad (3)$$

and ∇ denotes the gradient operator.

2.2 BCRLBs for Discrete-Time Dynamical Systems

Bayesian processors [1] (a.k.a. Bayesian *filters*) use a state-space representation of the system and are particularly useful to characterize uncertainty sources online. For this purpose, let us consider $\{X_k, k \in \mathbb{N}\}$ a first-order Markov process denoting an n_x-dimensional system state vector with initial distribution $p(x_0)$ and transition probability $p(x_k|x_{k-1})$. Also, let $\{Y_k, k \in \mathbb{N} \setminus \{0\}\}$ denote n_y-dimensional conditionally independent noisy observations. Then,

$$x_k = f(x_{k-1}, \omega_{k-1}) \qquad (4)$$

$$y_k = g(x_k, v_k), \qquad (5)$$

where ω_k and v_k denote independent, not necessarily Gaussian, random vectors.

Four different versions of the BCRLB can be used as a lower bound for the MSE in discrete-time dynamical systems [13]. These four versions are now defined. Let $x_{0:k} = \begin{bmatrix} x_0^T & x_1^T & \ldots & x_k^T \end{bmatrix}^T$ and $y_{1:k} = \begin{bmatrix} y_1^T & y_2^T & \ldots & y_k^T \end{bmatrix}^T$ denote a collection of augmented states and measurement vectors up to time k. The estimator of x_k is denoted as $\hat{x}_k(y_{1:k})$, which is a function of the measurement sequence $y_{1:k}$. Let also denote $\hat{x}_{0:k}(y_{1:k})$ the estimator of the whole state trajectory $x_{0:k}$. The inequalities regarding these versions of BCRLBs are summarized below.

A. Joint unconditional BCRLB

$$\mathbb{E}_{p(x_{0:k}, y_{1:k})}\{[\hat{x}_{0:k}(y_{1:k}) - x_{0:k}][\hat{x}_{0:k}(y_{1:k}) - x_{0:k}]^T\} \geq J_{0:k}^{-1} \quad (6)$$

$$J_{0:k}^{-1} = \mathbb{E}_{p(x_{0:k}, y_{1:k})}\{-\Delta_{x_{0:k}}^{x_{0:k}} \log p(x_{0:k}, y_{1:k})\} \quad (7)$$

B. Marginal unconditional BCRLB

$$\mathbb{E}_{p(x_k, y_{1:k})}\{[\hat{x}_k(y_{1:k}) - x_k][\hat{x}_k(y_{1:k}) - x_k]^T\} \geq J_k^{-1} \quad (8)$$

$$J_k^{-1} = \mathbb{E}_{p(x_k, y_{1:k})}\{-\Delta_{x_k}^{x_k} \log p(x_k, y_{1:k})\} \quad (9)$$

C. Joint conditional BCRLB

$$\mathbb{E}_{p(x_{0:k}, y_k | y_{1:k-1})}\{[\hat{x}_{0:k}(y_{1:k}) - x_{0:k}][\hat{x}_{0:k}(y_{1:k}) - x_{0:k}]^T\} \geq J_{0:k}(y_{1:k-1})^{-1} \quad (10)$$

$$J_{0:k}(y_{1:k-1})^{-1} = \mathbb{E}_{p(x_{0:k}, y_k | y_{1:k-1})}\{-\Delta_{x_{0:k}}^{x_{0:k}} \log p(x_{0:k}, y_k | y_{1:k-1})\} \quad (11)$$

D. Marginal conditional BCRLB

$$\mathbb{E}_{p(x_k, y_k | y_{1:k-1})}\{[\hat{x}_k(y_{1:k}) - x_k][\hat{x}_k(y_{1:k}) - x_k]^T\} \geq J_k(y_{1:k-1})^{-1} \quad (12)$$

$$J_k(y_{1:k-1})^{-1} = \mathbb{E}_{p(x_k, y_k | y_{1:k-1})}\{-\Delta_{x_k}^{x_k} \log p(x_k, y_k | y_{1:k-1})\} \quad (13)$$

These bounds can be classified according to two main criteria. On the one hand, the bound is said to be *joint* if it restricts the MSE of the whole state trajectory $x_{0:k}$, whereas if it solely limits the MSE of the state vector x_k, the bound is said to be *marginal*. On the other hand, if the bound considers measurements $y_{1:k-1}$ as a random vector, it is said to be *unconditional*, whereas if $y_{1:k-1}$ is a vector of known measurements, it is said to be *conditional*.

In [14] an elegant way for computing the marginal unconditional BCRLB J_k^{-1} (see Eq. (9)) was proposed without manipulating large matrices at each time instant k in the following manner:

$$J_{k+1} = D_k^{22} - D_k^{21}(J_k + D_k^{11})^{-1} D_k^{12} \quad (14)$$

where

$$D_k^{11} = \mathbb{E}\{-\Delta_{x_k}^{x_k} \log p(x_{k+1}|x_k)\} \quad (15)$$

$$D_k^{12} = \mathbb{E}\{-\Delta_{x_k}^{x_{k+1}} \log p(x_{k+1}|x_k)\} = (D_k^{21})^T \quad (16)$$

$$D_k^{22} = \mathbb{E}\{-\Delta_{x_{k+1}}^{x_{k+1}} [\log p(x_{k+1}|x_k) + \log p(y_{k+1}|x_{k+1})]\} \quad (17)$$

$$= D_k^{22,a} + D_k^{22,b}. \quad (18)$$

with expectations taken with respect to $p(x_{0:k+1}, y_{1:k+1})$. It is important to remark that the marginal unconditional BCRLB considers random measurement vectors. In [15], the marginal conditional BCRLB $J_k(y_{1:k-1})^{-1}$ (see Eq. (13)) was introduced and also developed in an elegant recursive way for its computation, in the following manner:

$$J_{k+1}(y_{1:k}) = B_k^{22} - B_k^{21}(J_k^A(Y_k) + B_k^{11})^{-1} B_k^{12}, \tag{19}$$

where

$$B_k^{11} = \mathbb{E}\{-\Delta_{x_k}^{x_k} \log p(x_{k+1}|x_k)\} \tag{20}$$

$$B_k^{12} = \mathbb{E}\{-\Delta_{x_k}^{x_{k+1}} \log p(x_{k+1}|x_k)\} = (B_k^{21})^T \tag{21}$$

$$B_k^{22} = \mathbb{E}\{-\Delta_{x_{k+1}}^{x_{k+1}} [\log p(x_{k+1}|x_k) + \log p(y_{k+1}|x_{k+1})]\} \tag{22}$$

$$= B_k^{22,a} + B_k^{22,b}, \tag{23}$$

with expectations taken with respect to $p(x_{0:k+1}, y_{k+1}|y_{1:k})$. On the other hand, $J_k^A(y_{1:k})$ is defined as the auxiliary BIM matrix for x_k, being its inverse equal to the $n_x \times n_x$ lower-right block of the inverse of the auxiliary BIM matrix $I_k^A(y_{1:k})$, where

$$I_k^A(y_{1:k}) = \mathbb{E}_{p(x_{0:k}|y_{1:k})}\{-\Delta_{x_{0:k}}^{x_{0:k}} \log p(x_{0:k}|y_{1:k})\}. \tag{24}$$

Now that BCRLBs have been properly defined, let us focus on how to use this concept in the context of the failure prognostic problem and, more specifically, as a criterion to measure the correctness of a given failure prognostic algorithm implementation.

3 Methodology for Prognostic Algorithm Design

Let us assume that it is required to implement a probability-based prognostic algorithm to measure the risk of future usage for failing equipment in real-time. Let $\theta \in \Theta \subseteq \mathbb{R}^{n_\theta}$ be a vector of hyper-parameters that allows to configure any implementation of this probability-based prognostic algorithm. A step-by-step methodology to tune these hyper-parameters will be defined, trying to maximize the efficacy of the prognostic algorithm while considering specific efficiency constrains (typically imposed by maximum processing time and/or computational cost). Now, it is important to note that some hyper-parameters will have a positive impact on the efficacy of the algorithm, while in other cases most of the impact can be measured in terms of an improvement on the efficiency. For this reason, the components of the vector θ are grouped in two sets of hyper-parameters: those that primarily affect the efficiency of the algorithm (conveniently arranged in the vector

$\theta_A \in \Theta_A \subseteq \mathbb{R}^{n_{\theta_A}}$, $n_{\theta_A} < n_\theta$), and those that primarily have impact on the quality of obtained results (arranged in the vector $\theta_B \in \Theta_B \subseteq \mathbb{R}^{n_{\theta_B}}$, where $\theta^T = [\theta_A^T \ \theta_B^T]$ and $n_{\theta_A} + n_{\theta_b} = n_\theta$). Parameter vector θ_A is typically tuned to meet efficiency constraints (for example, maximum processing time); however, the actual question is how to choose adequate values for the components of θ_B.

The Bayesian Cramér–Rao Lower Bound concept will help to measure the performance of the failure prognostic algorithm conditional to a realization of θ. In that regard, a design methodology that uses predictive BCRLBs can be implemented to determine fundamental limits for the predicted state MSE (at any future time instant) and a feasible region $\overline{\Theta_B} \subset \Theta_B$ for the hyper-parameter vector θ_B, assuming that θ_A is chosen to meet efficiency constraints. This feasible region is thereby characterized by all values of θ_B for which the predicted state MSE does not violate the corresponding predictive BCRLB (at any future time instant). This design methodology can be summarized as follows:

1) Choose θ_A such that efficiency specifications are met. Compute (recursively) Bayesian Cramér–Rao bounds for the predicted state MSE (also referred to as predictive BCRLBs), starting from time k_p and up to a time prediction horizon defined by $k_h > k_p$.
2) Choose realizations of the hyper-parameter vector $\theta_B \in \Theta_B$. You may use sampling schemes to obtain these realizations from a prior distribution.
3) Execute the prognostic algorithm, conditional to each one of the obtained realizations for θ. Discard realizations that generate predicted state MSEs smaller than the predictive BCRLB at any time $k_p < k < k_h$.
4) For all realizations of θ that were not discarded in Step 3) compute a weighted average of the ℓ^1 distances between MSE curves (per component of the state vector) and the corresponding BCRLB curves over time. Choose $[\theta_A^T \ \hat{\theta}_B^T]$ as the realization that minimizes the aforementioned weighted average. Compute the predicted ToF PMF conditional to $[\theta_A^T \ \hat{\theta}_B^T]$.
5) Explore the impact associated with a relaxation in soft efficiency constraints: Modify θ_A to allow less efficient algorithm implementations. Go through Steps 1)–4), and assess the impact on the resulting ToF PMF using a metric of choice. Iterate until the impact on the resulting ToF PMF is negligible.

This step-by-step design methodology requires a formal definition for Bayesian Cramér–Rao Lower Bounds for the predicted state MSE, as well as a feasible procedure to compute this bound recursively. A formal definition of BCRLB for the predicted state MSE can be obtained following the ideas presented in [15, 16], although a formal definition for the concept of the Time-of-Failure Probability Mass Function is required.

Fortunately, a formal definition for ToF PMF has been already presented in [3, 17] (Acuña's ToF PMF, see Eq. (26)). Indeed, let τ_F denote the ToF random variable:

Definition 1 (System Failure Function) A system *failure* is characterized by the function

$$F : \mathbb{R}^{n_x} \times \Omega \to \{0, 1\}$$

$$(x, \omega) \mapsto F(x, \omega) = \mathbb{1}_{\text{SystemFailurein} x}(\omega),$$

where $F(X_k)(\cdot) := F(X_k, \cdot)$ corresponds to a binary random variable indicating whether the system is in a failure condition or not, at the k-th time instant.

Theorem 1 (Acuña's Failure Probability Mass Function) *Considering the probability space $(\mathbb{N}, \sigma(\mathbb{N}), \mathscr{P})$ and given that $\mathscr{P}(\cup_{i=0}^{k-1}\{\tau_F = i\}|y_{1:k_p})$ as a function of time $k \in \mathbb{N}$ is always absolutely continuous with respect to the counting measure in $\mathbb{N}$, if the following conditions hold:*

- $\tau_F < +\infty$, $\mathscr{P}(\tau_F = \cdot|y_{1:k_p})$-a.s.
- $\tau_F > k_p$.

Then the mapping $\mathscr{P}(\tau_F = \cdot|y_{1:k_p}) : \mathbb{N} \to [0, 1]$ defines the Acuña's Time-of-Failure Probability Mass Function as

$$\mathscr{P}_A(k|y_{1:k_p}) := \mathscr{P}(\tau_F = k|y_{1:k_p}) \tag{25}$$

$$= \mathscr{P}(F(X_k) = 1|y_{1:k_p}) \prod_{j=k_p+1}^{k-1} \Big(1 - \mathscr{P}(F(X_j) = 1|y_{1:k_p})\Big), \tag{26}$$

with

$$\mathscr{P}(F(X_k) = 1|y_{1:k_p}) = \int_{\mathbb{R}^{n_x}} \mathscr{P}(F(x_k) = 1) p(x_k|y_{1:k_p}) dx_k. \tag{27}$$

*This probability measure is well-defined (satisfies the conditions of Probability Mass Function) and corresponds to the **unique** Bayesian probability function that can characterize the risk of future failures in discrete-time systems. Therefore, it holds that*

1) $\mathscr{P}_A(k|y_{1:k_p}) = 0$, $\forall k \in \mathbb{N}$, $k \leq k_p$.
2) $0 \leq \mathscr{P}_A(k|y_{1:k_p}) \leq 1$, $\forall k \in \mathbb{N}$.
3) $\sum_{i=0}^{+\infty} \mathscr{P}_A(k|y_{1:k_p}) = 1$.

3.1 Conditional Predictive Bayesian Cramér–Rao Lower Bounds

In actual prognostic algorithm implementations, measurements are always assumed to be available, because it is inadequate to prognosticate a failure even before the fault could be diagnosed. In this regard, a proper performance metric based on Bayesian Cramér–Rao lower bounds for the predicted state mean square error (MSE) needs to be conditional to measurement data.

Let $x_{k_p:k} = \begin{bmatrix} x_{k_p}{}^T & x_{k_p+1}{}^T & \ldots & x_k{}^T \end{bmatrix}^T$ and also $x^i, i = 1, 2, \ldots, (k - k_p + 1)n_x$, be the i-th component of the vector $x_{k_p:k}$. The initial focus is on finding a lower bound for the MSE associated with any estimator of $x_{k_p:k}$ (bound for the predictive state MSE). Afterwards, it is shown that a recursion could be used to compute the bound with ease.

Let $\hat{x}_{k_p:k}(y_{1:k_p})$ be an estimator of $x_{k_p:k}$ conditional to the set of measurements acquired until the prognostic time k_p, $k > k_p$. Besides, let $\tilde{x}_{k_p:k} \triangleq \hat{x}_{k_p:k}(y_{1:k_p}) - x_{k_p:k}$ be the estimation error and $p_k^{cp} \triangleq p(x_{k_p:k}|y_{1:k_p})$. The second-order derivative is denoted

$$\Delta_x^y = \nabla_x \nabla_y^T, \qquad (28)$$

where $\nabla_x = \begin{bmatrix} \frac{\partial}{\partial x_1}, \frac{\partial}{\partial x_2}, \ldots, \frac{\partial}{\partial x_{n_x}} \end{bmatrix}$ is a gradient operator of dimensions $1 \times n_x$.

Definition 2 The *Conditional Predictive Bayesian Information Matrix (CPBIM)* is defined as

$$I_{cp}(x_{k_p:k}|y_{1:k_p}) \triangleq \mathbb{E}_{p_k^{cp}} \left\{ \left[\nabla_{x_{k_p:k}}{}^T \log p_k^{cp} \right] \left[\nabla_{x_{k_p:k}} \log p_k^{cp} \right] \right\} \qquad (29)$$

Two theorems help to introduce both joint and marginal versions of the CP-BCRLB. The joint version represents a bound for the predictive state MSE associated with the whole state trajectory $x_{k_p:k}$, and requires to incur in a series of expensive matrix computations. In contrast, the marginal version allows to obtain a bound for the predicted state MSE related to x_k, $k > k_p$, which can be easily computed using a recursive expression.

Theorem 2 (Joint Conditional Predictive BCRLB) *Let us assume the following conditions about the density p_k^{cp}:*

1. p_k^{cp} is absolutely continuous and $\frac{\partial p_k^{cp}}{\partial x^i}$ is absolutely integrable with respect to $x_{k_p:k}$, this is

$$\int \left| \frac{\partial p_k^{cp}}{\partial x^i} \right| dx_{k_p:k} < +\infty \qquad (30)$$

2. For each x^i, with $i = 1, 2, \ldots, (k - k_p + 1)n_x$,

$$\lim_{x^i \to +\infty} x^i p(x_{k_p:k}) = \lim_{x^i \to -\infty} x^i p(x_{k_p:k}) = 0 \qquad (31)$$

The MSE associated with any estimator $\hat{x}_{k_p:k}(y_{1:k_p})$ of the state trajectory $x_{k_p:k}$ is lower bounded

$$\mathbb{E}_{p_k^{cp}}\{\tilde{x}_{k_p:k}\tilde{x}_{k_p:k}^T | y_{1:k_p}\} \geq I_{cp}^{-1}(x_{k_p:k}|y_{1:k_p}), \tag{32}$$

where $I_{cp}^{-1}(x_{k_p:k}|y_{1:k_p})$ is referred to as the *Joint Conditional Predictive BCRLB (JCP-BCRLB)*.

Finally, Theorem 3 presents a recursive formula that allows to compute the marginal version of the bound CP-BCRLB.

Theorem 3 (Marginal Conditional Predictive BCRLB) *Let us define*

$$S_{i+1}^i = \mathbb{E}\{-\Delta_{x_i}^{x_i} \log p(x_{i+1}|x_i)\} \tag{33}$$

$$S_{i+1}^{i,i+1} = \mathbb{E}\{-\Delta_{x_i}^{x_{i+1}} \log p(x_{i+1}|x_i)\} \tag{34}$$

$$S_{i+1}^{i+1} = \mathbb{E}\{-\Delta_{x_{i+1}}^{x_{i+1}} \log p(x_{i+1}|x_i)\} \tag{35}$$

with $S_{i+1}^{i+1,i} = S_{i+1}^{i,i+1\,T}$, $i = k_p, k_p+1, \ldots, k$. The MSE associated with x_k is lower bounded as

$$\mathbb{E}_{p_k^{cp}}\{\tilde{x}_k \tilde{x}_k^T | y_{1:k_p}\} \geq C_k^{22} \tag{36}$$

where C_k^{22} is named as Marginal Conditional Predictive BCRLB (MCP-BCRLB), and can be recursively computed as

$$[C_k^{22}]^{-1} = S_k^k - S_k^{k,k-1}[[C_{k-1}^{22}]^{-1} + S_k^{k-1}]^{-1} S_k^{k-1,k} \tag{37}$$

considering the initial condition $[C_{k_p}^{22}]^{-1} = S_{k_p}^{k_p} = \mathbb{E}\{-\Delta_{x_{k_p}}^{x_{k_p}} \log p(x_{k_p}|y_{1:k_p})\}$.

In the context of fault diagnosis (or system monitoring), Bayesian approaches assume that the system exogenous inputs as known (or even accurately measured). As a result, exogenous inputs (in this case, system operating profiles) are omitted in the notation that is used to describe either state vector prior or posterior PDFs. The failure prognostic problem, though, incorporates an additional source of uncertainty: it is not exactly known how the system is going to be operated in the future, i.e., future operating profiles can be characterized as a random process. The incorporation of future operating profiles in the analysis is, indeed, important because the bounds for maximum precision in long-term predictions depend on the quality of the characterization of system inputs. The procedure to perform the analysis requires to define an augmented state vector, including future exogenous inputs as additional states in the fault prognostic problem. An example to illustrate this procedure is presented in Sect. 4.5.

However, before continuing to the next section, it is important to note that from the standpoint of designers of prognostic algorithms, the methodology presented in Sect. 3 can be always applied assuming that the future operating profile of the system is known: The resulting algorithm will serve its purpose well, conditional to a good characterization of system inputs. The task of providing such good characterization of future operating profiles can be considered as a separate action that can be executed by an independent module.

3.2 Analytic Computation of MCP-BCRLBs

The computation of MCP-BCRLBs requires the computation of expectations over the predictive state probability density. This fact implies that, in the case of nonlinear systems, the designer may need to perform Monte Carlo simulations to tune the hyper-parameters of a given prognostic algorithm. If possible, it would be preferable to avoid that situation, because of the associated computational efforts. Fortunately, MCP-BCRLBs can be analytically calculated in systems where the state transition equation has additive noise and is linear with respect to the state vector, i.e.,

$$x_{k+1} = f(x_k, u_k) + \omega_k \qquad (38)$$

$$= A_k(u_k) \cdot x_k + B_k(u_k) + \omega_k, \qquad (39)$$

where u_k is the system input, $A_k(u_k)$ is an n_x-dimensional square matrix, $B_k(u_k)$ is an $n_x \times 1$ matrix, and ω_k is an n_x-dimensional zero mean Gaussian random vector. Indeed, consider the case where ω_k has covariance matrix Σ_k:

$$-\log p(x_{i+1}|x_i) = c + \frac{1}{2}[x_{i+1} - f(x_i, u_i)]^T \Sigma_i^{-1} [x_{i+1} - f(x_i, u_i)] \qquad (40)$$

$$-\nabla_{x_i} \nabla_{x_i}^T \log p(x_{i+1}|x_i) = A_i(u_i)^T \Sigma_i^{-1^T} \nabla_{x_i} f(x_i, u_i) \qquad (41)$$

$$= A_i(u_i)^T \Sigma_i^{-1} A_i(u_i) \qquad (42)$$

$$-\nabla_{x_{i+1}} \nabla_{x_i}^T \log p(x_{i+1}|x_i) = -A_i(u_i)^T \Sigma_i^{-1^T} \nabla_{x_{i+1}} x_{i+1} \qquad (43)$$

$$= -A_i(u_i)^T \Sigma_i^{-1} \qquad (44)$$

$$-\nabla_{x_{i+1}} \nabla_{x_{i+1}}^T \log p(x_{i+1}|x_i) = \Sigma_i^{-1^T} \nabla_{x_{i+1}} x_{i+1} \qquad (45)$$

$$= \Sigma_i^{-1} \qquad (46)$$

Note that $\Sigma_i^{-1^T} = \Sigma_i^{-1}$ since the covariance matrix is assumed to be symmetric and $\nabla_y \nabla_x^T = \Delta_x^y$. Besides, it is worth noting that as a general practice in systems engineering, the notation for $p(x_{i+1}|x_i)$ omits its dependency on the input u_k, as it is assumed that the latter is accurately and precisely measured. Although this may not be the case in prognostics, this dependency would remain implicit in the notation of this chapter. Therefore, from Eqs. (33)–(35), the recursion for MCP-BCRLBs has analytic expressions:

$$S_{i+1}^{i} = \mathbb{E}\{-\Delta_{x_i}^{x_i} \log p(x_{i+1}|x_i)\} = A_i(u_i)^T \Sigma_i^{-1} A_i(u_i) \quad (47)$$

$$S_{i+1}^{i,i+1} = \mathbb{E}\{-\Delta_{x_i}^{x_{i+1}} \log p(x_{i+1}|x_i)\} = -A_i(u_i)^T \Sigma_i^{-1} \quad (48)$$

$$S_{i+1}^{i+1} = \mathbb{E}\{-\Delta_{x_{i+1}}^{x_{i+1}} \log p(x_{i+1}|x_i)\} = \Sigma_i^{-1} \quad (49)$$

4 Case Study: End-of-Discharge Time Prognosis of Lithium-Ion Batteries

We now proceed to apply the proposed methodology for prognostic algorithm design and hyper-parameter tuning on an illustrative case study, which corresponds to the problem of End-of-Discharge (EoD) time prognostics in lithium-ion (Li-Ion) batteries. This case study assumes that a filtering stage is carried out by a particle-filtering algorithm, following the recommendations suggested in [5] (in terms of the number of particles utilized, among other implementation issues). In that regard, it is assumed that posterior estimates for both the battery State-of-Charge (SoC, defined as the ratio between the actual available energy and the maximum battery storage capacity E_{crit}) and internal polarization resistance are always available at the time where the prognostic algorithm is executed. The failure condition in this case is characterized by SoC levels below 10%.

On the one hand, Sect. 4.4 presents the implementation of the design methodology presented in Sect. 3, assuming that the exogenous input (future discharge current profile) is known along the whole prediction window. In this case, uncertainty sources are confined to the initial condition (state posterior PDF at the beginning of prognostic window) and the process noise embedded within the dynamics of the hidden system states. On the other hand, Sect. 4.5 explores the impact of system input uncertainty on the computed bounds for the predictive MSE.

4.1 State-Space Model

Filtering and prognostic stages use a state-space model to represent the evolution in time of the Li-Ion battery voltage as a function of (1) the SoC, (2) the battery internal impedance, and (3) the discharge current (exogenous system input). The objective in this case study is to prognosticate the moment in which the battery energy has depleted below 10%. As it has been already mentioned in Sect. 1, and as in any other prognostic problem, the "ground truth" failure PMF (in this case, the EoD time PMF) can be computed offline using Monte Carlo simulations for future trajectories of the state vector.

In actual implementations of failure prognostic algorithms, it is also necessary to characterize the future evolution of exogenous inputs to the state-space model (future operating profiles). Particularly, for EoD time prognostic purposes, Pola et al. [5] proposes to use a probabilistic characterization of the battery discharge current, via Markov Chains. However, it is important to note that, without loss of generality, both the performance assessment of a given prognostic algorithm and of the exogenous input characterization can be conducted separately [18]. Indeed, it is always possible to evaluate the performance of the prognostic algorithm conditional to a specific realization of the future usage profile, and then use the Law of Total Probability to incorporate the uncertainty associated with exogenous inputs. For this reason, this study will now assume that the future battery usage profile is known, solely focusing on computing the EoD time PMF conditional to that given profile.

For most of the battery operating range, the relationship between SoC and the *Open Circuit Voltage* (OCV) curve can be well characterized by an affine function, see "*zone 2*" in Fig. 1. However, the state-space model proposed in [5] is used instead, allowing to characterize the nonlinear behavior present in "*zone 1*" and "*zone 3*." Also, a structure proposed in [19] has been adopted to model the dependency between the polarization resistance and the battery discharge current.

State Transition Model

$$x_{k+1} = x_k - v_{oc}(x_k) \cdot u_k \cdot \frac{T_s}{E_{crit}} + \omega_k \tag{50}$$

Measurement Model

$$y_k = v_{oc}(x_k) - u(k) \cdot R_{int}(x_k, u_k) + \eta_k, \tag{51}$$

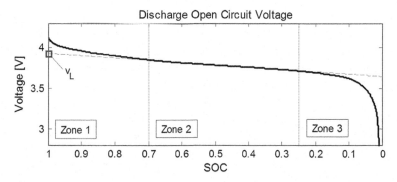

Fig. 1 OCV curve of a Li-Ion cell (black line) and the projection of its linear operational range (dashed gray line) as a function of SoC [5]

with

$$v_{oc}(x_k) = v_L + (v_0 - v_L) \cdot e^{\gamma \cdot (x_2(k)-1)} + \alpha \cdot v_L \cdot (x_2(k) - 1) \ldots$$
$$\ldots + (1 - \alpha) \cdot v_L \cdot (e^{-\beta} - e^{-\beta \cdot \sqrt{x_2(k)}}) \quad (52)$$

and

$$R_{\text{int}}(x_k, u_k) = r_0(u_k) + r_1(u_k) \cdot x_k + r_2(u_k) \cdot x_k^2. \quad (53)$$

In this representation, the input to the system $u_k = i_k[A]$ is defined as the discharge current, while $y_k = v_k[V]$ is the voltage at the battery terminals. The state x_k is the battery SoC measured with respect to E_{crit}, the expected total energy delivered by the battery, whereas the absolute value of the internal impedance is represented by the function $R_{\text{int}}(x_k, u_k)$. The process noise ω_k and the measurement noise η_k assume a zero mean Gaussian distribution. Finally, $T_s[s]$ is the sample time, and v_0, v_L, α, β, and γ are model parameters to be estimated offline (see [5] for more details).

Since a faulty condition is defined in this case by SoC values below a 10%, then Eq. (27) (required for computing the ToF PMF) becomes:

$$\mathcal{P}(F(x_k) = 1) = \mathbb{1}_{\{x \in \mathbb{R}: x < 0.1\}}(x_k). \quad (54)$$

4.2 Prognostic Algorithm

A particle-filtering-based prognostic algorithm [4] is selected to illustrate the design methodology. This algorithm uses, as initial condition, an empirical state posterior distribution that results from a PF implementation. It also considers that prognostic stage begins at time k_p, and that the state posterior distribution at that time instant is denoted by

$$p(x_{k_p}|y_{1:k_p}) = \sum_{i=1}^{N_p} w_{k_p}^{(i)} \delta_{x_{k_p}^{(i)}}(x_{k_p}), \quad (55)$$

where N_p is the amount of samples used by the PF implementation.

0) Resample $p(x_{k_p}|y_{1:k_p})$ and get a set of N_θ equally weighted particles.

Then, for each future time instant k, $k > k_p$, perform the following steps:

1) Compute the expected state transitions $x_k^{*(i)} = \mathbb{E}\{f(x_{k-1}^{(i)}, u_{k-1}, \omega_{k-1})\}$, $\forall i \in \{1, \ldots, N_\theta\}$, and calculate the empirical covariance matrix

$$\hat{S}_k = \frac{1}{N_\theta - 1} \sum_{i=1}^{N_\theta} [x_k^{*(i)} - \overline{x}_k^*][x_k^{*(i)} - \overline{x}_k^*]^T, \qquad (56)$$

with $\overline{x}_k^* = \frac{1}{N_\theta} \sum_{i=1}^{N_\theta} x_k^{*(i)}$.

2) Compute $\hat{D}_k$ such that $\hat{D}_k \cdot \hat{D}_k^T = \hat{S}_k$.
3) Update the samples as

$$x_k^{(i)} = x_k^{*(i)} + h_\theta \cdot \hat{D}_k \cdot \varepsilon_k^{(i)}, \; \varepsilon_k^{(i)} \sim \mathscr{E}, \qquad (57)$$

where $\mathscr{E}$ is the Epanechnikov kernel and h_θ corresponds to its bandwidth.

Thus, in this case, the hyper-parameters vector for the prognostic algorithm is defined as $\theta^T = \begin{bmatrix} N_\theta & h_\theta \end{bmatrix}$.

4.3 Avoiding Monte Carlo Simulations in EoD Prognostic Algorithms

The battery discharge model can be easily approximated by a structure that holds the necessary requirements to obtain an analytic expression of MCP-BCRLBs (see Sect. 3.2). The state transition equation for the SoC is

$$x_{k+1} = x_k - v_{\text{oc}}(x_k) \cdot u_k \cdot \frac{T_s}{E_{\text{crit}}} + \omega_k, \qquad (58)$$

which is nonlinear with respect to x_k because of the term $v_{\text{oc}}(x_k)$. However, from Fig. 1 it is possible to recognize a wide operating zone in which $v_{\text{oc}}(x_k)$ is linear with respect to x_k. Indeed, if $v_{\text{oc}}(x_k)$ is linearized around $x_o = 0.5$, then it is possible to write:

$$v_{\text{oc}}(x_k) \approx v_{\text{oc}}(x_o) + \left.\frac{\partial v_{\text{oc}}(x_k)}{\partial x_k}\right|_{x_k = x_o} \cdot (x_k - x_o) \qquad (59)$$

And thus, the state transition equation can be approximated by:

$$\begin{aligned} x_{k+1} = &\underbrace{\left(1 - \left.\frac{\partial v_{\text{oc}}(x_k)}{\partial x_k}\right|_{x_k=x_o} \cdot u_k \cdot \frac{T_s}{E_{\text{crit}}}\right)}_{A_k(u_k)} \cdot x_k \ldots \\ &\ldots + \underbrace{\left(\left.\frac{\partial v_{\text{oc}}(x_k)}{\partial x_k}\right|_{x_k=x_o} \cdot x_o - v_{\text{oc}}(x_o)\right) \cdot u_k \cdot \frac{T_s}{E_{\text{crit}}}}_{B_k(u_k)} + \omega_k, \end{aligned} \qquad (60)$$

expression that has the required form $x_{k+1} = A_k(u_k) \cdot x_k + B_k(u_k) + \omega_k$.

4.4 Prognostic Algorithm Design: Known Future Operating Profiles

The methodology presented in Sect. 3 is now utilized to tune hyper-parameters N_θ and h_θ of the prognostic algorithm proposed by Orchard and Vachtsevanos [4] (see Sect. 4.2), when this algorithm is used to solve the problem of EoD time prognosis (see Sect. 4.1). Please note that the parameter N_θ directly affects the computational effort of the method (i.e., $\theta_A = N_\theta$), while $\theta_B = h_\theta$ is more related to the capability of the algorithm to appropriately represent probability densities.

The performance of the prognostic algorithm is tested when predicting the EoD time at different moments during the battery discharge process. In this regard, and given that a full discharge takes approximately 11,628 [s], the prognostic routine is executed at 4000 [s] of operation. As it was previously mentioned in Sect. 4.1, the future discharge current is assumed to be known without loss of generality, since the aim is to assess the performance of the algorithms and characterize the evolution in time of the uncertainty associated with the state vector. Additionally, no simplifications in the model are considered in this case (as those described in Sect. 4.3), and thus the system is simulated embracing its nonlinear behavior illustrated in Fig. 1. The discharge current profile is generated from random realizations of a random walk model with an initial condition of 12.5 [A] (see Fig. 2).

We now proceed to apply the proposed methodology to this case study, step-by-step.

Step 1: Generate MCP-BCRLBs The first step of the methodology is to compute the sequence of MCP-BCRLBs, which in turn requires computation of an initial condition for the recursion. The procedure to achieve this goal has been

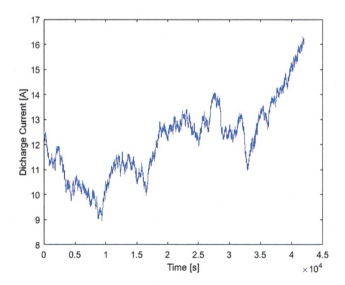

Fig. 2 Illustration of battery discharge current profile

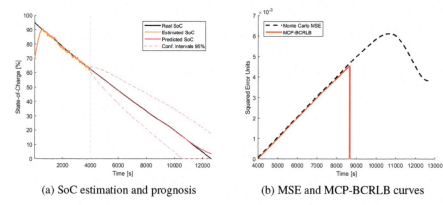

(a) SoC estimation and prognosis (b) MSE and MCP-BCRLB curves

Fig. 3 Example results for $k_p = 4000$ [s]. (**a**) Shows the results for battery SoC filtering and prediction stages. The estimation stage assumes an incorrect initial condition of 70% for the SoC, and is executed using a PF with 100 particles [5]. Long-term predictions are built simulating 10^5 random state trajectories. These predictions are used to compute MSE and MCP-BCRLB curves in (**b**)

reported in [13]. The hyper-parameter N_θ is set to 100 particles, following the recommendations stated in [5]. As the recursion requires to compute expectations over predicted state probability density functions, the MCP-BCRLB cannot be computed analytically. To overcome this difficulty, we simulate 10^5 random trajectories for the evolution of the battery SoC using Eq. (50). Figure 3 shows the results obtained when state predictions are computed at $k_p = 4000$ [s].

Step 2: Choose candidates for algorithm hyper-parameters The bandwidth h_θ of each Epanechnikov kernel has a theoretical optimal value h_{opt} when particles are sampled from Gaussian distribution with unity covariance matrix (see Eq. (61)) [4]. Although this is seldom the case, given that the underlying hypothesis behind this property is rarely valid in nonlinear systems (i.e., samples are not independent and identically distributed in the algorithm presented in Sect. 4.2), particle-filtering-based prognostic algorithm implementations use this value as an educated guess. Since in this case study $n_x = 1$, we should use $h_{\text{opt}} = 0.8529$.

$$h_{\text{opt}} = A \cdot N_\theta^{-\frac{1}{n_x+4}}, \quad A = \left(8 \cdot c_{n_x}^{-1} \cdot (n_x + 4) \cdot (2 \cdot \sqrt{\pi})^{n_x}\right)^{\frac{1}{n_x+4}} \qquad (61)$$

However, if we set $h_\theta = h_{\text{opt}}$, the implementation of the particle-filtering-based prognostic algorithm performs poorly in terms of predicted state MSE (far greater than corresponding MCP-BCRLB). This fact motivates to search for other hyper-parameter candidates. For illustrative purposes, we will analyze the following options for the bandwidth for the Epanechnikov kernel: $h_{\theta,1} = 0.0093$, $h_{\theta,2} = 0.0090$, $h_{\theta,3} = 0.0087$, $h_{\theta,4} = 0.0084$, and $h_{\theta,5} = 0.0081$.

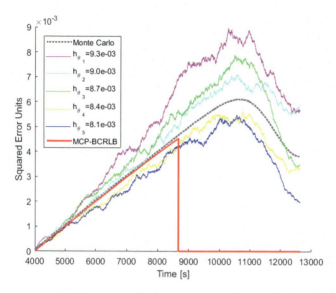

Fig. 4 Predictive state MSE and MCP-BCRLB curves computed at $k_p = 4000\,[\text{s}]$

Step 3: Discard hyper-parameter candidates related to implementations that violate MCP-BCRLBs Figure 4 shows the resulting predictive state MSE curves for realizations of the particle-filtering-based prognostic algorithm ($N_\theta = 100$) that used the proposed candidates for the hyper-parameter h_θ; all of them executed at time $k_p = 4000\,[\text{s}]$. It can be noticed that predictive MSE curves associated with candidates $h_{\theta,4}$ and $h_{\theta,5}$ violate the MCP-BCRLB curve. Although the candidate $h_{\theta,2}$ also generates a MSE curve that violates the bound, it must be noted that this situation occurs solely for a small set of time instants. We will not discard this candidate yet, since an increment in the number of particles may help alleviate the situation described above.

Step 4: Use the ℓ^1-norm to select the most appropriate hyper-parameter candidate After executing Step 3), two candidates remain: $h_{\theta,1}$ and $h_{\theta,3}$. It is proposed to use the ℓ^1-norm to measure the distance between predicted state MSEs and MCP-BCRLB curves, hoping that this information could be useful to discriminate between these two candidates:

$$\left\|\text{MSE}_{h_{\theta,1}} - \text{MCP} - \text{BCRLB}\right\|_1 = 33.1985 \tag{62}$$

$$\left\|\text{MSE}_{h_{\theta,3}} - \text{MCP} - \text{BCRLB}\right\|_1 = 26.3207. \tag{63}$$

It is important to note that candidate $h_{\theta,3} = 0.0087$ represents an appropriate choice. Intuition indicates that it is worthwhile to explore which could be the best choice for h_θ if efficiency constraints are relaxed and a larger number of particles is allowed in the implementation.

Table 1 Dissimilarity between predicted state MSE and MCP-BCRLB curves (ℓ^1 distance)

	$h_{\theta,1}$	$h_{\theta,2}$	$h_{\theta,3}$	$h_{\theta,4}$	$h_{\theta,5}$
$N_\theta = 100$	33.1985	25.5914✗	26.3207✓	19.5358✗	18.2365✗
$N_\theta = 500$	25.4781	23.9873✓	20.7479✗	19.5817✗	19.6815✗
$N_\theta = 1000$	24.3008	24.8117✓	21.4785✗	21.2526✗	19.0523✗
$N_\theta = 5000$	26.0585	23.1602	21.9363✓	20.2669✗	19.2155✗
$N_\theta = 10{,}000$	26.0338	23.8571	21.2822✓	20.5395✗	19.7839✗

Candidates that were discarded in Step 3) are marked with a ✗ symbol. Candidates associated with minimum distances are marked with a ✓ symbol

Step 5: Relax efficiency soft-constraints We now proceed to relax soft-constraints associated with efficiency criteria (in this case, the number of particles N_θ). This procedure helps to understand the cost (in terms of algorithm efficacy) that is associated with computational effort constraints. For this purpose, let us increase the hyper-parameter value to $N_\theta = 500$. After going through Steps 1)–4) once more, it is interesting to note that in this new scenario, the most appropriate choice for the hyper-parameter vector would have been (see Table 1):

$$\theta^T = \begin{bmatrix} N_\theta & h_{\theta,2} \end{bmatrix} = \begin{bmatrix} 500 & 0.0090 \end{bmatrix}. \tag{64}$$

The aforementioned steps allow to choose hyper-parameter candidates in terms of the quality of predicted state MSE curves. However, to be fair, the quality of the true outcome of probability-based prognostic algorithms (the ToF PMF) should also be included in the analysis. To do this, it is necessary to use a metric that aims at discriminating between two ToF PMFs. In this case, an ℓ^1 distance is suggested as a measure of changes between two ToF PMFs.

Let us consider, then, the case illustrated in Fig. 5, which shows the ToF PMFs obtained when implementing a PF-based algorithm for the EoD time prognostics with two choices for the hyper-parameters vector: $\theta_1^T = [100\ \ 0.0087]$ and $\theta_2^T = [500\ \ 0.0090]$. The latter choice, θ_2, assumes a relaxation of efficiency constraints (i.e., $N_\theta = 500$ instead of 100 particles). The ℓ^1 distance between these two PMFs indicates that the impact associated with the increment in the size of the particle population is not negligible. Furthermore, a direct comparison with the "ground truth" EoD PMF, approximated by a million simulations of the state transition model, indicates that this increment in N_θ allows a better characterization of the left tail of the ToF PMF.

Considering all of the above, we may now proceed to explore the impact (in terms of algorithm efficacy) of a larger increase in the number of particles N_θ. For that purpose, we propose to use the following sequence of hyper-parameters candidates, indexed according to an efficiency criterion:

$$\{\theta_n\}_{n=1}^5 = \left\{ \begin{bmatrix} N_{\theta_n} \\ h_{\theta_n} \end{bmatrix} \right\}_{n=1}^5 \tag{65}$$

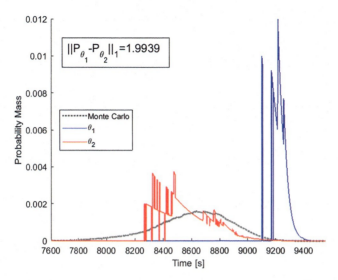

Fig. 5 Time-of-Failure PMFs for two choices of algorithm hyper-parameters $\theta_1^T = [100\ \ 0.0087]$ and $\theta_2^T = [500\ \ 0.0090]$. Prognosis is executed at time $k_p = 4000\,[\text{s}]$. Gray-dashed line shows the "ground truth" EoD PMF, which is approximated by a million simulations of the state transition model

$$= \left\{ \begin{bmatrix} 100 \\ 0.0087 \end{bmatrix}, \begin{bmatrix} 500 \\ 0.0090 \end{bmatrix}, \begin{bmatrix} 1000 \\ 0.0090 \end{bmatrix}, \begin{bmatrix} 5000 \\ 0.0087 \end{bmatrix}, \begin{bmatrix} 10000 \\ 0.0087 \end{bmatrix} \right\}. \tag{66}$$

It is evident from a direct comparison between Figs. 5 and 6 that the quality of the prognostic result (ToF PMF) increases as you also increase the number of particles. The above, particularly in regard to the characterization of the left tail of the PMF (the most useful source of information to quantify operational risk in prognostics). Indeed, when comparing with respect to the "ground truth" EoD PMF, evidence indicates that efficacy does not increase significantly when $N_\theta = 10,000$. Thus, in terms of the final design for this specific case study, it would be recommendable to use $\theta_4^T = [5000\ \ 0.0087]$. This final hyper-parameter choice aims at a combination that provides reasonable results in terms of a truthful characterization for the risk of failure, using the least computational resources.

The choice of $\theta_4^T = [5000\ \ 0.0087]$ as the definitive value for the hyper-parameter vector of the prognostic algorithm represents the whole purpose of the methodology that was described in Sect. 3: To provide the tools that are required to assess the quality of the results provided by a specific prognostic algorithm implementation, so that the final design balances the efficacy of the method with the efficiency of the code. In this case, it is unnecessary to increase the number of particles to 10,000, because the quality that is obtained with 5000 particles is already adequate.

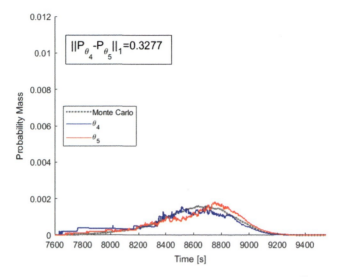

Fig. 6 Time-of-Failure PMFs for two choices of algorithm hyper-parameters $\theta_4^T = [5000\ 0.0087]$ and $\theta_5^T = [10{,}000\ 0.0087]$. Prognosis is executed at time $k_p = 4000\,[s]$. Gray-dashed line shows the "ground truth" EoD PMF, which is approximated by a million simulations of the state transition model

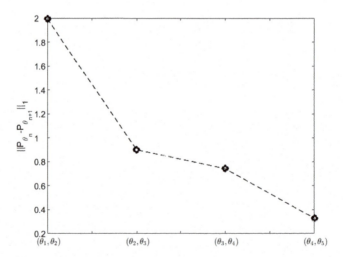

Fig. 7 Summary of ℓ^1 distances between EoD PMFs. The n-th iteration of the design procedure is related to hyper-parameters vector θ_n. Prognostics executed at time $k_p = 4000\,[s]$

To complement the previous analysis, Fig. 7 includes a graph that shows the evolution of ℓ^1 distances between obtained ToF PMFs as we compare the performance associated with candidates $\theta_1, \ldots, \theta_5$ (where θ_n is the candidate selected in the n-th iteration of the design procedure). It is interesting to note that the relaxation of efficiency constrains (N_θ, in this case) entails decreasing differences regarding EoD PMFs results in terms of the proposed metric (ℓ^1 distance between ToF PMFs). Eventually, as the prognostic algorithm approximates classic Monte Carlo simulations (by increasing the number of particles), it can be noted that the efficacy increases (although, obviously, efficiency decreases).

4.5 Prognostic Algorithm Design: Statistical Characterizations of Future Operating Profiles

Unfortunately, in actual implementations of failure prognostic algorithms, the future operating profile of the system is unknown and, therefore, we are forced to incorporate the uncertainty related to exogenous inputs into the computation of Bayesian lower bounds for the predictive MSE. We now proceed to explain how to proceed in these cases, computing MCP-BCRLBs in the light of the illustrative application previously shown in Sect. 4.4, where future operating profiles are now assumed to be uncertain.

To include future exogenous input uncertainty into the analysis, it is first required to augment the state vector that defines the system during the prognostic stage. Thus,

$$\check{x}_k = \begin{bmatrix} x_k \\ u_k \end{bmatrix}. \tag{67}$$

The augmentation of the state vector explicitly recognizes the random nature behind future inputs to the system, as well as the impact that this uncertainty may have on its future condition. Although the state transition equation has been previously defined, now it is required to model the evolution in time of u_k as an stochastic process. In our case study (battery End-of-Discharge time), it is known that the system input u_k corresponds to the discharge current of the Li-Ion battery and that its future evolution in time can be characterized as a random walk for most practical situations (electric bicycles, unmanned aerial vehicles). As a consequence, the augmented system results:

$$\underbrace{\begin{bmatrix} x_{k+1} \\ u_{k+1} \end{bmatrix}}_{\check{x}_{k+1}} = \underbrace{\begin{bmatrix} x_k - v_{\text{oc}}(x_k) \cdot u_k \cdot \frac{T_s}{E_{\text{crit}}} \\ u_k \end{bmatrix}}_{\check{f}(\check{x}_k)} + \underbrace{\begin{bmatrix} \omega_k \\ v_k \end{bmatrix}}_{\check{\omega}_k} \tag{68}$$

where the process noise v_k assumes a zero mean Gaussian distribution. A complete description of the new system would require, in addition, to define measurement equations. However, as these equations have no impact on prior distributions of the state in long-term predictions, we have not included those expressions explicitly in this book chapter.

According to Theorem 3, and considering all of the above, the recursion elements of the augmented system become:

$$\check{S}_{i+1}^i = \mathbb{E}\left\{-\Delta_{\check{x}_i}^{\check{x}_i}\log p(\check{x}_{i+1}|\check{x}_i) = \begin{bmatrix} -\frac{\partial^2 \log p(\check{x}_{k+1}|\check{x}_k)}{\partial x_k^2} & -\frac{\partial^2 \log p(\check{x}_{k+1}|\check{x}_k)}{\partial x_k \partial u_k} \\ -\frac{\partial^2 \log p(\check{x}_{k+1}|\check{x}_k)}{\partial u_k \partial x_k} & -\frac{\partial^2 \log p(\check{x}_{k+1}|\check{x}_k)}{\partial u_k^2} \end{bmatrix}\right\} \quad (69)$$

$$\check{S}_{i+1}^{i,i+1} = \mathbb{E}\left\{-\Delta_{\check{x}_i}^{\check{x}_{i+1}}\log p(\check{x}_{i+1}|\check{x}_i) = \begin{bmatrix} -\frac{\partial^2 \log p(\check{x}_{k+1}|\check{x}_k)}{\partial x_k \partial x_{k+1}} & -\frac{\partial^2 \log p(\check{x}_{k+1}|\check{x}_k)}{\partial x_k \partial u_{k+1}} \\ -\frac{\partial^2 \log p(\check{x}_{k+1}|\check{x}_k)}{\partial u_k \partial x_{k+1}} & -\frac{\partial^2 \log p(\check{x}_{k+1}|\check{x}_k)}{\partial u_k \partial u_{k+1}} \end{bmatrix}\right\} \quad (70)$$

$$\check{S}_{i+1}^{i+1} = \mathbb{E}\left\{-\Delta_{\check{x}_{i+1}}^{\check{x}_{i+1}}\log p(\check{x}_{i+1}|\check{x}_i) = \begin{bmatrix} -\frac{\partial^2 \log p(\check{x}_{k+1}|\check{x}_k)}{\partial x_{k+1}^2} & -\frac{\partial^2 \log p(\check{x}_{k+1}|\check{x}_k)}{\partial x_{k+1} \partial u_{k+1}} \\ -\frac{\partial^2 \log p(\check{x}_{k+1}|\check{x}_k)}{\partial u_{k+1} \partial x_{k+1}} & -\frac{\partial^2 \log p(\check{x}_{k+1}|\check{x}_k)}{\partial u_{k+1}^2} \end{bmatrix}\right\} \quad (71)$$

Now, the only remaining step towards the computation of MPC-BCRLBs is the definition of proper initial conditions to start the recursion. This initial condition requires knowledge about the precision with which you can characterize the error related to either posterior estimates of the state or inputs to the system. On the one hand, from Sect. 4.4 a MCP-BCRLB for the MSE associated with the posterior estimate of x_{k_p} is already available. On the other hand, the discharge current applied at time k_p (i.e., I_{k_p}) (exogenous input in this case study) can be measured before executing the prognostic routine. Moreover, it is known that I_{k_p} is not correlated to x_{k_p} because the system is causal. If it is assumed that I_{k_p} is measured with a perfect sensor, theoretically the lower bound for the associated MSE would be null. However, as sensors in practice are not perfect, it is correct to assume measurement uncertainty for I_{k_p}. Thus, the initial condition for the MCP-BCRLBs recursion can be defined as,

$$[\check{C}_{k_p}^{22}]^{-1} = \begin{bmatrix} [C_{k_p}^{22}]^{-1} & 0 \\ 0 & \varepsilon_I^{-1} \end{bmatrix}, \quad (72)$$

with $\varepsilon_I > 0$. Note the matrix $[\check{C}_{k_p}^{22}]^{-1}$ is defined as an inverse and, in consequence, must be non-singular. This property must be enforced by a proper choice for the constant ε_I. This constant (arbitrary small and positive value) represents a lower bound for the precision of the sensor you are using to measure the system input at time k_p, i.e., I_{k_p}.

Figure 8 shows the results obtained once all these concepts are applied to the problem of End-of-Discharge time prognosis, and a comparison between the MCP-BCRLBs that determine the maximum achievable precision when future operating profiles are assumed known or random. As expected, both the predictive MSE and the associated MCP-BCRLBs curves show an increment when incorporating uncertainty on future inputs of the system. Not only represents this result an empirical validation of the concepts described in this book chapter, but also demonstrates that it is actually possible to theoretically quantify the impact related to different

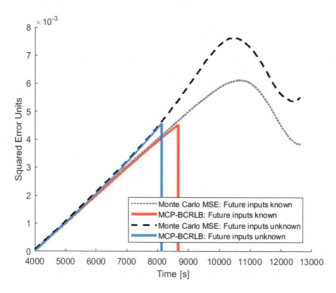

Fig. 8 Impact on MSE and MCP-BCRLB curves associated with SoC prognostics after assuming unknown future operating profiles characterized by a stochastic model. Prognostics executed at time $k_p = 4000\,[\text{s}]$

probabilistic characterizations of future input profiles. Moreover, whereas it seems as a trivial fact that errors should increase as uncertainty is added, throughout this procedure it is shown that this uncertainty can be formally included in this rigorous methodology for prognostic algorithm design. It is expected that these concepts will prove useful to the readers, providing guidelines about how to proceed in the design and implementation of modules that aim to generate a probabilistic representation of the operational risk in which a faulty system incurs when operated in uncertain environments.

An important aspect that we want to emphasize before we finish is that, from a design standpoint, it is absolutely valid to study and adjust the prognostic algorithm performance in a case where the future system operating profile is known. Obviously, in this case the procedure needs to be executed offline, so that the designer could incorporate this knowledge in the analysis and, as it was mentioned before, the quality of the characterization of future system inputs is then managed as a separate and independent problem. However, the computation of MCP-BCRLBs when modeling these future exogenous inputs with specific stochastic models provides a measure of the impact that these profiles may have on the operational risk of the system. Indeed, testing different stochastic characterizations of exogenous inputs (either changing structures or (hyper-)parameters) may provide a measure of robustness of MCP-BCRLBs as a standpoint for prognostic algorithm design.

5 Conclusions

This book chapter presents a prognostic performance metric based on the concept of Bayesian Cramér–Rao Lower Bounds (BCRLBs) for the predicted state mean square error (MSE), which is conditional to measurement data and model dynamics. This metric allows to implement a formal step-by-step design methodology to tune prognostic algorithm hyper-parameters, which allows to guarantee that obtained results do not violate fundamental precision bounds. Both the metric and the proposed design methodology are verified and validated using the problem of End-of-Discharge time prognosis as a case study.

The design methodology distinguishes between hyper-parameters that affect the efficiency of the implementation and those that have impact on the efficacy of obtained results, providing a structured procedure to explore different combinations that could improve the characterization of the ToF PMF.

From the standpoint of designers of prognostic algorithms, this methodology can be always applied assuming that the future operating profile of the system is known: The resulting algorithm will serve its purpose well, conditional to a good characterization of system inputs. The task of providing such good characterization of future operating profiles can be considered as a separate action that can be executed by an independent module. Nevertheless, we also provide an illustrative example to demonstrate the impact that exogenous input uncertainty may have on the bounds for the predictive MSE.

We expect that these results will motivate researchers to use formal and rigorous design procedures when tuning hyper-parameters of prognostic algorithms. In the humble opinion of the authors of this book chapter, the whole PHM community could benefit from this change of paradigm.

Acronyms

BCRLB	Bayesian Cramér–Rao Lower Bound
BIM	Bayesian Information Matrix
CPBIM	Conditional Predictive Bayesian Information Matrix
CRLB	Cramér–Rao Lower Bound
EoD	End-of-Discharge
JCP-BCRLB	Joint Conditional Predictive Bayesian Cramér–Rao Lower Bound
Li-Ion	Lithium-ion
MC	Monte Carlo
MCP-BCRLB	Marginal Conditional Predictive Bayesian Cramér–Rao Lower Bound
MSE	Mean squared error
OCV	Open Circuit Voltage
PDF	Probability density function
PF	Particle filter
PHM	Prognostics and Health Management

PMF	Probability Mass Function
SoC	State-of-Charge
ToF	Time-of-Failure

References

1. Doucet, A., de Freitas, N., Gordon, N. (eds.): Sequential Monte Carlo Methods in Practice. Springer, New York (2001)
2. Liu, D., Luo, V., Peng, Y.: Uncertainty processing in prognostics and health management: an overview. In: 2012 IEEE Conference on Prognostics and System Health Management (PHM) (2012)
3. Acuña, D.E., Orchard, M.E.: Particle-filtering-based failure prognosis via sigma-points: application to Lithium-Ion battery State-of-Charge monitoring. Mech. Syst. Signal Process. **85**, 827–848 (2017)
4. Orchard, M.E., Vachtsevanos, G.: A particle-filtering approach for on-line fault diagnosis and failure prognosis. Trans. Inst. Meas. Control **31**, 221–246 (2009)
5. Pola, D.A., Navarrete, H.F., Orchard, M.E., Rabié, R.S., Cerda, M.A., Olivares, B.E., Silva, J.F., Espinoza, P.A., Pérez, A.: Particle-filtering-based discharge time prognosis for lithium-ion batteries with a statistical characterization of use profiles. IEEE Trans. Reliab. **64**(2), 710–721 (2015)
6. Engel, S., Gilmartin, B., Bongort, K., Hess, A.: Prognostics, the real issues involved with predicting life remaining. In: 2000 IEEE Aerospace Conference Proceedings, vol. 6, pp. 457–469 (2000)
7. MacKay, D.J.C.: Introduction to Monte Carlo methods. In: Jordan, M.I. (ed.) Learning in Graphical Models. MIT Press, Cambridge (1998)
8. Orchard, M.E., Kacprzynski, G., Goebel, K., Saha, B., Vachtsevanos, G.: Advances in uncertainty representation and management for particle filtering applied to prognostics. In: Proceedings of International Conference on Prognostics and Health Management, pp. 1–6 (2008)
9. Acuña, D.E., Orchard, M.E., Saona, R.J.: Conditional predictive Bayesian Cramér-Rao Lower Bounds for prognostic algorithms design. Appl. Soft Comput. **72**, 647–665 (2018)
10. Cramér, H.: Mathematical Methods of Statistics. Princeton University Press, Princeton (1946)
11. Rao, C.R.: Information and the accuracy attainable in the estimation of statistical parameters. Bull. Calcutta Math. Soc. **37**, 81–89 (1945)
12. Van Trees, H.L.: Detection, Estimation and Modulation Theory. Wiley, New York (1968)
13. Fritsche, C., Ozkan, E., Svensson, L., Gustafsson, F.: A fresh look at Bayesian Cramér-Rao bounds for discrete-time nonlinear filtering. In: 17th International Conference on Information Fusion (2014)
14. Tichavský, P., Muravchik, C.H., Nehorai, A.: Posterior Cramér-Rao bounds for discrete-time nonlinear filtering. IEEE Trans. Signal Process **46**(5), 1386–1396 (1998)
15. Zuo, L., Niu, R., Varshney, P.: Conditional posterior Cramér-Rao lower bounds for nonlinear sequential Bayesian estimation. IEEE Trans. Signal Process. **59**, 1–14 (2011)
16. Šimandl, M., Královec, J., Tichavský, P.: Filtering, predictive, and smoothing Cramér-Rao bounds for discrete-time nonlinear dynamic systems. Automatica **37**, 1703–1716 (2001)
17. Acuña, D.E., Orchard, M.E.: A theoretically rigorous approach to failure prognosis. In: Annual Conference of the Prognostics and Health Management Society (2018)
18. Olivares, B.E., Cerda, M.A., Orchard, M.E., Silva, J.F.: Particle-filtering-based prognosis framework for energy storage devices with a statistical characterization of state-of-health regeneration phenomena. IEEE Trans. Instrum. Meas. **62**(2), 364–376 (2013)
19. Burgos, C., Orchard, M.E., Kazerani, M., Cárdenas, R., Sáez, D.: Particle-filtering-based estimation of maximum available power state in Lithium-Ion batteries. Appl. Energy **161**, 349–363 (2016)

Performance Degradation Monitoring and Quantification: A Wastewater Treatment Plant Case Study

Iñigo Lecuona, Rosa Basagoiti, Gorka Urchegui, Luka Eciolaza, Urko Zurutuza, and Peter Craamer

1 Introduction

Wastewater treatment plants (WWTPs) are responsible for treating the wastewater produced by cities and industries, removing the biological and chemical contaminants from the discharge water in order to protect both the public health and the environment. To get the best quality on the effluent water, the wastewater passes through a complex process, which includes physical and biological processes, divided into different phases [19].

The first step of the process consists on removing the largest solid waste. This waste is separated using an automatic sieve. After that, the wastewater is stored in tanks to remove the soaps and oils it contains by creating air bubbles that make these pollutants come up to the surface. In this phase it is also reduced the water circulation speed, so the sand particles are settled at the bottom of the tank.

To remove the waste that is dissolved in the water, a biological treatment is carried out. With this treatment the nitrogen, phosphorus and organic matters are removed. The tanks where this process is carried out had a huge bacteria concentration. These bacteria are fed with the waste that is dissolved in the water, removing the contaminants from the water. As a result of this process a biological sludge is produced from the waste. In the last step of the process, this sludge is separated from the water using a settling system. The sludge is collected to produce biogas and the clean water is poured back to the river.

I. Lecuona (✉) · R. Basagoiti · L. Eciolaza · U. Zurutuza
Faculty of Engineering, Mondragon Unibertsitatea, Arrasate - Mondragon, Spain
e-mail: ilecuona@mondragon.edu; rbasagoiti@mondragon.edu; leciolaza@mondragon.edu; uzurutuza@mondragon.edu

G. Urchegui · P. Craamer
MSI Grupo, Andoain, Spain
e-mail: gurchegui@msigrupo.com; pcraamer@msigrupo.com

© Springer Nature Switzerland AG 2019
E. Lughofer, M. Sayed-Mouchaweh (eds.), *Predictive Maintenance in Dynamic Systems*, https://doi.org/10.1007/978-3-030-05645-2_13

During the last years different technologies had been implemented with the aim of improving the wastewater treatment process [2] and complying with the limit values fixed by the law, which have become more and more restrictive. The main concern for the WWTP managers has always been to satisfy the water quality standards.

However, the growing awareness on environmental issues, sustainable living and limited energy supply has prompted an increase in energy saving research in the WWTP domain [37]. The equipment of the WWTPs (blowers, pumps, etc.) is operating 24 h per day. For the plant managers, suppose a challenge to check that the equipment is working under healthy conditions, due to the non-linear changes in the performance generated by effects such as, wear, tear and clogging situations. For the correct operations of the plant, it is very important to detect these changes in the performance, in order to prevent a failure and also not increase the operational costs.

1.1 Energy Consumption on WWTPs

The process of treating the wastewater is energy intensive, making the WWTPs one of the main energy consumers inside a community. According to a study carried out by Gude [17], in developed countries, the electrical consumption of these plants is in the order of 1% of the overall national consumption. According to Eurostat data [12], in the year 2016, the overall European consumption by the WWTPs was approximately of 27 TWh/year, equivalent to the global energy consumption of a country like Serbia [31].

With the aim of discovering the percentage of efficient WWTPs, Hernández-Sancho et al. [20] analysed the data of 177 WWTPs from the Valencian Community, Spain. The results of this work show that only 10% of the plants could be considered energy efficient [32].

To quantify the potential energy and cost savings of inefficient WWTPs, Castellet and Molinos-Senante analysed the data of 49 WWTPs in [5], also in the Valencian Community. The results of this study show that 29 of the 49 WWTPs were inefficient, and if these plants were to become efficient, they could collectively save more than 22 million Euro annually, where 25% will be saved on energy costs.

It is also important to remark that energy demand in this industry will grow over time due to a number of factors, such as population growth and the corresponding growth in the contaminant load to be treated, as well as increasingly stringent regulatory and environmental protection standards for effluent quality and residual water reuse. These changes are expected to result in more energy intensive processes [20]. The estimations for the European market of wastewater treatment predict a growing compounding rate of 4.1% per year [10], which adds more value to the energy savings on this industry.

In order to quantify the potential energy savings of the different processes that take place during wastewater treatment, multiple studies have been carried out.

Goldstein and Smith stated in [16] that 80% of the electricity use goes towards moving and treating the wastewater, whose main consumers are the pumping and blower systems [35].

The Water Environment Federation published a manual [34] where they quantify the energy consumption for the different processes. They found that the main energy consumer of a WWTP is the aeration process, which accounts for 54.1% of the energy consumption, followed by the pumping systems, with 14.3% of the consumption. Therefore, by applying energy savings on these systems, bigger savings can be achieved.

1.2 Energy Savings Through Maintenance

From the energy efficiency point of view, the performance of the equipment is a key feature. The WWTP is a very challenging scenario, working 24 h per day and with very changing characteristics of the wastewater. Its components suffer effects such as wear, tear and clogging situations, which reduce the performance of the equipment and increase the energy consumption. Therefore, by checking that the equipment of the WWTP is working on its best performance range, energy consumption can be reduced and also improved the maintenance task scheduling.

As Torregrossa et al. highlighted in [32], energy savings can be achieved by improving the maintenance strategies of the different WWTP equipment; despite the maintenance does not directly provide measures for energy performance, the detection of anomalies at an early stage can indirectly produce energy savings.

In 1999, the Department of Energy of the USA published a sourcebook [11] for industry with the most common maintenance strategies for efficiency in wastewater pumping systems. The simplest approach is a fixed interval scheduled maintenance program; whereas to apply a more advanced approach, a predictive or conditional maintenance, the use of information provided by vibration sensors, temperature and energy consumption, is necessary.

A method for early failure detection for wastewater pumping systems was developed by Berge et al. [4] based on condition monitoring. They used an array of sensors including: pump vibration, motor winding temperature, motor current, motor bearing temperature and pump inflow.

In the case of the pumps, the performance is affected by the wear and tear of its internal mechanic components, such as the impeller, shaft, bearings and the sealing.

According to a study [13] carried out by the European Commission, due to these effects, the performance of the pump is reduced between 10 and 15%. Hence, preventive maintenance tasks are periodically performed by replacing worn key internal components. Figure 1 shows schematically how proper maintenance insures a high performance of the pump over time.

In some pumping systems, the concentration of suspended solids within the wastewater can be especially high, which together with unfavourable operational conditions, can create partial or complete clogging situations. When clogging

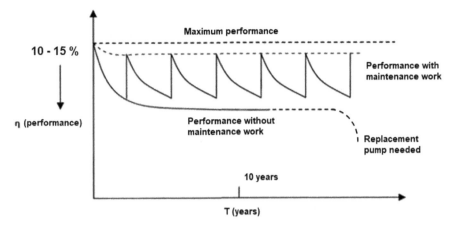

Fig. 1 Typical pump performance over time. Derived from [13]

situations arise, the flow-rate generated by a pump is reduced and its energy consumption increased, reducing its hydraulic and energetic performances. Small clogging situations often disappear without any action needs to be taken, in the case the pump is able to pump any of the accumulated solids over time. However, in harder clogging situations, a maintenance task may be needed, such as disassembling the pump and manually cleaning the clogged parts.

In summary, there are two main causes that influence the performance of a pump: the wear/tear of its components and the clogging situations. Typically, clogging situations affects the performance of a pump in the short term—from few minutes to several days—as depends on the characteristics of the wastewater, while the wear/tear of components change the performance of a pump in the medium or long term—from weeks to several months.

In the case of the blowers of the aeration system, its performance is also affected by the wear and tear of its internal mechanic components, mainly by the filter. The filter guarantees the air quality for the process. However, they are very likely to be clogged, reducing the performance and the energetic efficiency of the process and increasing the wear of other components (e.g. of the motor and oil).

Despite the importance of the wear/tear effects and the clogging situations and its consequences over the performance of the WWTP equipment, most of researches on the field did not take it into account in their works [9, 35, 37, 38]. The main reason of removing this effect from the analysis is to reduce the complexity. The clogging situations that arise in the pumping systems had a random nature, as they depend on the wastewater characteristics. The assumptions made in this analysis limit the application of the obtained results on a real case.

The literature shows a research that addresses this issue. Torregrossa et al. [33] proposed a daily data-driven approach for a detailed pump efficiency analysis using KPIs (key performance indicators). This approach is based on signal decomposition to identify short- and long-term performance fluctuations. The information provided by this approach is useful for the plant managers, as it allows them to take decisions based on real performance information. However, this approach has also some limitations, as it only works for single-speed pumping systems.

Therefore, a need for a new approach that can be useful for a wider industrial application has been identified. The aim of this research is to present a methodology to quantify and monitor the degradation of different industrial equipment, and show its applicability on two WWTP processes: the pumping systems and the blowers of the aeration process.

2 Methodology

As previously described, the process of treating the wastewater is complex and has different phases [19]. The research carried out with the aim of optimizing the global process of a WWTP, like the one performed by Moles et al. [23], obtained poor results as they were not able to take into account all the complexities of the process. Moreover, the assumptions that made on it usually limit the applicability of the results in industrial environments.

Due to these limitations, this methodology is going to be based on a divide and conquer approach. The idea of this approach is to break a complex problem into smaller and easier to solve problems. Therefore, when applying this methodology to analyse a process, like the pumping or the aeration processes, they are going to be divided into individual components of the system, pumps in the case of the pumping system and blowers in the aeration system, as these systems are made of multiple components.

It is necessary to analyse individually the components of a system, as the degradation does not affect in the same way to all the equipment of a system. Applying this approach to the problem will allow to analyse and compare with each other the individual components of the system.

Once identified the elements to be analysed, the individual components of the system, the next step is to define a modelling approach. The first modelling approaches investigated for WWTP equipment [1, 3] have been based predominantly on physics and mathematical programming [22].

The problem of these modelling approaches is that numerous assumptions are made during the process of building the model, making them rather abstract and limiting the applicability of these models in industrial environments [36]. Comas et al. had a similar conclusion in [8], where they suggested that it was necessary to apply other modelling techniques rather than the classic ones for the WWTP processes by their heuristic reasoning ability and work in conditions of uncertainty or qualitative information.

To model complex, dynamic and non-linear scenarios, the literature shows successful applications of data mining approaches in scenarios such as business administration [24], medical informatics [25] and also WWTP processes [30, 36].

For this reason, a data mining modelling approach has been selected to build the models for the individual components of the WWTP systems.

When Hand et al. wrote the book Principles of Data Mining [18], they defined data mining as the analysis of observational datasets to find unknown relationships and to summarize the data in novel ways that are both understandable and useful to the data owner. The relationships and summaries derived through a data mining exercise are referred to as models or patterns.

To build models using data mining approaches, data is needed. Nowadays, the availability of SCADA (supervisory control and data acquisition) systems in the WWTP domain opens the door to the use of the science of data mining to develop models in this scenario. The SCADA systems, besides controlling the process, provide synchronized and reliable data of the process, an essential feature to analyse and build the models.

However, as it has been highlighted in Sect. 1.2 of this chapter, the performance of these systems changes over time. Due to the effects such as wear, tear and clogging situations, the performance of the components is reduced. Therefore, the data used to build the models must be from a representative period of its operation, in this case, the period with the best performance. This period will be identified using data mining techniques.

Once selected the data from period with the best performance, the next step is to build the model using these data. The model will be built using a regression algorithm, as the aim is to predict how it is going to work the component under healthy status conditions, when the equipment is being efficient.

The output of this model will be the observed best performance of the component for the actual operational conditions. The predicted output value can be different from the observed one if the performance of the equipment has been reduced due to the wear/tear effects, the clogging situations or any other factor that could reduce its performance. Hence, comparing the predicted value and the observed one, it is possible to estimate the degradation or loss in performance that had suffered the component.

To analyse the performance of the component over time, observations are grouped by day. For each day a metric that will be used as a key performance indicator (KPI) is calculated. Here, the used metric is the relative mean error (RME). The relative mean error measures the average, daily, relative offset between the observations and the prediction based on a healthy state of the component and has the following formula:

$$\text{RME} = \frac{100\%}{n} \sum_{j=1}^{n} \left(\frac{y_j - \widehat{y}_j}{\widehat{y}_j} \right) \quad (1)$$

where n is the number of observations made for 1 day, y_j is the observed value of the dependent variable and $\widehat{y}_j$ is the predicted value.

By looking at the values of this metric is possible to quantify the loss in performance or degradation that had suffered the component being analysed, as the metric provides the percentage value of this degradation.

3 Results

To show how this approach is applied in a real environment, in this section three WWTP processes are going to be analysed, two pumping systems and one aeration system. Figure 2 shows the locations of these equipment inside the WWTP.

The data used to test is real operational data of two WWTPs located in the Basque Country, a highly industrialized region from northern Spain.

To work with these data the R programming environment [26] and the IDE RStudio [27] are going to be used.

3.1 External Recirculation Pumping System

This pumping system is located after the clarifier of the biological treatment and its work is to move the sludge collected by the clarifier to the sludge treatment line and back to the secondary treatment, as it is shown in Fig. 2. This pumping system is likely to suffer of clogging situations, as it moves big amounts of sludge. Actually, a periodic preventive maintenance strategy, which includes cleaning, damaged part replacement and adjustment tasks, is scheduled in order to prevent its breakdown. Therefore, it is an interesting testing scenario for this approach.

The pumping system uses two identical pumps to move the wastewater and has a flow metre at the end of the system that measures the flow generated by the two pumps. Usually this pumping system operates with one pump; always the same one, as it does not use a rotational scheduling. But when the flow it needs to generate cannot be reached with only one pump, both pumps work at the same time and synchronized, at the same speed.

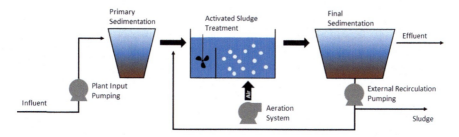

Fig. 2 Diagram of a WWTP process with the locations of the equipment analysed

This work is going to be focused on one pump of the system, the one that is programmed to operate standalone. In this pump is possible to isolate the relation between the variables. The configuration of the systems, where there is a common flow metre for both pumps, forces to find situations where only one pump is working, in order to have data to measure the performance of a pump.

3.1.1 Experimental Setup

The dataset of this pumping system consists on values of the frequency (Hz) of the two pumps and the flow-rate (m^3/h) at the exit of the pumping system. The dataset contains the mean value calculated every 15 min of these three signals for almost 2 years, from 2016-01-01 to 2017-12-15.

The data must be splitted for the individual pumps of the system, in order to find the relation between variables for each pump. This is done by isolating the pumps, finding the cases where the system has been working only with one pump being analysed.

3.1.2 Application of the Methodology

As described in the Methodology (Sect. 2) this approach relies on defining the best performance of the component. The performance of the pump will be checked over time by monitoring how the relation between the frequency of the pump and the generated flow-rate evolves over time. In Fig. 3 the observed flow-rate versus the applied pump frequency is shown for the whole 2-year period. From the figure it can be seen, for instance, that at 30 Hz the pump has generated flow-rates between roughly 50 and 100 m^3/h.

Therefore, a time interval needs to be found for which the pump performs optimally. For this pump, the period with the highest performance is found between May and June 2016. In Fig. 3 the dots in blue show the observation made within the time interval that has been defined as the pumps optimal performance.

The data from this period is used to fit a regression model, which defines for each frequency the mean flow-rate to expect under healthy conditions of the pump. After having built different models using multiple regression algorithms with a cross-validation approach [21, 28], a linear regression has been chosen as the best model for this pump. This modelling also allows an easy comparison between different pumps by comparing the intercept and slope coefficients of the model. Figure 4 shows the results of the analysis of the residuals.

The predictions of this model will show the theoretical output of the pump under healthy conditions. By calculating the daily value of the RME metric using the models predictions and the observed values, the mean percentage degradation of the pump for the day is estimated.

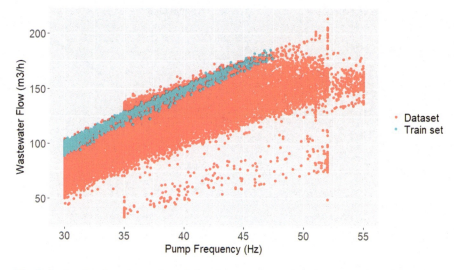

Fig. 3 Scatterplot that shows the relation between the pumps frequency (X axis) and the wastewater flow (Y axis) generated by the external recirculation pump. The dots in red show the whole 2-year period data and in blue the data which represents the optimal performance of the pump

3.1.3 Results and Discussion

Figure 5 shows the daily RME value with time for the results of all the frequencies, as the degradation affects in the same way to the entire operational frequency range. It also shows a smoothed value of this indicator, calculated using a locally weighted regression method.

A locally weighted regression, or *loess*, is a way of estimating a regression surface through a smoothing procedure, fitting a function of the independent variables locally and in a moving fashion analogous to how a moving average is computed [7]. It is used to enhance the visual information on a scatterplot [6].

In Fig. 5, when the RME value is close to zero, this implies that the pump is working close to its optimal performance. When the value increases negatively, means that the performance of the pump is decreasing.

From Fig. 5, it can be seen that for Period 1 (May–June 2016) the value of RME is close to zero, as this is the period used to fit the linear model (optimal performance of the pump). Immediately after this period, the performance of the pump decreases almost instantly by more than 30% (Period 2). After Period 2, the performance increases over 1 month and the pump performs optimally again at the beginning of July 2016, after which another drop in performance occurs over a 3 month period (final 2 months referred to as Period 3).

After a long period of various variability from May 2017, with severe performance losses like the one from Period 4, the performance of the pump is optimal

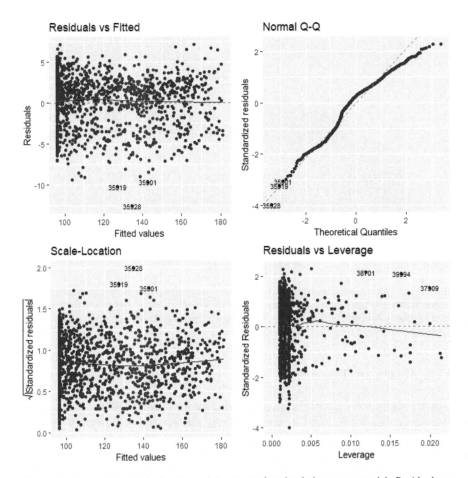

Fig. 4 Analysis of the residual values of the external recirculation pump model. *Residuals vs Fitted* plot shows that the residuals do not have non-linear patterns, as they are equally spread around a horizontal line. *Normal Q-Q* plot shows that the residuals are normally distributed, since they are lined on the straight dashed line. *Scale-Location* plot shows a horizontal line with randomly spread points, which is an indicator of equal data variance (homoscedasticity). *Residuals vs Leverage* plot shows that no outlying data have been used to build the model

again (Period 5). Actually, at the beginning of year 2017 the minimum work frequency of the pump is increased from 30 to 35 Hz, which helps to prevent clogging situations in the pump.

Between August 2017 and the end of the year 2017 two large clogging situations occur. The first takes place at the beginning of September (Period 6) for which the performance of the pump decreases by more than 50% instantly. After manual cleaning the clogging, the pump performance was immediately recovered. The second clogging (Period 7), taking place in December, is a steadily degradation resulting in a loss of more than 30% of its performance.

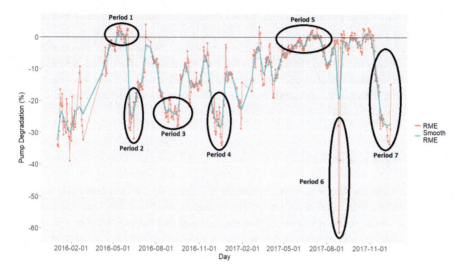

Fig. 5 Line chart that shows in red the daily values of the RME metric and in blue the smoothed trend of this metric for the external recirculation pump. Each value represents the percentage degradation of the pump compared with a healthy condition status

Checking the results with the plant manager, it has been seen that the large performance variations shown by the RME indicator are mainly related to clogging situations in the pump. However, despite all the clogging occurrences the pump suffered during these 2 years, it is observed, for instance, that the performances within the Periods 1 and 5 are the same. So, during this one year and a half period, the periodical maintenance works carried out to the internal components of the pump have insured that the performance of the pump has not been reduced due to mechanical degradation of its components. In case of permanent mechanical degradation of the pump, the RME indicator would not reach again values close to zero.

3.2 Plant Input Pumping System

This pumping system is located at the entrance of the WWTP. It collects the wastewaters from a nearby town and it pumps them to the plant, which is 5 km away, as it is shown in Fig. 2.

The pumping system uses two pumps of different size, a big one and a small one, and has a flow metre at the end of the system that measures the flow generated by the pumps. The elevation level of this pumping well is adjustable. The pumping system is configured to maintain this level at a constant value by adjusting the output flow-rate. Actually, a periodic preventive maintenance strategy, which includes cleaning, damaged part replacement and adjustment tasks, is scheduled in order to prevent its breakdown.

This work is going to be focused on the small pump of the system, as it is the one with the highest work load, operating for 90% of the time from the analysed dataset. A big working load will speed up the wear/tear effects of its internal mechanic components. Therefore, the small pump of the system is more likely of suffering these effects.

3.2.1 Experimental Setup

The dataset of this pumping system consists on values of the frequency (Hz) of the two pumps, the elevation level (%) of the pumping well and the flow-rate (m^3/h) at the exit of the pumping system. The dataset contains the mean value calculated every 15 min of these signals for more than a year and a half, from 2016-04-12 to 2017-12-15.

The data must be splitted for the individual pumps of the system, in order to find the relation between variables for each pump. This is done by isolating the pumps, finding the cases where the system has been working only with one pump being analysed.

3.2.2 Application of the Methodology

The first step of this approach is to define how it is measured the performance for this scenario. The performance of this pump will be checked over time by monitoring how the relation between the frequency of the pump and the elevation level of the pumping well changes the generated flow-rate over time. In Fig. 6 the observed flow-rate versus the applied pump frequency is shown for the whole year and a half period. The reason for looking at this relation is the higher relation between these variables (correlation value is 0.97), compared to the relation between the wet elevation level and the flow-rate (correlation value is 0.52).

From Fig. 6 it can be seen, for instance, that at 40 Hz the pump has generated flow-rates between roughly 20 and 65 m^3/h. Despite the elevation level causes variations in the flow-rate generated by the pump, it is not the fact that explains all this variability of the flow-rate.

Therefore, a time interval needs to be found for which the pump performs optimally. For this pump, the period with the highest performance is in June 2016, weeks 24 and 25. In Fig. 6 the dots in blue show the observation made within the time interval that has been defined as in which the pump performs optimally.

The data from this period is used to fit a regression model, which defines for each frequency the mean flow-rate to expect under healthy condition of the pump. After having built different models using multiple regression algorithms with a cross-validation approach, a multivariate adaptive regression splines (MARS) algorithm has been chosen as the best model for this pump. MARS is a non-parametric

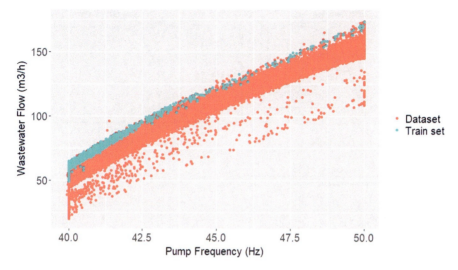

Fig. 6 Scatterplot that shows the relation between the pumps frequency (X axis) and the wastewater flow (Y axis) generated by the plant input pump. The dots in red show the whole year and a half period data and in blue the data which represents the optimal performance of the pump

regression approach that makes no assumption about the underlying functional relationship between the dependent and independent variables [14, 15]. Figure 7 shows the results of the analysis of the residuals.

The predictions of this model will show the theoretical output of the pump under healthy conditions. By calculating the daily value of the RME metric using the models predictions and the observed values, the mean percentage degradation of the pump for the day is estimated.

3.2.3 Results and Discussion

Figure 8 shows the daily RME value with time and a smoothed value of this indicator, calculated using a local regression method. When the RME value is close to zero, this implies that the pump is working close to its optimal performance. When the value increases negatively, means that the performance of the pump is decreasing.

From Fig. 8, it can be seen that for Period 1 (2 weeks from June 2016) the value of RME is close to zero and positive, as this is the period used to fit the model (optimal performance of the pump). After this period, around September 2016 (Period 2) the performance of the pump continues being positive, thus the actual performance is better than the modelled optimal performance. However, the data from this period has been discarded to build the model since it has been working at a constant frequency (40 Hz). To build a representative model data from the whole operational range is needed.

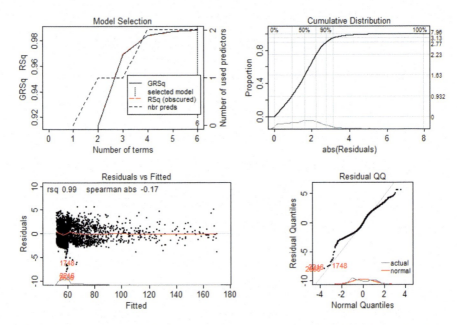

Fig. 7 Analysis of the residual values of the plant input pump model. *Model Selection* plot shows that the best results of the model are obtained using six terms and the two predictors, and that RSq and GRSq (normalized values of the residual sum of squares (RSS) and the generalized cross-validation (GCV), respectively) lines run together. *Cumulative Distribution* plot shows that the median absolute residual is 1.63 and that 95% of the absolute values of residuals are less than 3.13. *Residuals vs Fitted* plot shows that the residuals do not have non-linear patterns, as they are equally spread around a horizontal line. *Residual QQ* plot shows that the residuals are almost normally distributed, since most of them are lined on the straight dashed line

From Period 2 onwards a continuous degradation of the pump is observed until Period 3, more than 9 months, and the pump is degraded by more than 15%. This degradation is not constant, as it has a variability where it changes from positive to negative degradation.

After Period 3, the pump starts to get closer to its optimal performance point. However, after 2 months, its performance starts to decrease, losing more than 25% of its optimal performance (Period 4). Once again the pump recovers performance, but without reaching the optimal performance point. And after a short period, the pump starts to deteriorate, losing more than 35% of its performance (Period 5).

During this year and a half of operation, the performance of the pump has been constantly changing. The periods where the pumps recover performance represent periods where maintenance task has been done, mainly cleaning the clogging situations. However, the pump never reaches again to the optimal performance point. This fact could be produced by the wear/tear of its internal mechanic components. To reach again this optimal performance point, it will be necessary to fix or replace the internal mechanic components of pump.

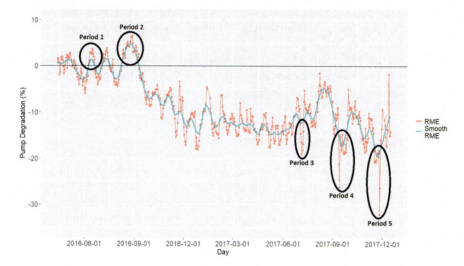

Fig. 8 Line chart that shows in red the daily values of the RME metric and in blue the smoothed trend of this metric for the plant input pump. Each value represents the percentage degradation of the pump compared with a healthy condition status

3.3 Aeration System Blowers

The aeration system is located in the biological treatment phase of the process, as it is shown in Fig. 2. It provides the oxygen needed in the biological process, where micro-organisms consume the waste dissolved in the water.

This aeration system uses four identical air blowers to provide the air to the system. Each blower has a diffuser at the entrance to regulate the amount of air flow that enters to the blower, and therefore, control the air flow at the output of the system. The amount of air flow it needs to generate is controlled by the pressure of the system, which needs to be at a constant value. This system uses the four blowers, with a rotational schedule. Actually, a periodic preventive maintenance strategy, which includes cleaning, damaged part replacement and adjustment tasks, is scheduled in order to prevent its breakdown.

This work is going to analyse the four blowers of the system. The application of a divide and conquer approach allows to model the different blowers of the system and compare the degradation that had suffered each one.

3.3.1 Experimental Setup

The dataset of the aeration system consists on values of the diffuser position (%) of the four air blowers, where 0% represents the minimum open position, but not closed. It also has the values of energy consumption of each blower, measured by the current (A) consumption, and the value of the pressure of the system. The dataset contains the mean value calculated every 15 min of these signals for almost 2 years, from 2016-03-01 to 2017-12-31.

The data must be splitted for the individual blowers of the system, in order to build the relation between its variables. This is done by isolating the blowers by their operational conditions.

3.3.2 Application of the Methodology

The first step of this approach is to define how it is measured the performance for this scenario. In this case, the performance of the blowers will be checked over time by monitoring how the relation between the diffusers position and the pressure of the system changes the current consumption of the blower over time. In Fig. 9 the observed current consumption and the diffusers position is shown for one of the four blowers. Blowers are named from A to D, the figures show the values of the B blower. The reason for looking at this relation is the higher relation between these variables (correlation value is 0.98), compared to the relation between the systems pressure and the current consumption (correlation value is −0.33).

From Fig. 9 it can be seen, for instance, that when the diffusers position is at 0%, the current consumption of the blower ranges between 100 and 125 A. This fact also affects the rest of blowers of the system.

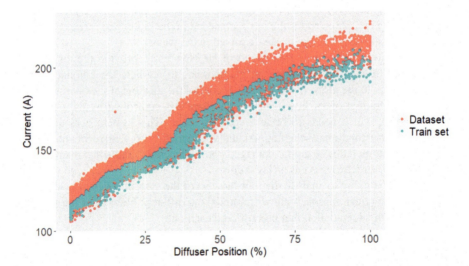

Fig. 9 Scatterplot that shows the relation between blowers diffuser position (X axis) and the current consumption (Y axis) for the B blower of the aeration system. The dots in red show the whole 2-year period data and in blue the data which represents the optimal performance of the blower

Therefore, a time interval needs to be found where the blower has been operating optimally. In this case, the optimal performance of the blower is understood as the cases where the current consumption has been minimal for the operating diffuser positions. For this blower, the period with the highest performance, where the blower has been more efficient, is in June 2017. In Fig. 9 the dots in blue show the observations made within the time interval that has been defined as optimal performance for the blower.

The data from this period is used to fit a regression model, which defines for each diffuser position and pressure value the current consumption to expect under healthy conditions of the blower. After having built different models using multiple regression algorithms with a cross-validation approach, a MARS algorithm has been chosen as the best model for this blower. Figure 10 shows the results of the analysis of the residuals.

The same process has been done with the rest of the blowers of the system (A, C and D), getting very similar models for each one. The predictions of these models will show the theoretical output of each blower under healthy conditions. By calculating the daily value of the RME metric using the models predictions and the observed values, the mean percentage degradation of the blower for the day is estimated.

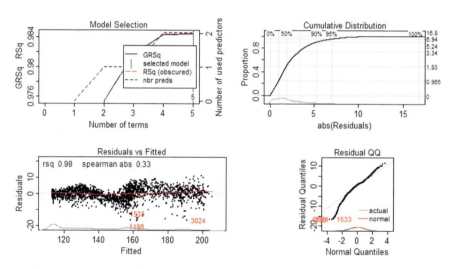

Fig. 10 Analysis of the residual values of the aeration system B blower model. *Model Selection* plot shows that the best results of the model are obtained using five terms and the two predictors, and that RSq and GRSq (normalized values of the residual sum of squares (RSS) and the generalized cross-validation (GCV), respectively) lines run together. *Cumulative Distribution* plot shows that the median absolute residual is 1.83 and that 95% of the absolute values of residuals are less than 7. *Residuals vs Fitted* plot shows that the residuals do not have non-linear patterns, as they are equally spread around a horizontal line. *Residual QQ* plot shows that the residuals are almost normally distributed, since most of them are lined on the straight dashed line

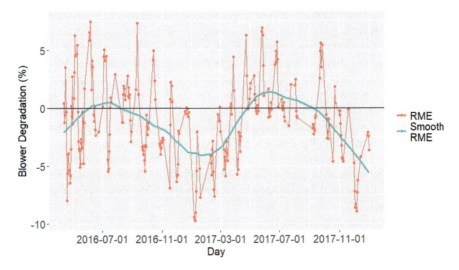

Fig. 11 Line chart that shows in red the daily values of the RME metric and in blue the smoothed trend of this metric for the B blower of the aeration system. Each value represents the percentage degradation of the blower compared with a healthy condition status

3.3.3 Results and Discussion

Figure 11 shows the daily RME value with time and a smoothed value of this indicator, calculated using a local regression method. When the RME value is close to zero, this implies that the blower is working close to its optimal performance. When the value increases negatively, means that the performance of the blower is changing, in this case by increasing its current consumption.

From Fig. 11 it can be seen that the value of this indicator has a big variability over time. However, looking at the smoothed value of the indicator, a seasonal performance variation effect can be identified. The effects of the seasonal performance variations are present in winter months, where the current consumption of the blower is increased compared with the summer month ones. Looking at the values of the RME indicator, this difference can reach values of 10%.

The same effect has been detected for the rest of the blowers of the system. Figure 12 shows the smoothed value of the RME indicator for the four blowers of the system, identified with names from A to D. For all the blowers, the current consumption is increased in winter months, whereas in summer months the blowers reduce their current consumption.

Figure 12 also shows the smoothed daily mean value of the air temperature at the WWTP. This signal has been added to the analysis to check if the seasonal performance variation is related to the air temperature. After calculating the correlation value between the blowers performance variation and the air temperature, the obtained results (correlation value is between 0.7 and 0.83 for the different blowers) indicate that the air temperature has direct influence over the performance of this equipment.

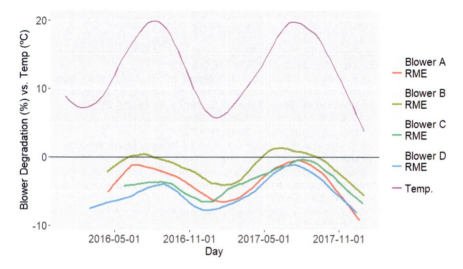

Fig. 12 Line chart that shows the smoothed air temperature at the WWTP in the upper side of the plot and the smoothed trend of the RME indicator for the flour blowers of the aeration system. It is visible how the temperature and RME indicators are correlated, therefore the temperature has direct impact on the blowers' performance

4 Conclusions and Future Works

The presented methodology has been applied over different WWTP equipment, the blowers of an aeration systems and the pumps from two pumping systems, and the obtained results have been validated by the WWTP managers.

In the blowers, a seasonal performance variation effect has been identified applying this methodology, related with the air temperature, which has direct impact on the current consumption of the blowers, as it is shown in the work of Stasyshan [29]. In this work, the author shows how with higher air temperature the power consumption of a blower is reduced. The next step is to add the air temperature variable to the blowers model, in order to predict correctly the current consumption variations caused by the air temperature and detect and analyse other variations in the current consumption.

In the pumping systems, this methodology has allowed to detect and quantify the effects of clogging situations, which can reduce the performance of a pump in more than 20% in less than a year of operations. This fact has not been taken into account in most of the works on the field, as it is a complex effect to model, due to the randomness of its nature. The next step will be to build models for the different pumping systems of the plant using the individual pump models.

Acknowledgements This work has been developed by the Intelligent Systems for Industrial Systems group supported by the Department of Education, Language policy and Culture of the Basque Government. It has been partially funded by the Basque Government.

References

1. Aizenchtadt, E., Ingman, D., Friedler, E.: Quality control of wastewater treatment: a new approach. Eur. J. Oper. Res. **189**(2), 445–458 (2008)
2. Ayesa, E., De la Sota, A., Grau, P., Sagarna, J., Salterain, A., Suescun, J.: Supervisory control strategies for the new WWTP of Galindo-Bilbao: the long run from the conceptual design to the full-scale experimental validation. Water Sci. Technol. **53**(4–5), 193–201 (2006)
3. Barán, B., von Lücken, C., Sotelo, A.: Multi-objective pump scheduling optimisation using evolutionary strategies. Adv. Eng. Softw. **36**(1), 39–47 (2005)
4. Berge, S., Lund, B., Ugarelli, R.: Condition monitoring for early failure detection. Frognerparken pumping station as case study. Proc. Eng. **70**, 162–171 (2014)
5. Castellet, L., Molinos-Senante, M.: Efficiency assessment of wastewater treatment plants: a data envelopment analysis approach integrating technical, economic, and environmental issues. J. Environ. Manag. **167**, 160–166 (2016)
6. Cleveland, W.S.: Robust locally weighted regression and smoothing scatterplots. J. Am. Stat. Assoc. **74**(368), 829–836 (1979)
7. Cleveland, W.S., Devlin, S.J.: Locally weighted regression: an approach to regression analysis by local fitting. J. Am. Stat. Assoc. **83**(403), 596–610 (1988)
8. Comas, J., Dzeroski, S., Sànchez-Marrè, M.: Applying Machine Learning Methods to Wastewater Treatment Plant Data. Universitat Politècnica de Catalunya, Spain (2000)
9. da Costa Bortoni, E., de Almeida, R.A., Viana, A.N.C.: Optimization of parallel variable-speed-driven centrifugal pumps operation. Energ. Effic. **1**(3), 167–173 (2008)
10. Di Lorenzo, M., Scott, K., Curtis, T.P., Katuri, K.P., Head, I.M.: Continuous feed microbial fuel cell using an air cathode and a disc anode stack for wastewater treatment. Energy Fuel **23**(11), 5707–5716 (2009)
11. DOE, U.: Improving pumping system performance: a sourcebook for industry. Prepared for the US Department of Energy, Motor Challenge Program by Lawrence Berkeley National Laboratory (LBNL) and Resource Dynamics Corporation (RDC) (1999)
12. Eurostat: Simplified energy balances - annual data. Technical report, European Commission (2017)
13. Falkner, H., Reeves, D.: Study on improving the energy efficiency of pumps. European Commission (2001)
14. Friedman, J.H.: Multivariate adaptive regression splines. Ann. Stat. **19**, 1–67 (1991)
15. Friedman, J., Hastie, T., Tibshirani, R.: The Elements of Statistical Learning, vol. 1. Springer Series in Statistics. Springer, New York (2001)
16. Goldstein, R., Smith, W.: Water & Sustainability (Volume 4): US Electricity Consumption for Water Supply & Treatment-The Next Half Century. Electric Power Research Institute, Palo Alto (2002)
17. Gude, V.G.: Energy and water autarky of wastewater treatment and power generation systems. Renew. Sust. Energ. Rev. **45**, 52–68 (2015)
18. Hand, D.J., Mannila, H., Smyth, P.: Principles of Data Mining. MIT press, Cambridge (2001)
19. Henze, M., van Loosdrecht, M.C., Ekama, G.A., Brdjanovic, D.: Biological Wastewater Treatment. IWA publishing, London (2008)
20. Hernández-Sancho, F., Molinos-Senante, M., Sala-Garrido, R.: Energy efficiency in Spanish wastewater treatment plants: a non-radial DEA approach. Sci. Total Environ. **409**(14), 2693–2699 (2011)
21. Kohavi, R., et al.: A study of cross-validation and bootstrap for accuracy estimation and model selection. In: IJCAI, vol. 14, pp. 1137–1145, Montreal, Canada (1995)
22. Kusiak, A., Zeng, Y., Zhang, Z.: Modeling and analysis of pumps in a wastewater treatment plant: a data-mining approach. Eng. Appl. Artif. Intell. **26**(7), 1643–1651 (2013)
23. Moles, C., Gutierrez, G., Alonso, A., Banga, J., et al.: Integrated process design and control via global optimization-a wastewater treatment plant case study. Chem. Eng. Res. Des. **81**(5), 507–517 (2003)

24. Ngai, E.W., Xiu, L., Chau, D.C.: Application of data mining techniques in customer relationship management: a literature review and classification. Expert Syst. Appl. **36**(2), 2592–2602 (2009)
25. Palaniappan, S., Awang, R.: Intelligent heart disease prediction system using data mining techniques. In: IEEE/ACS International Conference on Computer Systems and Applications, 2008. AICCSA 2008, pp. 108–115. IEEE, Piscataway (2008)
26. R Core Team: R: A Language and Environment for Statistical Computing. R Foundation for Statistical Computing, Vienna (2018). https://www.R-project.org
27. RStudio Team: RStudio: Integrated Development Environment for R. RStudio, Inc., Boston (2015). http://www.rstudio.com/
28. Shao, J.: Linear model selection by cross-validation. J. Am. Stat. Assoc. **88**(422), 486–494 (1993)
29. Stasyshan, R.: How inlet conditions impact centrifugal air compressor performance. https://www.airbestpractices.com/technology/air-compressors/how-inlet-conditions-impact-centrifugal-air-compressor-performance
30. Thunberg, A., Sundin, A., Carlsson, B.: Energy optimization of the aeration process at Kappala wastewater treatment plant. In: 10th IWA Conference on Instrumentation, Control & Automation, pp. 14–17 (2009)
31. Torregrossa, D., Schutz, G., Cornelissen, A., Hernández-Sancho, F., Hansen, J.: Energy saving in WWTP: daily benchmarking under uncertainty and data availability limitations. Environ. Res. **148**, 330–337 (2016)
32. Torregrossa, D., Hansen, J., Hernández-Sancho, F., Cornelissen, A., Schutz, G., Leopold, U.: A data-driven methodology to support pump performance analysis and energy efficiency optimization in waste water treatment plants. Appl. Energy **208**, 1430–1440 (2017)
33. Torregrossa, D., Hansen, J., Hernández-Sancho, F., Cornelissen, A., Schutz, G., Leopold, U.: Pump efficiency analysis of waste water treatment plants: a data mining approach using signal decomposition for decision making. In: International Conference on Computational Science and Its Applications, pp. 744–752. Springer, Cham (2017)
34. WEF, W.E.F.: Manual of practice (mop) no. 32: energy conservation in water and wastewater facilities. Prepared by the Energy Conservation in Water and Wastewater Treatment Facilities Task Force of the Water Environment Federation (2009)
35. Zeng, Y., Zhang, Z., Kusiak, A., Tang, F., Wei, X.: Optimizing wastewater pumping system with data-driven models and a greedy electromagnetism-like algorithm. Stochastic Environ. Res. Risk Assess. **30**(4), 1263–1275 (2016)
36. Zhang, Z., Kusiak, A.: Models for optimization of energy consumption of pumps in a wastewater processing plant. J. Energy Eng. **137**(4), 159–168 (2011)
37. Zhang, Z., Zeng, Y., Kusiak, A.: Minimizing pump energy in a wastewater processing plant. Energy **47**(1), 505–514 (2012)
38. Zhang, Z., Kusiak, A., Zeng, Y., Wei, X.: Modeling and optimization of a wastewater pumping system with data-mining methods. Appl. Energy **164**, 303–311 (2016)

Fuzzy Rule-Based Modeling for Interval-Valued Data: An Application to High and Low Stock Prices Forecasting

Leandro Maciel and Rosangela Ballini

1 Introduction

Prediction of asset prices movements in financial markets plays a key role in asset allocation, portfolio management, risk assessment, technical analysis strategies implementation, and derivatives pricing [35, 43]. The temporal evolution of assets prices, stock indices, and exchange rates is observed as single-valued time series [3]. If only the opening (or closing) price is observed daily, the intraday variability and important information is neglected [12].

Besides the forecasting of intraday time series appears as a solution to this problem, high-frequency data characteristics such as irregular temporal spacing, strong diurnal patterns, and complex dependence result in obstacles to traditional time series models. Additionally, in real-world situations, the precise prediction of the sequence of intraday prices for 1 day ahead, for example, is almost impracticable. On the other hand, these limitations can be alleviated if the highest and the lowest values of prices are measured daily, what originates interval time series (ITS) [16]. One must notice that variables of similar nature include electricity prices, power load and generation, meteorology, production rates, and traffic flows [45].

L. Maciel (✉)
São Paulo School of Politics, Economics and Business, Federal University of São Paulo, Osasco, SP, Brazil
e-mail: maciel.leandro@unifesp.br

R. Ballini
Institute of Economics, University of Campinas, Campinas, SP, Brazil
e-mail: ballini@unicamp.br

The modeling and forecasting of financial ITS have received considerable attention in the literature recently.[1] Analysis of large data sets, or namely big-data, such as high-frequency data, based on the ITS framework is a new domain to study statistically detectable patterns, attracting researchers from economics and finance [36]. The daily high and low financial prices can be seen as references values for investors in order to place buy or sell orders. Furthermore, these prices are related with the concept of volatility. Authors in [1] show that the difference between the highest and lowest log prices over a fixed sample interval, also known as the log range, is a highly efficient volatility measure.[2] In this context, works of [6] and [40] indicate that the volatility range appears robust to microstructure noise such as bid-ask bounce, due to the limitations of traditional single-valued volatility models that fall to use the information contents inside the reference period of the prices, resulting in inaccurate forecasts [11].

The literature addresses ITS modeling through extensions of traditional data analytics and prediction techniques [17]. For instance, authors in [17] use a vector error correction model (VECM) to forecast low, high, and closing values of three foreign exchange rates: Dollar-Yen, Pound-Dollar, and Mark-Dollar. They indicate that the data dynamics are different for high and low, and their prediction improves forecasting models based on closing prices. Also using a VECM framework, the work of [10] indicates that the use of high and low values of the Dow Jones Industrial, S&P 500, and NASDAQ increases forecasting accuracy.

Authors in [24] proposed an interval-valued linear model for stock market forecasting. Interval-valued data are represented by higher and lower equity prices as interval bounds. Using interval midpoints, the method is estimated by ordinary least squares. Using the same approach, research provided in [21] estimated the lower and upper interval bounds of the series separately. Both approaches are compared in [22], which the authors indicate that the former provide more accurate forecasts.

For stock index and exchange rate ITS forecasting, Arroyo et al. [3] considered methods such as exponential smoothing, autoregressive integrated moving average (ARIMA), artificial neural networks, and k-nearest neighbors. Univariate and multivariate predictions are evaluated using the interval bounds as the interval's midpoints (center) and half-lengths (radius). The results suggested that less accurate results are observed when the lower and upper bounds time series are predicted independently. Similar analysis is also provided in [36], revealing the potential of univariate threshold models for S&P 500 index ITS forecasting.

Considering intraday data, Yang et al. [50] evaluated the forecasting performance of an interval-valued linear regression model for Dow Jones, Nasdaq, and S&P 500 interval time series, which accurate forecasts are observed in comparison with

---

[1] The literature has introduced several interval time series forecasting methods. Examples include [18, 28, 44, 47].

[2] The literature that considers the high–low range prices as a proxy for volatility dates back to the 1980s with the work of [34].

single-valued prediction time series methods. More recently, using computational intelligence techniques, Xiong et al. [46] applied a support vector machine model to predict simultaneously the highest and lowest prices of S&P 500, Nikkei 225, and FTSE 100 indices.

Summarizing, the literature has provided evidence on the high potential of using ITS framework in economics and finance data forecasting, as it is able to deal with large data sets, e.g., high-frequency data, and also accounts the information related to prices variability, neglected in traditional time series methods that are based only on closing prices. Further, most of the techniques assumed a linear structure of the data dynamics even the strong evidence of nonlinearities in financial market data, e.g., [15, 20, 23, 36].

The current literature advocates the use of ITS framework in economics and finance, since it provides appropriate mechanisms to analyze large data sets such as high-frequency data, and also supplements the information extracted by the time series of the closing values considering high and low prices in terms of a proxy measure for volatility. However, most of the current approaches, even the ones based on interval-valued data, assume a linear structure in representing time series process dynamics. Several studies indicate that nonlinear models do provide a richer understanding of the dynamics of variables of interest [36]. Dueker et al. [15], Guidolin et al. [20], and Henry et al. [23] are examples that suggest evidences of threshold nonlinearities in exchange rates, bond and stock markets, respectively. In addition, the relevance of considering data nonlinearities in ITS analysis is also stressed out by Maia et al. [31], Rodrigues and Salish [36], and Roque et al. [37].

ITS prediction approaches have been proposed in the literature to model financial data: traditional statistical techniques such as interval exponential smoothing methods [3], VAR model [19], and VECM [10].[3] The authors showed that when ITS under study are linear and stationary, better forecasts are achieved [48]. Due to the intrinsic complexity and volatility of ITS (e.g., interval-valued stock prices), they appear nonlinear and nonstationary. To overcome this limitation, machine learning techniques such as interval multilayer perceptrons (iMLP) [37] and multi-output support vector regression [46] have shown a relevant nonlinear modeling capability for ITS in real world.

Using both linear and nonlinear concepts, Xiong et al. [48] suggest a "linear and nonlinear" modeling framework based on VECM and multi-output support vector regression to forecast agricultural commodity future interval prices. Using Chinese future market data, the research indicates that the method is a promising tool for forecasting interval-valued agricultural commodity futures prices, as the model able to capture both linear and nonlinear patterns exhibited in future prices.

Although the potential of machine learning approaches to model ITS, they do not consider the uncertainty inherited in financial data. It is well known that financial markets are often affected by news, expectations, and investors psychological states [32]. Thus, the uncertainty among agents plays a key role when modeling

---

[3] Arroyo et al. [3] provide a survey on ITS forecasting methodologies in finance and economics.

and forecasting financial time series. Further, machine learning approaches such as artificial neural networks models, due to their black-box nature, also lack on interpretability, in linguistic terms. This issue is very important for traders when performing their strategies based on forecasts, for example. Therefore, an interpretable model is demanding.

To overcome these limitations, this chapter proposes a fuzzy rule-based model for interval-valued data (iFRB). Fuzzy rule-based (FRB) models have achieved success in business and engineering applications, mostly in the field of fuzzy control, in response to the need of flexible and robust decisions in the face of rapid change, imperfect information, uncertainty, and ambiguous objectives [9, 52]. Fuzzy set theory enables the processing of imprecise information by means of membership functions, in contrast to Boolean characteristic mappings [51]. The capability of fuzzy systems to approach nonlinear functions is often stressed in the literature. Additionally, the advantage of fuzzy systems is its linguistic interpretability [5, 39]. The relationship between the input variable(s) and the output variable(s) is represented by means of fuzzy if–then rules of the following general form: if "antecedent proposition," then "consequent proposition." iFRB assumes Takagi-Sugeno (TK) type models, where the consequent part is expressed as a (non)linear relationship between the input variable and the output variable [42].

Most of the FRB models consider the data described by real-valued variables. When handling real-world complex data, these models are very restrictive, since they do not take into account variability and/or uncertainty inherent to the data.[4] In iFRB, instead of real numbers, variables are represented by intervals. The identification of iFRB concerns the identification of the antecedents and consequents of the fuzzy rules. Rules antecedents are identified using the fuzzy clustering approach for symbolic interval-valued data based on the participatory learning (PL) paradigm, iPL, suggested by Maciel et al. [30]. Participatory learning provides a paradigm for learning that emphasizes the pervasive role of what is already known or believed in the learning process. However, this work extends iPL with adaptive Hausdorff distances, as a mechanism to compute the (dis)similarity between intervals. The advantage of these adaptive distances is that the clustering algorithm is able to recognize clusters of different shapes and sizes. Rules consequent parameters are estimated using traditional least squares techniques taking advantage of the intervals midpoints.

Hence, this chapter suggests iFRB modeling for financial ITS forecasting. ITS are constructed by minimum and maximum stock prices. Experiments are conducted using the main stock index of the Brazilian financial market, the IBOVESPA, for the period from January 2000 to December 2015, focusing on one step ahead forecasts. iFRB is compared against univariate time series methods such as random walk,

---

[4]For instance, the work of Leite et al. [26] proposes an interval-based evolving modeling (IBeM) approach that recursively adapts both parameters and structure of rule-based models. In IBeM the clusters are represented by intervals, i.e., granular local models, such that the consequents of the rules are also represented by intervals. Therefore, the model is able to produce interval outputs but is not designed to process interval-valued data.

exponential smoothing, ARIMA, and threshold autoregressive (TAR), and with multivariate approaches such as VECM and the interval neural network (iMLP). In addition to the use of accuracy measures and statistical tests to compute models performance, this work also discusses the results using quality measures designed for the interval time series framework, as well as in terms of additional forecast descriptive statistics such as efficiency and coverage rates.

iFRB is able to simultaneously handle nonlinear patterns and the uncertainty exhibited in financial time series on an interval-valued data framework, which is essential to capture all the relevant information in intraday market price variation. Additionally, the interpretability of the system in linguistic terms improves planning and making decisions for traders' strategies. Further, the model is able to handle huge data sets when the data is summarized by means of symbolic data, i.e., through the use of interval-valued variables. Each interval is then a summarization of data variability in a period, in our case, in a day of trading. The contributions of this work are outlined as follows. First, most of the methods are based on single-valued price series. In comparison with interval-valued time series-based models, the suggested method also incorporates the nonlinearities and uncertainty of financial markets and provides an interpretable modeling framework. Second, the possibility of forecasting the lower and upper bounds of interval-valued stock index series simultaneously by an interval fuzzy model is examined. The third contribution is that not only statistical accuracy but also quality measures designed for interval time series are used to assess the practicability of the suggested model for interval-valued stock index forecasting as empirical application. Finally, the literature still demands works related to ITS forecasting in finance, mostly for emergent economies like Brazil, due to growing availability of high-frequency data as well as its applications (high-frequency trading), in which informative forecasts play an essential role.

This chapter is organized as follows. After this introduction, Sect. 2 gives a brief reminder of the interval arithmetic adopted in this work. Section 3 describes the interval fuzzy rule-based model. The empirical application on the IBOVESPA index is discussed in Sect. 4. Finally, Sect. 5 concludes the work and suggests issues for further investigation.

2 Interval Arithmetic

Let an interval x be a closed bounded set of real numbers:

$$x = [x^L, x^U] \in \Im, \tag{1}$$

where $\Im = \{[x^L, x^U] : x^L, x^U \in \Re, x^L \leq x^U\}$, x^L and x^U are the lower and upper bounds of the interval, respectively. An m-dimensional interval vector $\mathbf{x}$ is an ordered m-tuple of intervals $\mathbf{x} = [x_1, x_2, \ldots, x_m]^T$, where $x_j = [x_j^L, x_j^U] \in \Im$, $j = 1, \ldots, m$.

The midpoint of an interval x, $\bar{x}$, is calculated as:

$$\bar{x} = \frac{x^L + x^U}{2}. \tag{2}$$

Therefore, a sequence of intervals at time steps $t = 1, 2, \ldots, T$ is an interval time series (ITS).

The arithmetic operations used in this chapter are [33]:

$$\begin{aligned}
x + y &= [x^L + y^L, x^U + y^U], \\
x - y &= [x^L - y^U, x^U - y^L], \\
xy &= [\min\{x^L y^L, x^L y^U, x^U y^L, x^U y^U\}, \max\{x^L y^L, x^L y^U, x^U y^L, x^U y^U\}], \\
x/y &= x(1/y), \text{ with } 1/y = [1/y^U, 1/y^L].
\end{aligned} \tag{3}$$

The interval fuzzy rule-based model (iFRB) suggested in this work also requires a metric table to measure the (dis)similarities between intervals. In this work we adopt the Hausdorff distance, as suggested by Carvalho et al. [8]. The Hausdorff distance between two vectors of intervals, $\mathbf{x}$ and $\mathbf{y}$, denoted by $dH(\mathbf{x}, \mathbf{y})$, is then calculated as[5]:

$$dH(\mathbf{x}, \mathbf{y}) = \sum_{j=1}^{m} \left(\max\{|x_j^L - y_j^L|, |x_j^U - y_j^U|\} \right). \tag{4}$$

The next section addresses iFRB, the fuzzy rule-based method for interval-valued data.

3 Interval Fuzzy Rule-Based Model

The interval fuzzy rule-based model suggested in this work is composed by a set of fuzzy rules with affine interval consequents:

$$\mathscr{R}_i : \text{ IF } \mathbf{x} \text{ is } \mu_i \text{ THEN } y_i = \theta_{i0} + \theta_{i1} x_1 + \cdots + \theta_{im} x_m, \tag{5}$$

where $\mathscr{R}_i$ is the i-th fuzzy rule, $i = 1, 2, \ldots, c$, c is the number of fuzzy rules, $\mathbf{x} = [x_1, x_2, \ldots, x_m]^T$, $x_j \in \mathfrak{I}$, $j = 1, \ldots, m$, comprises the input, μ_i is the fuzzy set of

---

[5] One must notice that the subtraction of intervals can produce intervals with negative extremes. However, in all operations related to the current approach, the subtraction operation between intervals is not needed, which is also one of the advantages by using a distance metric such as the Hausdorff distance.

the antecedent of the i-th fuzzy rule with membership function $\mu_i(\mathbf{x}) : \Im \to [0, 1]$, $y_i \in \Im$ is the i-th rule output, and θ_{i0} and $\theta_{ij} \in \Re$, $j = 1, \ldots, m$, are the parameters of the consequent of the i-th rule, which are represented by single-valued variables.

Fuzzy inference in iFRB is similar to the traditional Takagi–Sugeno model, but the operations correspond to interval operations, as described in Sect. 2. Therefore, the output is calculated as:

$$y = \sum_{i=1}^{c} \left(\frac{\mu_i(\mathbf{x}) y_i}{\sum_{j=1}^{c} \mu_j(\mathbf{x})} \right). \tag{6}$$

In terms of normalized degrees of activation, the output in (6) is

$$y = \sum_{i=1}^{c} \lambda_i y_i = \sum_{i=1}^{c} \lambda_i \mathbf{x}_e^T \theta_i, \tag{7}$$

where

$$\lambda_i = \frac{\mu_i(\mathbf{x})}{\sum_{j=1}^{c} \mu_j(\mathbf{x})}, \tag{8}$$

is the normalized firing level correspondent to the i-th rule, $\theta_i = [\theta_{i0}, \theta_{i1}, \ldots, \theta_{im}]^T$ the vector of parameters, and $\mathbf{x}_e = [1 \ \mathbf{x}^T]^T$ the expanded input vector.

The TS model uses parametrized fuzzy regions and associates each region with a local affine (linear) model. The nonlinear nature of the model originates from the fuzzy weighted combination of multiple linear models. The contribution of a local model to the model output is proportional to its degree of activation.

iFRB modeling requires: (1) learning the antecedent part of the model via an interval fuzzy clustering algorithm, and (2) estimation of the parameters of the interval affine consequents. The i-th fuzzy cluster defines μ_i, the antecedent of the i-th fuzzy rule.

3.1 Interval Participatory Learning Fuzzy Clustering with Adaptive Distances

To identify the antecedents of the fuzzy rules, this chapter adopts the interval participatory learning fuzzy clustering algorithm (iPL), suggested by Maciel et al. [29, 30]. Further, this work extends iPL using adaptive Hausdorff distances and also incorporates in a fuzzy inference system to evaluate its capability in interval-valued time series forecasting. Distances are adaptive in the sense that is calculated differently for each cluster due to its intra-class structure at each iteration. Therefore, it is able to model data structures represented by clusters with distinct shapes, sizes, and orientation as highlighted by Carvalho et al. [8].

The clustering objective is to divide a data set $\mathbf{X} = \{\mathbf{x}_1, \ldots, \mathbf{x}_T\}, t = 1, 2, \ldots, T$, in $c, 2 \leq c \leq T$, fuzzy subsets, standing c for the number of classes or clusters and T the number of patters. The main difference here is that the data $\mathbf{x}_t$ are interval-valued variables.

iPL is based on the participatory learning (PL) paradigm [49] in which the model learning is based on the current knowledge of the model, i.e., the current model is part of the learning process and contributes to its self-organization due to new information. The principle of PL is that the new observation represented by the data influences self-organization or model revision. This influence is based on the compatibility of the data to the current model structure (cluster structure) [30, 41].

In clustering, clusters are defined by its centers or prototypes. In this chapter, $\mathbf{V} = \{\mathbf{v}^1, \ldots, \mathbf{v}^c\}, \mathbf{v}^i \in [0,1]^m, i = 1, \ldots, c$ represents the cluster centers in a cluster structure. The learning process corresponds to the learning of this variable. In this case, observations $\mathbf{x}_t \in [0,1]^m$ correspond to the knowledge related to the learning process of the variable $\mathbf{V}$, i.e., each data $\mathbf{x}_t, t = 1, 2, \ldots, T$, is used to learn about $\mathbf{v}^i$. Using the PL paradigm, the learning process consists on how observations $\mathbf{x}_t$ are compatible or not with the current estimates of the cluster centers $\mathbf{v}^i$.

In a current estimate of $\mathbf{v}_t^i$ after $t-1$ observations, a data $\mathbf{x}_t$ contributes to the current knowledge about the system if it is compatible (close) to $\mathbf{v}_t^i$, i.e., the compatibility of a new data is measured with all the current c clusters. If that is the case, $\mathbf{v}_t^i$ is updated according to [30, 41]:

$$\mathbf{v}_t^i = \mathbf{v}_{t-1}^i + G_t^i (\mathbf{x}_t - \mathbf{v}_{t-1}^i), \tag{9}$$

where

$$G_t^i = \alpha \rho_t^i \tag{10}$$

with $\alpha \in [0, 1]$ standing for the learning rate and ρ_t^i for the compatibility degree between $\mathbf{x}_t$ and $\mathbf{v}_{t-1}^i$.

The compatibility ρ_t^i is calculated as:

$$\rho_t^i = 1 - d_t^i, \tag{11}$$

where d_t^i is a (dis)similarity measure. For interval-valued data, the interval participatory learning algorithm (iPL) proposed by Maciel et al. [30] used the Hausdorff distance for intervals as in Eq. (4). This work extends iPL with adaptive Hausdorff distances which associates a distance d^i to each cluster i (and its prototype $\mathbf{v}^i$) such that the sum of the distances $d^i(\mathbf{x}, \mathbf{v}^i)$ between objects $i \in c$ and the prototype $\mathbf{v}^i$ is as small as possible. The compatibility measure in iFRB is computed as:

$$\rho_t^i = 1 - \gamma^i d H_t^i = 1 - \sum_{j=1}^m \gamma_j^i \left(\max\{|x_{j,t}^L - v_{j,t-1}^{i,L}|, |x_{j,t}^U - v_{j,t-1}^{i,U}|\} \right), \tag{12}$$

where $\gamma^i = (\gamma_1^i, \gamma_2^i, \ldots, \gamma_m^i)$ is the vector of weights measuring the adaptivity of the distances, with $\gamma_j^i > 0$ and $\prod_{j=1}^{m} \gamma_j^i = 1$, and m is the dimension of the data.

An arousal mechanism is also considered in order to monitor the compatibility of the current cluster structure with the new observations. Higher values of the arousal indicate less confidence of the current system with the new observation, i.e., indicates the incompatibility of the observation with the current system. In clustering, this idea retains as a case in which a data is far enough for the current cluster centers, indicating the need of creating a new cluster. Thus, the arousal index $a_t^i \in [0, 1]$ of cluster i at t is computed as:

$$a_t^i = a_{t-1}^i + \beta(1 - \rho_t^i - a_{t-1}^i), \tag{13}$$

where $\beta \in [0, 1]$ controls the rate of change of arousal. When β is closer to one, the faster the system is to sense compatibility variations.

The rule for creating a new cluster is: if $a_t^i \geq \tau \in [0, 1]$, a new cluster is created with center initiated as $\mathbf{v}_t^{i+1} = \mathbf{x}_t$. Otherwise, the most compatible cluster with $\mathbf{x}_t$ is updated using (9).

By incorporating the arousal mechanism (13) into (10) we have:

$$G_t^i = \alpha(\rho_t^i)^{1-a_t^i}. \tag{14}$$

When $a_t^i = 0$, $G_t^i = \alpha \rho_t^i$, which is the procedure with no arousal. If the arousal index increases, the similarity measure has a reduced effect. The arousal index can be interpreted as the complement of the confidence we have in the truth of the current belief, the rule base structure [29, 30]. Thus, if a new data is incompatible with the current cluster structure the arousal will indicate the need of creating a new cluster.

Finally, iFRB clustering also verifies the creation of redundant clusters. A cluster is declared redundant if its similarity to another cluster is greater than or equal to a threshold value $\lambda \in [0, 1]$. Thus, the compatibility index among cluster centers, i and j, can be computed as:

$$\rho_t^{i,j} = 1 - \gamma^i d H_t^{i,j}. \tag{15}$$

If $\rho_t^{i,j} \geq \lambda$, the cluster i is declared redundant and replaced by the average between the new data and the current cluster center.

3.2 Rules Consequent Parameters Identification

After antecedents learning, the next step of iFRB identification is the estimation of the parameters of the consequent linear models. To take advantage of the standard form of the least squares algorithm, the procedure in this work uses the midpoint of the intervals, $\bar{x}$, as in (2), also suggested by Maciel et al. [29]. The expression (7) can be rewritten as:

$$\bar{y} = \Lambda^T \Theta, \tag{16}$$

where $\Lambda = [\lambda_1 \bar{\mathbf{x}}_e^T, \lambda_2 \bar{\mathbf{x}}_e^T, \ldots, \lambda_c \bar{\mathbf{x}}_e^T]^T$ is the fuzzily weighted extended input, $\bar{\mathbf{x}}_e = [1 \ \bar{x}_1 \ \bar{x}_2 \ \ldots \ \bar{x}_m]^T$ is the expanded data vector, and $\Theta = [\theta_1^T, \theta_2^T, \ldots, \theta_c^T]^T$ is the parameter matrix, $\theta_i = [\theta_{i0}, \theta_{i1}, \ldots, \theta_{im}]^T$.

The weighted recursive least squares (wRLS) algorithm is considered to update the consequent parameters using the locally optimal error criterion wRLS[6]:

$$\min E_L^i = \min \sum_{t=1}^{T} \lambda_i \left(\bar{y}_t - \bar{\mathbf{x}}_{et}^T \theta_{it} \right)^2. \tag{17}$$

Parameters are then updated as [27]:

$$\theta_{i,t+1} = \theta_{it} + \Sigma_{it} \bar{\mathbf{x}}_{et} \lambda_{it} \left(\bar{y}_t - \bar{\mathbf{x}}_{et}^T \theta_{it} \right), \quad \theta_{i0} = 0, \tag{18}$$

$$\Sigma_{i,t+1} = \Sigma_{it} - \frac{\lambda_{it} \Sigma_{it} \bar{\mathbf{x}}_{et} \bar{\mathbf{x}}_{et}^T \Sigma_{it}}{1 + \lambda_{it} \bar{\mathbf{x}}_{et}^T \Sigma_{it} \bar{\mathbf{x}}_{et}}, \quad \Sigma_{i0} = \Omega I, \tag{19}$$

where Ω is a large number (usually $\Omega = 1000$), and Σ is the dispersion matrix.

3.3 iFRB Identification Procedure

The interval fuzzy rule-based (iFRB) identification steps are summarized in this section. Initialization of clusters centers initial values $\mathbf{V}^0$ are based on the selection of two random points of $\mathbf{X}$: $\mathbf{V}^0 = \{\mathbf{v}^1 \ \mathbf{v}^2\}$. Thus the fuzzy partition matrix $\mathbf{U} \in \Re^{(T \times c)}$ of $\mathbf{V}^0$ is computed, whose element $u_{it} \in [0, 1]$, $i = 1, 2, \ldots, c$, is the membership degree of the t-th data point $\mathbf{x}_t$ to the i-th cluster, the one with center $\mathbf{v}^i$. Membership degrees are calculated as:

$$u_{it} = \left(\sum_{j=1}^{c} \left(\frac{d H_t^i}{d H_t^j} \right)^{\frac{2}{\eta-1}} \right)^{-1}, \tag{20}$$

where η is the fuzzification parameter, using $\eta = 2$ as default value.

---

[6]Brandt and Diebold [2] show that global optimization does not guarantee locally adequate behavior of the sub-models that form the TS model.

Parameters α, β, τ, and λ are defined by the user. Using a data set with T observations, for $k = 1$, where k stands for the number of iterations, the antecedents are determined by the fuzzy clustering algorithm based on a Picard iterative process. The convergence criterion is based on an error metric (E) defined as:

$$E = \max_i |dH^i(\mathbf{v}_k^i, \mathbf{v}_{k-1}^i)|, \ i = 1, \ldots, c, \tag{21}$$

which measures the variations on the prototypes (cluster centers) representatives.

Thus, if a maximum number of iterations, $k_{\max}$, is reached or if the error is smaller than a threshold ϵ, $E \leq \epsilon$, the algorithm stops. At each iteration the vector of weights, γ^i, which defines the adaptability of the distances, is updated as follows [7, 8]:

$$\gamma_j^i = \frac{\left\{\prod_{j=1}^{m}\left[\sum_{t=1}^{T}(u_{it})^{\eta}\left(\max\{|x_{tj}^L - v_{tj}^L|, |x_{tj}^U - v_{tj}^U|\}\right)\right]\right\}^{1/m}}{\sum_{t=1}^{T}(u_{it})^{\eta}\left(\max\{|x_{tj}^L - v_{tj}^L|, |x_{tj}^U - v_{tj}^U|\}\right)}, \tag{22}$$

$j = 1, 2, \ldots, m$.

Carvalho [7] showed that (22) locally minimizes the fitting between the clusters and their representatives (prototypes) based on adaptive Hausdorff distances for interval-valued data clustering.[7]

With fixed antecedents, rules consequent parameters are obtained using wRLS algorithm. Figure 1 describes the iFRB identification steps.

4 Computational Experiments

The iFRB approach introduced in this chapter gives a flexible modeling procedure and can be applied to a range of problems such as process modeling, time series forecasting, classification, system control, and novelty detection. The performance of iFRB for financial interval time series forecasting is compared against univariate models such as random walk (RW), exponential smoothing (ES), ARIMA, and threshold autoregressive (TAR). Note that univariate models do not process the data as intervals. In this work, ITS are constructed by minimum and maximum stock prices. Thus, univariate techniques predict intervals attributes, minimum and maximum stock prices values, independently. Further, multivariate approaches such as the vector error correction model (VECM) and the interval multilayer perceptron neural network (iMLP) [37] are also considered as alternatives. Multivariate models account for the interdependence on the data without the need of specifying the relationship between the series. As follows the accuracy measures and statistical tests used to access models performance are described, followed by the empirical results considering the interval IBOVESPA stock index forecasting.

---

[7] The proof of (22) can be found in [7].

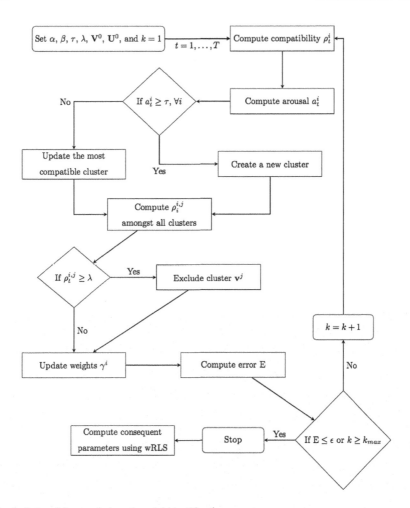

Fig. 1 Interval fuzzy rule-based model identification

4.1 Performance Assignment

In order to access models performance, traditional time series error measures such as root mean squared error (RMSE) and symmetric mean absolute percentage error (SMAPE) are considered:

$$\text{RMSE} = \sqrt{\frac{1}{T}\sum_{t=1}^{T}(y_t - \hat{y}_t)^2}, \quad (23)$$

$$\text{SMAPE} = \frac{1}{T} \sum_{t=1}^{T} \frac{|y_t - \hat{y}_t|}{(|y_t| + |\hat{y}_t|)/2}, \qquad (24)$$

where $\hat{y}_t$ is the t-th forecasted value, y_t the t-th actual value, and T is the sample size.

To translate the performance metrics to intervals, let an ITS, $\{Y_t\}_{t=1}^{T}$, be a sequence of intervals observed in successive instants in time $t = 1, \ldots, T$. Each interval is represented by:

$$Y_t = [y_t^L, y_t^U] \in \Im. \qquad (25)$$

In the context of financial ITS, y_t^L and y_t^U correspond to the minimum and maximum stock price values at t, respectively.

To measure the forecasting error considering the ITS attributes (lower and upper bounds) the error for each series attributes is computed separately. This work also employs the Diebold–Mariano [14] statistic test to evaluate the null hypothesis of equal predictive accuracy.

Since the data has an interval structure, it implies that both characteristics (lower and upper bounds) that describe intervals have to be taken into consideration jointly. Therefore, the overall accuracy of the fitted and forecasted ITS is also measured by the mean distance error (MDE) of intervals:

$$\text{MDE} = \frac{\sum_{t=1}^{T} D(Y_t, \hat{Y}_t)}{T}, \qquad (26)$$

where Y_t and $\hat{Y}_t$ are the observed and forecasted ITS and $D(\cdot)$ is an interval distance. As suggested in [3] and [36], the Euclidian distance for intervals is selected for lower and upper bounds representation:

$$D_E(Y_t, \hat{Y}_t) = \sqrt{\left(y_t^L - \hat{y}_t^L\right)^2 + \left(y_t^U - \hat{y}_t^U\right)^2}. \qquad (27)$$

The normalized symmetric difference (NSD) of intervals is also considered [36]:

$$D_{\text{NSD}}(Y_t, \hat{Y}_t) = \frac{w(Y_t \cup \hat{Y}_t) - w(Y_t \cap \hat{Y}_t)}{w(Y_t \cup \hat{Y}_t)}, \qquad (28)$$

where $w(\cdot)$ indicates the width of the interval.

Thus the mean distance error of intervals using both Euclidian (MDE^E) and NSD (MDE^{NSD}) as distance function is computed. The advantage of NSD distance is that it is a normalized distance measure not influenced by data magnitude as the Euclidian distance.

The computation of descriptive statistics for ITS is also conducted as suggested by Rodrigues and Salish [36]. This chapter calculates the coverage rate

$$R^C = \frac{1}{T} \sum_{t=1}^{T} \frac{w(Y_t \cap \hat{Y}_t)}{w(Y_t)}, \qquad (29)$$

and the efficiency rate

$$R^E = \frac{1}{T} \sum_{t=1}^{T} \frac{w(Y_t \cap \hat{Y}_t)}{w(\hat{Y}_t)}. \qquad (30)$$

Both rated give additional information on what part of the observed ITS is covered by its forecasts (coverage) and what part of the forecast covers the observed ITS (efficiency), which have to be considered jointly. Better forecasts are identified when coverage and efficiency rates are reasonably high and the difference between them is small [36].

4.2 Empirical Results

In terms of financial time series, an interval forecast representing the minimum and maximum prices for the next period of a given stock can be used to estimate its volatility in that period. Moreover, this forecast provides valuable information to the investor and can play an important role in establishing investment strategy [3]. This work evaluates the suggested iFRB modeling framework using as empirical application the Brazilian stock market. Data comprise daily minimum and maximum values of IBOVESPA for the period from January 2000 to December 2015.[8] Minimum and maximum values of IBOVESPA are used as intervals lower and upper bounds representatives.

Forecasting methods are adjusted by dividing all series in two periods: training and validation. The training period, using data from January 2000 to December 2011, is used for the initialization and adjustment of the methods, while the validation period, or out-of-sample set, concerning all remaining data, is used to assess the performance of the adjusted method using new data. Minimizing the training error has been the criterion followed to estimate the parameters for all the methods, apart from the ARIMA and VECM models where the Schwarz information criterion [38] and the residual autocorrelation function have also been considered.

It is worth mentioning that, in the case of models such as ARIMA and VECM, it is tested whether the time series are stationary or not by means of the augmented Dickey–Fuller test (ADF) [13]. As for the bivariate approach, the Johansen test [25]

---

[8]Data were collected in Economatica.

Table 1 Descriptive statistics of lower and upper bounds of IBOVESPA ITS for the period from January 2000 to December 2015

Statistic	Lower bound	Upper bound
Mean	40,098.48	40,984.33
Maximum	45,401.00	73,920.00
Minimum	8224.00	8513.00
Std. dev.	19,493.96	19,836.37
Skewness	−0.1877	−0.1998
Kurtosis	1.5688	1.5649
Jarque-Bera	361.31	366.60
p-Value	1.16E−16	1.22E−16
ADF test	−0.1066	−0.1720
p-Value	0.5889	0.6129

is used to determine whether the lower and upper bound time series are cointegrated or not. If that is the case, it makes sense to forecast them using a VECM model [3].

Table 1 shows the descriptive statistics for IBOVESPA ITS lower and upper bounds. As expected, time series of IBOVESPA intervals lower and upper bounds are very similar, mainly in terms of mean and standard deviations. These series have heavy left-side tails as indicated by the negative skewness coefficients, and also lower values of kurtosis. The Jarque-Bera [4] statistics indicate that the series are non-normal with a 99% confidence level. Concerning the augmented Dickey–Fuller (ADF) [13] unit root test statistics, the lower and upper bounds time series are integrated.[9]

Models specifications were based on simulations using training data. According to the Schwarz information criterion and the residual autocorrelation function, ARIMA(2,1,3) and TAR(2) were selected for both lower and upper bounds of IBOVESPA ITS, whereas a VECM(2) was set for IBOVESPA ITS. The iMLP and the interval fuzzy rule-based model (iFRB) take the following formulation:

$$\hat{y}_t = f(y_{t-1}, y_{t-2}, \ldots, y_{t-p}). \tag{31}$$

The number of lags, p, was also selected based on simulations in order to reach the best performance in terms of RMSE and SMAPE. An iMLP was chosen with 5 neurons in the hidden layer and 2 lagged values of the series as input. Further, for iFRB experiments indicate $p = 1$, with control parameters: $\alpha = 0.07$, $\beta = 0.22$, $\tau = 0.15$, and $\lambda = 0.16$. The estimated iFRB model achieved three fuzzy rules.[10]

Table 2 shows the accuracy measures, RMSE and SMAPE, from all forecasting approaches for IBOVESPA ITS. Results in this chapter concern the out-of-sample data set, i.e., the period from January 2012 to December 2015. Forecasts are one step

---

[9] According to the Johansen test, the lower and upper bounds time series are cointegrated.

[10] iFRB control parameters depend on the data and could be selected based on simulations. An alternative to automatic parameters selection is, for instance, the use of smart grid search techniques.

Table 2 RMSE and SMAPE values based on one step ahead forecasts of lower and upper bounds IBOVESPA ITS using data in the period from January 2012 to December 2015

Models	RMSE		SMAPE	
	Lower bounds	Upper bounds	Lower bounds	Upper bounds
RW	655.78	669.84	0.00935	0.00931
ES	655.88	669.94	0.00935	0.00931
ARIMA	649.59	662.15	0.00922	0.00917
TAR	645.22	656.32	0.00916	0.00912
VECM	419.33	476.20	0.00781	0.00729
iMLP	342.91	367.12	0.00487	0.00548
iFRB	335.87	349.61	0.00465	0.00533

Table 3 Diebold–Mariano test statistics for lower and upper bounds IBOVESPA ITS forecasting using data in the period from January 2012 to December 2015

Method	ES	ARIMA	TAR	VECM	iMLP	iFRB
Lower bound						
RW	0.237	0.536	1.254	2.762*	3.414*	3.987*
ES	–	−0.761	0.672	2.554*	3.220*	3.888*
ARIMA	–	–	0.556	2.653*	−3.664*	4.140*
TAR	–	–	–	−2.543*	4.561*	3.987*
VECM	–	–	–	–	2.131*	2.456*
iMLP	–	–	–	–	–	0.116
Upper bound						
RW	0.265	−0.453	−1.377	3.315*	3.615*	4.315*
ES	–	0.543	0.515	−2.873*	−2.981*	−4.110*
ARIMA	–	–	−0.891	2.699*	3.220*	3.976*
TAR	–	–	–	2.981*	3.562*	3.561*
VECM	–	–	–	–	−3.009*	−3.253*
iMLP	–	–	–	–	–	−0.315

*Significant at 5% level

ahead. Lower RMSE and SMAPE values indicate better accuracy. Considering both RMSE and SMAPE metrics, the multivariate and interval approaches, i.e., VECM, iMLP, and iFRB models, performed better for all lower and upper bounds time series in comparison with the univariate econometric benchmarks. The suggested approach, iFRB, showed the lowest RMSE and SMAPE values in all cases (Table 2). Among the univariate methods, the threshold autoregressive achieved better accuracy, since it is able to model time series nonlinearities. It is interesting to remark that iMLP and iFRB improved the results in comparison with the VECM method, since they account for the interdependence of interval bounds as VECM does but they are also designed to process interval-valued data naturally.

In addition to goodness of fit, as mirrored by forecast error, the models were evaluated statistically. The Diebold–Mariano [14] test statistics for upper and lower bounds IBOVESPA ITS are summarized in Table 3. The test is performed for each pair of models. The null hypothesis of equal predictive accuracy is rejected with 5%

confidence level, that is if |DM| > 1.96. From this point of view, for both lower and upper bounds of IBOVESPA ITS, concerning the univariate methods (RW, ES, ARIMA, and TAR), they can be considered equally accurate (|DM| < 1.96), but they provide statistically inferior forecasts when compared against VECM, iMLP, and iFRB. Further, the interval approaches, iMPL and iFRB, are equally accurate in statistical terms but achieved statistically superior performance than the VECM model for both lower and upper bounds time series. Nonetheless, in order to evaluate the better interval representation it is necessary to consider the interval representatives forecasts jointly.

In order to access the interval structure of the time series, IBOVESPA ITS forecasts were compared in terms of the mean distance error (MDE) of intervals using both the Euclidian distance (MDE^E) and the normalized symmetric difference (NSD) distance of intervals (MDE^{NSD}). Table 4 reports the results from MDE^E and MDE^{NSD} accuracy metrics of the actual and forecasted IBOVESPA ITS. Again, the lowest the MDE values, the highest the models accuracy. The better results are from the VECM, iMLP, and iFRB, since they are able to capture the interdependence between IBOVESPA ITS bounds. Among the benchmark techniques, TAR method showed slightly lower interval error values. iFRB modeling approach reports the lowest MDE values, using both Euclidian and normalized symmetric difference distance metrics (Table 4). It worth to note that the interval techniques, iMLP and iFRB, again provided the better results.

Additionally, intervals descriptive statistics, coverage (R^C) and efficiency (R^E) rates, are reported in Table 5. They provide additional information about the adequacy of the forecasts. In this case, these statistics reveal the percentage of the actual ITS is covered by its forecast (coverage), and what part of the forecast covers

Table 4 Mean distance error (MDE) values using Euclidian distance, MDE^E, and the normalized symmetric difference (NSD), MDE^{NSD}, based on one step ahead forecasts of IBOVESPA ITS using data in the period from January 2012 to December 2015

Models	MDE^E	MDE^{NSD}
RW	765.81	0.5901
ES	765.91	0.5902
ARIMA	761.30	0.5899
TAR	757.83	0.5482
VECM	354.12	0.3887
iMLP	307.60	0.3211
iFRB	298.51	0.3004

Table 5 Coverage (R^C) and efficiency (R^E) rates values based on one step ahead forecasts of IBOVESPA ITS using data in the period from January 2012 to December 2015

Models	R^C	R^E
RW	0.5747	0.5565
ES	0.5746	0.5564
ARIMA	0.5809	0.5654
TAR	0.6173	0.5783
VECM	0.7001	0.6879
iMLP	0.7895	0.7763
iFRB	0.8356	0.8211

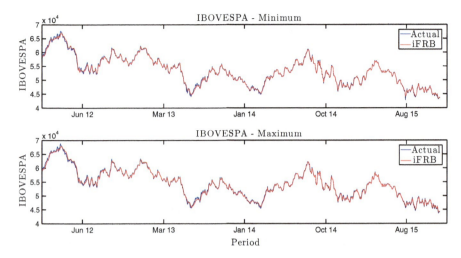

Fig. 2 IBOVESPA minimum and maximum actual and predicted values by iFRB for the period from January 2012 to December 2015

the realized ITS (efficiency). To evaluate the models, these rates must be considered jointly and the higher their values the better is the forecast. Further, the closeness of the results of these two statistics can be taken as an indicator of the quality of the forecasts.

For both coverage and efficiency rates, iFRB model showed the highest values, indicating more accurate predictions (Table 5). VECM and iMLP also provided similar results. Notice that these two statistics values are very close, which corroborate the adequacy of the models. Again, the econometric forecasting methods performed worst.

Summing up, the empirical results in this chapter indicate the adequacy of iFRB for IBOVESPA interval time series forecasting. According to traditional and interval quality measures, in general, the interval approaches iMLP and iFRB showed better performance, even when compared against the multivariate VECM method. Further, iFRB also improves the results from iMLP, since besides the neural network method being a nonlinear and interval technique, iFRB also considers the uncertainty inherited in data due to its fuzzy nature. Figure 2 illustrates the actual and forecasted values by the iFRB model of IBOVESPA lower (minimum) and upper (maximum) bounds for the testing data, covering the period from January 2012 to December 2015. Note that the high capability of the model to deal with nonlinear and time-varying dynamics. In line with [36], the results indicate that the contribution of nonlinear models to a good forecast performance is significant, also for the Brazilian equity market, and moreover, an accuracy improvement is obtained when interval-valued based methods are considered.

It is worth to remark that the iFRB model is an interpretable modeling framework in linguistic terms, which may be employed to improve decision-making process by labeling fuzzy rules in order to give some insight about the system or even about trading rules, for example. Using the IBOVESPA ITS training data the model found a rule base with three rules. The rule base is described as follows:

$$R^1 : \text{IF } y_t \text{ is } v^1 = [0.2314, 0.3452] \text{ THEN } y^1_{t+1} = 0.0232 - 0.2534 y_t$$

$$R^2 : \text{IF } y_t \text{ is } v^2 = [0.4652, 0.5517] \text{ THEN } y^2_{t+1} = 0.0242 + 0.0175 y_t$$

$$R^3 : \text{IF } y_t \text{ is } v^3 = [0.4982, 0.6620] \text{ THEN } y^3_{t+1} = 0.0315 - 0.2519 y_t$$

Therefore, for example, antecedents membership functions, $v^i = [v^L, v^U]$, $i = 1, 2, 3$, may be interpreted as "low," "medium," and "high" prices, for example, improving the model interpretability.

5 Conclusion

Interval time series (ITS) are time series where each period in time is described by an interval. In finance, ITS can be described as the evolution of the maximum and minimum prices of an asset throughout time. These price ranges are related to the concept of volatility. Hence, their accurate forecasts play a key role in risk management, derivatives pricing, and asset allocation, as well as supplement the information extracted by the time series of the closing price values for investors to place their sell and buy orders.

This work suggests a fuzzy rule-based model for interval-valued data (iFRB). The identification of iPFM concerns the identification of the antecedents and consequents of the fuzzy rules. Rules antecedents are identified using an interval fuzzy clustering approach with adaptive Hausdorff distances. Rules consequent parameters are estimated using traditional least squares techniques taking advantage of the intervals midpoints. As empirical application, iFRB is applied for financial ITS forecasting, in which the intervals are constructed by minimum and maximum stock prices, using the main stock index of the Brazilian financial market, the IBOVESPA, for the period from January 2000 to December 2015. iFRB is compared against univariate time series methods such as random walk, exponential smoothing, ARIMA, and threshold autoregressive (TAR), and with multivariate approaches such as VECM and the interval neural network (iMLP). In addition to the use of accuracy measures and statistical tests to compute models performance, this work evaluated the results using quality measures designed for the interval time series framework, as well as in terms of additional forecast descriptive statistics such as efficiency and coverage rates.

The results evidence the predictability of IBOVESPA ITS by the iFRB method in comparison with the alternative methods. A significant forecast contribution of interval approaches, iMLP and iFRB, is achieved. The interval quality measures also suggested that methodologies considering the interdependence of interval limits provide significant improvements in forecasting accuracy. iFRB appears as a potential tool for interval time series forecasting since the method is able to process interval-valued data naturally, capture data nonlinearities, take into account the uncertainty inherited to the data by its fuzzy nature, and also do provide an interpretable model in linguistic terms. Future works shall include the use of other clusters features to perform the construction of the cluster structure, the use of an interval technique to compute iFRB consequents, the evaluation of iFRB for different markets and economies as well as its application in trading strategies and in risk management by using range-based volatility estimators.

Acknowledgements The authors thank the Brazilian Ministry of Education (CAPES) and the São Paulo Research Foundation (FAPESP) for their support.

References

1. Alizadeh, S., Brandt, M.W., Diebold, F.X.: Range-based estimation of stochastic volatility. J. Financ. **57**(3), 1047–1091 (2002)
2. Angelov, P., Filev, D.: An approach to online identification of Takagi-Sugeno fuzzy models. IEEE Tran. Syst. Man Cybern. B **34**(1), 484–498 (2004)
3. Arroyo, J., Espínola, R., Maté, C.: Different approaches to forecast interval time series: a comparison in finance. Comput. Econ. **27**(2), 169–191 (2011)
4. Bera, A., Jarque, C.: Efficient tests for normality, homoscedasticity and serial independence of regression residuals: Monte Carlo evidence. Econ. Lett. **7**, 313–318 (1981)
5. Bisdorff, R.: Logical foundation of fuzzy preferential systems with application to the electre decision aid methods. Comput. Oper. Res. **27**(7–8), 673–687 (2000)
6. Brandt, M.W., Diebold, F.X.: A no-arbitrage approach to range-based estimation of return covariances and correlations. J. Bus. **79**(1), 61–74 (2006)
7. Carvalho, F.A.T.: Fuzzy c-means clustering methods for symbolic interval data. Pattern Recogn. Lett. **28**(4), 423–437 (2007)
8. Carvalho, F.A.T., Souza, R.M.C.R., Chavent, M., Lechevallier, Y.: Adaptive Hausdorff distances and dynamic clustering of symbolic interval data. Patter Recogn. Lett. **27**(3), 167–179 (2006)
9. Chang, P., Wu, J., Lin, J.: A Takagi–Sugeno fuzzy model combined with a support vector regression for stock trading forecasting. Appl. Soft Comput. **38**, 831–842 (2016)
10. Cheung, Y.W.: An empirical model of daily highs and lows. Int. J. Financ. Econ. **12**(1), 1–20 (2007)
11. Chou, R.Y.: Forecasting financial volatilities with extreme values: the conditional autoregressive range (CARR) model. J. Money Credit Bank. **37**(3), 561–582 (2005)
12. Degiannakis, S., Floros, C.: Modeling CAC40 volatility using ultra-high frequency data. Res. Int. Bus. Financ. **28**, 68–81 (2013)
13. Dickey, D.A., Fuller, W.A.: Distribution of the estimators for autoregressive time series with a unit root. J. Am. Stat. Assoc. **74**(266), 427–431 (1979)
14. Diebold, F.X., Mariano, R.S.: Comparing predictive accuracy. J. Bus. Econ. Stat. **13**(3), 253–265 (1995)
15. Dueker, M.J., Sola, M., Spangnolo, F.: Contemporaneous threshold autoregressive models: estimation, testing and forecasting. J. Econ. **141**(2), 517–547 (2007)
16. Engle, R.F., Russel, J.: Analysis of high frequency data. In: Aït-Sahalia, Y., Hansen, L.P. (eds.) Handbook of Financial Econometrics, vol. 1: Tools and Techniques, pp. 383–346. Elsevier, Amsterdam (2009)

17. Fiess, N.M., MacDonald, R.: Towards the fundamentals of technical analysis: analysing the information content of high, low and close prices. Econ. Model. **19**(3), 353–374 (2002)
18. Froelich, W., Salmeron, J.L.: Evolutionary learning of fuzzy grey cognitive maps for the forecasting of multivariate, interval-valued time series. Int. J. Approx. Reason. **55**(6), 1319–1335 (2014)
19. García-Ascanio, C., Maté, C.: Electric power demand forecasting using interval time series: a comparison between VAR and iMLP. Energy Policy **38**(2), 715–725 (2010)
20. Guidolin, M., Hyde, S., McMillan, D., Ono, S.: Non-linear predictability in stock and bond returns: when and where is it exploitable? Int. J. Forecast. **25**(2), 373–399 (2009)
21. He, L.T., Hu, C.: Impacts of interval measurement on studies of economic variability: evidence from stock market variability forecasting. J. Risk Financ. **8**(5), 489–507 (2008)
22. He, L.T., Hu, C.: Impacts of interval computing on stock market variability forecasting. Comput. Econ. **33**(3), 263–276 (2009)
23. Henry, Ó., Olekaln, N., Summers, P.M.: Exchange rate instability: a threshold autoregressive approach. Econ. Rec. **77**(237), 160–166 (2001)
24. Hu, C., He, L.T.: An application of interval methods to stock marketing forecasting. Reliab. Comput. **13**(5), 423–434 (2007)
25. Johansen, S.: Estimation and hypothesis testing of cointegration vectors in Gaussian vector autoregressive models. Econometrica **59**(6), 1551–1580 (1991)
26. Leite, D., Costa, P., Gomide, F.: Interval approach for evolving granular system modeling, pp. 271–300. Springer, New York (2012)
27. Ljung, L.: System identification: theory for the user. Prentice-Hall, Upper Saddle River (1988)
28. Lu, W., Chen, X., Pedrycz, W., Liu, X., Yang, J.: Using interval information granules to improve forecasting in fuzzy time series. Int. J. Approx. Reason. **57**, 1–18 (2015)
29. Maciel, L., Ballini, R., Gomide, F.: Evolving granular analytics for interval time series forecasting. Granul. Comput. **1**(4), 213–224 (2016)
30. Maciel, L., Ballini, R., Gomide, F., Yager, R.R.: Participatory learning fuzzy clustering for interval-valued data. In: Proceedings of the 16th International Conference on Information Processing and Management of Uncertainty in Knowledge-Based Systems (IPMU 2016), Eindhoven, pp. 1–8 (2016)
31. Maia, A.L.S., de Carvalho, F.A.T.: Holt's exponential smoothing and neural network models for forecasting interval-valued time series. Int. J. Forecast. **27**(3), 740–759 (2011)
32. Miwa, K.: Investor sentiment, stock mispricing, and long-term growth expectations. Res. Int. Bus. Financ. **36**, 414–423 (2016)
33. Moore, R.E., Kearfott, R.B., Cloud, M.J.: Introduction to interval analysis. SIAM Press, Philadelphia (2009)
34. Parkinson, M.: The extreme value method for estimating the variance of the rate of return. J. Bus. **53**(1), 61–65 (1980)
35. Pettenuzzo, D., Timmermann, A., Valkanov, R.: Forecasting stock returns under economic constraints. J. Financ. Econ. **144**(3), 517–553 (2014)
36. Rodrigues, P.M.M., Salish, N.: Modeling and forecasting interval time series with threshold models. Adv. Data Anal. Classif. **9**(1), 41–57 (2015)
37. Roque, A.M., Maté, C., Arroyo, J., Sarabia, A.: iMLP: applying multi-layer perceptrons to interval-valued data. Neural Process. Lett. **25**(2), 157–169 (2007)
38. Schwarz, G.: Estimating the dimension of model. Ann. Stat. **6**(2), 461–464 (1978)
39. Setnes, M., Babuska, R., Verbruggen, H.B.: Rule-based modelling: precision and transparency. IEEE Trans. Syst. Man Cybern. C **1**, 165–169 (1998)
40. Shu, J.H., Zhang, J.E.: Testing range estimators of historical volatility. J. Futur. Mark. **26**(3), 297–313 (2006)
41. Silva, L., Gomide, F., Yager, R.: Participatory learning in fuzzy clustering. In: IEEE International Conference on Fuzzy Systems, Reno, NV, pp. 857–861 (2005)
42. Takagi, T., Sugeno, M.: Fuzzy identification of systems and its applications to modeling and control. IEEE Trans. Syst. Man Cybern. **15**(1), 116–132 (1985)

43. Ustun, O., Kasimbeyli, R.: Combined forecasts in portfolio optimization: a generalized approach. Comput. Oper. Res. **39**(4), 805–819 (2012)
44. Wang, L., Liu, X., Pedrycz, W.: Effective intervals determined by information granules to improve forecasting in fuzzy time series. Expert Syst. Appl. **40**(14), 5673–5679 (2013)
45. Weron, R.: Electricity price forecasting: a review of the state-of-the-art with a look into the future. Int. J. Forecast. **30**(4), 1030–1081 (2014)
46. Xiong, T., Bao, Y., Hu, Z.: Multiple-output support vector regression with a firefly algorithm for interval-valued stock price index forecasting. Knowl. Based Syst. **55**, 87–100 (2014)
47. Xiong, T., Bao, Y., Hu, Z., Chiong, R.: Forecasting interval time series using a fully complex-valued RBF neural network with DPSO and PSO algorithms. Inform. Sci. **305**, 77–92 (2015)
48. Xiong, T., Li, C., Bao, Y., Hu, Z., Zhang, L.: A combination method for interval forecasting of agricultural commodity futures prices. Knowl. Based Syst. **77**, 92–102 (2015)
49. Yager, R.: A model of participatory learning. IEEE Trans. Syst. Man Cybern. **20**(5), 1229–1234 (1990)
50. Yang, W., Han, A., Wang, S.: Forecasting financial volatility with interval-valued time series data. In: Vulnerability, Uncertainty, and Risk, pp. 1124–1233 (2014)
51. Zadeh, L.A.: Fuzzy sets. Inf. Control **8**, 338–353 (1965)
52. Zhao, J., Wang, L.: Pricing and retail service decisions in fuzzy uncertainty environments. Appl. Math. Comput. **250**, 580–592 (2015)

Part III
Diagnosis, Optimization and Control

Reasoning from First Principles for Self-adaptive and Autonomous Systems

Franz Wotawa

1 Introduction

Systems that are able to automatically adapt their behavior in cases of faults have always been very much appealing for research and practice. Because of the increasing importance of applications like autonomous vehicles, the internet of things (IoT), or industry 4.0 dealing with increased autonomy, self-adaptation increases importance as well. For example, consider a truly autonomous vehicle transporting passengers from one location to another. If there is a system fault occurring during operation, there is no human driver working as fallback mechanism for assuring that the vehicle goes to a safe state, e.g., driving to an emergency lane of a highway and stopping there. In case of autonomous driving the system itself is responsible for any action after detecting a failure. This is one of the most significant differences to ordinary cars even if they have implemented automated functions like lane assist.

It is also worth noting that in many situations coming to a safe state is not that simple even for today's cars. For example, an emergency break as consequence of a fault in the control system of the car's engine might cause an accident if this is done on a high way with another car behind. Or another example is stopping a car in a tunnel without an emergency lane. Therefore, a car that is not moving is not necessarily in a safe state. As a consequence faults during operation should be handled in a smart way either via compensating or repairing faults during operation requiring self-adaptive systems. Of course it is worth mentioning that such a self-adaptive behavior does not allow to compromise safety. According to the IEEE

F. Wotawa (✉)
Technische Universität Graz, Institute for Software Technology, Graz, Austria
e-mail: wotawa@ist.tugraz.at

Systems and Software Engineering Vocabulary [29] fail-safe refers to a system or component that automatically places itself in a safe operating mode in the event of a failure.

Self-adaptive systems that assure fail-safe behavior in the context of autonomous vehicles are also often referred to fail-operational system, i.e., system that still remain operational even in case of faults. In order to implement fail-operational behavior we have to use methods that assure also the behavior to be fail-safe. We need methods that can be proven to work as expected never compromising safety. We therefore discuss methods relying on model-based reasoning in this chapter, because such methods guarantee to deliver all results that fulfill all properties specified in models. What remains is to prove that the models capture the important parts of the system and also its properties like safety.

The idea of using models for various purposes like diagnosis is not new and dates back to the early 1980s of the last century (see, for example, [11]). Model-based reasoning is characterized of using models directly to implement certain tasks without requiring to reformulate available knowledge. In case of diagnosis, the model is used to derive the expected behavior that can be compared with observations. If we see a deviation between the expected behavior and the observed one, model-based reasoning utilizes the model to identify the root cause of the detected misbehavior. The basic principles behind model-based reasoning are still very suitable for today's challenges like autonomous systems and driving. If a model is appropriately capturing its corresponding system, then all conclusions drawn from the model are reasonable and also appropriate. Therefore, validation and verification can focus on the model once the reasoning algorithms are tested.

In this chapter we discuss the basic principles of model-based reasoning including algorithms. We do not only focus on one available technique that makes use of models formalizing the correct behavior of components, but also abductive diagnosis where we use fault models for obtaining root causes in a similar way than medical doctors do when reasoning from observed symptoms back to hypotheses. In addition, we outline an approach for online repair of systems interacting with their environment using sensors and actuators. We discuss how to integrate diagnosis with repair and also the different types of repair. For the latter, we make use of a running example from the autonomous robotics domain. It is worth noting that the purpose of the chapter is mainly to give an overview of model-based reasoning for self-adaptive systems. Therefore, we also discuss related research and previous work that has been published.

This chapter is organized as follows: We first introduce the application domain of autonomous mobile robots in Sect. 2. Afterwards, in Sect. 3 we introduce the basic concepts of model-based reasoning including model-based diagnosis and abductive diagnosis. In Sect. 4 we discuss issues of modeling for model-based reasoning, followed by Sect. 5 where we introduce the basic architectures behind a self-adaptive system. There we make use of three examples outlining the different repair actions necessary to bring the system back into an operational state. Finally, we discuss related literature in Sect. 6 and conclude the chapter in Sect. 7.

2 Example

In this section, we introduce the example of a mobile robot having a differential drive for moving from one point in a plain to another. We will use this example in our chapter for introducing the basic concepts behind model-based reasoning and in an extended form for showing how self-adaptive behavior can be implemented using models of the system directly.

A differential drive comprises two wheels with varying speed. Depending on the speed of the wheels the robot either rotates, moves on a straight line, or on a curve. In the following, we discuss a kinematics model of a mobile robot with a differential drive. For more details we refer the interested reader to [17]. In Fig. 1 we show the underlying ingredients. We assume that the robot is at its current position (x_R, y_R) heading in a direction specified by the angle θ from the x-axis. We further assume that the distance between the wheels is d. Depending on the speed of the right or left wheel v_R, v_L, respectively, the robot rotates about its instantaneous center of curvature (ICC) with a rotational speed ω. The ICC lies on a straight line between the axis of the wheels and its distance from the center of the robot (lying on the same line) is R.

Obviously, there must be a relationship between ω and the speed of the wheels because both wheels are on the same line connected with ICC and thus have to have the same rotational speed. We are able to formalize this relationship as follows:

$$\begin{aligned} \omega(R + d/2) &= v_R \\ \omega(R - d/2) &= v_L \end{aligned} \tag{1}$$

Fig. 1 Mobile robot with differential drive

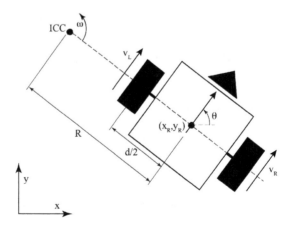

From Eq. (1) we are able to obtain R and ω if knowing v_R and v_L.

$$R = \frac{d}{2} \frac{v_R + v_L}{v_R - v_L}; \quad \omega = \frac{v_R - v_L}{d} \tag{2}$$

From Eq. (2) we can distinguish 3 corner cases of movement for a differential drive robot, which are usually used as available actions when planning a route for the robot from its current position to its finally expected position:

1. If $v_R = v_L > 0$, then ω becomes zero, and R infinite. Hence, the robot is moving on a straight line.
2. If $v_R = -v_L$, then R becomes zero, and we obtain a rotation around the center of the robot. The direction of the rotation in this case is clockwise (assuming $v_L > 0$) and counter clockwise, otherwise.
3. If $v_L = 0$ and $v_R > 0$ ($v_L > 0$ and $v_R = 0$), then $R = \frac{d}{2}$ ($R = -\frac{d}{2}$) and the robot rotates counter clockwise on its left wheel (clockwise on its right wheel).

Based on the above equations, we are also able to come up with a forward kinematics of the differential drive robot. In this case we assume that the robot is at a specific position (x_R, y_R) and direction with angle θ. The speed v_R and v_L are the control parameters to bring the robot to a new position. Using Eq. (2) we first obtain the ICC location:

$$\text{ICC} = (x - R\sin(\theta), y + R\cos(\theta)) \tag{3}$$

Assuming the robot is at its location at time t we are now able to state its new position at time $t + \Delta t$ where ICC_x and ICC_y references are the ICC location of the x- and y-axis, respectively:

$$\begin{pmatrix} x'_R \\ y'_R \\ \theta' \end{pmatrix} = \begin{pmatrix} \cos(\omega \Delta t) & -\sin(\omega \Delta t) & 0 \\ \sin(\omega \Delta t) & \cos(\omega \Delta t) & 0 \\ 0 & 0 & 1 \end{pmatrix} \begin{pmatrix} x_R - \text{ICC}_x \\ y_R - \text{ICC}_y \\ \theta \end{pmatrix} + \begin{pmatrix} \text{ICC}_x \\ \text{ICC}_y \\ \omega \Delta t \end{pmatrix} \tag{4}$$

Using Eq. (4) we are able to predict the movements of a robot with a differential drive over time providing that we know the speed of the wheels v_R and v_L. In control engineering someone would also be interested to compute values for v_R and v_L in order to reach a certain goal location. Searching for such values is also known as inverse kinematics problem. In case of a differential drive we are not able to compute such velocities. Instead what we can do is to separate this problem. We are able to move on a straight line and we are also able to rotate the robot on its current place. Hence, the problem of reaching an arbitrary location can be solved rotating the robot such that there is only a straight movement necessary to reach the goal, and afterwards move forward.

Without any doubt the given equations provide a model that explains the kinematics of a differential drive robot providing that the relevant information like the wheel speeds, the robot's dimensions, and the current location of a robot together with its direction is known. However, in the following, we will not use those equations directly. Instead we will focus on an abstraction of the behavior for diagnosis and also for implementing self-healing behavior. The abstraction we are going to use can be easily obtained from the cases we distinguished for the robot's movement providing the wheel speeds. For example, we might consider a finite number of values for speed, i.e., either the speed is 0 or the positive or negative nominal value. In this case the domain would be $\{v_n^-, 0, v_n^+\}$. Using this domain we are able to formalize the ordinary behavior of a differential drive robot using first order logic (FOL) as follows:

$$val(v_L, v_n^+) \wedge val(v_R, v_n^+) \rightarrow motion(straightline)$$

$$val(v_L, v_n^+) \wedge val(v_R, v_n^-) \rightarrow motion(rotate clockwise)$$

$$val(v_L, v_n^-) \wedge val(v_R, v_n^+) \rightarrow motion(rotate counter clockwise)$$

$$val(v_L, 0) \wedge val(v_R, 0) \rightarrow motion(stop)$$

In the rules we use the predicate $val/2$ stating that a particular speed given as first argument has the value given in the second argument. The predicate $motion/1$ is for establishing that a certain motion pattern is valid, i.e., either following the straight line or rotating clockwise or counter clockwise. The first rule formalizes the case where a robot moves on a straight line. The second is for indicating the case of clockwise rotation. The third one is for rotating counter clockwise, and the last one specifies the case where the robot stops moving. Note that this formalization does not comprise the case where the speed of one wheel is set to a value unequal 0, and the one of the other wheel is set to 0. Such a setting would also lead to a rotation and can be easily added to the abstract model if required.

The control problem of reaching a certain location can be represented using abstract values for the speed of the wheels. The following sequence assures that the robot first rotates and afterwards moves straightforward.

$$(v_L = v_n^+, v_R = v_n^-)_0, (v_L = v_n^+, v_R = v_n^+)_1$$

In this representation, we use $(\ldots)_i$ to indicate given values to be used at time i. Note that we do not consider a specific time. Instead each element indicates a state occurring at a particular point in time and lasting for a certain period. Hence, the presentation abstracts not only the values but also time. In the following section we show how such abstract models can be used for identifying the cause of a detected misbehavior. For example, we will consider the case of a differential drive robot that follows a wrong trajectory. In Fig. 2 we depict such a case, where a robot is expected to rotate clockwise followed by moving on a straight line but follows a curve. Obviously in this case either the speed of the left wheel is too low or the one of the right wheel too high.

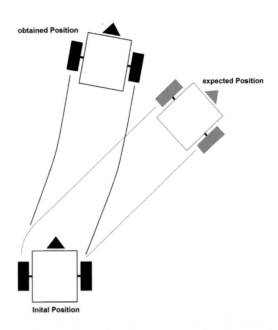

Fig. 2 A small mobile robot driving the wrong trajectory

3 Model-Based Reasoning

The underlying idea behind model-based reasoning is to use a model of a system directly to reason about the system. In one instance, i.e., model-based diagnosis, the system's model is used for identifying root causes in case of an observed behavior that contradicts the expected one. A model in model-based diagnosis (MBD) comprises the system's structure including its components and interconnections, as well as the component models. The health state of components, i.e., a predicate indicating whether a component is working as expected or not, is used to indicate a root cause. In this terminology an incorrectly working component maybe an explanation for the detected unexpected deviation in the observed behavior. Davis [11] was one of the first outlying basic principles behind MBD that Reiter [46] and De Kleer and Williams [13] further formalized and extended.

It is worth noting that in classical MBD the component models only describe the correct behavior. This makes the theory general applicable even in cases where there is no knowledge about faults and their consequences available. De Kleer et al. [14] later presented an extension incorporating models of faulty behavior into the theory including some theoretical consequences. For example, in MBD without fault models every superset of a diagnosis itself is a diagnosis, which is not the case when using fault models. Note that there is a close relationship between MBD and other diagnosis theories like abductive diagnosis [21]. In abductive diagnosis, symptoms are explained based on hypotheses, which—more or less—represent known faulty behavior. Console and Torasso [7] showed how to integrate also correct behavior into abductive reasoning, and later Console et al. [8] showed that abductive diagnosis is MBD using models of faulty behavior.

In this section, we recall the basic foundations behind MBD and also abductive diagnosis. For this purpose, we make use of the differential drive robot as a running example. In contrast to the mathematical model of the kinematics outlined in Sect. 2, we will use an abstract representation, which we initially discussed in the same section. Because of the fact that diagnosis relies on systems comprising interconnected components, we first start with such a component-oriented model for a differential drive robot. In Fig. 3 we depict such a robot where each wheel has a wheel encoder attached and is connected to an electric motor that drives the wheel. The wheel encoder is for giving feedback to a controller that supplies the motors of the left and the right wheel with their expected voltage level.

In Fig. 4 we summarize the component-oriented representation of the differential drive robot. The control component C is connected to motor M_L and M_R. The motors are connected to their corresponding wheels W_L and W_R, respectively. With attached wheel encoders E_L and E_R the rotational speed of the wheels is given back to the controller C. Hence, the motors work as actuators whereas the wheel encoders as sensors.

The behavior of each component can be specified again in an abstract way using the value domain $\{v_n^-, 0, v_n^+\}$ where distinguish a negative nominal value, zero, and

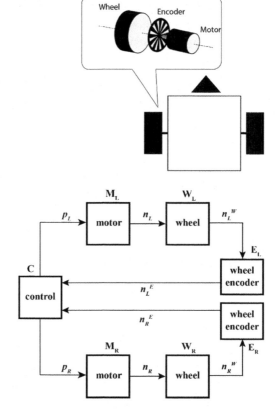

Fig. 3 A small mobile robot comprising a differential drive with two motors and two wheel encoders for obtaining the wheels' rotation

Fig. 4 The component-oriented model of the differential drive robot

a positive nominal value, respectively. For example, if the input to the motor is a positive nominal value, then its output, i.e., its rotational speed, is also a positive nominal value. We can similarly define the behavior of a wheel and the wheel encoders. We will further formalize the components' behavior when introducing MBD and afterwards abductive diagnosis in the following subsections.

3.1 Model-Based Diagnosis

In this subsection we outline the basic definitions of MBD from Reiter [46] in slightly adapted form. According to Reiter, MBD allows to reason directly from models and is therefore also called reasoning from first principles. A system model itself comprises components, their interconnections, and the components' behavior. All components of the system that might cause a misbehavior are assumed to be element of a set $COMP$. The structure and behavior has to be specified in a formal form in SD. Formally, a system (model) according to Reiter is defined as follows:

Definition 1 (Diagnosis System) A pair $(SD, COMP)$ is a diagnosis system providing that SD is a system description comprising a model of the system, and $COMP$ a set of system components.

Using Definition 1 we are able to represent the differential drive robot as diagnosis system as follows: We start with the components. In the representation we only take care of components, which we want to classify as faulty or correct. Hence, in this example, we only consider the motors and the wheel encoders to be faulty, ignoring the health state of the control component and the wheels, so that:

$$COMP_R = \{M_L, M_R, E_L, E_R\}$$

Despite this design decision, we have to formulate a model of the wheels as well. Basically, we have to state that if there is a rotation applied at the axis, it is also provided to the wheel encoder. Using FOL, we are able to express this behavior as follows:

$$\forall X : wheel(X) \rightarrow (\forall Y : dom_A(Y) \rightarrow (val(in(X), Y) \leftrightarrow val(out(X), Y)))$$

In the above rule we make use of a predicate $value/2$ to assign a value to a port of a component, where we assume that a wheel has only one input and one output. We further restrict the values using the predicate $dom_A/1$ to three values as follows representing the abstract value domain introduced in Sect. 2.

$$dom_A(v_n^-) \wedge dom_A(0) \wedge dom_A(v_n^+)$$

Note that in this domain we only consider nominal speed and speed 0. If needed in an application scenario, we might either use more abstract values or to use even models based on the continuous domain. In the former case, it is important to consider all abstract values that allow us to distinguish the different behavior of a system we want to diagnose. In the latter case, we might have to use a different underlying reasoning system for diagnosis.

In addition to the domain description, we add the information that we have two wheels to the model stating:

$$wheel(W_L) \land wheel(W_R).$$

The models for the motors and the encoders can be similarly formalized. In both cases there is only one input and one output. If the input is zero, then the output has also to be zero. If it is a nominal value, the value is propagated to the output. However, in contrast to the wheel, we now have the situation that a component might fail. In this case we do not know its behavior. In order to distinguish the health state of a component, we use a new predicate $Ab/1$ for each element of $COMP_R$ that if true, states that the component is faulty. For MBD we only specify the correct behavior requiring to formalize a rule in case $\neg Ab$ is true. For the motor and the encoder, we use the following rules for this purpose:

$$\forall X : motor(X) \rightarrow (\forall Y : dom_A(Y) \rightarrow (\neg Ab(X)$$
$$\rightarrow (val(in(X), Y) \leftrightarrow val(out(X), Y))))$$
$$\forall X : enc(X) \rightarrow (\forall Y : dom_A(Y) \rightarrow (\neg Ab(X)$$
$$\rightarrow (val(in(X), Y) \leftrightarrow val(out(X), Y))))$$

Again, we also represent the structure of the system formally:

$$motor(M_L) \land motor(M_R) \land enc(E_L) \land enc(E_R)$$

What is missing to finalize the FOL model, is a representation of the structure. This can be easily done, connecting the ports of the components:

$$val(out(M_L), X) \leftrightarrow val(in(W_L), X) \land val(out(W_L), X) \leftrightarrow val(in(E_L), X)$$
$$val(out(M_R), X) \leftrightarrow val(in(W_R), X) \land val(out(W_R), X) \leftrightarrow val(in(E_R), X)$$

Although we do not model the control component directly, we add some rules stating what should be the case for given motion patterns like stop, going straight, or rotating. The following rules formalize the motion patterns. There we only state the expected values of the different ports for the motors so that they would provide the expected motion pattern, which should also be visible at the outputs of the encoders.

$$stop \rightarrow (val(in(M_L), 0) \land val(in(M_R, 0)))$$
$$straight \rightarrow (val(in(M_L), v_n^+) \land val(in(M_R, v_n^+)))$$
$$rotclk \rightarrow (val(in(M_L), v_n^+) \land val(in(M_R, v_n^-)))$$
$$rotinvclk \rightarrow (val(in(M_L), v_n^-) \land val(in(M_R, v_n^+)))$$

The set of all the discussed FOL rules constitute the system description for the robot example SD_R.

After formalizing a diagnosis system, we have to state a diagnosis problem. Diagnosis is necessary if an observed behavior is in contradiction with the expected behavior, which we derive from the system description, i.e., the model. Hence, a diagnosis problem has two ingredients: (1) the diagnosis system and (2) a set of observations stating values for connections between components or equally good the ports of components. Formally, a diagnosis problem is defined as follows:

Definition 2 (Diagnosis Problem) A tuple $(SD, COMP, OBS)$, where $(SD, COMP)$ is a diagnosis system and OBS a set of observations, is a diagnosis problem.

For our robot example, we can easily state a diagnosis problem. For example, let us assume the faulty behavior we depict in Fig. 2. In case of the expected straight-line movement, we see a curve going to the left. Let us further assume that the encoder of the left wheel does not give us back the nominal value but the one of the right wheel does. Note that this assumption explains the actual behavior assuming that the left motor has less number of revolutions than expected. For this diagnosis problem, we can easily state the observations OBS_R:

$$OBS_R = \{straight, \neg val(out(E_L), v_n^+), val(out(E_R), v_n^+)\}$$

Note that predicates or rules given in a set are assumed to be true. Thus all elements of such sets can be considered as being connected using logic conjunctions, i.e., $\wedge$. Given a diagnosis problem, we are now interested in finding diagnoses. We first define a diagnosis formally.

Definition 3 (Diagnosis) Given a diagnosis problem $(SD, COMP, OBS)$. A set $\Delta \subseteq COMP$ is a diagnosis if and only if $SD \cup OBS \cup \{Ab(C)|C \in \Delta\} \cup \{\neg Ab(C)|C \in COMP \setminus \Delta\}$ is satisfiable.

In this definition of diagnosis, we are searching for an assignment of health states to all components, which eliminates all contradictions with the given observations. For example, $(SD_R, COMP_R, OBS_R)$ would lead to a contradicting logical sentence when assuming all components to be correct, i.e., setting their corresponding negated predicate $\neg Ab$ to true, because in this case we would expect $val(out(E_L), v_n^+)$ to be true, which contradicts OBS_R. When setting $Ab(M_L)$ to true and all other components to be working as expected, we are able to eliminate the contradiction. Hence, $\{M_L\}$ is a diagnosis. Unfortunately, assuming all components to be faulty would also be a diagnosis according to Definition 3. Therefore, we need a stronger definition of diagnosis, which focuses on parsimonious explanations.

Definition 4 (Minimal Diagnosis) Given a diagnosis problem $(SD, COMP, OBS)$. A diagnosis Δ for $(SD, COMP, OBS)$ is a minimal diagnosis if and only if there exists no diagnosis Δ' that is a subset of Δ.

The definition of minimal diagnosis assures that only the smallest diagnoses in terms of the subset relation are considered. In case of our robot example, we obtain two minimal diagnoses $\{M_L\}$ and $\{E_L\}$ indicating that either the left motor or the left encoder is not working as expected. From here on, we assume if not otherwise stated that we are interested in minimal diagnoses only.

Computing minimal diagnoses is computationally expensive. In particular, the problem of searching for a minimal diagnosis is NP-complete providing that a theorem prover or any other reasoning algorithm can check satisfiability in linear time, which is not even the case for propositional logic where we know that the satisfiability problem (SAT) is itself NP-complete. However, in practice this is not so much a big deal, because we are mostly not interested in finding all minimal diagnoses but also the smallest ones with respect to cardinality. Moreover, in most cases there are enough single fault or double fault diagnoses, and the challenge is more to reduce the number of diagnosis candidates.

In literature there are many diagnosis algorithms discussed in detail. Greiner et al. [25] discussed a corrected version of Reiter's diagnosis algorithm outlined in [46], which makes use of conflicts for computing diagnoses. For more details about conflicts and further issues regarding MBD we refer to [46]. More recently, Felfernig et al. [18] presented the *FastDiag* algorithm, which allows computing diagnoses directly from the model. Nica et al. [41] presented an empirical evaluation of different diagnosis algorithm focusing on runtime. To be self-contained we briefly outline a simple algorithm for diagnosis, which makes use of a theorem prover like *Prover9* [35] for checking satisfiability of the system description, the observations together with the health assumptions. The theorem prover is called using the function **TP** and takes a set of rules and facts as input. It returns consistent if the given argument is satisfiable and inconsistent, otherwise.

The **CompMBD** algorithm (Algorithm 1) computes minimal diagnoses up to a cardinality n. Minimality is assured because in Line 4 we remove all supersets of components that are already diagnoses. The algorithm obviously computes all

---

Algorithm 1 CompMBD $((SD, COMP, OBS), n)$

Input: *Given a diagnosis system $(SD, COMP)$, a set of observations OBS, and a number $n \in \{1, \ldots, |HYP|\}$.*
Output: *A set of minimal abductive diagnoses up to cardinality n*
1: Let $\Delta_S = \emptyset$
2: **for** $i = 1$ to n **do**
3:     Let C be the set of all combinations of elements of HYP of size i.
4:     Remove from C all elements where there exists a subset in Δ_S.
5:     **for** Δ in C **do**
6:         **if TP**$(SD \cup OBS \cup \{Ab(x)|x \in \Delta\} \cup \{\neg Ab(x)|x \in COMP \setminus \Delta\})$ is consistent **then**
7:             Add Δ to Δ_S.
8:         **end if**
9:     **end for**
10: **end for**
11: **return** Δ_S

diagnoses because we consider all combinations of components of a particular cardinality in each iteration. The elements in Δ_S have to be diagnoses according to Definition 3 because of the theorem prover call in Line 6. The runtime complexity is in the worst case $O(2^{|COMP|})$ ignoring the complexity of the theorem prover call because we search for diagnosis considering all combinations of components. **CompMBD** should therefore only be used in a small value of n preferable smaller or equivalent to 3, which seems to be sufficient for practical applications, where we are mainly interested in a small number of diagnoses as already discussed previously.

But how to obtain a small number of diagnoses? How can we select the most likely diagnoses and do not need to consider all computed diagnosis candidates? In literature we find two practicable reasonable answers to these questions. De Kleer and Williams [13] introduced a probabilistic framework for MBD. There the authors discuss how to assign probabilities to diagnoses. For this purpose, we need the probability that a given component $C \in COMP$ fails, i.e., $p_F(C)$. If we know this probability for each component, we are able to assign a probability to each diagnosis $\Delta \subseteq COMP$ as follows:

$$p(\Delta) = \prod_{C \in \Delta} p_F(C) \prod_{C \in COMP \setminus \Delta} (1 - p_F(C)) \qquad (5)$$

Hence, when using the probability of a diagnosis we can come up with a ranking of diagnoses presenting the most likely diagnosis first. Note that either the fault probabilities of components can be obtained due to the availability of reliable empirical data, e.g., from experiments, or we assign probabilities correspondingly to expectations. In the latter case, we might consider one component to be more likely to fail than another.

Alternatively to using probabilities, De Kleer and Williams suggested to make use of probing. For example, if we know that the encoder of our differential drive robot delivers the correct result given the motor's speed, we can conclude that only the motor has to be responsible for the wrong behavior. Hence, measuring values at certain connections between components can restrict the number of diagnoses. In their paper, De Kleer and Williams presented an optimal probing strategy that reduces the number of diagnoses with the least number of additional measurements. When using such a strategy, we are able to finally come up with a single fault diagnosis.

An alternative strategy for reducing the number of computed diagnoses is based on the introduction of fault models. There we make use of the faulty behavior of components to restrict the number of valid diagnoses given the observations. Struss and Dressler [57] discussed the integration of fault models into a general diagnostic framework, and—as already mentioned—De Kleer et al. [14] introduced a theory of diagnosis with fault models. Because of the increase in the search space, i.e., we do not only need to check all subsets of the set of components but also each possible mode, using fault models is not feasible for larger systems. Friedrich et al. [22]

presented an alternative way of reducing the number of diagnoses without the need for fault models. The authors suggested to specify logical rules representing physical necessities. If such a rule is violated during diagnosis, there is contradiction and the corresponding behavior is physically impossible. Using physical impossibilities does not change the overall computational complexity and effectively reduces the number of computed diagnoses.

3.2 Abductive Diagnosis

In contrast to MBD where we model only the correct behavior of components, abductive reasoning deals with models of the faulty behavior. From a logical perspective abductive reasoning makes use of hypotheses to explain certain symptoms. We have rules of the form *hypothesis* → *symptom*. Abduction is reasoning in the opposite direction of the implication →. In case of abductive diagnosis the hypotheses represent different health states of components, which cause a certain behavior.

The following formalization is extension of the definition of Friedrich et al. [21] where the authors focused on propositional horn clause abduction problems, i.e., abductive diagnosis based on propositional logic dealing with implication rules and facts only. We first define abductive diagnosis systems similar to diagnosis systems in MBD.

Definition 5 (Abductive Diagnosis System) A pair (SD, HYP) is an abductive diagnosis system where SD is a logical model, and HYP a finite set of hypotheses.

The definition of abductive diagnosis system is similar to Definition 1. The difference is that in case of MBD we have a set of components whereas in the case of abduction we use hypotheses. This change is due to the fact that in case of MBD we only consider the Ab and $\neg Ab$ health state for each component but we may have more health states when dealing with abduction.

To illustrate the definitions, we make again use of our differential drive robot example but simplify the model. Instead of using all components of the robot (see Fig. 4), we focus on the motors only. We first specify what is going to happen in case the motor is running as expected, too slow, or too fast. Such a behavior can be formalized in a general way as follows:

$$\forall x : (motor(x) \land expected(x)) \rightarrow nominalspeed(x)$$
$$\forall x : (motor(x) \land tooslow(x)) \rightarrow reducedspeed(x)$$
$$\forall x : (motor(x) \land toofast(x)) \rightarrow increasedspeed(x)$$

In these rules the predicates $expected/1$, $tooslow/1$, and $toofast/1$ represent the health states of the motors. We state that our system has two motors:

$$motor(m_L) \land motor(m_R)$$

Furthermore, we introduce three rules stating that we can only have one speed for a motor at each time.

$$\forall x : \neg(nominalspeed(x) \land reducedspeed(x))$$
$$\forall x : \neg(nominalspeed(x) \land increasedspeed(x))$$
$$\forall x : \neg(increasedspeed(x) \land reducedspeed(x))$$

These three rules are for assuring that only one of the predicates representing speed can be true at a time. To finalize the model for abductive diagnosis, we introduce rules deriving a motion pattern for a mobile robot. The goal is to classify the observed behavior like given in Fig. 2 using a single predicate. Depending on the speed of the left and right motor, we obtain the following motion patterns:

$$(reducedspeed(m_L) \land nominalspeed(m_R)) \rightarrow leftcurve$$
$$(reducedspeed(m_L) \land increasedspeed(m_R)) \rightarrow leftcurve$$
$$(nominalspeed(m_L) \land increasedspeed(m_R)) \rightarrow leftcurve$$
$$(nominalspeed(m_L) \land reducedspeed(m_R)) \rightarrow rightcurve$$
$$(increasedspeed(m_L) \land reducedspeed(m_R)) \rightarrow rightcurve$$
$$(increasedspeed(m_L) \land nominalspeed(m_R)) \rightarrow rightcurve$$
$$(nominalspeed(m_L) \land nominalspeed(m_R)) \rightarrow straight$$

In this example, we distinguish three motion patterns, which we want to explain using abductive diagnosis. The introduced rules are element of SD_A and $HYP_A = \{expected(m_L), tooslow(m_L), toofast(m_L), expected(m_R), tooslow(m_R), toofast(m_R)\}$. (SD_A, HYP_A) states an abductive diagnosis system for our differential drive robot example.

Before defining abductive diagnosis formally, we introduce the definition of an abductive diagnosis problem.

Definition 6 (Abductive Diagnosis Problem) A tuple (SD, HYP, OBS) is an abductive diagnosis problem where (SD, HYP) is a abductive diagnosis system and OBS is a set of observations.

This definition of a diagnosis problem is similar to the model-based diagnosis definition. For our running example the tuple $(SD_A, HYP_A, \{leftcurve\})$ represents an abductive diagnosis problem. A solution to this problem is an explanation for the observations, where an explanation is a conjunction of hypotheses that allow to derive the given observations.

Definition 7 (Abductive Diagnosis) Given an abductive diagnosis problem (SD, HYP, OBS). A set $\Delta \subseteq HYP$ is a diagnosis if and only if

1. $SD \cup \Delta \models OBS$, and
2. $SD \cup \Delta$ is satisfiable, i.e., $SD \cup \Delta \not\models \bot$

In this definition, we make use of logic reasoning to define abductive diagnosis. In the first part, we require that the given observations, i.e., the observed symptoms, can be logically derived ($\models$) from the model SD together with the hypotheses in Δ. The second part assures that we cannot trivially obtain OBS, which would be the case if SD together with Δ is inconsistent, because we can derive anything from inconsistent theories.

We furthermore define minimal abductive diagnoses as follows:

Definition 8 (Minimal Abductive Diagnosis) Given an abductive diagnosis problem (SD, HYP, OBS). An abductive diagnosis $\Delta \subseteq HYP$ for the given diagnosis problem is a minimal diagnosis if and only if there is no other diagnosis Δ' that is a subset of Δ.

For the differential drive robot example and the abductive diagnosis problem $(SD_A, HYP_A, \{leftcurve\})$ we are able to compute three different minimal abductive explanations:

$$\{tooslow(m_L), expected(m_R)\}$$
$$\{tooslow(m_L), toofast(m_R)\}$$
$$\{expected(m_L), toofast(m_R)\}$$

Either the left motor m_L is too slow, providing that m_R is running as expected or faster, or the right motor is running faster providing m_L being slower or running at nominal speed. This result might be further reduced measuring the speed of the motors and comparing it with the expected values. It is worth noting that we are also able to make use of a slightly changed definition of diagnosis probability from Eq. (5). In case of abductive diagnosis we assume that we know the probabilities for each health state, i.e., each element in HYP. Hence, the definition of diagnosis probability can be simplified.

$$p_A(\Delta) = \prod_{x \in \Delta} p(x)$$

In case of our robot example, we might state that the expected behavior is much more likely. In this case we would prefer the first and the last of the three diagnoses.

Similar to MBD abductive diagnosis is at least NP-complete. In the following, we outline a basic algorithm for computing all abductive diagnosis up to a specific size. The algorithm is not optimized. For other algorithms we refer the interested reader to [39] and [31].

Algorithm 2 CompAD $((SD, HYP, OBS), n)$

Input: *Given an abductive diagnosis system (SD, HYP), a set of observations OBS, and a number $n \in \{1, \ldots, |COMP|\}$.*
Output: *A set of minimal diagnoses up to cardinality n*
1: Let $\Delta_S = \emptyset$
2: **for** $i = 1$ to n **do**
3:     Let C be the set of all combinations of elements of $COMP$ of size i.
4:     Remove from C all elements where there exists a subset in Δ_S.
5:     **for** Δ in C **do**
6:         **if** TP$(SD \cup HYP)$ is consistent **then**
7:             **if** TP$(SD \cup HYP \cup \neg OBS)$ is inconsistent **then**
8:                 Add Δ to Δ_S.
9:             **end if**
10:         **end if**
11:     **end for**
12: **end for**
13: **return** Δ_S

Algorithm 2 implements the abductive diagnosis algorithm **CompAD** where we go through all subsets of the hypothesis set and check for diagnosis. In Line 6 we first check whether the hypotheses are consistent with the system description. If this is true, we check whether we are able to derive OBS using SD together with HYP. This has to be the case if assuming that the observations are not valid ($\neg OBS$) together with SD and HYP leads to an inconsistency. The algorithm obviously terminates and computes all minimal abductive diagnoses up to a size n. Minimality is assured because we remove all supersets of already detected diagnoses in Line 4.

3.3 Summary on Model-Based Reasoning for Diagnosis

The presented diagnosis approaches, i.e., MBD and abductive diagnosis, have been successfully used in practice. Whereas MBD makes use of a model capturing the correct behavior of components for computing diagnoses, abductive diagnosis relies on fault models. Both approaches rely on models that can be formally represented and where a reasoning mechanism is available for checking satisfiability. It is worth noting that we do not necessarily rely on FOL, propositional logic or any other logic formalism. We may also make use of constraints to represent models and constraint solving for checking satisfiability. For an introduction into constraints and constraint solving we refer to Rina Dechter's seminal book [16].

Although both diagnosis approaches are computationally demanding, current work, e.g., [41] and [31], has shown that MBD and abductive diagnosis can be used for practical applications and that their worst case computational complexity is not a limiting factor in practice. The only more severe restrictions of course are the necessity to have models that can be used for diagnosis. Obtaining such models in practice is not always that simple, but due to the increasing importance of models for system development, this challenge seems to be solvable. There has been more recent work dealing with obtaining models from available development artifacts, see, e.g., [64].

4 Modeling for Diagnosis and Repair

Model-based reasoning requires that we have a model of the system we want to diagnose. In particular, we need a model that can be feed into a reasoning engine in order to determine consistency. In the previous sections we made use of FOL or other logics as underlying modeling language. However, model-based reasoning is not restricted to logic as a formalism for modeling. Beside the use of a certain modeling language a model to be used for model-based reasoning has also to provide means for setting or characterizing the health state of components or any other assumption we want to reason about.

In order to come up with models for diagnosis and repair, we first focus on some modeling principles that may be used. Wherever necessary, we distinguish modeling for MBD from abductive diagnosis. When starting modeling of systems the first part comprises coming up with the system's architecture, i.e., its parts and their interconnections together with the system's environment and interface. The interface of the system and its context is important in order to allow systems communicating with their environments. For modeling, we have to know this information as well. The architecture of a system can be seen as component-connection model where connections are interfaces between the ports of components. A connection is for exchanging information and data. From an abstract point of view each connection has a name and a type, e.g., a natural number or an array of reals.

After identifying the components, their interfaces, i.e., ports, and the connections, we have to represent the behavior of each component. In case of digital circuits such a behavior can be expressed using Boolean logic whereas in case of physical components differential equations would be a more appropriate form if we want to closely describe the real behavior of components over time. It is worth noting that the data types of the connections of course are also the same for connected components and should be used to come up with a component model. This component model has to capture the correct behavior in case of MDB and the fault behavior in case of abductive diagnosis. Before discussing the differences in modeling for MBD and abductive diagnosis, we first have a closer look at the underlying data types.

As already said, in principle formalisms used to describe models range from logic sentences to differential equations or even programs formulating a certain behavior. In case of diagnosis in general such fine-grained representations of reality are not needed and some form of abstraction is usually sufficient. Instead of using real valued connections we might be able to distinguish some finite number of values that allow representing the behavior sufficiently. For example, let us consider again the robot example. Instead of dealing with the exact values of voltage and current used to control the number of rotations of a motor, it is sufficient to consider the case where there is voltage applied and thus the motor starts rotating, and the case where there is no voltage and the motor stops rotating. We may also want to distinguish a case where the voltage is too low or too high causing a decreased or increased speed. However, this depends on the current application.

For modeling the behavior we therefore require to search for the right degree of abstraction. Right in this context is informally speaking a set of values that can be distinguished leading to different behaviors. Using abstraction has the advantage of not requiring sophisticated simulation and requiring less computational resources. The use of abstraction for modeling is not new and has been proposed in the context of artificial intelligence almost 30 years ago. In qualitative reasoning (QR)[1] [62] researchers have been working on coming up with different kind of abstractions and also different underlying modeling principles. For the latter Kuiper's qualitative simulation [33], Forbus's qualitative process theory [20], and De Kleer et al.'s work on confluences [12] are worth mentioning. In qualitative simulation, ordinary differential equations are mapped to their abstract corresponding qualitative differential equations, which can be used to obtain all possible behaviors of a system without knowing the exact values. In addition, qualitative simulation also allows specifying new qualitative values in certain cases. Qualitative process theories make use of processes for specifying the abstract behavior of systems. There not components are of importance but processes determining the value of variables. There are still modeling environments, e.g., [4], available that are mainly used for formulating from biology and sustainable engineering.

In qualitative reasoning, we distinguish two possible cases of abstraction. Either there is a mapping of quantities directly to their corresponding qualitative values [10] or we represent deviations [56]. For example, in the case of an electrical motor we might be only interested in the case where the motor is stopping, rotating clockwise or anticlockwise. When knowing the electrical characteristics we may come up with the following mapping:

$$
\begin{aligned}
[-12\ V, -0.5\ V] &\mapsto -\omega \\
]-0.5\ V, 0.5\ V[&\mapsto 0 \\
[0.5\ V, 12\ V] &\mapsto +\omega
\end{aligned}
$$

---

[1]In qualitative reasoning variable values are abstract representations of their original domain. Instead of quantities like real numbers, qualitative representations are used for various purposes like simulation or diagnosis. Because of using qualitative values the name qualitative reasoning was established.

In this mapping, the assumption is that the motor can take voltages from -12 V to $+12$ V and that the rotational speed can be either anticlockwise ($-\omega$), zero (0), or clockwise ($+\omega$). The abstract domain $\{-\omega, 0, +\omega\}$ is totally ordered and we are also able to introduce abstract operations for such domains to be used further on for modeling the behavior. Of course because of abstraction we are losing information and sometimes the operators cannot distinguish potential outcomes. In this case, the question comes up to increase the abstract domain and to introduce new abstract values. This process can also be automated. Sachenbacher and Struss [48] presented a solution for automated domain abstraction.

In contrast to the direct representation of quantities as elements of an abstract domain, deviation models only consider as the name suggested deviations from nominal or expected behavior. Instead of stating that a value is 0.9 and therefore lower than the expected value of 1.0, the deviation, e.g., the value is small, is used. Hence, in case of deviation models we do not have a mapping of values to their qualitative representation but a mapping of deviations to their representation, e.g., "$<$" for stating a value to be smaller as expected. Deviation models have been used successfully in diagnosis, e.g., [56]. For more information about qualitative models have a look at [58, 59].

In the following, we discuss providing models for MBD and abductive diagnosis separately. Let us start with *modeling for MBD*. As already outlined we have a component-connection model and (possible abstract) data types for the connections, ports, and interfaces to the system environment and context. What we need now is the behavior of components. What we do first is to come up with certain types of components like a logical and with two inputs and one output. For each component type we need to specify a relationship between the different ports in case the component is working as expected. Hence, for each component C of type $type$ we have to come up with rules of the form $\forall C : (type(C) \rightarrow (\neg Ab(C) \rightarrow Behav))$ where $Behav$ specifies the correct behavior of component C.

In case of our mobile robot example explained in Sect. 2 and later in Sect. 3, we defined the behavior of a motor as $(val(in(X), Y) \leftrightarrow val(out(X), Y))$ stating that every value Y on the input port $in(X)$ has to be transferred to its output $out(X)$ and vice versa. Hence, we do not only specify one direction of data flow but formalize the model in a relational way ignoring the information whether a certain port works as input or output.

The behavior might also be given as equation like $v = R \cdot i$ representing a model for an electrical resistor. In such an equation we also do not have a data flow direction. We are only specifying relations that constraint given quantities or qualities. Of course it is necessary that the underlying reasoning mechanism is able to handle the given models. When relying on FOL we would not be able to specify the resistor model as given. In this case we may use a different kind of logic or a different (abstract) representation.

In addition, to the component models, we might also want to add rules stating physical impossibilities [22]. For example, let us consider an analog circuit comprising two bulbs in parallel coupled with a battery. A simple model would state that bulbs are lighting if there is voltage provided, and that a battery provides voltage.

In case one bulb is lighting but the other is not, there would be two diagnoses, i.e., the bulb that is not lighting and also the battery. The latter diagnosis is of course physically impossible, because an empty or broken battery cannot provide voltage. We can solve this issue via stating the impossibility: If a bulb is lighting, then there has to be a voltage. When adding such a rule, the battery cannot longer be a diagnosis.

Modeling for abductive diagnosis starts with the same input, i.e., the structure of the system. However, instead of defining the correct behavior, we are interested in the faulty cases and their consequences. Hence, we adopt a form of cause–effect reasoning, where causes are (faulty) health states of components and there effect are the symptoms we observe and want to explain. In this setting symptoms are deviations from the expected behavior. This type of modeling goes beyond MBD where we only have the health states correct or faulty, which we represent using $\neg Ab$ and Ab, respectively. In case of abductive diagnosis we have one or more health states for each component, which can also be represented using predicates. For each of these health states we present their consequences formally in the model. For this purpose, we would usually use rules of the form $\forall C : type(C) \to (cause_i(C) \to effects_i(C))$ where $type(C)$ is the type of the component, e.g., a Boolean and gate, $cause_i$ represents the health state, and $effects_i$ the consequences following from the given cause.

Such knowledge can be easily obtained from a failure mode and effect analysis (FMEA) [6, 26], which is regularly used in the context of safety critical systems in order to analyze the consequences of a fault and its corresponding risks. From the FMEA we obtain a table of the form:

Failure mode	Effects	Risk
⋮	⋮	⋮

This table can be almost directly mapped to cause–effect rules. For more details we refer to Wotawa [64] explaining the transformation in detail.

For practical applications it would be a significant advantage to have general modeling languages available that allows for writing models for model-based reasoning, like for simulation where we have beside Matlab/Simulink[2] other languages like Modelica [24]. Note that in contrast to simulation where all boundary conditions have to be known, in diagnosis we need models that allow to specify also the unknown behavior. Modeling languages for diagnosis have to deal with this specific requirement. Fleischanderl et al. [19] were one of the first introducing a modeling language for MBD considering the component-connection modeling principles and also different data types. The only limitation was the lack of considering time information in the models. Bonus et al. [5] presented an extension

---

[2] See https://de.mathworks.com/products/simulink.html.

of Modelica that can be used of modeling for MBD. Most recently, Nica and Wotawa [40] presented a language that also allows to come up with models considering time.

It is also worth mentioning modeling approaches that make use of physical models written in Modelica for fault localization using MBD and abductive diagnosis. De Kleer et al. [15, 36] discussed an approach for extracting diagnosis system models from Modelica models using explicit fault modes. Sterling et al. [55] presented an approach for mapping Modelica programs to diagnosis systems directly considering abstractions, and Peischl et al. [43] outlined the use of Modelica programs for obtaining cause–effect models for abductive diagnosis.

In summary, (1) modeling starts with identifying the boundaries of systems and their internal structure comprising components and their interconnections. In the second step (2) we have a look at the underlying data types of the connections and component ports as well as the interfaces of the system to its environment. For the data types, we have to elaborate on abstractions that are strong enough to distinguish the important behavioral aspects of components. If this is done, we formalize the component behavior in step (3) where we distinguish the case of MBD from abductive diagnosis. If we only know the correct behavior, we have to rely on MBD and define the component behavior as relations over the component ports. In case of abductive reasoning, we define rules of the form *causes imply effects* where a cause is a certain fault of a component. For abductive reasoning, it is also possible to determine the model from FMEAs, which are often used in engineering practice to determine the system's risk.

Until now, we have discussed the modeling steps for diagnosis. But are there any specific parts when dealing with *repair* or self-repair? Friedrich et al. [23] were one of the first considering repair as part of diagnosis. They formalized the repair process including diagnosis as one part. Basically, the general diagnostic process continues the diagnosis and probing loop until one candidate can be obtained, which should be replaced. Hence, in the simplest form repair is only a replacement of a faulty component. In case of fail-operational behavior or self-adaptive systems this would be close to the spare components that can be invoked whenever the component itself becomes faulty. However, as motivated in the introduction, we do not always have a spare component. Here we would require to change the system in a way such that the important functionality can still be guaranteed.

In the first case, i.e., using spare components for replacing broken components during runtime, we do not need an additional model. For this purpose, it is sufficient to know which component is faulty and should be replaced. Hence, both diagnosis approaches can be used directly. This is not possible, when wanting a system to behave as expected but maybe in a degraded mode. In such a case we need to model the degradation. We can do this via modeling which parameters have to be changed and how in case of a fault. Hence, we would need to specify also what can be done in case of a specific fault. In the next section, we discuss repair in much more detail and also provide examples for each of the different repair cases.

5 Self-adaptation Using Models

In this section, we discuss the use of model-based reasoning for self-adaptation. According to Weyns et al. [63] self-adaptation is the ability of a system to adapt to dynamic and changing operating conditions autonomously, i.e., without requiring human intervention. Self-adaptation can be seen as form of self-healing behavior, which is also known as autonomic computing where a system can detect, diagnose, and repair localized faults originating from software or hardware (see [30]). In the previous sections of this chapter, we already discussed the use of models for diagnosis purposes. Hence, we have to extend our framework to the capabilities of repair and in particular self-repair [9].

Repair itself can be classified according to [60] in (1) attributive repair and (2) functional repair. In attributive repair the idea is to restore the system to its original state, whereas in functional repair the focus is only on restoring the (important) functionality but not necessarily bringing the system back to its fault-free state. Hence, in functional repair we are also satisfied with a degraded behavior. Nevertheless in both types of repair we require a certain redundancy in order to bring back the system in a desired state. For example, in software-based self-repair [49], the reconfiguration is done on side of the user program to allow execution on the available processors.

It is worth noting that in order to achieve self-adaptation or self-healing the system itself has to monitor its health-state and react appropriately over time in case the actual behavior of the system is in contradiction with its expected behavior. After detecting such a relevant behavioral deviation, the system has to perform diagnosis using its underlying knowledge and later on self-repair. Hence, a self-healing system has to follow a certain system architecture. For example, IBM suggested such a reference model for autonomic control loops [28], which is also called the MAPE-K (Monitor, Analyze, Plan, Execute, Knowledge) loop. In this section, we introduce a simplified architecture, which requires monitoring, diagnosis (which is a kind of analysis), and repair (which comprises planning and execution). For all these three steps we rely on a model, i.e., knowledge of the system.

In Fig. 5 we see on the left the classic architecture of a system that interacts with its environment. For this purpose, we have a sensor and an actuator level for obtaining information from the environment and interacting with the environment. The system and their in particular the control component makes use of the measurements coming from the sensor level and its internal state to compute values for the actuators. For example, our differential drive robot may localize an object at a certain position via a laser range sensor or a computer vision system and computes the commands for the motors for moving to detected object. In case of a fault in any parts of the system, the robot will not work as expected anymore because it cannot detect the misbehavior and also not react accordingly.

This situation is different in case of the system architecture on the right side of Fig. 5. There we have an additional smart diagnostics component, which takes the information provided by the sensors together with the current state of the control

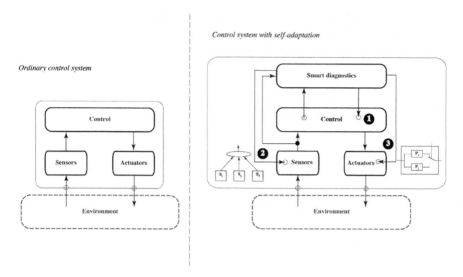

Fig. 5 Control system without (left) and with (right) a smart diagnostic component

component to derive a diagnosis and afterwards repair actions, which might change the control component itself (1), the sensor information (2), or the actuators (3). For all three cases, we are going to outline examples later in this part of the chapter. Before, we discuss the structure of the smart diagnostics and how model-based reasoning can be integrated.

A smart diagnostics has to monitor the current system and based on this information to draw conclusions about the health state of the system over time. Hence, in every step at a certain point in time t, we have to evaluate the system's behavior with respect to deviations from the expected behavior. For this purpose, we can use the health state of the system obtained at the immediately previous time step $t - 1$ to predict a behavior. If this behavior is equivalent to the observed value, we know that the system is still in the same health state at time t and no further action is required. Otherwise, we have to run diagnosis to explain the deviation and in case of identifying the root cause to repair the system.

It is worth noting that there might be the case that a fault occurring at a time step is only visible after more than one time step. In this case, diagnosis might give us back a wrong result. In order to overcome this issue, the time span between two time steps used for monitoring has to be defined as being large enough for all faults to be visible in the observations. If this is not possible, e.g., because of underlying physical processes requiring substantially different time spans for being observed, we have to introduce separated monitoring/diagnosis/repair cycles for all of these processes.

In Fig. 6 we illustrate time step t using information from the previous step $t - 1$. Note that the observations are extracted from the sensor data and the internal state of the control block. When implementing the smart diagnostics there has to be

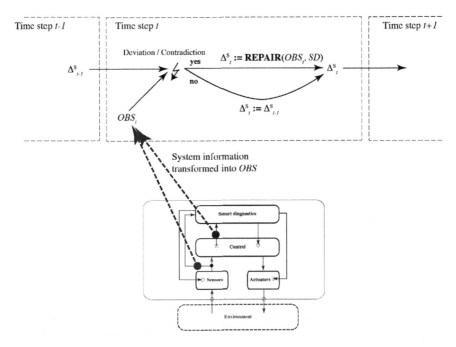

Fig. 6 A monitoring/diagnosis/repair cycle at time step t

a program mapping the monitored variables to their corresponding observations OBS_t. This mapping does not necessarily need to set all potential observations. There may be observations that cannot be obtained at all points in time. But this is not a problem for model-based reasoning because there computation is done on available information only. It is also worth noting that in case of FOL as underlying modeling language, the mapping has to set the truth values of predicates. All observations that can be obtained must be true at the certain point in time t.

In Fig. 6 we also see that in case of a contradiction we call a method **REPAIR** and not diagnosis. This method calls a diagnosis algorithm for identifying candidates, selects the best candidate, and computes necessary repair steps either to bring the system into a correct state again or in a degraded mode where the system still can deliver its required functionality. Depending on the repair step the health state of a system Δ may or may not change. In case of using spare parts the system after repair should incorporate only healthy components. In a degraded mode, the system has some components still not working as expected but which have been compensated. Formally, we can define a system health state as mapping of components to their health states.

Definition 9 (System Health State) A system health state Δ^S for a system with a set of components $COMP$ is a sequence of pairs $\langle (c_1, h_1), \ldots, (c_{|COMP|}, h_{|COMP|}) \rangle$ where for all components $c \in COMP$ we have a corresponding (component) health state h.

Algorithm 3 REPAIR (SD, OBS)

Input: *Given the model of a SD and a set of observations OBS at the current time t*
Output: *A system health state Δ^S*
1: Compute the set of diagnosis Δ_S using either **CompMBD** or **CompAd** depending on the diagnosis method used.
2: Select the one diagnosis Δ from Δ_S.
3: Apply repair mechanisms using Δ.
4: Let Δ^S be the old system health state at time $t-1$.
5: **if** Spare part repair **then**
6:     Change the health states in Δ^S for all components in Δ to their correct state, i.e., $\neg Ab$.
7: **end if**
8: **if** Compensating action repair **then**
9:     Change the health states in Δ^S for all components in Δ with their corresponding health state provided in Δ.
10: **end if**
11: **return** Δ^S

In a smart diagnostics the system health state is continuously adapted over time using the current observations and the past system health state. In the following, we first summarize the **REPAIR** algorithm and afterwards introduce three possible cases how such a smart diagnostics maybe react and why.

Algorithm 3 implements the necessary steps for repairing a system. The way a system is repaired is not explicitly stated in the algorithm and depends on the kind of possible action. However, we distinguish two cases either repairing via using new components or compensating actions. A compensating action in the context of this work is any change of parameters or health states of the system that still allow the system to behave as close as possible to its expectations. Compensating actions are very much application specific. Furthermore, in **REPAIR** we also do not specifically describe how to select a diagnosis. In practice, we either make use of probabilities for obtaining the most likely diagnosis, or we may make a random selection. Ideally, the diagnosis itself should only provide us with one diagnosis. However, that cannot be assured always. Moreover, if we select a wrong diagnosis, this will be visible later in time when there is again a deviation with the expected values. In such a situation we may consider changing the probabilities of faults and finally improve reasoning and selection over time. De Kleer and Williams [13] suggested such a process. The only issue that has to be assured is that the selection of the diagnosis does not violate certain properties like safety. If this can be the case we further are able to use such properties during diagnosis selection.

We now discuss the three cases of faults with corresponding repair that may occur during operation of a system. In Fig. 5, we see three different locations of a fault: ❶ indicates a fault in the control block, ❷ states a fault in the sensor block, and ❸ represents the case of a fault on side of the actuators. For all three cases, we discuss examples.

Control (❶):   Let us consider the differential drive robot from Fig. 3 and its architecture depicted in Fig. 4. Instead of modeling the behavior using qualitative domains, we now make use of equations and variables to specify the behavior. In particular for motors we say that the voltage applied is proportional to their number of rotations. The following rule states this behavior where s_X represents the speed of the motor and v_X the applied voltage. In this rule we set the constant stating the proportion between speed and voltage to 1.0.

$$\forall X : motor(X) \rightarrow (\neg Ab(X) \rightarrow (s_X = 1.0 \cdot v_X))$$

Encoders can be similarly modeled. Instead of speed a rotational encoder (or wheel encoder) returns pulses. In the following rule we assume that there are 36 pulses delivered for one full rotation. Hence, the number of pulses t_X is a function $36 \cdots s_X$ where s_X is the speed that should be measured.

$$\forall X : enc(X) \rightarrow (\neg Ab(X) \rightarrow (t_X = 36 \cdot s_X))$$

In addition, we have to model the structure, i.e., the components and connections, of the system:

$$motor(m_L) \wedge motor(m_R)$$

$$enc(e_L) \wedge enc(e_R)$$

$$s_{m_L} = s_{e_L} \wedge s_{m_R} = s_{e_R}$$

For this example, the above rules and equations represent the model SD and $COMP = \{m_L, m_R, e_L, e_R\}$. Let us further assume that we have the following observations, where the voltage comes from the control block and the number of pulses from the encoders.

v_{m_L}	v_{m_R}	t_{e_L}	t_{e_R}
2.0 V	2.0 V	72	72

In this case, obviously everything is fine and there is no contradiction with the expectations. Let us now consider the case of a robot following the wrong trajectory depicted in Fig. 2. For this case, we would receive observations like the following, where the number of pulses on the left side is small than expected:

v_{m_L}	v_{m_R}	t_{e_L}	t_{e_R}
2.0 V	2.0 V	54	72

Using MBD, we are able to compute the two minimal diagnoses: $\{m_L\}$ and $\{e_L\}$. This result can be further improved when using physical impossibilities or other known properties, e.g., stating that a broken encoder would not deliver any pulses, i.e., $\forall X : enc(X) \rightarrow (\neg Ab(X) \vee t_X = 0)$. In addition, we might extend the model also to *compensate* the fault. In this example, the left motor is not running the right speed. If extending the voltage, we may speed up the motor. Hence, we may come up with a component representing the proportionality factor between the voltage and the speed. The new model would look like:

$$\forall X : motor(X) \rightarrow (\neg Ab(X) \rightarrow (s_X = c_X \cdot v_X))$$

$$\forall X : const(X) \rightarrow (\neg Ab(X) \rightarrow (c_X = 1.0))$$

$$const(c_L) \wedge const(c_R)$$

$$c_{m_L} = c_{c_L} \wedge c_{m_R} = c_{c_R}$$

The set of components would be $COMP = \{m_L, m_R, e_L, e_R, c_L, c_R\}$. Using this model together with the rule stating the property that broken encoders do not deliver pulses, we obtain again two minimal diagnoses but this time: $\{m_L\}$ and $\{c_L\}$. The encoder is not a diagnosis anymore because we observe pulses. Let us now take a closer look at diagnosis $\{c_L\}$. When the constant is wrong, any value for variable c_{c_L} can be inserted. Using an equation solver a value of 0.75 for c_{c_L} explains the lower value of t_{e_L}. Assuming that this reduction of speed when providing the same voltage is—more or less—equally distributed over the working range of the input voltage, we would multiply the voltage with $\frac{1}{0.75}$ which equals to 1.33 before using it as input for the left motor.[3]

In this example, the compensation would be valid until new observations contradict again the behavior, maybe leading to further adaptations of the input voltage. Moreover, the diagnosis $\{c_L\}$ with its corresponding compensating value would be part of the system health state of the differential drive robot. The repair rule for this diagnosis would state how to compute the compensating value and how to apply it. The application may be either part of the component block, or before or in the actuator block of the smart diagnostics architecture (see Fig. 5 right picture).

Sensor (❷): In the second example, we illustrate how smart diagnostics can be used for sensor fusion and there in particular on identifying contradictions in sensor information and to react in an appropriate manner. For illustration purposes let us consider a mobile robot comprising the three sensors, inertial measurement unit (IMU), computer vision system (CVS), and the wheel encoders. All of this sensor can be used to identify the direction a robot is moving. From the IMU we would get an acceleration vector in the direction of movement when starting. The CVS can provide among other things an optical flow indicating the direction, and from the encoders we receive pulses where the difference indicates the direction. In the following we formalize the behavior of the different sensors:

---

[3]In practice, someone might make use of a more sophisticated process after detecting such a fault. A search procedure might be used to find the right voltage levels for the left motor in order to guarantee moving on a straight line.

$$\forall X : imu(X) \to (\neg Ab(X) \to (\forall Y : accdir(X,Y) \leftrightarrow dir_I(X,Y)))$$

$$\forall X : cvs(X) \to (\neg Ab(X) \to (\forall Y : optflowdir(X,Y) \leftrightarrow dir_V(X,Y)))$$

$$\forall X : encs(X) \to \left(\neg Ab(X) \to \begin{pmatrix}(equiv(X) \leftrightarrow dir_E(X, straight)) \wedge \\ (rightgreater(X) \leftrightarrow dir_E(X, left)) \wedge \\ (leftgreater(X) \leftrightarrow dir_E(X, right)) \wedge\end{pmatrix}\right)$$

Note that $dir_I/2$, $dir_V/2$, and $dir_E/2$ are predicates for representing the current direction of the IMU, the CVS, and the encoders, respectively. In this example, we do not distinguish the encoders for the different wheels. We only assume that we know their number of pulses and are able to compare them leading to the respective truth values for the corresponding predicates.

In order to check consistency we further have to come up with consistency properties stating that all the different direction values should be the same.

$$\forall X : \forall Y : \forall Z : \forall V : (dir_I(X,V) \wedge dir_V(Y,V) \wedge dir_E(Z,V)).$$

Moreover, we have to state that each sensor is only allowed to deliver one value at a particular time.

$$\forall X : \forall V : \forall W : dir_I(X,V) \wedge dir_I(X,W) \to V = W$$

$$\forall X : \forall V : \forall W : dir_V(X,V) \wedge dir_V(X,W) \to V = W$$

$$\forall X : \forall V : \forall W : dir_E(X,V) \wedge dir_E(X,W) \to V = W$$

After specifying the structure of the system comprising three components $COMP = \{imu1, cvs1, encs1\}$ using the following rules, we obtain a complete model for sensor fusion:

$$imu(imu1) \wedge cvs(cvs1) \wedge encs(encs1)$$

Let us assume now that we obtain the following observations:

$$OBS = \{accdir(imu1, straight), optflowdir(cvs1, right),$$

$$equiv(encs1)\}.$$

From SD and OBS and assuming all components to work correctly we get a contradiction because the CVS is returning a different value. From the model, we are able to compute two minimal diagnoses, i.e., $\{cvs1\}$ and $\{imu1, encs1\}$. When focusing on the smallest diagnosis first, we would disable the CVS, which is the repair action for this example. Note that in this example we do not have a compensating action. Instead we simply turn off a component, which would not influence the rest of the system because the other two sensors determine the value used for further controlling the robot.

Actuator (❸): In the last example, we show the case of spare parts. Assume that we have the differential drive robot but this time we add power circuits for providing enough voltage and current for the motors. When using a similar model than in Sect. 3, we can determine which motor is faulty and therefore, which power circuit should be replaced. If we have a hardware architecture with integrated spare parts that can be enabled, the repair action would simply enable the new power circuit and disable the old one.

The presented approach for repairing detected and localized faults based on models has the advantage of relying on system models and therefore always delivering the best possible diagnoses for the given observations and the system model. We do not rely on data for learning diagnosis, therefore the approach can be used even in situations where such data is not available, e.g., after the development of systems where no feedback from its use is available. The approach allows for implementing fail-operational systems. We can assure that the proposed diagnosis fulfills given properties before applying it. Furthermore, the underlying algorithms deliver all minimal diagnoses as explained in this chapter. In order to prove that the approach is working as expected, we have to validate the model and there in particular their capabilities of specifying the behavior of the system. In this section, we further discussed how to use diagnosis for coming up with repair suggestions and how to integrate them in a smart diagnostics.

6 Related Research

The use of model-based reasoning for self-adaptation is not new. Rajan et al. [45] discussed the use of model-based reasoning in a combination with planning to form a new control system for an automated space probe, which was successfully tested in space. Later Hofbauer et al. [27] and Brandstötter et al. [3] introduced a model-based adaptive system that allows a robot adapting its drive in case of a fault in a motor. There the authors suggested to use hybrid automata for modeling where each state of the automata represents a certain health state of the robot. Besides diagnosis the approach also allowed to adapt the kinematics autonomously.

Besides reacting on hardware faults, there is also work on dealing with software faults occurring during operation. Steinbauer et al. [54] discussed a model-based approach for a mobile robot control system that is able to reduce the number of software processes that have to be restarted in case of a crash. The presented approach relies on an abstract model of the software considering the dependencies between software components. For an overview of the use of model-based reasoning to self-adaptive behavior, we refer the interested reader to [53].

Krenn and Wotawa [32] introduced another interesting approach to implement self-healing behavior based on models. There the model captures the ordinary behavior using rules that have to fulfill a certain goal, e.g., obtaining sensor measurements and sending them to a particular server. By specifying alternative rules to achieve a certain sub-goal, it is possible to easily add redundancy. The execution system searches for those rules that can be executed at a particular point in time considering such alternative rules.

In the context of self-healing system research there is a lot of research dealing with self-adaptation and self-repair. Some like [9] discuss changes in traditional design practice in order to implement self-repair for producing systems comprising hardware and software components. For reconfiguration of electronic devices we refer, for example, to [2, 37, 47, 61]. For self-adapting software have a look at [38, 51] and more recently [44, 49, 52, 63]. Such software systems have to have the ability to modify itself in response to a change in its operating environment.

In the domain of automotive systems Seebach et al. [50] presented an approach for implementing self-healing behavior. There the authors made use of the adaptive cruise control (ACC) system as example application. In their approach the ACC system is deactivated and restarted after reconfiguration. Especially, in case of automotive systems fail-safe properties have to be fulfilled even after reconfiguration. Barbosa et al. [1] introduced the use of the formal modeling language Lotos to check monitored quantities over time. In particular, the presented tool monitors the execution traces generated by a self-adaptive system and annotates the probabilities of occurrence of each system action on their respective transition on the system model, created at design time as labelled transition system (LTS) that is used for checking properties.

In the context of cyber-physical systems there has also been research published. Niggemann and Lohweg [42] discussed the state of the art of diagnosis of cyber-physical systems and the research questions but focusing more on production systems. Mahadevan et al. [34] introduced the application of causal diagrams for fault localization. The approach presented in this work relies in contrast to these papers on model-based reasoning based on formal models.

7 Conclusions

In this chapter, we discussed the basic principles and foundations behind model-based reasoning in detail. We motivated why reasoning based on models is of particular interest for self-adaptive systems using a running example from the autonomous mobile robot domain. We illustrated different challenges that occur in this domain and discussed possible solutions. Regarding diagnosis we introduced two different methods, i.e., (1) model-based diagnosis and (2) abductive diagnosis. The first relies on models of system components only considering the correct behavior. The latter requires knowledge about faults and their corresponding behavior. For both approaches we outlined a simple algorithm allowing for computing minimal diagnoses from models and observations.

Because modeling is the important part, when introducing model-based reasoning, we also explained how to model for diagnosis. We outlined a process that starts with a system architecture comprising components and connections. From this discussed how to come up with abstract models that map potentially infinite domains into a small set. For this purpose, we referred to qualitative reasoning and the abstractions discussed there. Furthermore, we introduced concepts for extracting models from other development artifacts like FMEAs or simulation models.

After presenting the basic concepts, we focused on the integration of diagnosis for self-adaptive systems. There we introduced a system combining a smart diagnostics with an ordinary control system interacting via sensors and actuators with its environment. We discussed the basic repair cycle and an algorithm for combining diagnosis with repair. In contrast to previous work we did not only rely on simple repair actions, e.g., replacing one component with its spare part, but also discussed how compensating actions can be integrated and used in the proposed setting. For the latter it is worth noting that the system model itself can be used to obtain sufficient information about the compensation.

The presented approach can be integrated into cyber-physical systems like autonomous vehicles in order to implement fail-operational behavior where compensating repair as well as automated repair using redundant hardware that can be enabled during operation is required.

Acknowledgements The research was supported by ECSEL JU under the project H2020 737469 AutoDrive—Advancing fail-aware, fail-safe, and fail-operational electronic components, systems, and architectures for fully automated driving to make future mobility safer, affordable, and end-user acceptable. AutoDrive is funded by the Austrian Federal Ministry of Transport, Innovation and Technology (BMVIT) under the program "ICT of the Future" between May 2017 and April 2020. More information on https://iktderzukunft.at/en/ bm. The author wants to thank Dr. Iulia Nica for providing an initial survey on self-healing systems and application inspiring parts of the related research section.

References

1. Barbosa, D.M., Lima, R.G.D.M., Maia, P.H.M., Costa, E.: Lotus@runtime: a tool for runtime monitoring and verification of self-adaptive systems. In: IEEE/ACM 12th International Symposium on Software Engineering for Adaptive and Self-Managing Systems (SEAMS), pp. 24–30. IEEE, Buenos Aires (2017)
2. Boesen, M., Madsen, J.: eDNA: a bio-inspired reconfigurable hardware cell architecture supporting self-organisation and self-healing. In: Proceedings of the 2009 NASA/ESA Conference on Adaptive Hardware Systems, pp. 147–154. IEEE Computer Society Press, Moscone Convention Center, San Francisco (2009). https://doi.org/10.1109/AHS.2009.22
3. Brandstötter, M., Hofbaur, M., Steinbauer, G., Wotawa, F.: Model-based fault diagnosis and reconfiguration of robot drives. In: IEEE/RSJ International Conference on Intelligent Robots and System, pp. 1203–1209. IEEE, San Diego (2007)
4. Bredeweg, B., Bouwer, A., Jellema, J., Bertels, D., Linnebank, F., Liem, J.: Garp3—a new workbench for qualitative reasoning and modelling. In: Proceedings of the 20th International Workshop on Qualitative Reasoning (QR-06), pp. 21–28. Dartmouth College, Hanover (2006)
5. Bunus, P., Isaksson, O., Frey, B., Münker, B.: RODON—a model-based diagnosis approach for the DX diagnostic competition. In: Proceedings of the International Workshop on Principles of Diagnosis (DX) (2009)

6. Catelani, M., Ciani, L., Luongo, V.: The FMEDA approach to improve the safety assessment according to the IEC61508. Microelectron. Reliab. **50**, 1230–1235 (2010)
7. Console, L., Torasso, P.: Integrating models of correct behavior into abductive diagnosis. In: Proceedings of the European Conference on Artificial Intelligence (ECAI), pp. 160–166. Pitman Publishing, Stockholm (1990)
8. Console, L., Dupré, D.T., Torasso, P.: On the relationship between abduction and deduction. J. Log. Comput. **1**(5), 661–690 (1991)
9. Coyle, E., Maguire, L., McGinnity, T.: Self-repair of embedded systems. In: Proceedings of the Engineering Applications of Artificial Intelligence, vol. 17, pp. 1–9 (2004). https://doi.org/10.1016/j.engappai.2003.11.009
10. Dague, P.: Qualitative reasoning: a survey of techniques and applications. AI Commun. **8**(3/4), 119–192 (1995)
11. Davis, R.: Diagnostic reasoning based on structure and behavior. Artif. Intell. **24**, 347–410 (1984)
12. de Kleer, J., Brown, J.S.: A qualitative physics based on confluences. Artif. Intell. **24**, 169–203 (1984)
13. de Kleer, J., Williams, B.C.: Diagnosing multiple faults. Artif. Intell. **32**(1), 97–130 (1987)
14. de Kleer, J., Mackworth, A.K., Reiter, R.: Characterizing diagnoses and systems. Artif. Intell. **56**(2–3), 197–222 (1992)
15. de Kleer, J., Janssen, B., Bobrow, D.G., Kurtoglu, T., Saha, B., Moore, N.R., Sutharshana, S.: Fault augmented Modelica models. In: 24th International Workshop on Principles of Diagnosis (DX), pp. 71–78 (2013)
16. Dechter, R.: Constraint Processing. Morgan Kaufmann, San Francisco (2003)
17. Dudek, G., Jenkin, M.: Computational Principles of Mobile Robotics, 2nd edn. Cambridge University Press, New York (2010)
18. Felfernig, A., Schubert, M., Zehentner, C.: An efficient diagnosis algorithm for inconsistent constraint sets. AI EDAM **26**(1), 53–62 (2012). https://doi.org/10.1017/S0890060411000011
19. Fleischanderl, G., Schreiner, H., Havelka, T., Stumptner, M., Wotawa, F.: DiKe—a model-based diagnosis kernel and its application. In: Proceedings of the Joint German/Austrian Conference on Artificial Intelligence (KI), Vienna (2001)
20. Forbus, K.D.: Qualitative process theory. Artif. Intell. **24**, 85–168 (1984)
21. Friedrich, G., Gottlob, G., Nejdl, W.: Hypothesis classification, abductive diagnosis and therapy. In: Proceedings of the International Workshop on Expert Systems in Engineering. Lecture Notes in Artificial Intelligence, vol. 462. Springer, Vienna (1990)
22. Friedrich, G., Gottlob, G., Nejdl, W.: Physical impossibility instead of fault models. In: Proceedings of the National Conference on Artificial Intelligence (AAAI), Boston, pp. 331–336 (1990). Also appears in Readings in Model-Based Diagnosis (Morgan Kaufmann, 1992)
23. Friedrich, G., Gottlob, G., Nejdl, W.: Formalizing the repair process. In: Proceedings of the European Conference on Artificial Intelligence (ECAI), pp. 709–713. Wiley, Chichester (1992). Also appeared in the Proceedings of the Second International Workshop on Principles of Diagnosis, Milano (1991)
24. Fritzson, P.: Object-Oriented Modeling and Simulation with Modelica 3.3—A Cyber-Physical Approach, 2nd edn. Wiley-IEEE Press, Piscataway (2014)
25. Greiner, R., Smith, B.A., Wilkerson, R.W.: A correction to the algorithm in Reiter's theory of diagnosis. Artif. Intell. **41**(1), 79–88 (1989)
26. Hawkins, P.G., Woollons, D.J.: Failure modes and effects analysis of complex engineering systems using functional models. Artif. Intell. Eng. **12**, 375–397 (1998)
27. Hofbaur, M.W., Köb, J., Steinbauer, G., Wotawa, F.: Improving robustness of mobile robots using model-based reasoning. J. Intell. Robot. Syst. **48**(1), 37–54 (2007)
28. IBM: An architectural blueprint for autonomic computing (2003)
29. ISO/IEC/IEEE: Systems and software engineering—vocabulary. 24765:2010(E), pp. 1–418 (2010). https://doi.org/110.1109/IEEESTD.2010.5733835
30. Kephart, J., Chess, D.: The vision of autonomic computing. Comput. Mag. **36**(1), 41–52 (2003)

31. Koitz, R., Wotawa, F.: On the feasibility of abductive diagnosis for practical applications. In: 9th IFAC Symposium on Fault Detection, Supervision and Safety of Technical Processes (2015)
32. Krenn, W., Wotawa, F.: Intelligent, fault adaptive control of autonomous systems. In: Madrid, N.M., Seepold, R.E.D. (eds.) Intelligent Technical Systems. Lecture Notes in Electrical Engineering, vol. 38, pp. 175–188. Springer, Berlin (2009). https://doi.org/10.1007/978-1-4020-9823-9_13
33. Kuipers, B.: Qualitative simulation. Artif. Intell. **29**, 289–388 (1986)
34. Mahadevan, N., Dubey, A., Karsai, G., Srivastava, A., Liu, C.C.: Temporal causal diagrams for diagnosing failures in cyber-physical systems. In: Proceedings of the Annual Conference of the Prognostics and Health Management Society (PHM). PHM Society (2014). https://www.phmsociety.org/node/1439
35. McCune, W.: Prover9 and mace4 (2005–2010). http://www.cs.unm.edu/~mccune/prover9/
36. Minhas, R., de Kleer, J., Matei, I., Saha, B.: Using fault augmented modelica models for diagnostics. In: Proceedings of the 10th International Conference on Modelica, Lund, pp. 437–445 (2014)
37. Moreno, J., Madrenas, J., Faura, J., Canto, E., Cabestany, J., Insenser, J.: Feasible evolutionary and self-repairing hardware by means of the dynamic reconfiguration capabilities of the FIPSOC devices. In: Proceedings of the Evolvable Systems: From Biology to Hardware (ICES 1998), pp. 345–355. Springer, Berlin (1998)
38. Musliner, D., Goldman, R., Pelican, M., Krebsbach, K.: Self-adaptive software for hard real-time environments. Intell. Syst. **14**(4), 23–29 (1999)
39. Ng, H.T., Mooney, R.J.: An efficient first-order horn-clause abduction system based on the ATMS. In: Proceedings of the Ninth National Conference on Artificial Intelligence (AAAI-91), pp. 494–499. MIT Press, Anaheim (1991)
40. Nica, I., Wotawa, F.: The SiMoL modeling language for simulation and (re-)configuration. In: Proceedings of the International Conference on Current Trends in Theory and Practice of Computer Science, vol. 7147, pp. 661–672. Springer, Berlin (2012)
41. Nica, I., Pill, I., Quaritsch, T., Wotawa, F.: The route to success—a performance comparison of diagnosis algorithms. In: Proceedings of the Twenty-Third International Joint Conference on Artificial Intelligence (IJCAI), Beijing (2013)
42. Niggemann, O., Lohweg, V.: On the diagnosis of cyber-physical production systems: state-of-the-art and research agenda. In: Proceedings of the Twenty-Ninth AAAI Conference on Artificial Intelligence (AAAI), pp. 4119–4126. Association of the Advancement of Artificial Intelligence, Menlo Park (2015)
43. Peischl, B., Pill, I., Wotawa, F.: Abductive diagnosis based on modelica models. In: 27th International Workshop on Principles of Diagnosis (DX) (2016).
44. Pilgerstorfer, P., Pournaras, E.: Self-adaptive learning in decentralized combinatorial optimization—a design paradigm for sharing economies. In: 2017 IEEE/ACM 12th International Symposium on Software Engineering for Adaptive and Self-Managing Systems (SEAMS), pp. 54–64. IEEE, Buenos Aires (2017)
45. Rajan, K., Bernard, D., Dorais, G., Gamble, E., Kanefsky, B., Kurien, J., Millar, W., Muscettola, N., Nayak, P., Rouquette, N., Smith, B., Taylor, W., Tung, Y.: Remote agent: an autonomous control system for the new millennium. In: Proceedings of the 14th European Conference on Artificial Intelligence (ECAI), Berlin (2000)
46. Reiter, R.: A theory of diagnosis from first principles. Artif. Intell. **32**(1), 57–95 (1987)
47. Rincon, F., Teres, L.: Reconfigurable hardware systems. In: Proceedings of the International Conference on Semiconductor, New York, vol. 1, pp. 45–54 (1998)
48. Sachenbacher, M., Struss, P.: Task-dependent qualitative domain abstraction. Artif. Intell. **162**(1–2), 121–143 (2005). https://doi.org/10.1016/j.artint.2004.01.005
49. Schölzel, M., Koal, T., Müller, S., Scharoba, S., Röder, S., Vierhaus, H.T.: A comprehensive software-based self-test and self-repair method for statically scheduled superscalar processors. In: 17th Latin-American Test Symposium (LATS), Foz do Iguacu, pp. 33–38 (2016). https://doi.org/10.1109/LATW.2016.7483336

50. Seebach, H., Nafz, F., Holtmann, J., Meyer, J., Tichy, M., Reif, W., Schäfer, W.: Designing self-healing in automotive systems. In: Xie, B., Branke, J., Sadjadi, S.M., Zhang, D., Zhou, X. (eds.) Proceedings of the 7th International Conference on Autonomic and Trusted Computing (ATC'10), pp. 47–61. Springer, Berlin (2010)
51. Seltzer, M., Small, C.: Self-monitoring and self-adapting operating systems. In: The Sixth Workshop on Hot Topics in Operating Systems, California, pp. 124–129 (1997)
52. Shevtsov, S., Weyns, D., Maggio, M.: Handling new and changing requirements with guarantees in self-adaptive systems using SimCA. In: 2017 IEEE/ACM 12th International Symposium on Software Engineering for Adaptive and Self-Managing Systems (SEAMS), pp. 12–23. Buenos Aires (2017)
53. Steinbauer, G., Wotawa, F.: Model-based reasoning for self-adaptive systems—theory and practice. In: Assurances for Self-Adaptive Systems: Principles, Models, and Techniques, pp. 187–213. Springer, Berlin (2013). https://doi.org/10.1007/978-3-642-36249-1_7
54. Steinbauer, G., Mörth, M., Wotawa, F.: Real-time diagnosis and repair of faults of robot control software. In: RoboCup. Lecture Notes in Computer Science, vol. 4020, pp. 13–23. Springer, Berlin (2005)
55. Sterling, R., Struss, P., Febres, J., Sabir, U., Keane, M.M.: From modelica models to fault diagnosis in air handling units. In: Proceedings of the 10th International Conference on Modelica. Linköping University Press, Lund (2014)
56. Struss, P.: Deviation models revisited. In: Working Papers of the 15th International Workshop on Principles of Diagnosis (DX-04) (2004)
57. Struss, P., Dressler, O.: Physical negation—Integrating fault models into the general diagnostic engine. In: Proceedings 11th International Joint Conference on Artificial Intelligence, Detroit, pp. 1318–1323 (1989)
58. Travé-Massuyès, L., Ironi, L., Dague, P.: Mathematical foundations of qualitative reasoning. AI Mag. **24**(4), 91–106 (2004)
59. Travé-Massuyès, L., Prats, F., Sánchez, M., Agell, N.: Relative and absolute order-of-magnitude models unified. Ann. Math. Artif. Intell. **45**(3–4), 323–341 (2005)
60. Umeda, Y., Tetsuo, T., Hiroyuki, Y.: A design methodology for self-maintenance machines. J. Mech. Des. **117**, 41–53 (1995)
61. Villasenor, J., Hutchings, B.: The flexibility of configurable computing. Signal Process. Mag. **15**(5), 67–84 (1998)
62. Weld, D., de Kleer, J. (eds.): Readings in Qualitative Reasoning about Physical Systems. Morgan Kaufmann, San Mateo (1989)
63. Weyns, D., Holvoet, T.: An architectural strategy for self-adapting systems. In: Proceedings of the 2007 International Workshop on Software Engineering for Adaptive and Self-Managing Systems (SEAMS '07). IEEE Computer Society, Washington (2007). http://dx.doi.org/10.1109/SEAMS.2007.3
64. Wotawa, F.: Failure mode and effect analysis for abductive diagnosis. In: Proceedings of the International Workshop on Defeasible and Ampliative Reasoning (DARe) (2014)

Decentralized Modular Approach for Fault Diagnosis of a Class of Hybrid Dynamic Systems: Application to a Multicellular Converter

Moamar Sayed-Mouchaweh

1 Learning from Data Streams

1.1 Basic Definitions and Motivation

A fault can be defined as a nonpermitted deviation of at least one characteristic property of a system or one of its components from its normal or intended behavior. Fault diagnosis is the operation of detecting faults and determining possible candidates that explain their occurrence. Most of real systems are hybrid dynamic systems (HDS) [1] in which the discrete and continuous dynamics cohabit. The discrete dynamics are described by discrete state variables, while the continuous dynamics are described by continuous state variables.

The general principle of model-based diagnosis approaches [2] is based on the use of a model of the system normal and/or fault behaviors. Discrete-event model-based diagnosis approaches [3, 4] describe the system as discrete mode changes in response to the occurrence of discrete events. Therefore, they ignore the continuous dynamics of the system. Continuous model-based diagnosis approaches [5] represent the system dynamics as a continuous time evolution using differential equations. However, they do not take into account the discrete changes of the system discrete modes or configurations.

Consequently, both approaches cannot be used to perform the fault diagnosis of HDS since in the latter both continuous and discrete dynamics and the interactions between them must be taken into account. Hence, fault diagnosis of HDS must deal with the evolution of continuous dynamics in each discrete mode in order to

---

M. Sayed-Mouchaweh (✉)
Institute Mines-Telecom Lille Douai, Douai, France
e-mail: moamar.sayed-mouchaweh@mines-douai.fr

© Springer Nature Switzerland AG 2019
E. Lughofer, M. Sayed-Mouchaweh (eds.), *Predictive Maintenance in Dynamic Systems*, https://doi.org/10.1007/978-3-030-05645-2_16

construct a diagnosis module (called diagnoser). The latter may be modeled in HDS by introducing parameters into the system model or using explicit fault events or/and fault modes.

Discretely controlled continuous systems (DCCS) [3] are a special class of HDS widely used in the literature. They are composed by a plant with continuous dynamics and supervisory discrete control. The latter generates discrete control events in order to change regulator set points or the plant configuration. Their behavior is described through a set of discrete operation modes and a set of algebraic differential equations within each discrete mode. Transitions from one discrete mode to another one are achieved upon discrete control events, e.g., open or close a switch, based on continuous state conditions. This paper focuses on the fault diagnosis of DCCS.

1.2 State of the Art

Several approaches have been proposed in the literature for fault diagnosis of HDS. They can be divided into two main categories according to how they model the system's hybrid dynamics. In the first category [6–8], the hybrid model is an extension of the continuous model by adding the system discrete modes. The fault-free continuous behavior is defined in each discrete mode by relations over observable variables. These relations are used in order to generate residuals sensitive to a certain subset of faults. A fault is diagnosed when the value of the sensitive residuals to this fault is different from zero.

In the second category, the discrete model is extended or enriched by adding events generated by the abstraction of system's continuous dynamics. In [9–12], a set of residuals or guards are defined to represent the fault-free continuous dynamics in each discrete mode. The occurrence of unobservable discrete fault generates unpredicted transition from one discrete mode to another one. In this case, the residuals or guards, defined for the discrete mode before the fault occurrence, are different from zero in the discrete mode after the fault occurrence. This is due to the fact that these residuals or guards describe the continuous dynamics' conditions in the normal operation conditions. This change of residuals or guards' values in a discrete mode indicates the occurrence of a discrete fault leading the system to reach unpredicted discrete mode. The continuous dynamics can also be integrated in the discrete model by using the occurrence time of events [13, 14]. In this case, the occurrence of faults does not change events ordering but only alters their timing characteristics. Therefore, a fault is diagnosed when predicted events occur too late or too early or they do not occur at all during their predefined time intervals.

Finally, the events referring to the continuous dynamics can be generated by the use of a bond graph. The authors in [15] construct temporal causal graphs (TCG) for each normal and fault discrete mode based on the use of a global hybrid bond graph. When measurement deviations, caused by fault occurrence, are observed through residuals, TCG are used to determine the effects that faults have on the measurements as well as the temporal order in which they deviate. Then, fault signature is defined for each fault as the qualitative values of the magnitude and the first nonzero derivative change which can be observed in the residuals. The online diagnosis is performed by building the current signature of the system based on the measurements' deviations and comparing it to each of the different fault signatures.

The common drawback of these approaches is the fact that they do not scale to large systems with multiple discrete modes. Indeed, the fault diagnosis in these approaches is performed based on the use of a global model representing both the discrete and continuous dynamics. The global model can be too large to be physically constructed for systems with a large number of discrete modes. As an example, for telecommunication networks, as the one studied in [16], the number of states of the global model is of the order of 210×4300. Therefore, constructing the global model is physically unfeasible.

1.3 Contribution of the Proposed Approach

In this chapter, an approach to perform the fault diagnosis of HDS, in particular DCCS, without the use of a global model is proposed. This approach exploits the modularity of the system by dividing it into several discrete components. Then, the local model for each of the latter is built. This local model includes the normal discrete modes as well as the ones reached in response to the occurrence of faults impacting the discrete behavior of this component.

The continuous dynamic behavior in each discrete mode is defined by a set of analytical redundancy relations (ARRs). An ARR is a constraint defined over the observable continuous variables in a discrete mode [17]. They are used to generate residuals. The latter allow evaluating the consistency between the system model (fault-free behavior) and the real observation. A fault occurs when ARRs' value is different from zero. The abstraction of these residuals generates events that are used to enrich the local discrete models. These events can turn unobservable transitions, due to unobservable fault events, into observable transitions leading to infer the discrete fault occurrence.

A local diagnoser is built for each discrete component based on the use of the corresponding enriched local model. The residuals are defined in such a way that they are sensitive only to the faults impacting the dynamic behavior of one local component. Therefore, the events defined by the abstraction of these residuals are generated when a fault in this local component occurs. The local diagnoser of a component diagnoses only the faults occurring in its associated component (see Fig. 1). This can be useful for the diagnosis of multiple faults. When faults occurred in several components, the associated local diagnosers announce mutually the occurrence of these faults. Apart that the diagnosis is performed without the use of a global model, another advantage is related to the improvement of the diagnosis robustness in the sense that when one local diagnoser is failed, the other local diagnosers remain operational and they continue to perform the fault diagnosis in their respective discrete components.

The paper is structured as follows. Section 2 details the proposed approach by explaining each of its steps presented in Fig. 1. In Sect. 2, the complexity for constructing the local diagnosers is demonstrated to be polynomial in the size of local models. In addition, the diagnosis obtained by the proposed decentralized structure is demonstrated to be equivalent to a centralized diagnosis structure. The evaluation of this approach using several simulation scenarios and the obtained results are discussed in Sect. 3. Section 4 ends the paper by some concluding remarks. A three-cellular converter is used throughout the paper as an example in order to illustrate and validate the proposed approach. Power converters are electrical systems used in order to turns the DC (direct current) power into AC (alternating current) power, AC power into AC power, DC power into DC power, and AC power into DC power. The three-cell converter is adapted for DC/DC and DC/AC power conversion. It is based on the combination of three switches (cells of commutation) allowing the current flowing from the voltage source toward the output load.

2 Proposed Approach

The proposed approach performs the fault diagnosis based on the steps illustrated in Fig. 1. These steps are detailed in the following subsections.

2.1 System Decomposition

DCCS consist of continuous components (Ccs) whose operation modes are switched according to the configuration or discrete mode of its discrete components (Dcs). In order to illustrate the proposed approach, the three-cellular converters [18], depicted in Fig. 2, are used. The continuous dynamics of the system are described by state vector $X = [Vc_1 \quad Vc_2 \quad I]^T$, where Vc_1 and Vc_2 represent, respectively, the floating voltage of capacitors C_1 and C_2 and I represents the load current flowing from source E toward the load (R, L) through three elementary switching cells $S_j, j \in \{1, 2, 3\}$. The latter represent the system discrete dynamics. Each discrete switch S_j has two discrete states: S_j opened ($h_q^j = 0$) or S_j closed ($h_q^j = 1$), where h_q^j is the discrete output of S_j. The control of this system has two main tasks: (1) balancing the voltages between the switches and (2) regulating the load current to a desired value. To accomplish that, the controller changes the switches' states from opened to closed or from closed to opened by applying discrete commands "CS_j" or

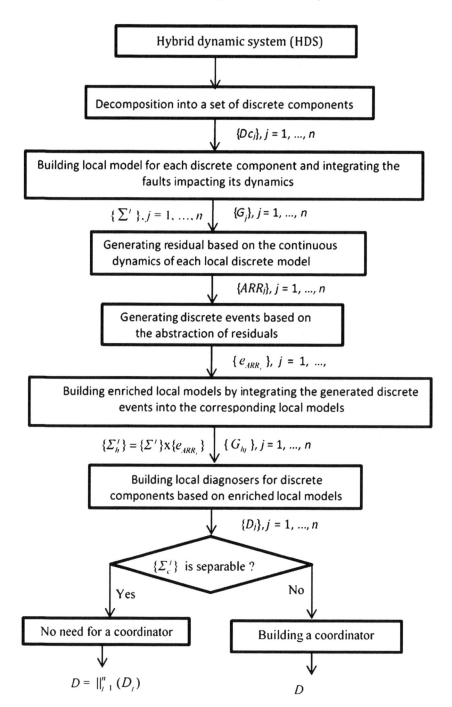

Fig. 1 Steps of the proposed decentralized modular approach

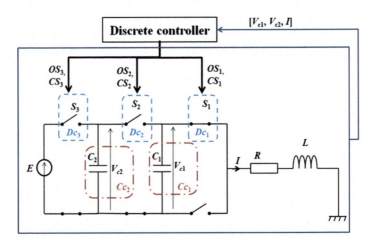

Fig. 2 Three-cell converter description and decomposition

"OS_j" to each discrete switch $S_j, j \in \{1, 2, 3\}$ (see Fig. 2) where CS_j refers to "close switch S_j" and OS_j to "open switch S_j." Thus, the considered example is a DCCS. The latter is decomposed into three discrete components S_1, S_2, and S_3 representing the three switches of the three-cellular converter (see Fig. 2).

Let G be the system global model and Σ be a finite set of events produced by G. It includes observable Σ_o and unobservable Σ_u events. $\Sigma_c \subseteq \Sigma_o$ where Σ_c is the set of controllable events, e.g., OS_1, CS_1. Σ^+ is the set of all event sequences over Σ. Let Σ^* be equal to $\Sigma^+ \cup \{\varepsilon\}$ where ε denotes the empty event sequence. A subset $L \subseteq \Sigma^*$ is called a language over Σ. G observes the system by one global observation mask $M : \Sigma^* \to \Sigma_o^*$. Thus, M erases the unobservable events in an event sequence. The inverse global observation mask is defined as: $M^{-1}(u) = \{s \in L : M(s) = u\}$. Let Σ_f be the set of fault events which can occur in the system. Σ_f is partitioned into different fault types. Each fault type requires the identification not of the fault event itself but of the type of fault when such an event occurs in the system. Let $\Sigma_f = \Sigma_{F_1} \cup \cdots \cup \Sigma_{F_r}$, where $\Sigma_{F_j}, j \in \{1, \ldots, r\}$, denotes disjoint sets of fault events corresponding to different fault types. It consists of the set of faults which have the same effect according to either the configuration or maintaining procedure.

For the three-cellular converters, eight faults can be considered for the diagnosis [19, 20] as depicted in Table 1. This paper focuses on the fault diagnosis of discrete faults because they are more frequent and their consequences are more destructive. For instance, in open-circuit (stuck-off) failure, the system operates in degraded performance. However, unstable load may lead to further damage on the system. Therefore, the diagnosis of these faults is necessary to ensure the system safety and quality.

Table 1 Faults for the diagnosis of three-cell converters

Fault types	Fault labels	Fault event—fault description
Discrete faults	F_1	f_{s1so}—S_1 stuck opened
		f_{s1sc}—S_1 stuck closed
	F_2	f_{s2so}—S_2 stuck opened
		f_{s2sc}—S_2 stuck closed
	F_3	f_{s3so}—S_3 stuck opened
		f_{s3sc}—S_3 stuck closed
Parametric faults	F_4	f_{C1}—Abnormal change in the nominal values of C_1 due to C_1 aging
	F_5	f_{C2}—Abnormal change in the nominal values of C_2 due to C_2 aging

2.2 Discrete Component Modeling

The nominal and faulty behaviors of each discrete component $Dc_j, j \in \{1, \ldots, n\}$, is modeled using a finite state automaton G_j defined by:

$$G_j = \left(Q^j, \Sigma^j, q_0^j, \delta^j \right)$$

where $Q^j = \left\{ q_k^j \right\}$ is a finite set of discrete states (modes) of Dc_j. It includes normal and discrete failure states.

G_j describes the component dynamical behavior by sequences of states (representing the component discrete modes) and discrete events entailing the transition from one state to another. G_j represents the normal behavior that this component can execute in response to a control command event as well as faulty behaviors in response to the occurrence of a set of predefined fault events in the system.

The output of q_k^j is characterized by real discrete output vector $h_q^j \in \{0, 1\}$ and nominal discrete output vector $\tilde{h}_q^j \in \{0, 1\}$. At normal discrete mode (state) $\tilde{h}_q^j = h_q^j$, while in faulty modes $\tilde{h}_q^j \neq h_q^j$. $\Sigma^j = \Sigma_o^j \cup \Sigma_u^j$ is the event set of Dc_j. It includes observable events Σ_o^j corresponding to control command events and unobservable events Σ_u^j. Σ_u^j includes discrete fault events Σ_f as well as normal but unobservable events. $q_0^j \in Q^j$ is the initial state. $\delta^j : Q^j \times \Sigma^j \to Q^j$ is the state transition function. A transition $\delta^j(q^j, e) = q^{j+}$ corresponds to a change from state q^j to state q^{j+} after the occurrence of event $e \in \Sigma^j$. K is the set of control command event sequences generated by the controller and sent to the different Dcs.

$L_j \subseteq \Sigma_j^*$ is the local language of Dc_j representing its local behavior. Therefore, Σ, Σ_o, and Σ_u are equal, respectively, to $\Sigma_1 \cup \cdots \cup \Sigma_n$, $\Sigma_{1o} \cup \cdots \cup \Sigma_{no}$, and

$\Sigma_{1u} \cup \cdots \cup \Sigma_{nu}$. Similarly, a local observation mask can be defined for each local model G_j as: $M_j : \Sigma_j^* \to \Sigma_{jo}^*$. The natural projection from Σ^* to Σ_j^* is defined by: $P_j : \Sigma^* \to \Sigma_j^*$.

G can be obtained by achieving the synchronous composition [21] between its local models: $G = G_1 \parallel G_2 \parallel \ldots \parallel G_n$, where the symbol $\parallel$ is the parallel or synchronous composition operator. It builds the system global model from its individual interconnected component models. In this parallel or synchronous composition, a common event between two components can only be executed if both components execute it simultaneously. However and contrary to product composition, the private events which can be executed by only one component can be executed whenever possible. Therefore, in this type of interconnection, a component can execute its private events without the participation of the other components. The synchronous composition between local languages produces the global behavior of the system: $L = L_1 \parallel L_2 \parallel \ldots \parallel L_n$.

The set of control command event sequences executed by Dc_j is represented by K^j. It is worth to mention that $K = \parallel_{j=1}^{j=n} (K_j)$ if and only if $\{\Sigma^j\}$ is separable [22]. This is due to the fact that because of components' partial observation, the parallel or synchronous composition operator between K_j may generate event sequences that do not belong to K. Therefore, a coordinator is required in order to delete these event sequences thanks to its limited global observation. For more details, reader can refer to [23] and the references therein.

Figure 3 shows G_1 of Dc_1 of the three-cellular converter of Fig. 2. It is worth to mention that the states with N label cannot be distinguished from states with fault labels ($F1$, $F2$) since the real output h_q^j, $j \in \{1, \ldots, n\}$ is not observable or measurable.

Fig. 3 Local model G_1 for switch S_1 of the three-cell converter

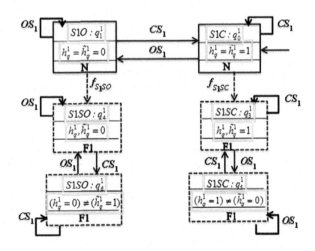

2.3 Residual Generation Based on System Continuous Dynamics

The continuous dynamics of the DCCS can be described in each discrete state q as follows:

$$\dot{X} = A_{(q)} X + B_{(q)} u$$

where X is the state vector and u is the input vector. In the case of linear systems, $A_{(q)}$ and $B_{(q)}$ are constant matrices of appropriate dimensions.

For each discrete mode q, an analytical redundancy relation (ARR) can be defined as a constraint based on the continuous dynamics in q. ARR contains observable, or measurable, continuous variables and can be determined off-line. Then, it can be evaluated online to test the consistency between the observed behavior and the predicted one. Therefore, an ARR can be expressed as a residual r. When ARR is satisfied, this means that the observed behavior matches the predicted (the model) one and in this case $r = 0$. While in the opposite case, r is different of zero which indicates the occurrence of a fault.

For instance, the multicellular converter with p cells or switches and a main voltage source E has 2^p discrete modes and $p - 1$ reference voltage (Vc_{jref}) of the floating capacitors as follows:

$$Vc_{jref} = j\frac{E}{p}, j = 1, \ldots, p - 1$$

The discrete controller ensures the simultaneous regulation of the load current I and Vc_j in order to be at their reference values. The dynamics of multicellular converter with $p = 3$ and a load consisting of a resistor R and inductance L can be described by [24]:

$$\begin{cases} \dot{V}c_1 = -h_q^1 \frac{1}{C_1} I + h_q^2 \frac{1}{C_1} I \\ \dot{V}c_2 = -h_q^2 \frac{1}{C_2} I + h_q^3 \frac{1}{C_2} I \\ \dot{I} = -\frac{R}{L} I + h_q^1 \frac{1}{L} Vc_1 + h_q^2 \frac{1}{L} (Vc_2 - Vc_1) + h_q^3 \frac{1}{L} (E - Vc_2) \end{cases}$$

where $u(t) = E$.

The continuous dynamics of the three-cellular converter are thus described by state vector $X = \begin{bmatrix} Vc_1 & Vc_2 & I \end{bmatrix}^T$ and $Vc_{1ref} = \frac{E}{3}$, $Vc_{2ref} = \frac{2E}{3}$.

The output voltage V_s for a multicellular converter of p cells can take $p + 1$ levels as follows:

$$V_s = Eh_p + \sum_{j=1}^{p-1} Vc_j (h_j - h_{j+1})$$

In this paper, it is assumed that only V_s can be measured.
r_q can take the following expression in each q:

$$r_q = (x_m)_q - \tilde{x}_q$$

where x_m is the measured continuous variable and $\tilde{x}_q$ its nominal or reference value. $\tilde{x}_q$ can be defined as a function of the nominal discrete states or modes of the DCCS as follows:

$$\tilde{x}_q = f\left(\tilde{q}^1, \tilde{q}^2, \ldots, \tilde{q}^j, \ldots, \tilde{q}^n\right)$$

Since each discrete component state q is represented by its output h_q^j, the previous equation can be rewritten as follows:

$$\tilde{x}_q = f\left(\tilde{h}_q^1, \tilde{h}_q^2, \ldots, \tilde{h}_q^j, \ldots, \tilde{h}_q^n\right)$$

For the example of the three-cellular converter, the residual r_q can be defined as follows:

$$r_q = (V_{sm})_q - \left(\tilde{V}_s\right)_q$$

$$\left(\tilde{V}_s\right)_q = E\tilde{h}_q^3 + \frac{E}{3}\left(\tilde{h}_q^1 - \tilde{h}_q^2\right) + \frac{2E}{3}\left(\tilde{h}_q^2 - \tilde{h}_q^3\right)$$

Let the occurrence of faults impacting a discrete component $Dc_j, j \in \{1, \ldots, n\}$ be indicated by the fault label F_j, while their absence is indicated by the label N. The occurrence of a fault in Dc_j entails an unanticipated transition to a discrete state or mode q^j different from the predicted or normal one $\tilde{q}^j$. Therefore, $\tilde{h}_q^j$ is different from h_q^j.

In order to define a residual sensitive to the faults that can impact only one discrete component Dc_j, the residual is decomposed as follows:

$$r_j = (x_m)_q - \left(\tilde{x}_q\right)^j, \, j \in \{1, \ldots, n\},$$
$$\left(\tilde{x}_q\right)^j = f\left(h_q^1, \ldots, \tilde{h}_q^j, \ldots, h_q^n\right)$$
$$r_j = (x_m)_q - f\left(h_q^1, h_q^2, \ldots, \tilde{h}_q^j, \ldots, h_q^n\right), \, j \in \{1, \ldots, n\}$$

Indeed, when a fault impacts a discrete component Dc_j, $\tilde{h}_q^j$ is different from h_q^j. Therefore, r_j is a residual sensitive to the faults that can occur only in the discrete component Dc_j.

For the three-cellular converter, three residuals can be defined where each residual $r_j, j \in \{1,2,3\}$ is sensitive to faults impacting one discrete component $Dc_j, j \in \{1, \ldots, n\}$:

$$r_1 = (V_{sm})_q - \left(Eh^3 + \frac{E}{3}\left(\tilde{h}^1 - h^2\right) + \frac{2E}{3}\left(h^2 - h^3\right)\right)$$

$$r_2 = (V_{sm})_q - \left(Eh^3 + \frac{E}{3}\left(h^1 - \tilde{h}^2\right) + \frac{2E}{3}\left(\tilde{h}^2 - h^3\right)\right)$$

$$r_3 = (V_{sm})_q - \left(E\tilde{h}^3 + \frac{E}{3}\left(h^1 - h^2\right) + \frac{2E}{3}\left(h^2 - \tilde{h}^3\right)\right)$$

2.4 Enriched Local Models Building

As it is explained before, the occurrence of a discrete fault entails an unanticipated change of the continuous dynamics in a discrete state or mode. This unanticipated change is inferred by the mean of residuals measuring the difference between the nominal (predicted) continuous behavior and the real one observed through sensors. These residuals can be zero, above zero, or below zero. Therefore, their abstraction can generate three events as follows:

$$\forall j \in \{1, \ldots, n\}, \{e_{ARR_j}\} = \begin{bmatrix} r_j = 0 \Rightarrow e_j^0 \\ r_j > 0 \Rightarrow e_j^+ \\ r_j < 0 \Rightarrow e_j^- \end{bmatrix}$$

These events may turn unobservable transitions into observable ones leading to distinguish normal states (behavior) from fault ones. Therefore, these events are integrated into the local models $G_j, j \in \{1, \ldots, n\}$ in order to obtain the corresponding enriched ones $G_{hj}, j \in \{1, \ldots, n\}$.

G_{hj} is defined as Mealy machine (MM) model where the control command event (input event) is associated to an event generated by the abstraction of the residuals (output event related to the sensor reading). Formally, G_{hj} can be expressed as follows:

$$G_{hj} = \left(Q^j, \Sigma_h^j, q_0^j, \delta_h^j\right)$$

where $\Sigma_h^j = \Sigma^j \times \{e_{ARR_j}\}$ and $\delta_h^j : Q^j \times \Sigma_h^j \rightarrow Q^j$.

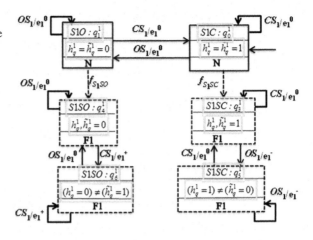

Fig. 4 Enriched local model G_{h1} for switch S_1 (Dc_1) of the three-cell converter

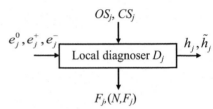

Fig. 5 Local diagnoser structure

For the three-cell converter, Σ_h^1, Σ_h^2, and Σ_h^3 for, respectively, Dc_1, Dc_2, and Dc_3 are defined as follows:

$$\Sigma_h^1 = \left(\Sigma^1 = \{OS_1, CS_1\}\right) \times \left(\{e_{ARR_1}\} = \left\{e_1^0, e_1^+, e_1^-\right\}\right)$$

$$\Sigma_h^2 = \left(\Sigma^2 = \{OS_2, CS_2\}\right) \times \left(\{e_{ARR_2}\} = \left\{e_2^0, e_2^+, e_2^-\right\}\right)$$

$$\Sigma_h^3 = \left(\Sigma^3 = \{OS_3, CS_3\}\right) \times \left(\{e_{ARR_3}\} = \left\{e_3^0, e_3^+, e_3^-\right\}\right)$$

Figure 4 shows the enriched local model G_{h1} of the discrete component Dc_1 of the three-cell converter of Fig. 2.

2.5 Local Hybrid Diagnoser Construction

For each discrete component Dc_j, a local diagnoser $D_j, j \in \{1, \ldots, n\}$ (see Fig. 5), is built based on the use of the corresponding enriched local model G_{hj}.

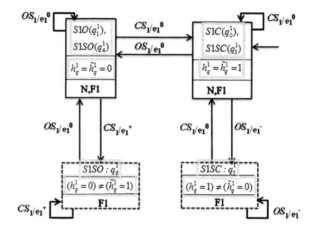

Fig. 6 Local diagnoser D_1 for the discrete component (switch) Dc_1 (S_1) of the three-cell converter

For the example of the three-cell converter, three local diagnosers, D_1, D_2, and D_3, are built for the three switches S_1 (Dc_1), S_2 (Dc_2), and S_3 (Dc_3).

Each local diagnoser integrates the pure discrete events (command control events) generated by the controller and the events generated by the abstraction of the residuals defined based on the continuous dynamics of the system. $D_j, j \in \{1, \ldots, n\}$, contains only the observable events. Therefore, the states linked by unobservable transitions are fused since they are not distinguishable. In this case, states with normal label (N) may be fused with states with a fault label. Hence, this fusion creates confusion between the normal and faulty states. This is due to the fact that the real output h_q^j, $j \in \{1, \ldots, n\}$, is not observable or measurable. The events generated by the abstraction of the continuous dynamics may lead a local diagnoser to reach a state where this confusion is removed in order to confirm the occurrence of a fault. Hence, these events allow the fault diagnosis by reaching certain states with one fault label within bounded time. This is due to the fact that they turn unobservable transitions into observable ones. In this case, the state real output h_q^j, $j \in \{1, \ldots, n\}$, can be inferred at these states with certain fault labels.

For the three-cell converter, Fig. 6 shows the local diagnoser D_1 built based on the enriched local model G_{h1} (see Fig. 4) of the discrete component Dc_1 (see Fig. 3). Likewise, the other local diagnosers, D_2 and D_3, can be built.

Figure 7 shows the scheme of the proposed decentralized fault diagnosis of HDS.

2.6 Equivalence Between Centralized and Decentralized Diagnosis Structures

Diagnosability property [25] ensures that a predefined set of faults can be diagnosed by a centralized diagnoser built using a global model of the system, while co-diagnosability [26] guarantees that these faults are diagnosed in decentralized

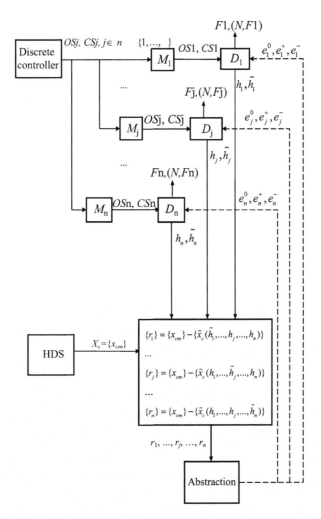

Fig. 7 Proposed decentralized fault diagnosis for HDS

manner using a set of local diagnosers. A fault must be diagnosed by at least one local diagnoser by using its proper local observation of the system. However, co-diagnosability property is stronger than diagnosability property. If a system is co-diagnosable, then it is diagnosable; while a diagnosable system does not ensure that it is co-diagnosable. Therefore, it is necessary to ensure the equivalence between centralized and decentralized diagnosis structures in the sense that if the system is diagnosable then it is co-diagnosable.

In the case that the controller command K is separable [22], then this equivalence is verified without the need of a coordinator. In this case, a fault diagnosed by a centralized diagnoser can be co-diagnosed by at least one local diagnoser. However, when the controller language (desired behavior) K is not separable, then a coordinator is required to ensure this equivalence. There exist several approaches that can build the coordinator without the need of a global model. Examples of these approaches can be found in [23] and the references therein.

For the three-cell converter, it is easy to verify that the controller command K is separable since $K = \|_{j=1}^{j=n} (K_j)$. Therefore, the global diagnoser D is equal to $D = \|_{j=1}^{j=n} (D_j)$. This can be demonstrated easily as follows. Let D be the centralized diagnoser that can infer the occurrence of any event sequence belonging to $L - K$. Indeed, any fault event sequence belongs to $L - K$ since the latter includes all the fault sequences that the system can generate but they do not belong to the desired behavior (represented by K). Since the controller command K is separable, then we can write:

$$L - K = \|_{j=1}^{n} L_j - \|_{j=1}^{n} K_j$$
$$L - K = \|_{j=1}^{n} (L_j - K_j)$$
$$D = \|_{j=1}^{n} D_j$$

2.7 Computation Complexity Analysis

Centralized diagnosis approaches are not suitable for large-scale systems as the telecommunication networks. Indeed, in the latter, the global model can contain a huge number of states. Therefore, constructing the global model is physically unfeasible.

Let $|G|$ be the number of states of G and $|\Sigma|$ be the number of events in G. The number of transitions for G is equal to $|G| \times |\Sigma|$. Let us assume that there are n local sites, i.e., n local diagnosers, in the system. The construction of the global diagnoser D requires computing G. Therefore, the computation complexity for constructing D is of the order $O(|G| \times |\Sigma|) = O(|G^j|^n \times n|\Sigma^j|)$. Therefore, the complexity computation of D using centralized diagnosis approach is exponential with the number of components n.

Let $|G_j|, |\Sigma_j|, j \in \{1, \ldots, n\}$, be the number of states in local model G_j and the number of events in Σ_j, respectively. The construction of local diagnoser D_j requires computing G_j. Therefore, the computation complexity of D_j is of the order $O(|G_j| \times |\Sigma_j|)$. The computation complexity of the n local diagnosers required to build the global diagnosis decision is of the order $O(n(|G_j| \times |\Sigma_j|))$. Consequently, the computation complexity of the proposed decentralized diagnosis approach is polynomial in the number of system components and the size of local models. This computation complexity does not depend on the number of states and transitions of the global model but it depends on the size of local models.

It is worth to mention that the proposed approach has the advantage to be scalable in the sense that adding or removing components does not require modifying any of the existing diagnosers but only adding new diagnosers for the new components or removing the diagnosers for the removed components. This is not the case for a centralized diagnosis structure where adding or removing components requires to rebuild the centralized diagnoser from scratch.

In the proposed approach, the set of local diagnosers are built using the local models of the system discrete components. Then, the global diagnoser can be constructed as a synchronous composition of these local diagnosers. Since the local diagnosers are computed using the local models, then the global diagnoser can be computed without the need for a global model but only local models. The global diagnoser can then be used to verify whether the predefined set of faults is diagnosable or not. Consequently, the diagnosability property can be verified using the proposed approach without the need for a global model.

3 Experimental Results

In order to evaluate the proposed approach, simulations were carried out for the three-cell converter using Matlab-Simulink™ environment and Stateflow™ toolbox. The parameters used in these simulations are: $E = 60$ V, $C_1 = C_2 = 40$ µF, $R = 200$ Ω, and $L = 0.1$ H. Figure 8 shows the proposed decentralized fault diagnosis structure for the three-cell converter.

Discrete controller commands are assured by a pulse width modulation (PWM) signal [24]. Figure 9 depicts the control of the three switches S_1, S_2, and S_3. When the triangular signal is below the reference signal (ref in Fig. 9), the associated switch is controlled to be opened. When the triangular signal is above the reference signal, the associated switch is controlled to be closed. This sequence of control is periodic with a period of 0.02 s.

Test scenarios are generated for single and multiple faults as follows (see Fig. 10). Each fault f, impacting each one of the switches (see Table 1), is generated starting at time t_{sf} and ending at time t_{ef}. Then, the system returns to normal operating conditions before generating a new fault for a certain time.

The output measured voltage V_{sm} in response to the generated fault scenarios of Fig. 10 is shown in Fig. 11. We can see that a stuck-on, respectively stuck-off, of a switch adds, respectively removes, a tension level of $E/3$ (20 V) to, respectively from, V_{sm}. In the case of multiple faults where one switch is stuck-on, respectively stuck-off, and another switch is stuck-off, respectively stuck-on, the impact of these

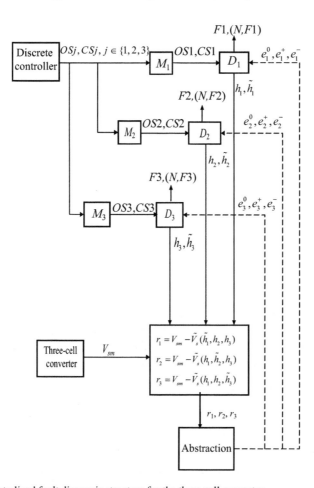

Fig. 8 Decentralized fault diagnosis structure for the three-cell converter

two faults on V_{sm} is masked. This is due to the fact that the consequence of one fault (adding or removing a tension level equal to $E/3$) compensates the consequence of the other fault. However, since it is impossible to have both faults at the same time and since the diagnosis delay is less than the time between two consecutive faults, then this multiple fault scenario can be diagnosed by the proposed approach where the first fault is diagnosed by one local diagnoser and then the second fault is diagnosed by a second local diagnoser.

The real state outputs, h^1, h^2, and h^3 of switches S_1, S_2, and S_3 according to the generated fault scenarios of Fig. 10 are shown in Fig. 12. We can see that when a fault occurs, the real state output does not change, react, anymore in response to the control command events.

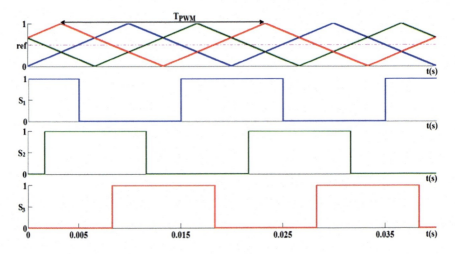

Fig. 9 PWM for control of three switches S_1, S_2, and S_3

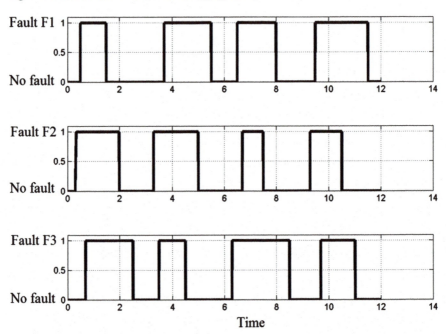

Fig. 10 Generated single and multiple fault scenarios

Figure 13 shows the residuals r_1, r_2, and r_3 generated in response to the fault scenarios of Fig. 10. We can see that each residual is sensitive to a fault impacting its corresponding discrete component (switch). Multiple faults can be diagnosed when two or more of the local diagnosers declare a fault.

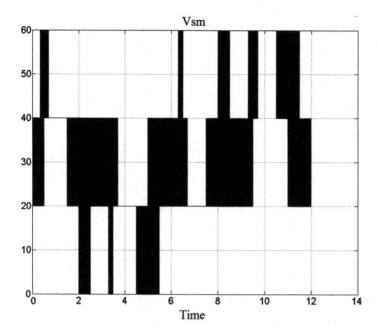

Fig. 11 Output measured voltage V_{sm} in response to the generated fault scenarios of Fig. 10

Figure 14 shows the local diagnosers' decisions in response to the fault scenarios of Fig. 10. We can see that each local diagnoser is able to diagnose with certainty and within bounded time the occurrence of a fault impacting its corresponding discrete component (switch).

The diagnosis delay corresponds to the time when the residuals are silent because the system is in a silent discrete state. In the latter, the continuous dynamics cannot allow the discrimination between normal and fault operation conditions. For instance, if the switch S_3 is in S_3O (S_3 opened) and the stuck-open fault occurs in S_3. In this discrete mode, it is impossible to distinguish between S_3O and S_3SO since the continuous dynamics, represented by V_{sm}, are the same. When the controller issues the control command event CS_3, S_3 is expected to change its discrete mode to S_3C. In this case, it is possible to distinguish between S_3C and S_3SO thanks to the continuous dynamics, represented by V_{sm} (see Fig. 6). The time between the occurrence of S_3 stuck-open fault event when S_3 is in S_3O and the occurrence of the control command event CS_3 represents the diagnosis delay to diagnose the S_3 stuck-open fault. However, it is worth mentioning that the control command signal changes quickly leading to change permanently the discrete state of switches. Therefore, the diagnosis delay is supposed to be very small.

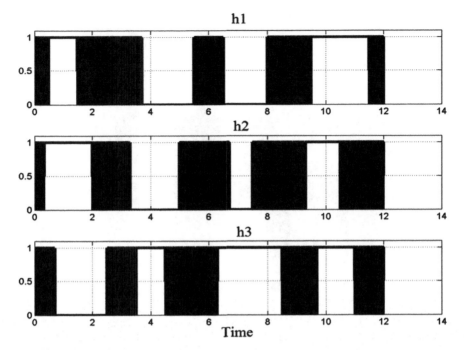

Fig. 12 Real state output for each discrete component (switch) in response to the generated fault scenarios of Fig. 10

4 Conclusion

In this paper, a decentralized modular approach to perform single and multiple discrete fault diagnosis of hybrid dynamic systems, in particular discretely controlled continuous systems, is proposed. This approach performs the fault diagnosis based on the use of a set of local diagnosers. The latter are built using the system components' local models. The system continuous dynamics is abstracted in order to generate events that are used to enrich the local models. These events allow improving the co-diagnosability of the local diagnosers by turning unobservable transitions into observable ones. There are two advantages for the proposed approach. Firstly, adding or removing components does not require modifying any of the existing diagnosers but only to add new diagnosers for the new components or to remove the diagnosers for the removed components. Secondly, the set of local diagnosers are built using the local models of the system discrete components. Therefore, the diagnosis is performed without the need for a global model.

Fig. 13 Local residuals, r_1, r_2, and r_3, generated in response to the fault scenarios of Fig. 10

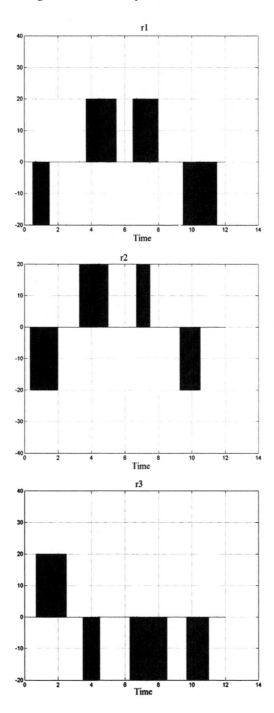

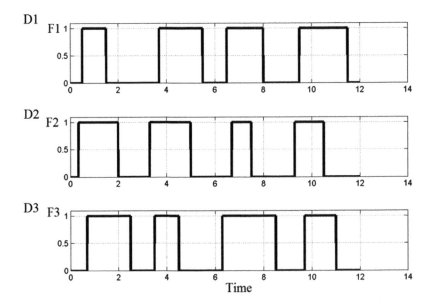

Fig. 14 Local diagnosers' decisions in response to the generated fault scenarios of Fig. 10

A future work is to develop the local diagnosers to be adaptive according to the changes in their environments. These changes are represented by the occurrence of new faults which are not known in advance. Therefore, the local diagnosers can adapt their inference engine to these changes by integrating online the new fault behaviors over time.

References

1. Van Der Schaft, A.J., Schumacher, J.M.: An Introduction to Hybrid Dynamical Systems, p. 251. Springer, London (2000)
2. Ding, S.X.: Model–based Fault Diagnosis Techniques: Design Schemes, Algorithms, and Tools, 1st edn. Springer, Berlin (2008)
3. Lunze, J.: Fault diagnosis of discretely controlled continuous systems by means of discrete-event models. Discret. Event Dyn. Syst. **18**(2), 181–210 (2008)
4. Sayed-Mouchaweh, M.: Discrete Event Systems: Diagnosis and Diagnosability. Springer, Berlin (2014)
5. Patton, R.J., Frank, P.M., Clark, R.N. (eds.): Issues of Fault Diagnosis for Dynamic Systems. Springer, London (2000)
6. Cocquempot, V., El Mezyani, T., Staroswiecki, M.: Fault detection and isolation for hybrid systems using structured parity residuals. In: Control Conference, 2004. 5th Asian, vol. 2, pp. 1204–1212. IEEE (2004)
7. Kamel, T., Diduch, C., Bilestkiy, Y., Chang, L.: Fault diagnoses for the Dc filters of power electronic converters. In: 2012 IEEE Energy Conversion Congress and Exposition (ECCE), pp. 2135–2141. IEEE (2012, September)

8. Van Gorp, J., Defoort, M., Djemai, M., Veluvolu, K.C.: Fault detection based on higher-order sliding mode observer for a class of switched linear systems. IET Control Theory Appl. **9**(15), 2249–2256 (2015)
9. Bayoudh, M., Travé-Massuyes, L., Olive, X.: Hybrid systems diagnosability by abstracting faulty continuous dynamics. In: Proceedings of the 17th International Principles of Diagnosis Workshop, pp. 9–15 (2006)
10. Bhowal, P., Sarkar, D., Mukhopadhyay, S., Basu, A.: Fault diagnosis in discrete time hybrid systems–a case study. Inf. Sci. **177**(5), 1290–1308 (2007)
11. Rahiminejad, M., Diduch, C., Stevenson, M., Chang, L.: Open-circuit fault diagnosis in 3-phase uncontrolled rectifiers. In: 2012 3rd IEEE International Symposium on Power Electronics for Distributed Generation Systems (PEDG), pp. 254–259. IEEE (2012, June)
12. Vento, J., Travé-Massuyès, L., Puig, V., Sarrate, R.: An incremental hybrid system diagnoser automaton enhanced by discernibility properties. IEEE Trans. Syst. Man. Cybern. Syst. **45**(5), 788–804 (2015)
13. Derbel, H., Alla, H., Hadj-Alouane, N.B., Yeddes, M.: Online diagnosis of systems with rectangular hybrid automata models. IFAC Proc. Vol. **42**(4), 954–959 (2009)
14. Meseguer, J., Puig, V., Escobet, T.: Fault diagnosis using a timed discrete-event approach based on interval observers: application to sewer networks. IEEE Trans. Syst. Man. Cybern. Part. A Syst. Hum. **40**(5), 900–916 (2010)
15. Daigle, M.J., Koutsoukos, X.D., Biswas, G.: An event-based approach to integrated parametric and discrete fault diagnosis in hybrid systems. Trans. Inst. Meas. Control. **32**(5), 487–510 (2010a)
16. Pencolé, Y.: Decentralized diagnoser approach: application to telecommunication networks. In: International Workshop on Principles of Diagnosis (DX'00), pp. 185–192 (2000)
17. Travé-Massuyes, L., Escobet, T., Olive, X.: Diagnosability analysis based on component-supported analytical redundancy relations. IEEE Trans. Syst. Man. Cybern. Part A Syst. Hum. **36**(6), 1146–1160 (2006)
18. Shahbazi, M., Jamshidpour, E., Poure, P., Saadate, S., Zolghadri, M.R.: Open-and short-circuit switch fault diagnosis for nonisolated dc–dc converters using field programmable gate array. IEEE Trans. Ind. Electron. **60**(9), 4136–4146 (2013)
19. Louajri, H., Sayed-Mouchaweh, M.: Modular approach for the diagnosis of a class of hybrid dynamic systems: application to three cell converters. In: Proceedings of the 25th International Workshop on Principles of Diagnosis dx'14. (2014)
20. Uzunova, M., Bouamama, B.O., Djemai, M.: Hybrid bond graphs for diagnosis of three cells converter. IFAC Proc. Vol. **45**(20), 162–167 (2012)
21. Cassandra, C.-G., Lafortune, S.: Introduction to Discrete Event Systems, 2nd edn. Springer, New York (2008)
22. Willner, Y., Heymann, M.: Supervisory control of concurrent discrete-event systems. Int. J. Control. **54**, 1143–1169 (1991)
23. Sayed-Mouchaweh, M., Lughofer, E.: Decentralized fault diagnosis approach without a global model for fault diagnosis of discrete event systems. Int. J. Control. **88**(11), 2228–2241 (2015)
24. Defoort, M., Djemai, M., Floquet, T., Perruquetti, W.: Robust finite time observer design for multicellular converters. Int. J. Syst. Sci. **42**(11), 1859–1868 (2011)
25. Sampath, M., Segupta, R., Lafortune, S., Sinnamohideen, K., Teneketzis, D.: Diagnosability of discrete event systems. IEEE Trans. Autom. Control. **40**(9), 1555–1575 (1995)
26. Sengupta, R., Tripakis, S.: Decentralized diagnosability of regular languages is undecidable. In: 40th IEEE Conference on Decision and Control, pp. 423–428 (2002)

Automated Process Optimization in Manufacturing Systems Based on Static and Dynamic Prediction Models

Edwin Lughofer, Alexandru-Ciprian Zavoianu, Mahardhika Pratama, and Thomas Radauer

1 Introduction

Optimized production processes and accurate predictive maintenance systems [29] have been identified as two of the most important drivers of innovation in modern industrial facilities. As such, these topics represent a key issue in several Horizon 2020 call objectives. An overall (and somewhat ambitious) goal would be to construct close-to-ideal production processes that simultaneously (1) maximize multiple product quality criteria, reduce waste and negative environmental impact while (2) having the option to undergo on-line manual [65] or, ideally, automatic [52] preemptive adjustments (with a high success ratio) when items or parts of these processes show (in advance) behaviors that are likely to result (at a later stage) in a downtrend in product quality, a degraded performance, or even in complete system failure (i.e., increased downtime) [45].

Considering the complexities inherent in modern manufacturing processes and the likely time-wise restrictions, one key part of both the optimization stage and the predictive maintenance stage is the ability to obtain high-quality static and dynamic

---

E. Lughofer (✉)
Fuzzy Logic Laboratorium Linz-Hagenberg, Department of Knowledge-Based Mathematical Systems, Johannes Kepler University Linz, Linz, Austria
e-mail: edwin.lughofer@jku.at

A.-C. Zavoianu
Department of Knowledge-Based Mathematical Systems, Johannes Kepler University Linz, Linz, Austria
e-mail: ciprian.zavoianu@jku.at

M. Pratama
School of Computer Science and Engineering, Nanyang Technological University, Singapore, Singapore

T. Radauer
Stratec Consumables, Anif, Austria

prediction models for the various relevant targets (i.e., objectives and constraints) of the production process. For example, in the case of a micro-fluidic chip production process (which is the main focus of our case study), typical targets are several quality control (QC) indicators that measure the shape, flatness, and internal composition of the resulting chips. The inputs of the prediction models can be classified in two broad classes:

1. *process parameters*—controllable (machine) settings that are kept fixed over a longer period during the production process. These parameters are expected to have a direct influence on all OC indicators and their optimization usually requires the construction of static prediction models (being used as surrogates during process optimization), as physically testing numerous parameter combinations is usually a very expensive endeavor. For example, in the case of micro-fluidic chip production, one would be typically interested in optimizing the parameters controlling the bonding liner and the injection molding machines.
2. *process values*—denote dynamic (i.e., high-volume and high-frequency) read-only/diagnostics information usually obtained via (numerous) sensors that monitor key parts of the production process. While this type of data can be inexpensively (and inherently) measured, its usage for creating meaningful dynamic prediction models for QC indicators is not straightforward as it requires appropriate combinations of advanced techniques like (incremental) dimension reduction and on-line model adaptation.

The above described characteristics of the two main input classes also influence the suitability of different (predictive) modeling and optimization paradigms. For example, when wishing to account for shifting major external factors (new product variants, raw material quality, supply chain disruptions, etc.), production parameters are usually set and adjusted manually by operators/domain experts using experience and rules of thumb (i.e., formal and informal expert knowledge) that do not necessarily deliver optimal results. While more advanced parametrization solutions based on design of experiments (DoE) approaches (e.g., [7, 18]) or data-driven models (e.g., [3]) have been proposed, comprehensive approaches that aim to integrate (1) DoE strategies with (2) valuable expert knowledge and (3) effective optimization methods remain scarce—a first attempt has been proposed by the authors of this chapter in [74]. There the concentration laid on the static case by only respecting process parameters (optimizing them towards to achieve ideal product quality), but not taking into account any process values, which are measured during production and thus are typically able to reflect changing system dynamics and environmental influences.

When considering dynamic process values, most state-of-the-art predictive maintenance systems use well-established techniques from the field of forecasting and prognostics centered on analytical [5, 31], knowledge-based [13], hybrid models [50], or purely data-driven models [30, 59]. Some others use model-based predictive control strategies to optimize the trajectories of the targets, see, e.g., [63]. The main disadvantage of analytical and knowledge-based models lies in the fact that they require longer derivation and development phases and are often restricted to very particular application settings [26]. Data-driven models based on robust paradigms (e.g., active learning [33]) can be generated much faster, more or less automatically and mostly independently of the application scenario. They can incorporate advanced strategies like on-the-fly evolution of structural components and incremental learning of parameters [38] in order to cope with the potentially changing dynamics of the production process. More importantly, this can be achieved *without* the need for classical (human) supervised model re-calibration and model maintenance cycles (like the one proposed in [69]).

1.1 Our Approach

In this chapter, we demonstrate a holistic approach for automated process optimization (HAPO) in manufacturing systems during off-line (static case) and on-line mode (dynamic case), comprising the following concepts:

- Gathering initial expert knowledge about which process parameters and process values might have an intrinsic influence on the final product quality and thus should be supervised and optimized; additional knowledge about possible ranges of the parameters and even some (causal) relations, dependencies could be helpful for improving the performance of the models (see Sect. 2).
- Design of experiments (DoE) for the purpose to define some (initial) settings which are most likely important for the process: these settings are necessary for some machine testing and initial data collection cycles in order to gather training data for the predictive model construction phase. We propose a new DoE which comes with a hybrid data-driven and expert knowledge-based form, typically in alternating manner in order to increase expected model certainty, robustness, and finally predictive performance (Sect. 3.1.2).
- Predictive mapping construction based on the recordings obtained from the design of experiments phase: linear and nonlinear regression model techniques in combination with statistical evaluation procedure (for checking significant outperformance) play a central role (Sect. 3.1).
- Dynamic time-series-based forecast models construction based on the process values permanently recorded during the actual production over a longer time frame (typically several months); these are expected to capture also the system dynamics which is not necessarily steered by the process parameters in advance, such as varying charge compositions or specific environmental influences. The inputs are time-series of process values trends and the targets are various quality criteria (typically continuously measured), leading to a very high-dimensional learning problem in a batch process modeling setting (Sect. 3.2).
- Self-adaptive forecast models which are able to evolve their structure and to update the input space transformation (used for dimension reduction) are necessary to cope with changes in the mapping relations between trends and quality criteria; our approach includes rule evolution, rule splitting, recursive parameter adaptation with dynamic forgetting, and incremental update of the transformed loading space to achieve sufficient model flexibility (Sect. 3.2.2).

- Off-line process optimization methods based on predictive mappings in order to guide the machinery and the production process in the correct direction in advance; the mappings are used as surrogate models in a multi-objective optimization problem, which has been reduced from a (typically resulting) many-objective problem by cross-correlation analysis and clustering.
- On-line process optimization based on dynamic, self-adaptive forecast models in order to balance out nonoptimal parameter solutions and/or to react on undesired system dynamics properly and early; the forecast models are used as surrogate models in a multi-objective optimization problem, where an appropriate reduction of the very high input dimensionality (whole time-series trends for different process values) is necessary.

The optimization problems mentioned in the last two itemization points will be more clearly formalized in the subsequent section (Sect. 2). The combination of these concepts will be highlighted and extensively evaluated for a real-world manufacturing process based on micro-fluidic chip production, whereby we concentrated mainly on the bonding lining stage (last stage of production influencing chip quality most), but also included some recent findings for the multi-stage case by performing process optimization for bonding lining stage based on (process parameters of) injection molding stage. Some challenging results obtained during a 3-year running project will be presented in Sect. 5.3 in order to underline the applicability of the proposed concepts.

2 Problem Statement

2.1 *Process Optimization Based on Parameters*

As process parameters are the adjustable settings that control the machining procedure of a production process [23], they are the main means through which operators can influence product quality (such as to meet customer expectations). A typical demand is that process parameters are to be adjusted properly for different product variants, charges, machine components, etc., in order to meet customer expectations on product quality. This requires intrinsic knowledge about the process and is often done manually in current industrial installations—an issue which we aim to automate with the methodologies described in this chapter for yielding (multi-criteria) optimized production processes that are supported on-line (during their run) by a highly accurate and automated predictive maintenance approach, see the motivation in Sect. 1.

Often, the operators themselves do not know the ideal settings, and simply adjust the parameters based on their own past experience using rules of thumb, which are often affected by vague and uncertain knowledge. Different process parameter settings typically affect the product quality (in positive and negative way), mostly within the range of set points, which, however, may be also exceeded. The exceed is a no-go for a product as it indicates a bad part/item which should not be delivered to customers. Moreover, conventionally used settings are widely accepted by the operators/experts and thus may have a supremacy over other settings.

There are thus two goals for an automatized process parameter optimization in order to achieve high qualitative products and to avoid time-intensive manual tuning efforts:

1. Changing the process parameters in a way to achieve expected QC (quality criteria) production values that are as close as possible to their ideal values (which are a priori known, defined by experts).
2. Changing the process parameters in a way to achieve expected QC (quality criteria) production values that are as close as possible to their ideal values, but simultaneously ensure that the new process parameters are also as close as possible to their standard, default values—which the operators/experts are most aware of usually prefer in light of their past experience.

In both of the aforementioned cases, constraints are normally set on the minimal and maximal allowed values of all the process parameters that can be varied during the optimization cycles. A typical example for chip production is shown in Fig. 1.

As there are often many quality target values Q in parallel to be optimized, the first case can be formalized as a many-objective optimization problem:

$$f_i(\mathbf{x}) = opt_i!, \quad 1 \leq i \leq q \qquad (1)$$

$$\text{subject to } x_j \in [l_j, u_j], \quad 1 \leq j \leq p$$

	Influence Factors:	Unit:	Default:	Min:	Max:
(X1)	Opening Ratio (Diffusor) (VAU)	[%]	10	5	10
(X2)	Chamber Temperature (VAU)	[°C]	31	28	34
(X3)	Upper Tool Temperature (BND)	[°C]	50	40	60
(X4)	Time Marriage to Bonding (Rob. Speed)	[%]	100	50	100
(X5)	Lower Tool Temperature (BND)	[°C]	50	40	60
(X6)	Activation Time (VAU)	[s]	21	16	26
(X7)	Bonding Pressure BND)	[kN]	50	20	50
(X8)	Press Time t1 (BND)	[s]	15	5	25
(X9)	Speed High (BND)	[mm/min]	2,5	1	15
(X10)	Speed Low (BND)	[mm/min]	0,15	0,15	5
(X11)	Touchpoint P2 (BND)	[mm]	6,9	6,5	7,2

Fig. 1 Example of influencing process parameters which are important for steering the production process, their standard values, and allowed ranges

where the optimal opt_i values of the quality criteria and the variation intervals of each process parameter (i.e., $[l_j, u_j]$) are defined by the operators/domain experts. Thereby, $\mathbf{x} = \{x_1, \ldots, x_p\} \in [l_1, u_1] \times \cdots \times [l_p, u_p]$ denotes the parameter combination (which is in essence a multi-dimensional point), and $\{f_1(\mathbf{x}), \ldots, f_q(\mathbf{x})\}$ the quality indicators for q different criteria, typically measured as continuous numerical values. This induces that $f_1, \ldots, f_q$ are models in a supervised regression context, which are able to predict quality criteria (QCs) based on the current parameter setting. The establishment of such (surrogate) models, ideally with sufficient high-quality and different input sources (knowledge and data), is thus a central issue that must be tackled before the automated optimization of parameters can be start. This modeling stage is comprehensively handled in Sect. 3.

In the second case, an additional objective is inserted into the previously formalized many-objective optimization problem:

$$\sum_{i=1}^{p} d(x_i, X_l) = \min_{x_1, \ldots, x_p} ! \qquad (2)$$

with $X_1, \ldots, X_p$ denoting the currently used process parameter values and d marking a certain distance function (e.g., quadratic). This objective may receive a different weight during optimization compared to each single $f_1, \ldots, f_q$-based objective, or it may be handled in a specific (e.g., elitist) form within the optimization process.

In many practical cases there are usually intrinsic conflicts between the various QC indicators (not all of their ideal values can be achieved synchronously) meaning that Eq. (1) defines a nontrivial many/multi-objective optimization problem (MOOP) for which there is no single solution (i.e., $\nexists \mathbf{x}^*$). Thus, the reduction of the dimensionality p of the variable space and/or of the dimensionality q of the objective space can become an important issue when wishing (1) to increase the likelihood of coming closer to high-quality compromise solutions and (2) also to speed up the whole optimization process.

2.2 Process Optimization Based on Process Values Trends

Process values are important system variables permanently recorded on-line during the production process that reflect the actual state of the system during its runtime. This also means that they may indicate (unexpected) differences in the machining and system states when considering expectations based on the (pre-optimized) process parameter settings. Such differences can be induced, for instance, by some unknown system dynamics or environmental influences unexpectedly arising during production. In addition, in some cases, the differences indicated by the process values stem from the fact that the process parameters have not been optimized for a new product type or the optimization attempt has not been successful. In light of these undesirable production scenarios, it is also important to track the behavior of on-line recorded process values and especially to supervise their influence on the (final) quality of the production parts. The latter can be achieved in form of time-series-based forecast models $f_1, \ldots, f_q$, which are able to predict QC values in the future based on current time-series trends of process values (used as inputs), see Sect. 3.2. Formally, these forecast models can be defined as:

$$QC_q(t) = f_q(\mathbf{x}(t - n_1), \mathbf{x}(t - n_1 - 1), \ldots, \mathbf{x}(t - n_2)) \quad \forall q = 1, \ldots, Q, \quad (3)$$

where $QC_q(t)$ marks the qth quality criterion measured at time instance t, $\mathbf{x} = [x_1, x_2, \ldots, x_J]$ (containing J process values), and f_q is the input/output forecast mapping (the prediction model). Further, $n_1 < n_2$, and n_1 denotes the prediction horizon, whereas $n_2 = n_1 + k$, with k denoting the size of the window of past samples to be considered for forecasting.

Whenever the process values forecast problematic QC values, an on-line adjustment of respective process values and/or parameters in the "right direction" is demanded in order to prevent significant quality damage or even system failures that eventually translate to monetary loss for the company. The two main problematic scenarios one should expect concern QC predictions that either suddenly drop out of bounds as shown in Fig. 2 or slowly drift towards potentially problematic levels as shown in Fig. 3.

On the one hand, the on-line adjustment of process values may again lead to a many-objective optimization problem, depending on the number of QC values that are out of bounds or drifting. Nevertheless, typically, the number of

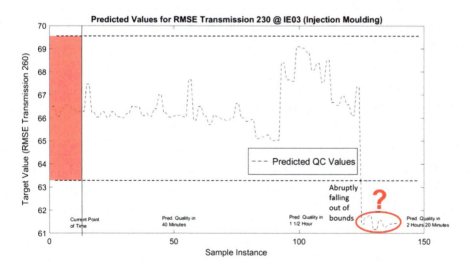

Fig. 2 Predicted QC values of a special transmission characteristics of a chip into the future; in around 2 h, there is a problem expected in terms of an abrupt change falling out of the allowed bounds of transmission values (shown as dashed lines)

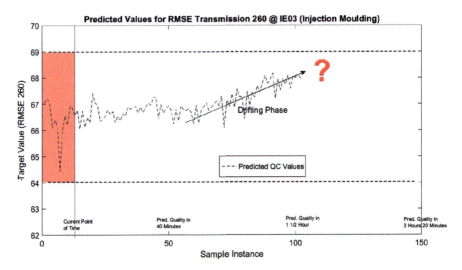

Fig. 3 Predicted QC values of a special transmission characteristics of a chip into the future; there is a drift phase expected in around 1 h (indicated by an arrow marker)

objectives requiring adjustment is expected to be much smaller than in the case of process parameter optimization (where *all* important QC values the experts want to supervise are always used as targets during the optimization procedure). On the other hand, the input dimensionality is typically much higher as one needs to consider entire trends of length k for a rather larger number of process values J (as is the case in our case study presented in Sect. 5) in order to construct meaningful prediction models. All these aspects should be carefully considered when designing the optimization procedure as they indicate a need for (1) pre and post-processing, (2) dimensionality reduction for both inputs and outputs, and (3) work-arounds based on a dynamic process parameter optimizations (as described in Sect. 4.2).

The on-line adjustment of the production process should be able to deliver fully automatic and/or expert-assisted corrective actions that aim to maintain QC values within their bounds when considering two types of application scenarios: (1) those concerning un-optimized production processes and (2) those concerning pre-optimized production processes.

Un-optimized production processes occur when, before starting production, the process parameters (1) were not or could not be optimized at all or (2) were not correctly optimized for the current product (e.g., due to the lack of good QC surrogate models). These cases lend themselves to a standard many/multi-objective optimization of the process values in which:

- the goal is to restore each problematic QC forecast back to its ideal (e.g., the midpoint between its allowed lower and upper bound);
- the forecast model of each problematic QC—defined in Eq. (3)—can be considered as the basis of a single-target surrogate fitness estimator;

- a dimensionally reduced collection of dynamic process values will form the solution encoding (notation $\mathbf{x^{agg}}$).

As dynamic production process values (DPVs) are in effect (sensor-based) data trends, dimensionality reduction should focus on using simple time-wise aggregators (i.e., average, slope, skew, variance, etc.) that are able to describe the behavior of an individual sensor (i.e., process value) during the given k-sized window of past samples that we use for prediction. Therefore, when considering:

1. a total number of $q' < q$ problematic QC indicators,
2. a dimensionally reduced number of p' sensor measurements (i.e., DPVs) that are deemed as most influencing for the predicted QC indicators,
3. a general DPV trend size of k,
4. the wish to describe trends using a compact summary description by statistical measurers, e.g., average value, slope, and variance,

the on-line process optimization tasks can be formalized as:

$$f_i(\mathbf{x^{agg}}) = opt_i!, \quad 1 \leq i \leq q' \quad (4)$$

$$\text{subject to } x_j^{agg} \in [l_j, u_j], \quad 1 \leq j \leq 3 \cdot p',$$

$$\text{where } \mathbf{x^{agg}} = [\mu_1, \Delta_1, \gamma_1, \ldots, \mu_{p'}, \Delta_{p'}, \sigma_{p'}],$$

and, for $1 \leq l \leq p' \wedge k \geq 2$, $\mu_l = trendAverage(DPV_l, k)$,

$$\Delta_l = trendSlope(DPV_l, k),$$

$$\sigma_l = trendVariance(DPV_l, k).$$

where f_i are the forecasting models as defined in (3) and obtained in advance; they are used as surrogate models for fitness evaluation during optimization: their predictions can be directly compared with the pre-defined, known optimal values of the QC indicators, i.e., a fitness function can be easily defined, which measures the distance between a predicted $\mathbf{f(x^{agg})}$ and the optimal QC indicator vector **opt**.

It is worthy to mention that for an un-optimized production process, Eq. (4) can be interpreted as a "free optimization" as there are no secondary constraints on the ranges of the ideal trend average, slope, and variance the many/multi-objective solver can propose. More importantly, a reverse mapping "optimized process values → optimized process parameters" should be generally available as the process parameters are the only means through which the overall process behavior can be influenced. This reverse mapping can be based on expert knowledge and/or purely data-driven methods (e.g., multiple linear regression between process value trends and process parameters).

In the case of pre-optimized production processes, problematic QC predictions are likely generated by (1) unknown system dynamics and (2) unexpected environmental influences. We have also identified two complementary strategies for restoring QC values within their desired bounds. The first one is based on (iterated) local optimizations in which we wish to restrict solver solutions to a close vicinity of the current DPV/(pre-optimized) process parameters settings by:

- reducing the lower and upper bound search intervals for each relevant x_j^{agg} from Eq. (4)
- introducing an extra optimization objective:

$$\sum_{i=j}^{3 \cdot p} d(x_j^{aggold}, x_j^{aggnew}) = \min! \qquad (5)$$

Thus, the closeness of optimized DPVs to the current trend of DPVs is respected as additional objective in order to regularize the free optimization problem by restricting it to smaller changes in the process values and associated parameters.

3 Establishment of Predictive Models

The subsequent sections describe the methodologies for a data-driven construction of static and dynamic surrogate mappings (being used in subsequent process optimization phases, Sect. 4), which we successfully applied for a particular production system, as described in Sect. 5.1.

3.1 Iterative Construction of Static Predictive Mappings (Parameters ⇒ Quality)

The goal is to establish data-driven predictive mappings that can predict QC values based on process parameter combinations. Therefore, various combinations need to be defined in advance and the resulting QC values after some production cycles observed when applying these settings to the machinery of the production (or to its "control wheel") → the observations can be collected as supervised data. It is thereby of utmost importance to carefully select process parameter value combinations 1.) which are known to have an essential effect on the quality of the chips at all and 2.) which are expected to induce a high generalization capacity of the predictive mappings (e.g., by a good coverage of the multi-dimensional parameter space).

Thus, we propose a knowledge-based construction strategy for QC predictive mappings that aims to combine:

1. operator/domain expert knowledge, especially for initial selection of parameters and valuable settings;
2. data-driven insights based on a design of experiments (DoE) methodology: we propose a new hybrid version combining model (parameter) uncertainty with parameter space coverage for sample (=parameter setting) selection.

3. linear and nonlinear modeling techniques for finally constructing the mappings based on the selected parameter settings; a reliable application of the latter certainly depends on the proportion between the dimensionality of the parameter space to the number of available samples for model training [21].

3.1.1 Expert Knowledge Initialization

Expert knowledge is mainly integrated in the mapping construction process during the initial data collection/generation stage. First, based on several discussions with experts, usually a so-called cause-effect (CE) diagram can be established in order to elicit the most effective parameters onto QC values, see, e.g., [74] for details. Such an analysis results in a parameter matrix as exemplarily shown in Fig. 1 and which is also required for setting up the concrete optimization problem as discussed throughout Sect. 2.1. Then, the first few process parameter combinations over the most effective parameters shown in the CE diagram and the associated parameter matrix that are to be tested are the ones recommended by production process operators. Apart from this, expert input is also used to restrict the individual domain of each parameter, i.e., limit the search space by defining appropriate maximal ranges of the parameter and to filter—either as a pre or post-processing step— invalid parameter combinations. Computer-aided design selection strategies such as the Taguchi L12 method or Full Factorial [46] may support the experts for choosing appropriate combinations based on these limitations. These initial settings are needed to be able to start with a reliable design of experiments (second step).

3.1.2 Hybrid Design of Experiments (HDoE)

The second step of the data collection stage is based on a domain-independent hybrid DoE-based strategy and aims to obtain further samples (=parameter settings)

1. that are well distributed in parameter space in order to assure a good coverage of the parameter space and thus to prevent extrapolation on new settings, and
2. that reduce the uncertainty of predictive mappings (and internal parameters) constructed from the initial samples selected by operators.

Regarding the latter, typical widely used choices in case of (predictive) regression problems are A-optimality, D-optimality, or E-optimality [14, 61], which can be calculated through the usage of the Fisher information matrix [15]. Often, it can be assumed that the initial number of samples defined by the experts is pretty low, a couple of samples (due to high effort for an appropriate selection and associated experiments on the machine for this selection). This means that a linear model is the most reliable option for representing the predictive mapping regression problem at the beginning. To approach well-distributed samples, a combination of space filling designs based on Latin hyper-cube sampling with minimax optimization [43] and corner points in the parameter space are promising choices.

Therefore, the skeleton of our suggested hybrid design of experiments (HDoE) is as follows:

1. $M + 2^p$ samples can be drawn from the parameter space, with p the dimensionality of the space. The M samples are obtained by Latin hyper-cube sampling. The remaining 2^p are the corner points of the parameter space in order to reduce the likelihood of extrapolation as much as possible. If p is small, typically $M >> p$; if p is large, an explicit selection of subsets of corner points should take place (as 2^p grows exponentially).
2. Then, in each odd iteration of sample selection, each of the remaining $M + 2^p - |S|$ constructed samples, with S and $|S|$ the set and number of samples selected so far, respectively, is checked how much it improves one of the following criteria:

   - *A-optimality* (variant 1): it seeks to minimize the trace of the inverse of the Fisher information matrix.
   - *D-optimality* (variant 2): it seeks to maximize the determinant of the Fisher information matrix.
   - *E-optimality* (variant 3): it seeks to maximize the minimum eigenvalue of the Fisher information matrix.

   Due to the linear model assumption, the Fisher information matrix is equivalent to the Hessian matrix $X_{ext}^T X_{ext}$, with X_{ext} always containing the initially collected samples from experts plus all the samples selected so far over the DoE iterations plus the new sample to be checked for improvement of the optimality criterion. Thus, in the first case that sample is selected which achieves minimal A-optimality over all samples when being joined with X_{ext} (so, A-optimality is calculated for all samples s_i, $[X_{ext}; s_i]$, $i \notin X_{ext}$).
3. In each even iteration of sample selection, each of the remaining $M + 2^p - |S|$ constructed samples is checked how much minimal distance it has to the already selected samples plus the initial ones. That one whose minimal distance is maximal is selected as it extends most/best the coverage of the current input space.

Due to the mixed selection strategy, we call our algorithm *hybrid design of experiment*, whose pseudo-code is provided in Algorithm 1.

Algorithm 1 Hybrid design of experiments for sample selection

Input: initial data matrix X containing initial process parameter combinations (no targets required at all); min-max values of all process parameters defining the corner points; number of Latin hyper-cube samples M (default 10,000).
Output: matrix S containing a pre-defined maximal number of selected samples N ordered by their importance for model quality improvement (most important first)

1. $S = \{\}$.
2. Generate M samples $\{\mathbf{x}_1, \ldots, \mathbf{x}_M\}$ according to Latin hyper-cube sampling with minimax operation.
3. Generate 2^p additional samples $\{\mathbf{x}_{M+1}, \ldots, \mathbf{x}_{M+2^p}\}$ as the corner points of the p-dimensional parameter space.
4. **For** $j = 1$ to N
5. **For** $i = 1$ to $M + 2^p$

    a. **If** $\mathbf{x}_i \in S$ (sample was selected before), continue;
    b. Erase vectors $Crit$ and $Dist$; $X_{ext} = [X; S; \mathbf{x_i}]$.
    c. **If** $mod(j, 2) = 0$ (odd number)
    d. Case Variant 1: $Crit(i) = trace([X_{ext}^T X_{ext}])$
    e. Case Variant 2: $Crit(i) = det([X_{ext}^T X_{ext}])$
    f. Case Variant 3: $Crit(i) = \min(eig([X_{ext}^T X_{ext}]))$
    g. **Else:**
    h. Compute Euclidean distance $dist_i(k)$ between $\mathbf{x}_i$ and all k samples in $[X; S]$;
    i. $Dist(i) = \min_{k=1,\ldots,|X \cup S|} dist_i(k)$
    j. **End If**

6. **End for** (inner loop)
7. **If** $mod(j, 2) = 0$ (odd number)
8. Sort $Crit$ in ascending order and store its associated samples $\{\mathbf{x}_s(1), \ldots, \mathbf{x}_s(M + 2^p - |S|)\}$;
9. Case Variant 1: Select first entry, thus $S = S \cup \mathbf{x}_s(1)$.
10. Case Variant 2+3: Select last entry, thus $S = S \cup \mathbf{x}_s(M + 2^p - |S|)$.
11. **Else**
12. Sort $Dist$ in ascending order and store its associated samples $\{\mathbf{x}_s(1), \ldots, \mathbf{x}_s(M + 2^p - |S|)\}$;
13. Select last entry, thus $S = S \cup \mathbf{x}_s(M + 2^p - |S|)$.
14. **End if**
15. **End for** (outer loop)

3.1.3 Predictive Mapping Models Construction

Like in most data-driven modeling tasks, a basic pre-processing step should be performed in order to at least detect and remove those process parameters that are expected to have a minor influence on the QC indicators. This helps to reduce the input dimensionality and thus to increase the generalization capabilities of the models. Process parameters can be seen as factors, which, due to expert knowledge input, can be divided into low, medium, and high levels. Thus, a multi-way ANOVA analysis [20] on the data collected can be applied to observe significant influences of factors and various groups of factors onto the QC values. Those factors having not any influence can be deleted → dimension reduction.

Given the limited number of data samples one is likely to have gathered after the two-stage data collection phase, linear regression models between process parameter settings (inputs) and QC (quality criteria) indicators (modeling targets) are likely to deliver the best (generalization) performance. Thus, training linear models is an intuitive first step. Nonlinear regression modeling paradigms like artificial neural networks (ANNs), support vector machines, or genetic programming should also be tested—especially for those QC indicators where linear models seem to fail in accuracy.

In light of a low sample count, over-fitting (especially for the nonlinear models) is highly likely. Therefore, we suggest to always apply an n-fold cross-validation strategy for assessing model performance (i.e., R^2) during training. Furthermore, since the number of cross-validation folds is limited for a low number of samples (i.e., $3 \leq n \leq 5$), a stability assessment of the most promising nonlinear regression models should be performed. The idea is to compare if, for a given QC indicator, the nonlinear model is able to generally deliver a superior modeling performance, when being compared to their linear counterparts, over several different cross-validation partitions. Only, when this is really significant, nonlinear models should be chosen (to avoid over-fitting as much as possible) [21, 57].

3.2 Time-Series-Based Forecast Models (Process Values ⇒ Quality) Learning and Adaptation

In manufacturing processes, we are often dealing with a specific type of continuous production process in which the quality measures/indicators are provided once for a whole process cycle (i.e., "batch"); in literature, this is termed *batch-processing* [3]. One batch can be associated with the period between the measurements of two consecutive QC info vectors (quality criteria vectors after N process cycles). This means that, for a particular period of time, one QC info vector (containing several important QC values to be supervised) is delivered that indicates the production quality for the whole period. This leads to the signal flow (of process values) and measurement frequency of QC info vectors shown in Fig. 4.

The production quality forecast problem can then be tackled by establishing a prediction model, the inputs to which are trends of the various (synchronously and permanently measured) process values within a particular time frame, and the targets are the quality criteria embedded in the QC-info vector. Ideally, the *prediction horizon* of such a model is as large as possible in order to be able to recognize downtrends in the quality as early as possible. This increases the likelihood of a successful predictive maintenance cycle, especially it increases the likelihood that process parameters can be optimized automatically in time whenever problems in the process values are recognized, see Sect. 4.2.

After an appropriate arrangement of the time-series data (for J process values) into a three-dimensional matrix and flattening this matrix accordingly, see [42] for details, we are ending up with the data-driven predictive model definition as in (3).

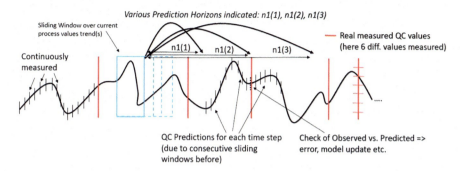

Fig. 4 Typical forecasting scheme in a batch-processing setting: the process signal (permanently sampled and measured) is shown as bold solid line, the QC info vector is indicated by vertical (red) lines and is recorded in B batches over time (here $B = 5$), not necessarily in equidistant manner—here only the case of $J = 1$, i.e., one process value time-series trends is shown, this can be generalized easily to J multiple process values by recording and processing them in parallel; short-, medium-, and long-term predictions of QC info in a batch process based on time-series trends of a process value measured in the past are indicated by different possible prediction horizons (double arrows), indexed with n1(1), n1(2), and n1(3)

In the following two subsections, we perform a summary of the basic aspects of our learning and (on-line) adaptation procedure for such types of models.

3.2.1 Learning by a Nonlinear (Fuzzy) Version of PLS (PLS-Fuzzy)

We are dealing with very high-dimensional data, as we have $J*k$ columns in sample matrix $X = [\mathbf{x}_1(t - n_1), \ldots, \mathbf{x}_1(t - n_2), \mathbf{x}_2(t - n_1), \ldots, \mathbf{x}_2(t - n_2), \ldots, \mathbf{x}_J(t - n_1), \ldots, \mathbf{x}_J(t - n_2)]$, where typically J reflects the variety of sensor recordings and $k = n_2 - n_1$ the size of the window containing the past trends; $\mathbf{x}$ is a column vector containing the various sample values. Consecutive values in a time-series can be expected to have similar information content when the dynamics between two or more measurements is not really high.

We thus develop a nonlinear variant of partial least squares regression (PLSR) [19], which (1) handles correlated "neighboring" inputs in an appropriate fashion (PLS thereby enjoyed a wide usage onto data learning problems in the past where such cases occurred, see, e.g., the field of chemometrics [64]), and (2) is able to resolve nonlinearities in the covariance structure between input matrix X and target $y = QC_q$. The latter provides us more flexibility than conventional PLSR (linear models) to model possible nonlinearities implicitly contained in the system, see Fig. 5 for a two-dimensional example, where one global component based on conventional PLSR would end up with an inaccurate representation of the data distribution and the (co)variance contained in it. Obviously, partitioning the data into two local regions and modeling the relationship between x and y for each region separately (dotted lines) is more appropriate.

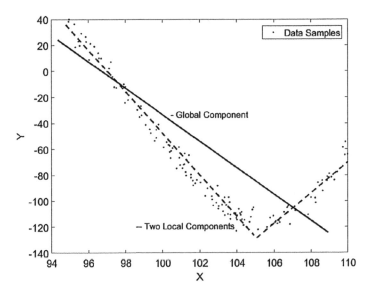

Fig. 5 A two-dimensional example of local data clouds showing two components/trends (as dashed lines), which cannot be sufficiently resolved by a global, linear model (solid line)

Our idea is thus to combine PLS with Takagi-Sugeno (TS) fuzzy systems, the latter having been applied in a wide range of applications fields in the past for supervised regression and prediction problems with numerical target values [1, 32, 51]. Furthermore, the architecture offers a partial partitioning of the input space into several regions in a natural way (sub-models represented by rules), for which partial direction vectors (latent variables) can be extracted in order to resolve locally varying covariance structures between X and y. We employ a specific variant of TS fuzzy systems, the *generalized* version of TS fuzzy systems, firstly introduced in [28], and which showed better predictive performance in several past studies [12, 36, 53, 54] by inducing more compact rule bases with similar or even less model errors than conventional TS fuzzy systems.

The connection between PLS and TS fuzzy systems is conducted on a global model level in the score space, because it reduces the input dimensionality to the most important variables in advance and then rule learning can start from scratch with an empty rule base, see below (the number of rules does not need to be defined in advance). Therefore, PLS is performed first on the data matrix X, which is then transformed to the score space by the resulting loading vectors (latent variables) P through multiplication, i.e., $X_S = X * P$. The learning of and prediction with TS fuzzy systems then operate fully on the score matrix X_S including score samples $\mathbf{x}_s$, which are typically of reduced space $p << J*k$. The reduced space can be elicited by using the p most important latent variables (with highest eigenvalues) explaining most of the covariance/variance structure in the data. Thus, in the equations above $\mathbf{x}_s$ is the projection of a single new sample onto the latent variable space, i.e., $\mathbf{x}_s =$

$\mathbf{x} * P$ with $\mathbf{x}$ containing the time-series trends of past k samples over all J process values: $\mathbf{x} = [x_1(t - n_1), \ldots, x_1(t - n_2), x_2(t - n_1), \ldots, x_2(t - n_2), \ldots, x_J(t - n_1), \ldots, x_J(t - n_2)]$.

As the nonlinearity degree of the learning problem is not a priori known, it is wise to estimate the appropriate number of fuzzy rules during the initial stage of fuzzy model construction. Therefore, we exploit the *Gen-Smart-EFS* algorithm [36] by passing all the available samples through its core learning engine in a single-pass manner (thus as pseudo-stream), after projecting all of these to the score space. By using *Gen-Smart-EFS* algorithm (see [36] for details) in combination with PLS, the *PLS-fuzzy* training approach results, as listed in Algorithm 2.

Algorithm 2 PLS-fuzzy employing robust Gen-Smart-EFS in batch mode (*PLS-fuzzy static*)

Input: Training matrix X, target vector $\mathbf{y} = QC_q$, parameter fac responsible for rule learning in the rule evolution criterion.

Output: p most important PLS projection directions stored in P, corresponding un-normalized projections stored in Q; PLS-fuzzy model stored in F.

1. Perform PLS with X and y to obtain the complete projection matrix P_{all} with components ranked along the columns.
2. Select the p most significant components from P_{all}, $P = p_1, \ldots, p_p$ and the corresponding un-normalized ones $Q = q_1, \ldots, q_p$. This can be done in a parameter grid search (iteratively increasing p) and performing the successive steps for each p or by using an accumulated trend of the explained variance in the data by the p components.
3. Perform *Gen-Smart-EFS* algorithm [36] using P, X, and y, thus obtaining C rules in the latent variable space.
4. Fine-tune the C rule centers $\mathbf{c}_i$ and shapes Σ_i^{-1}, $i = 1, \ldots, C$; this is done in several iterations over the whole data set X until convergence is met; thereby, centers are moved according to generalized vector quantization and the inverse covariance matrix is updated recursively by ($\mathbf{c} = \mathbf{c}_{win}$ with win denoting the index of the winning rule having highest membership degree in the current sample):

$$\Sigma^{-1}(k+1) = \frac{\Sigma^{-1}(k)}{1-\alpha} - \frac{\alpha}{1-\alpha} \frac{(\Sigma^{-1}(k)(\mathbf{x}_s - \mathbf{c}))(\Sigma^{-1}(k)(\mathbf{x}_s - \mathbf{c}))^T}{1 + \alpha((\mathbf{x}_s - \mathbf{c})^T \Sigma^{-1}(k)(\mathbf{x}_s - \mathbf{c}))} \quad (6)$$

with $\alpha = \frac{1}{k+1}$ and k the number of samples seen so far for which $\mathbf{c}$ has been the winning rule (cluster).

5. Estimate consequent parameters $\mathbf{w}$ by employing the elastic net formulation of weighted least squares objective (WLS) and employing LARS-EN algorithm [78] to solve it. Parameter α is set to 0.5 per default (equal influence of terms based on L_1 and L_2 norms), the regularization parameter λ is set according to the considerations in [35] (based on the condition of the Hessian matrix). The weighted version of LS is used because the parameters are estimated for each rule

separately, emphasizing the local learning spirit which has several robustness and interpretability advantages over global learning, as analyzed in [2].
6. Store extracted rules and consequent hyper-planes in F.

3.2.2 On-Line Model Adaptation with Increased Flexibility

Once the forecast models have been established (and fully evaluated) in batch mode, it is a challenge to keep the models up-to-date during further on-line production due to system dynamics, which is often a typical occurrence in today's production systems, e.g., because of varying charges, process setting, environmental conditions, etc. Only then, it can be guaranteed that the models stay reliable with a sufficient accuracy in such cases.

There are two possibilities for an update of fuzzy systems: (1) fully unsupervised not requiring any target values (QC criteria) to guide the model update and (2) supervised by using the (from time to time) measured (thus real) target information. The first variant typically leads to models with lower accuracy than the second option, as only the antecedent parts of the fuzzy systems (rule centers and shapes represented by inverse covariance matrices) and neither the consequents nor the latent variable space (input space of the fuzzy models) can be updated—as, e.g., analyzed and evaluated in [49] for a real-world production process (melamine resin). On the other hand, the first variant can be conducted permanently based on the permanent process values recordings, whereas the second one acts in a kind of post-adaptive manner, as only in case when new QC measurements are available a feasible update can be carried out; so the update is always delayed up to new QC measurements. So we propose to apply a mixture of both, whereas the update of the antecedents is achieved in the same way as in Step 4 of Algorithm 2 (for each new sample conducted in single-pass, non-iterating manner), and the update of the consequents (whenever a new QC measurement is available) through recursive fuzzily weighted least squares approach (RFWLS), which leads to an immediate convergence to the real optimal solution in each update step—see [36, 42] for formulas and the whole update algorithm.

To increase the flexibility of the update process in order to account for significant system dynamics, three basic functionalities in the model update are added, which are (compactly) described below:

- Dynamic forgetting of consequent and antecedent parameters over time in order to increase flexibility—this is especially required in case of upcoming system drifts [16, 25] or shifts [34].
- Rule splitting to a posteriori compensate potentially arising (gradual) drifts over time, which are not compensated by dynamic forgetting.
- Incremental update of the latent variable space to compensate changes in the covariance structure between input X and target y (leading to changes in associated importance of variables or (in our case) time-series points in X on y).

Dynamic Forgetting

Leaned on the approach demonstrated in [60], our aim is to dynamically adjust the forgetting factor as integrated (1) in the recursive fuzzily weighted least squares formulas and (2) in the resetting of the dynamic learning gain for updating the rule centers. The basic idea is that no forgetting is used as long as there is no explicit (local) drift occurring (and detected) in the system → thus the forgetting factor is set to 1. No forgetting leads to a more stable behavior when the process does not show any drift, so is in regular mode, as deeply analyzed in [60].

Here, we have to apply the local variant for drift detection, as the global one requires the model error trend line over time, which may be significantly delayed (as requiring the real measured value, see explanation above). The local variant relies on the weighted Kullback–Leibler divergence, as the forgetting factor is adjusted according to the trend line of Kullback–Leibler divergence values per rule over time: if there is a statistically significant up-trend (as can be checked through the Page-Hinkley (PH) test statistics applied on consecutive values [48]), the forgetting factor is decreased (thus the degree of forgetting increased) by a fraction of the intensity of the gradient of the PH statistics:

$$\lambda_t = \min(\max(\lambda_{t-1} - direction(drift\_ind)C_t, 0.9), 0.999) \quad (7)$$

$$C_t = \frac{drift\_accum_{t-1} - drift\_accum_t}{rmse_t \rho} \quad (8)$$

where λ_t is the forgetting factor at time block t, and C_t the amount of change in accumulated drift intensity, $drift\_accum_{t-1} - drift\_accum_t$ denotes the gradient in the PH statistics over two consecutive time points, $direction$ the direction of the drift and ρ a scaling factor typically set to a high value (around 1000); $rmse$ is an estimator of the root mean squared model error based on past samples (elicited during initial batch training of PLS-fuzzy).

Increasing the degree of forgetting in case of (local) drifts makes sense in order to enforce a rule movement towards the new data distribution while leaving the old one completely, and not to end up with a rule including a mixture of old and new local data distributions, which typically increases the model error ("blown-up rules").

Rule Splitting

The stronger rule movement can be only enforced when the drift becomes visible and thus "detectable" in the PH statistics. Sometimes a gradual drift arises over time which is not clearly seen in an abrupt fashion, but becomes more and more impacting on the model quality when being blindly integrated into the model update. Typically, it leads to an artificial sneaky blow-up of one or several rules—as shown in the left image in Fig. 6.

An a posteriori compensation of such a drift is beneficial to reduce the rule size and to decrease the model error. We therefore employed the concepts demonstrated in [39], which rely on two criteria for inducing a rule split: the local error of a rule and the rule size in terms of its volume [55, 56]. Both are compared with the other rules in the system based on the usage of statistical process control for automatic

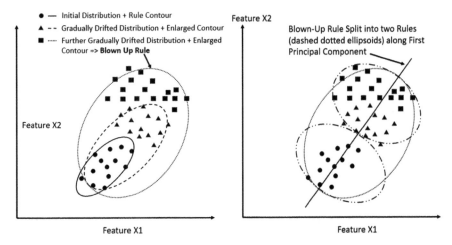

Fig. 6 A two-dimensional example for gradually drifting data distributions, leading to an artificial enlargement of the original rule (solid line) to a "blown-up" rule (dotted line); the right image shows how such larger rules are split: along the first principal component (with largest eigenvalue) = main axis of the ellipsoid and by halving it into two rules

thresholding. If they are estimated as extraordinary high, such a rule is split into two rules (1) by splitting its center according to:

$$\mathbf{c}_i(split1) = \mathbf{c}_i + a_i \frac{\sqrt{\lambda_i}}{2} \quad \mathbf{c}_i(split2) = \mathbf{c}_i - a_i \frac{\sqrt{\lambda_i}}{2} \qquad (9)$$

where λ_i corresponds to the largest eigenvalue of the covariance matrix of rule i, i.e., $\lambda_i = \max(\Lambda)$ and a_i to the corresponding eigenvector, which can both be obtained through classical eigen-decomposition of the covariance matrix; and (2) by splitting the covariance matrix according to:

$$\Sigma_i(split1) = \Sigma_i(split2) = A\Lambda^* A^T, \quad \Lambda^*_{jj} = \begin{cases} \Lambda_{jj} & j \neq 1 \\ \frac{\Lambda_{jj}}{4} & j = 1 \end{cases} \qquad (10)$$

where Λ_{11} is assumed to be the entry for the largest eigenvalue, and A the matrix of eigenvectors obtained through eigen-decomposition. The second line in (10) is because it shrinks the two split rules into the direction of the largest eigenvector (main components) of the original rule (and most responsible for the large rule size). This is achieved by taking the square-root of the largest eigenvalue as this then triggers the length of the corresponding ellipsoidal axis reduced to its half. An example of a rule split into two halves along its most significant eigenvector (first principal component shown as solid straight line) is visualized in Fig. 6.

Incremental Update of the Latent Variables (PLS Loading Space)

In particular cases, not only the characteristics of the (partial local) data distributions may change, but also the covariance relationship between process values (inputs) and QC indicators (targets); or, in other words, the main influencing directions (parts of the process values time-series) to best explain the variance in the target change over time. This can be, for instance, caused by a drift/shift in the composition of the material used during production [6]. It requires a rotation of the transformation space represented by the principal component directions obtained through partial least squares (see above).

Updating these directions builds upon on the concepts proposed in [75] and is based on the following:

- Recursive (exact) update of the first projection direction by exploiting the fact that, according to the NIPALS algorithm (also termed PLS1) [70], the first *unnormalized* direction can be represented as $q_1 = X^t y$, which is due to the maximization objective based on the covariance between Xp and y, with p denoting normalized directions, i.e., $J = \arg\max_{p^T p = 1}(Cov(Xp, y))$. Then, it is easy to see that

$$q_1(N+1) = q_1(N) - N\mu_y(N)\Delta(N+1) + y(N+1)\mathbf{x}_{N+1}(N+1), \quad (11)$$

where $\Delta(N+1) = \mu(N+1) - \mu(N)$, and $\mu(N+1)$ is the mean of the input features estimated from the first $N+1$ samples, μ_y is the mean of the target, and $\mathbf{x}_{N+1}(i)$ is the ith mean-centered sample using the mean $\mu(N+1)$ of all inputs.

- Obtaining the remaining projection directions $p_2, p_3, \ldots, p_p$ via the Krylov sequence, whose Gram–Schmidt orthogonalized form is given by $P = [q_1, Cq_1/q_1, \ldots, C^{k-1}q_1/\{q_1, Cq_1, \ldots, C^{k-2}q_1\}]$, where C is the covariance matrix, which can be recursively updated. This leads to the second PLS projection direction:

$$q_2(N+1) = Cq_1(N+1)$$

$$q_2(N+1) = q_2(N+1) - \left(q_2(N+1)^T \frac{q_1(N+1)}{\|q_1(N+1)\|}\right) \frac{q_1(N+1)}{\|q_1(N+1)\|}. \quad (12)$$

These two equations can easily be processed iteratively to obtain $q_3(N+1), q_4(N+1), \ldots, q_p(N+1)$. By normalizing these directions subject to the own L^2-norm, the updated projection directions $p_3(N+1), p_4(N+1), \ldots, p_p(N+1)$ are obtained.

Alternatively, instead of updating the global covariance matrix (which may be large and thus time-consuming whenever input dimensionality $J * k$ is huge), also the corresponding principal components can be updated by, e.g., using the well-known

CCIPCA approach as proposed in [67], and approximating the covariance matrix by the principal components directions. For further details and the full algorithm, see also [42].

4 Process Optimization with Predictive Models

The optimization problems as defined in Sect. 2.1 for the static and in Sect. 2.2 for the dynamic case can be solved by using the established predictive mappings and forecast models as surrogates within multi-objective optimization procedures. In the subsequent section, we provide a compact summary how we have achieved this.

4.1 Static Case (Mappings as Surrogates)

Having (static) mappings (surrogate models) for QC criteria enables the (virtually instant) cost-free evaluation of any process parameter combination and this, in turn, enables the search for that (those) process parameter combinations that are able to simultaneously optimize all the targeted QC criteria. In a general (real-world, larger-scale) production setting, there can be many QC criteria measured in parallel which characterize (measurements of the) final product items (e.g., the size, the shape, the appearance of various colors, etc.) [29], i.e., q in (1) is often higher than 3, 4, or 5 (denoting classical multi-objective problems); this then results in a *many-objective* optimization problem.

Nearly all classical techniques of solving such a problem are based on the central idea of reducing/restating the original many-objective optimization problem as one or more single-objective optimization problems. Obviously, the main advantage of this strategy is that the restated problem can be solved using single-objective optimization methods or smaller multi-objective problems each one operating on a subset of target values. The best known techniques that rely on transformations from many-objective to single-objective include the Tschebyscheff min-max criterion, the global criterion, the weighted sum method, goal programming, and the normal boundary intersection [44]. The disadvantage of reducing the many-objective problem to a single-objective one is that some articulation of preference (among the objectives) might be required before the start of the optimization and that the obtained single-objective optimization problem(s) can itself be very hard to solve (multi-modal, false global optima, small global solution attraction basins, etc.) by classical mathematical methods and require computationally intensive numerical methods that can produce very good results but are more "unorthodox" (i.e., heuristic-based and possibly less mathematically grounded) [24].

4.1.1 Evolutionary Algorithms for Solving Many-Objective Optimization Problems

Evolutionary algorithms (EAs) have proven extremely successful in tackling very complicated (real-life) nonlinear optimization problems by generally being able to discover acceptable solutions in reasonable time [8, 9]. The words "generally," "acceptable," and "reasonable" must be emphasized because, like natural evolution itself, an EA is a stochastic process that cannot provide any guarantees with regard to global solution optimality, success ratio, and time required for convergence. Nevertheless, in most cases, even after basic parameter tuning, the stochastic behavior of an EA can be controlled to a certain extent. The general structure of an EA is visualized in Fig. 7. Despite the drawback of not being guaranteed to find the optimal solutions, in many complex applications domains EAs have proven one of the most successful meta-heuristic global search methods. When considering multi-objective optimization problems (MOOPs), where one usually aims to find sets of Pareto non-dominated solutions that encompass the best trade-offs between several (i.e., 2–4) conflicting objectives, specialized EAs (like NSGA-II [11] and SPEA2 [77]) have become virtually canonical in the last 20 years. The main advantage of multi-objective evolutionary algorithms (MOEAs) is that, unlike classical optimization methods, by slightly adjusting the evolutionary model, these EAs are able to discover full Pareto non-dominated sets in a single run. For example, in the case of NSGA-II [11] and SPEA2 [77] (two of the most well-known MOEAs) the aforementioned adjustments to the evolutionary model are centered

---

Algorithm 1 The possible structure of a general Evolutionary Algorithm

```
1:  function EA(problem, popSize, genSize, stopCrit)
2:      P ← INITIALIZEPOPULATION(popSize, problem,)
3:      O ← Φ
4:      BestSolSet ← Φ
5:      EVALUATEFITNESS(P, problem)
6:      BestSolSet ← EXTRACTBESTSOLUTION(P, BestSolSet, problem)
7:      while stopCrit ≠ true do
8:          i ← 0
9:          while i ≤ genSize do
10:             S ← SELECTPARENTS(P)
11:             O' ← CREATEOFFSPRING(S)
12:             O ← O ∪ O'
13:             i ← i + |O'|
14:         end while
15:         EVALUATEFITNESS(O, problem)
16:         BestSolSet ← EXTRACTBESTSOLUTION(O, BestSolSet, problem)
17:         P ← SELECTFORSURVIVAL(P, O)
18:     end while
19:     return BestSolSet
20: end function
```

Fig. 7 General structure of an evolutionary algorithm (EA)

around the same selection for survival paradigm that focuses on an elitist approach to evolution based on a primary non-dominated ranking strategy and secondary (tie-breaker) objective space crowding strategy.

When considering MOOP that has more than 4 objectives, one is said to deal with a many-objective optimization problem. Specialized algorithms that deal with such scenarios have been recently proposed (e.g., NSGA-III adaptations [10]) but, given the very significant inherent difficulty of most many-objective optimization problems, a first step when solving them is to try and reduce the number of objectives one wishes to simultaneously optimize. There are basically two variants to accomplish such a reduction:

1. Explicit objective reduction strategies are based on discussing the relative importance of each objective with the decision maker (DM)—i.e., the process operator in our case. Apart from the obvious approach of simply removing the very low priority objectives from the problem formulation, one can also opt for the replacement of these objectives with a newly defined *synthetic objective* that aggregates them. The synthetic objective can be obtained using any of classical objective reduction techniques [44] that require no articulation of preference on behalf of the DM (e.g., Tschebyscheff min-max and global-criterion) or an a priori articulation of preference (e.g., weighted sum, lexicographic ordering, and goal programming).
2. Implicit objective reduction is data-driven and can be achieved by clustering objectives that are cross-correlated. Depending on the strength of intra-cluster cross-correlation, one could choose to reduce the entire cluster of objectives to:

    - one of its members that shall act as a "cluster representative" (very strong intra-cluster cross-correlation);
    - a new synthetic objective that aggregates all the members of the cluster (mild intra-cluster cross-correlation).

In our case study (Sect. 5.3), we have accomplished the former variant based on correlation checks within groups of QC indicators.

4.1.2 A New Efficient Method for Multi-Objective EA (DECMO2)

During the last 15 years since its proposal, the non-dominated sorting genetic algorithm II (NSGA-II) [11] has become one of the default (meta-heuristic) multi-objective solvers as its non-dominated sorting operator makes it highly robust and enables NSGA-II to discover high-quality *PN*s in many application domains [72]. Over the years, different paradigms have been introduced (such as differential evolution [27], cooperative co-evolution [71], and decomposition-based objective spacing [76]) which may help to further improve the convergence behavior of MOEAs.

Therefore, DECMO2 was designed by the authors of this chapter [73] as a co-evolutionary method to deliver fast average convergence and well-spaced PNs on a wide class of problems. The key feature of DECMO2 is that it tries to combine (and dynamically pivot between) three multi-objective search space exploration paradigms:

- *P*—one of the two equally sized sub-populations evolved in DECMO2 uses a SPEA2 [77] evolutionary model centered around the *environmental selection* operator which implements a two-tier selection for survival strategy that is very similar to the one of NSGA-II. Population *P* is also evolved using the SBX and PM operators.
- Sub-population *Q* adopts the GDE3 [27] search behavior that aims to benefit from the very good performance of differential evolution operators (e.g., *DE/rand/1/bin*) [62] on continuous optimization problems.
- The third multi-objective optimization paradigm is incorporated in DECMO2 via an archive *A* of well-spaced elite solutions that are maintained according to a (weighted Tschebyscheff) decomposition-based strategy similar to the one popularized by MOEA/D-DE [76]. Although *A* largely acts as a passive sub-population, from time to time (especially if the other search paradigms under-perform), a few individuals are evolved directly from *A* using differential evolution.

DECMO2 actively rewards the currently best-performing strategy by allowing the sub-population that implements it to generate a total of $m = \frac{2}{9}|P|$ more individuals than usual. A schematic overview of the search strategy proposed by the co-evolutionary solver is presented in Fig. 8.

4.2 Dynamic Case (Time-Series-Based Forecast Models as Surrogates)

4.2.1 Optimization Strategies

The particularities of the time-series-based forecasting and on-line optimization scenarios impose some extra restrictions on the used many/multi-objective solvers. Thus, for both pre-optimized and un-optimized production processes, one obvious option would be to filter state-of-the-art static solvers (i.e., those mentioned in Sect. 4.1.2) in order to discover those methods that can be parameterized to display both robustness and rapid convergence characteristics during very time limited optimization runs (e.g., 1 min/run) scheduled during two consecutive forecast model updates. In order to compensate for the short runtimes, especially in the case of un-optimized processes, parallel independent optimizations of static process parameters on the one side and on-line DPV trends (+ reverse mappings) on the other side can be carried out in order to maximize the chance of discovering and validating good solutions (e.g., through consensus).

Alternatively, one could also attempt to couple specialized dynamic many/multi-objective solvers (e.g., [17]) with the dynamic time-series-based forecast models

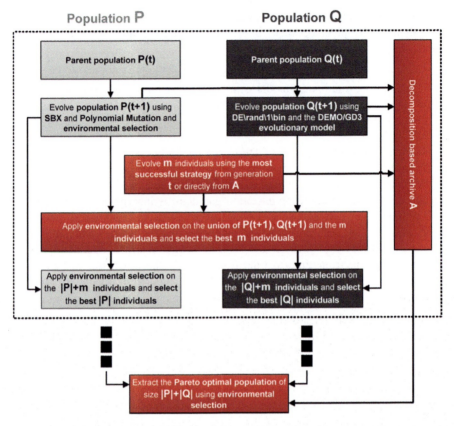

Fig. 8 The DECMO2 evolutionary model [73], where two populations are co-evolved with two different strategies and can exchange individuals through an archive A whenever requested

for the entire duration of the production process. Although still considerably less popular than their static counterparts, dynamic multi-objective solvers are very promising as they are designed to efficiently compensate for unexpected shifts in their fitness function and maintain competitive solutions throughout the entire run [22].

When considering the on-line optimization of pre-optimized production processes that are predicted to fall out of bounds, after including Eq. (5) in the optimization problem and applying the required reverse mappings (DPV trends → process parameters), the mild changes that could be suggested by the many/multi-objective solver in DPV trends should translate to minor changes in process parameters. This is a desired effect as the restricted approach is motivated in part by the fact that process operators might be reluctant to make large changes to process parameters that have been accepted (and so-far validated) as optimal but they might easily accept and test gradual shifts. Several repeats of restricted optimizations might be necessary to reverse a worrying QC trend. Depending on the complexity of the forecast models (i.e., fitness functions), fast-to-evaluate one-factor-at-a-time (OFAT) strategies might yield good results for restricted problems.

A complementary strategy we propose for pre-optimized production processes speculates the fact that preliminary static optimization results indicate that several (different) parameter configurations can deliver competitive QC results (see Table 1). Since some optimal parameter configurations might be far more robust than others with regard to unknown system dynamics and unexpected environmental influences, it would make sense to simply switch to a completely different (previously identified) best parameter configuration. In this case, Eq. (5) can also be used to filter alternative configurations based on their overall proximity to current settings.

4.2.2 Reducing Dimensionality of the Optimization Space

A final, but important issue concerns the dimensionality reduction of the input space of the optimization problem as defined in (4) (no matter whether in combination with (5) = restricted or not = unrestricted). When assuming a significant number of original process values $x_1, \ldots, x_J$ recorded during production (and often not pre-selected by experts), for which indeed several indicators are extracted to provide a compact information about their main time-series trends, it, however, soon may end up with a few hundreds of inputs (=genes in the individuals) for the heuristics-based solver. Thus, it is expected that the convergence and thus the speed of the multi-objective evolutionary algorithm is slowed down drastically. Hence, we suggest to perform an influence analysis between process values (trends) and (a subset of) quality criteria to be optimized: only those process values with a significantly higher influence compared to others are used in the optimization process—this also meets the expectations when performing the restricted optimization procedure employing (5), as more influencing process values can be modified more slightly than less influencing ones to achieve optimized QC values (as a slight change already has a significant impact on the QCs).

In our case, the influence of a process value can be calculated through its loadings obtained in all the latent variables finally used in the PLS-fuzzy models (see Sect. 3.2.1). Additionally, its impact in the consequent hyper-planes of the generalized TS fuzzy systems is important, as these indicate the regression trends in the rules (which are sub-models of local regions). As we perform dimension reduction in advance as a kind of filter stage before the optimization process starts, we have to calculate the influence globally (and not locally per actual sample); assuming that the fuzzy models contains C rules (evolving automatically during model training), we achieve this by:

$$influ_{i;r} = \frac{1}{k}\sum_{l=1}^{k}\sum_{j=1}^{p}\left(|q_{il,j}|\left(\frac{1}{C}\sum_{h=1}^{C}|w_{hj}|*range(j)\right)\right) \quad (13)$$

where $influ_i$ denotes the influence of the ith process value onto the rth target (quality criterion) to be optimized, p denotes the number of latent variables in the PLS-fuzzy model, and $q_{il;j}$ denotes the loading of the L^2-normalized jth principal component direction in the lth time-series sample of the ith process value (k samples in sum due to the sliding window size k); thus, the influences of all time-series samples of one process value are averaged. The L^2-normalization is important in order to assure the same range and thus the same impact in all loadings used as inputs to the PLS-fuzzy model (as also this has been trained by seeing all scores produced by all loadings as equally important). w_{hj} denotes the consequent parameter (=regression coefficient) of the jth component in the hth rule.

Then, the influence of a single process value can be checked versus the others, i.e., if

$$influ_{i;r} > \mu(\mathbf{influ}_r) + \sigma(\mathbf{influ}_r), \qquad (14)$$

with μ the mean and σ the standard deviation over the influences of all process values, the ith value has a significant influence onto quality criterion r and thus should be taken into account for optimization. The joined set of influencing variables over all q' quality criteria to be optimized is then taken as input parameter space. Further reduction could be achieved by using the average influences of each process value $i = 1, \ldots, p'$ over all QC indicators and applying (14) on the averaged $influ_i$ values. Furthermore, rule weights $\rho_h, h = 1, \ldots, C$ could be integrated into the last term in (13) in order to achieve a weighted average where more important rules (with higher weights) have a higher influence. A natural calculation of such a rule weight would be along the support of the rule in the training data based on which it was extracted: this can be measured in terms of the number training samples "falling" into the rule (for which the rule had highest membership degree among all rules).

The target space (number of f's in (4)) can be usually reduced in advance according to clustering and cross-correlation (see Sect. 5.2) or by some weighted objectives combining groups of quality indicators into one objective, or by selection—as discussed in Sect. 4.1.1.

5 Some Results from a Chip Production Process

5.1 Application Scenario

Our case study deals with the inspection of micro-fluidic chips used for sample preparation in DNA (deoxyribonucleic acid) sequencing. On the chip, the DNA and primers are packed into aqueous droplets in oil phase. Currently, they are checked by image inspection in a closed loop in the diagnostic instrument. Thus, an optical inspection determines whether particular events on the chips (peculiarities in image analysis) are erroneous or may even indicate severe errors in the production process. This is done in a posteriori manner, removing bad chips after production (in order not to deliver faulty components to customers), based on machine learning classifiers developed in a preliminary project, see [37, 66]. However, typically this does not prevent unnecessary waste and can even introduce greater complications and risks to the production system.

Predicting downtrends in the quality of the chips at an early stage, in order to decrease or even completely avoid waste and risks, is therefore an important challenge to be addressed. The idea is to supervise the continuously measured process values in order to see whether there are changes in the process that impact future quality. Two main process phases are essential for the final quality of the chips: injection molding (first stage in the production line) and bonding (third and last stage in the production line). Here, we concentrate on the last stage of the whole production line, because there the final chip quality is measured before the chips are delivered to the customers. We also checked the impact of injection molding process to the chip quality after bonding in a multi-stage context, where we conducted process optimization cycles to achieve ideal quality of chips shipped to customers.

5.2 Experimental Setup and Data Collection

Two different types of experiments were conducted over a longer time frame of several months:

- Supervision of the regular production process with little or no variations in the process parameters over a longer time frame: this was performed for the purpose to record time-series data permanently and to "track" any possible system dynamics (not caused by parameter settings, but maybe affecting QC values) over time.
- Explicit runs of specific production cycles based on several process parameter combinations of the most influential production parameters as shown in Fig. 1: this was performed for the purpose to see how parameter settings can affect final chip quality.

For the first type of experiments, the data of 17 process values, expected to be most important ones and most influencing final chip quality from experts' point of view, were permanently recorded and stored into a database server. The data necessary for modeling was provided via a set of custom database views, and it was made accessible via remote access to the database server. From there it could be extracted via a custom-made SQL (structured query language) client software package. 32 different QC criteria comprising the flatness values in six nest positions (the most important ones to supervised), *RMSE* values pointing to chip size and several types of *transmissions* were measured from time to time (only 2–3 times per day!). This sparsely taken QC measurements in combination with continuously measured process values lead to the batch forecast modeling problem as explained in Sect. 2. Over the whole period of July to December 2016, indeed a total of 79,716 time-series steps (of the process values) were recorded and distributed across the production of two different chip types. However, the QC information was recorded and stored only 524 times (approximately three times a day), and finally had to be reduced to 424 samples due to missing values and outliers. So, due to the unfolding process of a three-dimensional matrix as explained in Sect. 2, the final data matrix ended up with an input dimensionality of $17 * k = 850$, with $k = 50$ the default time-series length per process value. According to the low number of samples (i.e., even $N < p$), this in fact requires a significant reduction of the dimensionality (otherwise, the regression problem would be even under-determined), as can be well established with (linear or nonlinear) PLS (see Sect. 3.2.1). The first 166 samples (comprising the period from July–September 2016) were used as initial training (and evaluation) data and the remaining 258 samples as independent test set. The latter was used to determine the expected (future) errors of the batch off-line forecast models and as the basis for permanently updating the dynamic models (see Sect. 5.3).

For the second type of experiments the most essential process parameter combinations were obtained in the following way: (1) the experts proposed 12 (initial) parameter combinations that are expected to have a positive impact on the quality of the chips; (2) these 12 settings were used as input to our hybrid design of experiments (HDoE) strategy (as explained in Sect. 3.1.2), which then selected another N settings, which from the data-driven modeling viewpoint are expected to improve the robustness (and thus generalization capability) of the predictive mappings most when being trained on the joint (expert-based + data-driven selected) parameter settings. Therefore, we exploited the condition of the parameter covariance matrix as criterion to track the decrease in uncertainty when selecting more and more samples from a joint set (1) generated through edges of the parameter hyper-cube ($2^{11} = 2048$ samples) and (2) obtained by extended Latin hyper-cube sampling with min-max optimization (10,000 samples). The condition is defined by

$$cond(X) = \frac{\max(eig(X^T X))}{\min(eig(X^T X))}, \qquad (15)$$

a well-known and widely applied measure for parameter and model uncertainty in case of regression/mapping problems [20]. Figure 9 shows the trend lines of the condition of the parameter covariance matrix for different DoE criteria (y-axis) when selecting more and more samples (x-axis) up to $N = 100$. A-optimality can significantly outperform the other variants, either when being used stand-alone (which is state of the art [14]) or even more in the hybrid combination with a space filling approach (shown as dark bold dashed line). The latter also can significantly outperform the classical SoA A-optimality criterion, as the hybrid combination reduces model uncertainty faster, especially during the first 10–12 samples . The grey dashed line shows the case when only Latin hyper-cube samples are considered for space filling. Even though this strategy delivers slightly better performance after generating the first 7–8 samples when comparing with hybrid DoE (that also uses

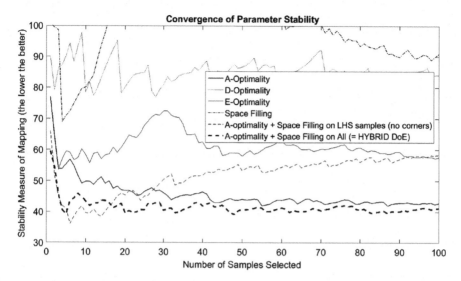

Fig. 9 Trace of parameter stability for various DoE-based sample selection criteria when incrementally (step-wise) adding up to 100 samples

corner points), the former is not able to ensure the stability of the expected model certainty further on. Hence, the A-optimality in combination with hybrid sampling was finally chosen to select another 11 samples (in addition to the 12 selected by the experts) to test at the bonding liner machine.

The joint 23 selected settings (12 expert-based + 11 from the data-driven hybrid DoE) were used at the production machine and the resulting real QC values were collected and stored, which could be used as basis for the predictive mapping construction phase. Thereby, 26 QC indicators were identified that characterize best the quality of a given process parameter setting. They belong to three main groups, namely *RMSE* (also used in the dynamic forecast modeling procedures), *skew=flatness* in the six nest positions (as also used in the dynamic forecast modeling procedures), and *void defects* (additional information to be optimized). These comprise the following indicators:

- for the two indicators (i.e., f_1 and f_2) in the *RMSE* group only a single value was computed for the whole production plate by averaging the individual QC values of every micro-chip in the plate;
- for the indicators regarding *skew=flatness*, the compliance of each of the 6 micro-chips in a production box was measured at two different time intervals (Cycles #1 and #4), resulting in a total of 12 QC indicators (f_3 to f_{14});
- the 12 QC indicators (f_{15} to f_{26}) regarding *void defects* were obtained in the same manner as in case of *skew*.

5.3 Results

Here, we summarize and show the most promising and attractive results we obtained during a project phase of about 3 years working with the data. These are based on optimal learning parameter configurations for the model training procedures, which have been automatically obtained through best parameter grid searches in combination with cross-validation and boot-strapping procedures [21], as well as default constructive parameters for optimization cycles using DECMO2 (which were basically tuned on other types of applications before). In all aspects, we show a fair comparison with related state-of-the-art methods, i.e., results from these when also being optimized over parameter grids.

5.3.1 Static Phase (Based on Process Parameter Settings)

In a few first trial-and-error test runs, we conducted a comparative performance between linear and best-performing MLP (nonlinear) predictive mappings on all 26 QC values when considering a given fivefold cross-validation partition. Thereby, we found that MLP can marginally improve the results for 19 out of 26 QC indicators; however, in most cases, the marginal MLP improvements were not stable across different random splits of the cross-validation folds.

As we had only a couple of (training) samples available (23 in sum), but 26 QC indicators in parallel to optimize (resulting in many-objective optimization problem), we could expect a significant bias when performing mapping construction and process optimization due to curse of dimensionality effects [4]. Therefore, we concentrated on the reduction of the input and output spaces:

- Performing a multi-way factor analysis, where the factors are the process parameters, in order to realize which factor has a statistically significant influence onto which QC indicator. This could be accomplished with the so-called M-way analysis of variance (N-way) ANOVA [47], where M should be as large as possible in order to take into account possible interactions between the factors, while still guaranteeing stable results for the respective data reference base at hand. p-values obtained from the results showed that only the first 8 of the 11 listed parameters in Fig. 1 are relevant for explaining any of the 26 QC indicators; the most important was parameter $X2$ (chamber temperature), which had an influence on most of the QC indicators, followed by parameters $X4$, $X5$, and $X6$ and finally by $X1$, $X3$, $X7$, and $X8$. We thus reduced the input parameter space to the first 8 input dimensions.
- Performing a cross-correlation analysis among all 26 QC indicators, which resulted in the cross-correlation matrix shown in Fig. 10. It clearly indicates that the original measurement-based classification of the QC indicators into three main groups is relevant as:

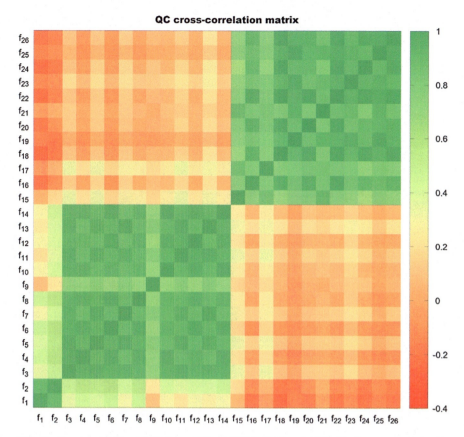

Fig. 10 Cross-correlation matrix between the 26 QC indicators, note the block-type structure with three clusters $c_1 = (f_1, f_2), c_2 = (f_3, \ldots, f_{14}), c_3 = (f_{15}, \ldots, f_{26})$

- QC indicators f_1 and f_2 are strongly intercorrelated and, from an optimization perspective, can be reunited in a *RMSE cluster*;
- QC indicators f_3 to f_{14} can form the *skew (flatness) cluster*;
- QC indicators f_{15} to f_{26} can form the *void defects cluster*.

Based on the strong inter-correlation inside the identified objective clusters, we could define a surrogate-based MOOP that contains one representative from each cluster with the reasoning that, by simultaneously aiming to minimize one of each group, we are in fact searching for process parameter settings that deliver Pareto optimal solutions (i.e., QC values) related to all 26 indicators. The three indicators from each group were selected by experts: f_1, f_{12}, and f_{18}.

Next, we analyzed the quality of the predictive mappings constructed on the reduced space of parameter settings for the indicators f_1, f_{12}, and f_{18}. In Fig. 11, we show the average R^2 values obtained by the linear and best-performing MLP regression models over 25 different cross-validation partitions. These partitions have

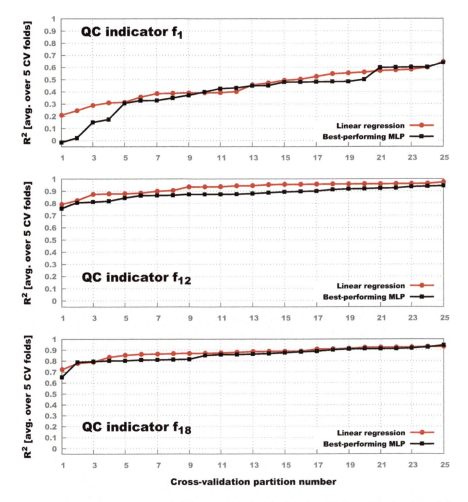

Fig. 11 Comparative performance of linear and best-performing MLP models for three QC indicators over 25 different cross-validation partitions

been carried out to omit randomness effects and to guarantee robust performance statements and conclusions. The plotted results clearly indicate that:

- some QC indicators (e.g., f_1) are generally difficult to model as R^2 values depend more on the cross-validation partition than on the used modeling method;
- some QC indicators (e.g., f_{12} and f_{18}) appear far easier to model but using advanced nonlinear methods does not seem to bring a consistent/stable advantage and can even deliver slightly worse results (e.g., f_{12}).

Based on the results in Fig. 11, we finally used the linear mappings as surrogates for the process parameter optimization phase, where, after normalization of the

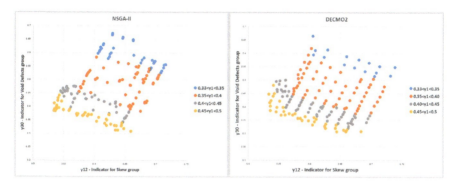

Fig. 12 Comparative f_{12} vs. f_{18} 2D Pareto fronts obtained using NSGA-II and DECMO2; the different colors mark different groups of f_1; each point denotes a multi-dimensional parameter vector (solution)

target values, the objective was to minimize all three (predicted) QC criteria towards 0. Given the fact that objective f_1 has proven harder to model (more unstable and lower R^2 values), considering the precise numerical values of this indicator in further analyses is not recommended. Therefore, in Fig. 12, which contains the final comparative optimization results obtained by NSGA-II and DECMO2, we show the obtained Pareto fronts over f_{12} and f_{18} when considering four broad quality groups with respect to f_1. It can be clearly seen that:

- the integration of the decomposition-based space exploration paradigm enables DECMO2 to maintain a better spread than NSGA-II across the entire Pareto front.
- NSGA-II also performs robustly as it is able to deliver a large number of (albeit more poorly spread) solutions in a key section of the Pareto front where the harder to model f_1 objective is also minimized.
- A good solution being closest to the original point in the Pareto front can be only achieved with a pretty high value of f_1 (lying in the fourth group); generally speaking, better values for f_1 also induce worse values for either f_{12} or f_{18}, and vice versa. Thus, a conflicting situation between the three QC groups could be found out when trying to optimize the essential process parameters, which means that the expert should ideally define her/his preferences (two out of these three groups).

Despite this observation of conflicting QC groups, two best parameter settings were selected from the Pareto fronts (based on expert preferences) and were tested on the real production machine to determine the real QC values (measured manually by experts) delivered by these settings. This on-site examination (validation) is necessary, as the predictive mappings are not perfect and thus may estimate wrong target values (for individuals = settings) during the optimization procedure. Surprisingly, while one of the new settings leads to improved QC results when comparing to the default parametrization (third row in Table 1) the bonding liner

Table 1 Possible process parameter combinations at the bonding liner; third row: default parametrization used by operators for several years, fourth and fifth rows: two parametrizations obtained by selecting promising solutions from the Pareto fronts as shown in Fig. 12, and the achieved QC values after a real test cycle phase with these settings—compared to the QC values achieved with the default parametrization, there is a clear improvement in skewness (flatness) and void defects

Process parameters											RMSE QCs		Skew QCs												Void defects QCs											
x_1	x_2	x_3	x_4	x_5	x_6	x_7	x_8	x_9	x_{10}	x_{11}	f_1	f_2	f_3	f_4	f_5	f_6	f_7	f_8	f_9	f_{10}	f_{11}	f_{12}	f_{13}	f_{14}	f_{15}	f_{16}	f_{17}	f_{18}	f_{19}	f_{20}	f_{21}	f_{22}	f_{23}	f_{24}	f_{25}	f_{26}
10	31	50	100	21	50.00	15.0	2.50	0.15	6.90	0.010	0.008	2	-13	11	-61	-25	3	3	-12	12	-60	-24	4	1	0	0	0	0	0	0	0	0	0	0	1
10	30	47	100	46	24	34.06	23.1	9.75	0.91	7.07	0.011	0.008	4	-68	-6	8	10	17	5	-67	-5	9	11	18	1	0	2	1	0	1	0	0	0	0	1	0
10	31	46	50	53	17	42.50	20.6	1.00	2.42	6.54	0.010	0.008	-4	-18	-18	-6	-7	-15	-3	-17	-17	-5	-6	-14	0	0	0	0	0	0	0	0	0	0	0	0

operators have been using for years for this specific product type (Gent-Brugge chips), the other tested setting also delivers competitive results. These two best parameter settings that were tested are presented in the last two lines of Table 1. It is noteworthy that, when comparing with the default parametrization, for the second setting (i.e., the last line in the table), the void defects could be reduced to 0 and the skewness (=flatness) of the chips is much closer to 0 in all nest positions in both cycles, while RMSE values (f_1) are equally good.

5.3.2 Dynamic Case (Based on Time-Series of Process Values)

Initial Batch Model Construction

First, an initial batch off-line modeling phase was conducted in order to optimize the learning parameters, especially fac in the rule evolution criterion [42], the number of latent variables to use after having been obtained by PLS as well as the ideal prediction horizon n_1 in (3), leading to a good trade-off between forecast accuracy and early problem recognition (ideally as early as possible, thus the horizon should be as long as possible). Various prediction horizons have been tested through in a trial-and-error modeling process: 1, 50, 100, 200, 300, 400, and 450 samples ahead. As the gap between two sample recordings is around 50 s, this belongs to timely horizons of around 50 s, 40 min, 80 min, 2 1/2 h, 4 h and 5 1/2 h, and 6 1/4 h into the future.

Finally, it turned out that a prediction horizon up to 300 samples (= around 4 h) did not decrease the accuracy of the forecast models significantly compared to the case when using a horizon of one single sample. However, when increasing the horizon to 400 and 450 samples, there were some significant downtrends for some QC targets. As a 4-h-ahead prediction was sufficient for the company, we concentrated on 300 samples horizon when doing further test with model adaptation and influence analysis, see below. More detailed results from the off-line modeling process are shown in Fig. 13, where the performance of *PLS-fuzzy* approach is compared to conventional (linear) PLS models: values in bold font indicate significantly better results (underlined by the Wilcoxon signed rank test [68]), which is achieved in four out of six nest positions for the flatness criterion, and this with errors up to maximal 15%, overshooting the 10% error bound wished by the company. This motivates the usage of dynamic adaptation of the forecast models based on newly recorded samples, see below.

Additionally, in Table 2 we show the optimal number of latent variables (=number of inputs) and the optimal number of rules to achieve a minimal cross-validation error, in order to identify a potential over-fitting over time for new samples drawn from the process. Thus, the number of inputs and the number of rules of the three right-most columns (for prediction horizons 200, 300, and 400) correspond to the PLS-fuzzy models achieving the errors as listed in Fig. 13. Obviously, both, the number of inputs and the number of rules are moderate for the more distant horizons, and even higher for shorter horizons. In these cases, the cross-validation procedure clearly produced overly optimistic estimates due to over-

	PLS –Fuzzy (robust-GenFIS) versus PLSR		
Prediction Horizon	200	300	400
Nest_01_Flatness	15.32 / 15.49	15.81 / 15.80	15.89 / 15.88
Nest_02_Flatness	13.15 / 13.68	13.39 / 13.38	13.83 / **12.97**
Nest_03_Flatness	**11.80** / 12.97	*11.75 / 13.11*	**11.80** / 13.02
Nest_04_Flatness	**10.77** / 12.85	*11.01 / 12.87*	12.35 / 12.75
Nest_05_Flatness	**10.65** / 11.82	*10.67 / 13.33*	**11.21** / 14.25
Nest_06_Flatness	**13.01** / 15.06	*13.01 / 14.80*	**12.92** / 15.49

Fig. 13 Mean absolute errors for the six flatness criteria achieved by PLS-fuzzy based on Gen-Robust-EFS (before the slashes) and conventional PLSR (after the slashes); bold values indicate significantly better results subject to a significance level of $\alpha = 0.05$ using Wilcoxon signed-rank test [68] applied on the residual vectors, red font values belong to the prediction horizons achieving best trade-offs with model accuracies

Table 2 Upper part: input dimensionality in terms of the number of latent variables in the forecast models for the various prediction horizons and the six flatness criteria; lower part: number of fuzzy rules in the forecast models for the various prediction horizons and the six flatness criteria

Prediction horizon	1	50	100	150	200	300	400
Number of inputs (latent vars)							
Nest_01_Flatness	4	1	4	3	2	2	2
Nest_02_Flatness	5	2	2	2	2	2	2
Nest_03_Flatness	2	2	2	2	2	2	2
Nest_04_Flatness	3	2	2	1	2	1	1
Nest_05_Flatness	2	2	2	2	2	2	2
Nest_06_Flatness	2	2	2	2	2	2	2
Number of fuzzy rules							
Nest_01_Flatness	10	4	20	5	3	1	1
Nest_02_Flatness	18	7	3	3	1	1	3
Nest_03_Flatness	3	1	3	2	3	2	3
Nest_04_Flatness	8	4	3	3	3	3	3
Nest_05_Flatness	1	1	1	3	1	1	3
Nest_06_Flatness	8	1	1	3	3	1	3

fitting. These errors may be decreased by using less complex models—however, such short-time horizons are less interesting from a practical perspective. Thus, for the more distant horizons, models that are relatively stable on new, unseen data can be expected; especially, for a prediction horizon of 300 samples (the preferred choice), only 2 latent variables were needed and a maximum of 3 rules for one flatness criterion, which is remarkably low.

Dynamic Model Adaptation

In order to increase model performance over time, especially to reach the desired model error bounds and to compensate (unforeseen, suddenly arising) system dynamics (e.g., due to environmental influences, new product charges, etc.), we tried to adapt the models with new incoming on-line samples on the fly. Therefore, we used the separate validation set (applied above for error elicitation and representing the period between September and December 2016) as pseudo-stream, as this has been drawn from the on-line process whenever new QC values were measured:

thus, always the latest 500 samples for each process values were stored in a ring-buffer, where, depending on the ideal prediction horizon of the initially generated model (300 in our case), the first part of it was used as input to model adaptation, and the measured QC values as targets to model adaptation. Thus, we achieved a fully supervised adaptation of the (fuzzy) models, which typically outperforms pure unsupervised adaptation [49, 58]. We tracked the model error development over time by using the interleaved test-and-then-train scenario, where upon a new QC measurement first the model error was updated (using the predictions of the QC as obtained 300 process cycles ago) and then the model was adapted.

Figure 14 shows the results in terms of the trend lines of the (over time) accumulated error for all six flatness nest position when using (1) no updating at all (static case), (2) model updating with all samples weighted equally, i.e., *PLS-fuzzy*, (3) model updating with forgetting embedded as discussed in Sect. 3.2.2, i.e.,*PLS-fuzzy adaptive + forgetting*, and (4) the full IPLS-GEFS method including incremental PLS loading space update. Obviously, error trends increased significantly after some time in the case of static models for nest positions 02–05, which can be compensated for using dynamically adaptive models. Furthermore, the forgetting strategy for increased model flexibility further helped to decrease the error trend lines, especially in the cases of nest positions 01, 02, and 06. For nest positions 03–06, a significant reduction of the MAE to below 10% was achieved, which was the goal defined by our industry partners, while for the first two nest positions (the two most difficult forecasting cases) the error converged from around 15% to close to 10%. The dynamically adaptive models, however, showed slightly poorer performance at the beginning of the stream for nest positions 01 and 02, which can be explained by adaptive forgetting causing too excessive model flexibility (due to an inappropriate forgetting factor setting).

The case in which the PLS space was updated in addition to the fuzzy model (= *PLS-fuzzy adaptive + dynamic PLS space = IPLS-GEFS*, by using the incremental PLS space concept in Sect. 3.2.2 (paragraph "Incremental Update of the Latent Variables (PLS space)") led to results resembling those without updating but using a forgetting factor: for nest positions 1, 2, and 6 the trend curves almost overlap, for nest positions 2 and 4 they are slightly better, and in the case of nest position 5 they become slightly worse. When taking into account that—unlike PLS-fuzzy models with forgetting, which require a forgetting factor Λ to be tuned—incremental updating of the PLS component space requires no additional parameter, *IPLS-GEFS* seems to be the better choice.

In all cases, incremental (linear) PLSR achieved similar performance as the adaptive PLS-fuzzy modeling variants until approximately sample number 300, but then deteriorated the performance significantly, in some cases even falling behind the static PLS-fuzzy model, which was not updated at all. This is another clear indicator of the process containing intrinsic nonlinearity: this was further underlined by the rule evolution trends, which showed an evolution of 1–2 rules at most over all six nest positions. Splitting was never activated, such that we can conclude that no (gradual) drift occurred in the on-line production phase during the period of 4 months, i.e., September to December 2016.

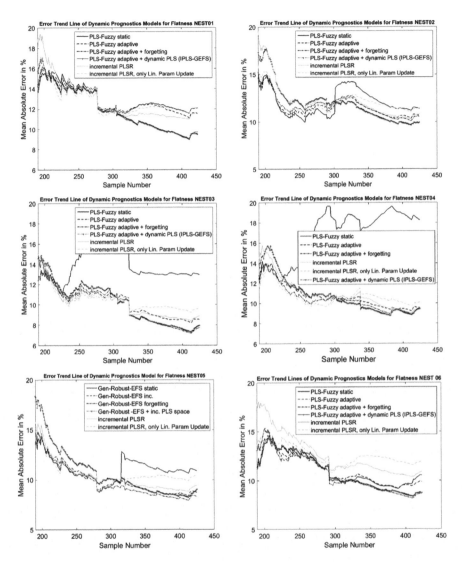

Fig. 14 Error trend lines over a time line of about 4 months for the **flatness in six nest positions** in case of no update (static PLS-fuzzy model), classical update of PLS-fuzzy model (rule evolution + parameter adaptation) in a life-long learning mode (PLS-fuzzy adaptive), update of PLS-fuzzy models with forgetting (using $\lambda = 0.95$ as forgetting factor setting), update by including incremental PLS (IPLS-GEFS), incremental PLSR (purely linear), and incremental PLSR by only updating the linear regression coefficients

Model Usage

Finally, we inspected the influence of the single process values onto the flatness criteria in order to see which process values have a higher influence and which ones have a negligible influence and thus can be eliminated before conducting any

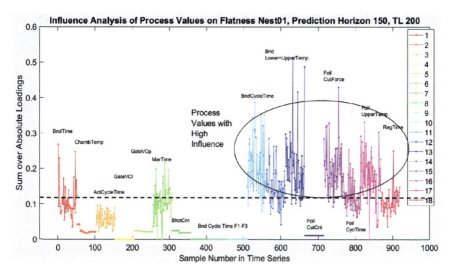

Fig. 15 Influence analysis plot for flatness NEST01 at bonding liner including 17 process values, each process value in a different color containing 50 time-series samples n_1 steps (=prediction horizon) before the QC value was measured; the dashed horizontal line indicates the threshold for a process value above which it becomes significant—as calculated by (14): the significant ones are highlighted by an ellipsis

optimization cycles—this then may lead to a reduction of the dimensionality of the solution space in the optimization problem, which in turn increases the likelihood to become solvable in a meaningful (or even real) time during the on-line process. The influence scores have been calculated as described in Sect. 4.2 (with the usage of the loadings of the variables in the most important latent variables used as inputs). Figure 15 depicts one example how the various process values influence the flatness criterion in nest position 01.

The three groups of process values with low, medium, and high influences are highlighted as such in the figure. Certainly, those with a low influence (close to 0) can be discarded for optimization as a change in them would not lead to any changes in the QC predictions, thus in the fitness calculations, anyway. Also the ones with a medium influence and still lying under the threshold (marked by the horizontal dashed line) may be discarded, especially when a co-optimization strategy with process parameters in parallel is carried out (then, most of the required variance can be steered by the process parameters). The ones inducing a high influence also have a high variance over the 50 time-series samples which are used as input to PLS. This should be respected during optimization, by, e.g., coding the mean values plus standard deviations for these process values in the individuals (solution vectors). In this example case, we ended up with a 21-dimensional solution space (mean, standard deviation, and slope for seven process values surrounded by the ellipsis). Similar drawings with similar influences have been obtained for other nest positions, such that the 21-dimensional solution space was not further expanded for optimization. Based on this reduced solution space, the on-line optimization (based on multi-objective solvers) as formulated and described in Sects. 2.2 and 4.2 can be started directly and efficiently whenever (some) flatness values fall out of their allowed bounds.

Process Optimization During Multi-Stage Production
Recent developments and experiments have been conducted for the multi-stage production process, where parameters and process values occurring during injection molding were optimized for achieving a better quality of chips after the end of bonding process. There, the same quality criteria were used in terms of three essential groups (RMSE, flatness, and void events) for optimization purposes. Long-term forecast models were established based on process values trends at injection molding to predict final chip quality at the end of bonding, where a cross-link for identifying the (time of) occurrence of the same chips across the production stages needed to be established for data collection—see [41] for details. The forecast models from injection process to bonding quality turned out to be of higher quality due to lower prediction errors (e.g., in the range of 5–8% for flatness) than the single stage forecast models from bonding process to bonding quality. According to the feedback of company experts, this is consistent to their expectations, especially for flatness and void events, which should be in large parts already ascertained after injection while bonding should have lower (but additional) influence on them.

Thus, process optimization was carried out for the multi-stage case as formulated and described in Sects. 2.2 and 4.2, where Pareto fronts with high quality (close to the origin of the three essential criteria) could be obtained. The fitness of the best individual after more than 1400 optimization cycles (populations), measured in terms of the distance to the origin, turned out to be 0.2; dimensionality reduction to 13 important process variables using the influence analysis described in Sect. 4.2.2 was essential, otherwise the fitness turned out to be much worse, i.e., 70, in the full dimensional space comprising 63 process variables at injection molding. Final Pareto fronts obtained when using optimization in the full space (left) and in the reduced space (right) are visualized in Fig. 16, where the fitness of the initial individuals are shown in dark dots and the optimized ones in lighter blue stars. For further details about the experiments and results including suggested process parameter settings for injection molding to optimize chip quality at the end of bonding, refer to [40].

6 Conclusion and Outlook

We discussed and demonstrated a novel holistic approach for off-line and on-line process optimization in manufacturing systems. The off-line phase relies on process parameters which can be steered as "control wheels" in order to obtain ideal machine settings leading to better production quality. Therefore, predictive mappings are constructed based on samples which are economically and efficiently

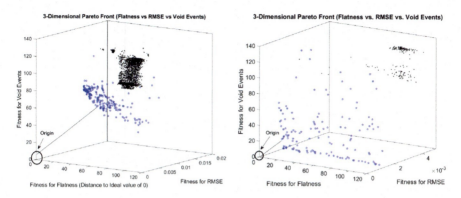

Fig. 16 Pareto fronts of individuals at the end of each process optimization run: left for the full space (63 process values, 189 dimensions), right for the reduced space (13 process values, 39 dimensions); dark dots mark fitness of individuals from initial population, blue stars the final individuals achieved through multi-objective optimization with NSGA-II

gathered from a new hybrid design of experiments approach, which are then used as surrogate models in an evolutionary-based multi-objective optimization solver (using various QC indicators as target criteria) to obtain ideal process parameter settings. The latter is based on a particular co-evolution strategy as embedded in the DECMO2 approach, with which the authors already enjoyed good experience on past multi-objective optimization applications. The on-line phase relies on process values permanently recorded during production, based on which undesired changes leading to worse product quality can be supervised. Time-series-based forecast models are constructed from on-line recorded data-based employing time-series transformation to reduce curse of dimensionality becoming apparent in a batch process modeling setting. They are able to self-adapt and evolve over time with sufficient flexibility (including rule splitting, dynamic forgetting, and updating the transformation space) and are used for permanent prediction of product quality, measured in terms of several QC indicators (with a prediction horizon of up to a few hours into the future). Based on early observed bad quality or (significantly) down-trending quality, on-line process optimization can start on a subset of QC indicators, again using an evolutionary-based multi-objective optimization solver, but with a reduced functionality, an incremental dynamic capability, and/or based on partial low-dimensional views on the objective space. Furthermore, reduction of the optimization space is required to ensure fast optimization cycles; this is achieved by a new influence analysis between process values trends and QC indicators within the context of PLS-fuzzy forecast models. The holistic approach has been implemented for a micro-chip production process in a larger manufacturing environment, comprising three stages of production stages (injection molding, oven and bonding liner), whereas the latter stage was extensively evaluated, for which some promising, successful results are shown. Additionally, process optimization in a multi-stage context from injection to bonding was achieved with nearly ideal fitness values of solutions in form of desired process values settings + trends.

Acknowledgement The authors acknowledge the Austrian research funding association (FFG) within the scope of the "IKT of the future" programme, project "Generating process feedback from heterogeneous data sources in quality control" (contract # 849962). The first author also acknowledges the support by the LCM-K2 Center within the framework of the Austrian COMET-K2 program.

References

1. Abonyi, J.: Fuzzy Model Identification for Control. Birkhäuser, Boston (2003)
2. Angelov, P., Lughofer, E., Zhou, X.: Evolving fuzzy classifiers using different model architectures. Fuzzy Sets Syst. **159**(23), 3160–3182 (2008)
3. Aumi, S., Corbett, B., Mhaskary, P.: Model predictive quality control of batch processes. In: 2012 American Control Conference, pp. 5646–5651. Fairmont Queen Elizabeth, Montréal (2012)
4. Carreira-Perpinan, M.: A review of dimension reduction techniques. Tech. Rep. CS-96-09, Dept. of Computer Science, University of Sheffield, Sheffield (1997)
5. Cauchi, N., Macek, K., Abate, A.: Model-based predictive maintenance in building automation systems with user discomfort. Energy **138**, 306–315 (2017)
6. Cernuda, C., Lughofer, E., Hintenaus, P., Märzinger, W., Reischer, T., Pawlicek, M., Kasberger, J.: Hybrid adaptive calibration methods and ensemble strategy for prediction of cloud point in melamine resin production. Chemom. Intell. Lab. Syst. **126**, 60–75 (2013)
7. Chockalingam, K., Jawahar, N., Ramanathan, K., Banerjee, P.: Optimization of stereolithography process parameters for part strength using design of experiments. Int. J. Adv. Manuf. Technol. **29**(1), 79–88 (2006)
8. Coello, C.C., Lamont, G.: Applications of multi-objective evolutionary algorithms. World Scientific, Singapore (2004)
9. Dasgupta, D., Michalewicz, Z.: Evolutionary Algorithms in Engineering Applications. Springer, Heidelberg (1997)
10. Deb, K., Jain, H.: An evolutionary many-objective optimization algorithm using reference-point-based nondominated sorting approach, part I: solving problems with box constraints. IEEE Trans. Evol. Comput. **18**(4), 577–601 (2014)
11. Deb, K., Pratap, A., Agarwal, S., Meyarivan, T.: A fast and elitist multiobjective genetic algorithm: NSGA-II. IEEE Trans. Evol. Comput. **6**(2), 182–197 (2002)
12. Dovzan, D., Logar, V., Skrjanc, I.: Implementation of an evolving fuzzy model (eFuMo) in a monitoring system for a waste-water treatment process. IEEE Trans. Fuzzy Syst. **23**(5), 1761–1776 (2015)
13. Fonseca, D.J.: A knowledge-based system for preventive maintenance. Expert Syst. **17**(5), 241–247 (2000)
14. Franceschini, G., Macchietto, S.: Model-based design of experiments for parameter precision: state of the art. Chem. Eng. Sci. **63**(19), 4846–4872 (2008)
15. Frieden, B., Gatenby, R.: Exploratory Data Analysis Using Fisher Information. Springer, New York (2007)
16. Gama, J., Zliobaite, I., Bifet, A., Pechenizkiy, M., Bouchachia, A.: A survey on concept drift adaptation. ACM Comput. Surv. **46**(4), article: 44 (2014)
17. Greeff, M., Engelbrecht, A.P.: Dynamic multi-objective optimization using PSO. In: Multi-Objective Swarm Intelligent Systems, pp. 105–123. Springer, Berlin (2010)
18. Gu, S., Ren, J., Vancso, G.: Process optimization and empirical modeling for electrospun polyacrylonitrile (PAN) nanofiber precursor of carbon nanofibers. Eur. Polym. J. **41**(11), 2559–2568 (2005)
19. Haenlein, M., Kaplan, A.: A beginner's guide to partial least squares (PLS) analysis. Underst. Stat. **3**(4), 283–297 (2004)

20. Harrel, F.: Regression Modeling Strategies. Springer, New York (2001)
21. Hastie, T., Tibshirani, R., Friedman, J.: The Elements of Statistical Learning: Data Mining, Inference and Prediction, 2nd edn. Springer, New York (2009)
22. Helbig, M., Engelbrecht, A.P.: Population-based metaheuristics for continuous boundary-constrained dynamic multi-objective optimisation problems. Swarm Evol. Comput. **14**, 31–47 (2014)
23. Jain, N., Jain, V., Debb, K.: Optimization of process parameters of mechanical type advanced machining processes using genetic algorithms. Int. J. Mach. Tools Manuf. **47**(6), 900–919 (2007)
24. Jong, K.D.: Evolutionary Computation: A Unified Approach. MIT Press, New York (2006)
25. Khamassi, I., Sayed-Mouchaweh, M., Hammami, M., Ghedira, K.: Discussion and review on evolving data streams and concept drift adapting. Evol. Syst. **9**(1), 1–23 (2017)
26. Kluska, J.: Analytical Methods in Fuzzy Modeling and Control, vol. 241. Springer, Berlin (2009)
27. Kukkonen, S., Lampinen, J.: GDE3: The third evolution step of generalized differential evolution. In: IEEE Congress on Evolutionary Computation (CEC 2005), pp. 443–450. IEEE Press, Piscataway (2005)
28. Lemos, A., Caminhas, W., Gomide, F.: Multivariable Gaussian evolving fuzzy modeling system. IEEE Trans. Fuzzy Syst. **19**(1), 91–104 (2011)
29. Levitt, J.: Complete Guide to Preventive and Predictive Maintenance. Industrial Press Inc., New York (2011)
30. Liao, W., Wang, Y.: Data-driven machinery prognostics approach using in a predictive maintenance model. J. Comput. **8**(1), 225–231 (2013)
31. Liu, Y.: Predictive modeling for intelligent maintenance in complex semiconductor manufacturing processes. Ph.D. thesis, University of Michigan, Ann Arbor (2008)
32. Lughofer, E.: Evolving fuzzy systems — fundamentals, reliability, interpretability and useability. In: P. Angelov (ed.) Handbook of Computational Intelligence, pp. 67–135. World Scientific, New York (2016)
33. Lughofer, E.: On-line active learning: a new paradigm to improve practical useability of data stream modeling methods. Inf. Sci. **415–416**, 356–376 (2017)
34. Lughofer, E., Angelov, P.: Handling drifts and shifts in on-line data streams with evolving fuzzy systems. Appl. Soft Comput. **11**(2), 2057–2068 (2011)
35. Lughofer, E., Kindermann, S.: SparseFIS: data-driven learning of fuzzy systems with sparsity constraints. IEEE Trans. Fuzzy Syst. **18**(2), 396–411 (2010)
36. Lughofer, E., Cernuda, C., Kindermann, S., Pratama, M.: Generalized smart evolving fuzzy systems. Evol. Syst. **6**(4), 269–292 (2015)
37. Lughofer, E., Weigl, E., Heidl, W., Eitzinger, C., Radauer, T.: Integrating new classes on the fly in evolving fuzzy classifier designs and its application in visual inspection. Appl. Soft Comput. **35**, 558–582 (2015)
38. Lughofer, E., Pollak, R., Zăvoianu, A.C., Meyer-Heye, P., Zorrer, H., Eitzinger, C., Haim, J., Radauer, T.: Self-adaptive time-series based forecast models for predicting quality criteria in microfluidics chip production. In: 2017 3rd IEEE International Conference on Cybernetics (CYBCONF), pp. 1–8. IEEE, Exeter (2017)
39. Lughofer, E., Pratama, M., Skrjanc, I.: Incremental rule splitting in generalized evolving fuzzy systems for autonomous drift compensation. IEEE Trans. Fuzzy Syst. **26**(4), 1854–1865 (2018)
40. Lughofer, E., Zavoianu, A., Pollak, R., Pratama, M., Meyer-Heye, P., Zörrer, H., Eitzinger, C., Radauer, T.: Autonomous supervision and optimization of product quality in a multi-stage manufacturing process based on self-adaptive prediction models. J. Process Control (2019, to appear)
41. Lughofer, E., Zavoianu, A.C., Pollak, R., Meyer-Heye, P., Zörrer, H., Eitzinger, C., Lehner, J., Radauer, T., Pratama, M.: Evolving time-series based prediction models for quality criteria in a multi-stage production process. In: Proceedings of the IEEE Evolving and Adaptive Intelligent Systems Conference (EAIS) 2018, Rhodos, pp. 1–10 (2018)

42. Lughofer, E., Zavoianu, A.C., Pollak, R., Pratama, M., Meyer-Heye, P., Zörrer, H., Eitzinger, C., Haim, J., Radauer, T.: Self-adaptive evolving forecast models with incremental PLS space update for on-line predicting quality of micro-fluidic chips. Eng. Appl. Artif. Intell. **68**, 131–151 (2018)
43. McKay, M., Beckman, R., Conover, W.: A comparison of three methods for selecting values of input variables in the analysis of output from a computer code. Technometrics **21**(2), 239–245 (1979)
44. Miettinen, K.: Nonlinear Multiobjective Optimization. Kluwer Academic Publishers, Boston (1999)
45. Mobley, R.: An Introduction to Predictive Maintenance, 2nd edn. Elsevier Science, Woburn (2002)
46. Montgomery, D.: Design and Analysis of Experiments. Wiley, New York (1991)
47. Montgomery, D.: Introduction to Statistical Quality Control, 6th edn. Wiley, New York (2008)
48. Mouss, H., Mouss, D., Mouss, N., Sefouhi, L.: Test of Page-Hinkley, an approach for fault detection in an agro-alimentary production system. In: Proceedings of the Asian Control Conference, vol. 2, pp. 815–818 (2004)
49. Nikzad-Langerodi, R., Lughofer, E., Cernuda, C., Reischer, T., Kantner, W., Pawliczek, M., Brandstetter, M.: Calibration model maintenance in melamine resin production: integrating drift detection, smart sample selection and model adaptation. Anal. Chim. Acta **1013**, 1–12 (featured article) (2018)
50. Paoletti, S., Juloski, A., Ferrari-Trecate, G., Vidal, R.: Identification of hybrid systems a tutorial. Eur. J. Control **13**(2–3), 242–260 (2007)
51. Pedrycz, W., Gomide, F.: Fuzzy Systems Engineering: Toward Human-Centric Computing. Wiley, Hoboken (2007)
52. Permin, E., Bertelsmeier, F., Blum, M., Bützler, J., Haag, S., Kuz, S., Özdemir, D., Stemmler, S., Thombansen, U., Schmitt, R., et al.: Self-optimizing production systems. Procedia CIRP **41**, 417–422 (2016)
53. Pratama, M., Anavatti, S., Angelov, P., Lughofer, E.: PANFIS: a novel incremental learning machine. IEEE Trans. Neural Netw. Learn. Syst. **25**(1), 55–68 (2014)
54. Pratama, M., Anavatti, S., Lughofer, E.: GENEFIS: towards an effective localist network. IEEE Trans. Fuzzy Syst. **22**(3), 547–562 (2014)
55. Pratama, M., Anavatti, S., Lu, J.: Recurrent classifier based on an incremental meta-cognitive scaffolding algorithm. IEEE Trans. Fuzzy Syst. **23**(6), 2048–2066 (2015)
56. Pratama, M., Lu, J., Anavatti, S., Lughofer, E., Lim, C.: An incremental meta-cognitive-based scaffolding fuzzy neural network. Neurocomputing **171**, 89–105 (2016)
57. Rhinehart, R.R.: Nonlinear Regression Modeling for Engineering Applications — Modeling, Model Validation, and Enabling Design of Experiments. Wiley, Chichester (2016)
58. Sayed-Mouchaweh, M., Lughofer, E.: Learning in Non-Stationary Environments: Methods and Applications. Springer, New York (2012)
59. Serdio, F., Lughofer, E., Zavoianu, A.C., Pichler, K., Pichler, M., Buchegger, T., Efendic, H.: Improved fault detection employing hybrid memetic fuzzy modeling and adaptive filters. Appl. Soft Comput. **51**, 60–82 (2017)
60. Shaker, A., Lughofer, E.: Self-adaptive and local strategies for a smooth treatment of drifts in data streams. Evol. Syst. **5**(4), 239–257 (2014)
61. Skrjanc, I.: Evolving fuzzy-model-based design of experiments with supervised hierarchical clustering. IEEE Trans. Fuzzy Syst. **23**(4), 861–871 (2015)
62. Storn, R., Price, K.V.: Differential evolution - a simple and efficient heuristic for global optimization over continuous spaces. J. Glob. Optim. **11**(4), 341–359 (1997)
63. Su, Z., Jamshidi, A., Núñez, A., Baldi, S., Schutter, B.D.: Multi-level condition-based maintenance planning for railway infrastructures — a scenario-based chance-constrained approach. Transp. Res. Part C Emerg. Technol. **84**, 92–123 (2017)
64. Varmuza, K., Filzmoser, P.: Introduction to Multivariate Statistical Analysis in Chemometrics. CRC Press, Boca Raton (2009)

65. Wang, L., Gao, R.X.: Condition Monitoring and Control for Intelligent Manufacturing. Springer, London (2006)
66. Weigl, E., Heidl, W., Lughofer, E., Eitzinger, C., Radauer, T.: On improving performance of surface inspection systems by on-line active learning and flexible classifier updates. Mach. Vis. Appl. **27**(1), 103–127 (2016)
67. Weng, J., Zhang, Y., Hwang, W.S.: Candid covariance-free incremental principal component analysis. IEEE Trans. Pattern Anal. Mach. Intell. **25**(8), 1034–1040 (2003)
68. Wilcoxon, F.: Individual comparisons by ranking methods. Biometrics **1**, 80–83 (1945)
69. Wise, B.M., Roginski, R.T.: A calibration model maintenance roadmap. IFAC PapersOnLine **48**(8), 260–265 (2015)
70. Wold, S., Sjöström, M., Eriksson, L.: PLS-regression: a basic tool of chemometrics. Chemom. Intell. Lab. Syst. **58**, 109–130 (2001)
71. Yang, Z., Tang, K., Yao, X.: Large scale evolutionary optimization using cooperative coevolution. Inf. Sci. **178**(15), 2985–2999 (2008)
72. Yusoff, Y., Ngadiman, M.S., Zain, A.M.: Overview of NSGA-II for optimizing machining process parameters. Procedia Eng. **15**, 3978–3983 (2011)
73. Zavoianu, A.C., Lughofer, E., Bramerdorfer, G., Amrhein, W., Klement, E.: DECMO2 — a robust hybrid multi-objective evolutionary algorithm. Soft Comput. **19**(12), 3551–3569 (2015)
74. Zavoianu, A.C., Lughofer, E., Pollak, R., Meyer-Heye, P., Eitzinger, C., Radauer, T.: Multi-objective knowledge-based strategy for process parameter optimization in micro-fluidic chip production. In: Proceedings of the SSCI 2017 Conference (CIES Workshop), Honolulu, pp. 1927–1934 (2017)
75. Zeng, X.Q., Li, G.Z.: Incremental partial least squares analysis of big streaming data. Pattern Recogn. **47**, 3726–3735 (2014)
76. Zhang, Q., Liu, W., Li, H.: The performance of a new version of MOEA/D on CEC09 unconstrained MOP test instances. Tech. rep., School of CS & EE, University of Essex (2009)
77. Zitzler, E., Laumanns, M., Thiele, L.: SPEA2: Improving the strength Pareto evolutionary algorithm for multiobjective optimization. In: Evolutionary Methods for Design, Optimisation and Control with Application to Industrial Problems (EUROGEN 2001), pp. 95–100. International Center for Numerical Methods in Engineering (CIMNE), Barcelona (2002)
78. Zou, H., Hastie, T.: Regularization and variable selection via the elastic net. J. R. Stat. Soc. Ser. B **67**, 301–320 (2005)

Distributed Chance-Constrained Model Predictive Control for Condition-Based Maintenance Planning for Railway Infrastructures

Zhou Su, Ali Jamshidi, Alfredo Núñez, Simone Baldi, and Bart De Schutter

1 Introduction

Maintenance is essential for the reliability, availability, and safety of a railway network, which is composed of various infrastructures like tracks, tunnels, stations, switches, overhead wiring, signaling systems, and safety control systems. In this paper we focus on track maintenance, which in general takes up a large portion of the annual maintenance budget of a railway infrastructure network, e.g., 40% for the Dutch railway network [1]. As shown in Fig. 1, a railway track contains different assets, e.g., rails, ballasts, sleepers, fastenings, welds, etc., that are interconnected and work together. These assets suffer from quality degradation over time due to regular usage. For example, the contact between wheel and rail leads to squats, a typical rolling contact fatigue that first appears on the rail surface and might cause rail breakage if not treated properly [2]. Early-stage squats can be effectively treated by grinding, while late-stage squats can only be addressed by rail replacement [3].

Due to the high cost of railway track maintenance interventions (e.g., over EUR 10,000 for one grinding operation), and the limited resource for track maintenance (e.g., limited track possession time for maintenance), how to plan maintenance interventions in a cost-efficient way without sacrificing the safety and reliability of the whole network has become a primary concern for railway infrastructure managers. This explains why most European countries have started a shift from reactive maintenance to proactive maintenance in recent years [4, 5]. Condition-based maintenance [6, 7], where maintenance interventions are planned based on

---

Z. Su (✉) · S. Baldi · B. De Schutter
Delft Center for Systems and Control, Delft, The Netherlands
e-mail: z.su-1@tudelft.nl; S.Baldi@tudelft.nl; B.DeSchutter@tudelft.nl

A. Jamshidi · A. Núñez
Section of Railway Engineering, Delft, The Netherlands
e-mail: A.Jamshidi@tudelft.nl; A.A.NunezVicencio@tudelft.nl

Fig. 1 Components of railway track

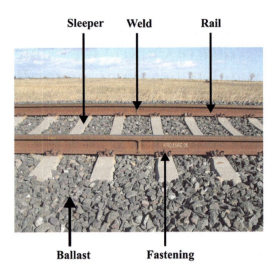

the "condition" of the asset, has been considered the most promising predictive maintenance strategy in various fields [8, 9], as most system failures are preceded by one or more indicative signals [10].

We consider condition-based maintenance optimization [11, 12], where the optimal planning of maintenance interventions is based on an explicit mathematical model describing the deterioration dynamics of the asset. This deterioration model can be either deterministic or stochastic. Examples of deterministic models include the linear model used in [13] to describe track quality degradation over tonnage, and the exponential model proposed in [14] for track geometry deterioration over time. The main advantage of deterministic model is that the resulting optimization problem is easier to solve than in case a more complex stochastic model is used. However, as a deterministic model only captures the nominal deterioration behavior of an asset, the resulting maintenance plan might not be robust enough in the presence of various sources of randomness like model uncertainties and measurement errors. In this case stochastic models are preferred. A bi-variant Gamma process is used in [15] to describe the evolution of longitudinal and transverse levels for a French high-speed line. A grey-box model is proposed in [16] to identify the stochastic aging process of track geometry using Monte Carlo simulation. Dagum probabilities are used in [17] to characterize the reduction of the standard deviation of the longitudinal level over time. In [18], a fuzzy Takagi-Sugeno internal model is applied to capture the most representative dynamics of squat evolution.

To make the proposed approach applicable to a wide range of defects in general railway infrastructures, we use a piecewise-affine model with bounded uncertain parameters as the deterioration model. The main contribution of this chapter is the development of a model-based, optimization-based approach for condition-based maintenance planning of railway infrastructures. The developed approach is robust but nonconservative, and the proposed distributed solution methods guarantee tractability even for large-scale infrastructure systems.

The paper is organized as follows: the theoretical background of the proposed approach is presented in Sect. 2, and the problem formulation is given in Sect. 3. Two distributed solution approaches are explained in Sect. 4. A numerical case study with computational experiments and comparison to other approaches is presented in Sect. 5. Finally, we conclude this work and provide future working directions in Sect. 6.

2 Preliminaries

We use Model Predictive Control (MPC) [19, 20] as the basic methodology for optimal condition-based maintenance planning for railway infrastructures. MPC follows a *receding horizon principle*. An optimization problem is solved at each sampling time step to predict the optimal sequence of maintenance actions for a given *prediction horizon*, based on the information (e.g., measurement data) available at the current time step. Only the first step of the maintenance action sequence is applied to the system, and a new optimization problem is solved at the next time step with new information. The prediction horizon is in general much shorter than the *planning horizon*, so the MPC optimization problem at each time step is much easier to solve than the correspondent optimization problem for the entire planning horizon. Although the MPC controller does not guarantee closed-loop optimality, in practice it usually gives a good control performance [21].

2.1 Hybrid and Distributed MPC

MPC has been applied to several real-world optimization problems like risk management [22] and supply chain management [23, 24]. If the system involved in these problems contains both continuous and discrete dynamics, we call it hybrid system. One way to address such a hybrid system is to transform it into a Mixed Logical Dynamical (MLD) system [25] and to solve a Mixed Integer Programming (MIP) problem at each time step. Another way is to adopt the concept of Time Instant Optimization (TIO) [26] and transform the MPC optimization containing both continuous and discrete decision variables into a non-smooth optimization problem with only continuous decision variables. Since both MIP problems and non-smooth optimization problems are NP-hard, hybrid MPC usually becomes computationally intractable for large-scale systems. In this case a distributed optimization scheme is usually adopted to improve the scalability of the MPC approach. In the control literature, most of the distributed optimization approaches are Lagrangian-based, e.g., Alternating Direction Method of Multipliers (ADMM) [27], and there is no guarantee of convergence to a global optimum for MIP problems. A continuous relaxation of binary variables is used in [28, 29], yielding a bound on the objective function value to warm-start the MIP problem. A practical approach is proposed in [30] for a class of networked hybrid MPC. This heuristic first determines the binary decision variables in the local problems, and then transforms the Mixed Integer Quadratic Programming (MIQP) problem into a set of Quadratic Programming (QP) problems via distributed coordination. One non-Lagrangian-based distributed method for MIP problems is the Distributed Robust Safe But Knowledgeable (DRSBK) algorithm [31], which adopts a constraint tightening technique.

In the operations research literature, Benders decomposition [32] and Dantzig-Wolfe decomposition [33] are the most well-known decomposition methods for large-scale Linear Programming (LP) and Mixed Integer Linear Programming (MILP) problems. Benders decomposition is designed for problems coupled through common variables, while Dantzig-Wolfe decomposition is for problems coupled through common constraints. Benders decomposition can provide global optimal solution for MILP problems in which the integer decision variables are only in the coupling variables. An up-to-date review on Benders decomposition is provided in [34]. Dantzig-Wolfe decomposition only solves an LP relaxation for MILP problems. One example of applying Dantzig-Wolfe decomposition to hybrid MPC is [35], which provides a suboptimal solution of the MILP problem via column generation.

2.2 Chance-Constrained MPC

Real-world problems like maintenance planning are influenced by various sources of randomness like model uncertainties, measurement error, and missing data. Robust control [36, 37], where control performance and constraint satisfaction are guaranteed when the uncertainties are within a specific range, might lead to a very conservative control strategy. In this case, the concept of chance-constrained optimization [38] can be adopted to achieve a balance between robustness and optimality. Chance-constrained MPC, where the probabilistic constraints are formulated as chance-constrained constraints and the objective is to optimize the expected value of the objective function, has been applied to various cases in industries like drinking water network management [39], hospital pharmacy stock management [40], and condition-based planning of railway infrastructures [41].

For chance-constrained optimization problems with known probability distributions of uncertainties, analytical approximation methods [42] are the most suitable solution approaches. When the probability distributions of uncertainties are unknown, scenario-based approaches [43] and sample average approximation methods [44] should be considered. Both approaches are based on randomization of uncertainties. The major difference is that scenario-based approaches have more restrictive assumptions on the convexity of the chance-constrained optimization problem, but require less randomized scenarios to obtain the same probabilistic guarantee as sample average approximation methods. On the other hand, sampling average approximation methods, which are based on Monte Carlo simulation, do not require convexity of the chance-constrained problem, but need a large number of scenarios to achieve an acceptable probabilistic guarantee.

Since most scenario-based approaches require the chance constraints to be convex with respect to the uncertain parameters, their applications to MILP chance-constrained problems are scarce. One notable example is [45]. However, the proposed bound in [45] on the number of scenarios is very conservative, and thus not suitable for large-scale chance-constrained problems. In this case, we choose a two-level approach [46] that lies between robust approach and scenario-based approach.

3 Problem Formulation

In this section, we first described the deterioration model in Sect. 3.1. The local chance-constrained MPC problem is formulated in Sect. 3.2, and the two-stage robust scenario-based approach to approximate the chance-constrained MPC problem is explained in Sect. 3.3. Finally, the centralized MLD-MPC problem that have to be solved at each time step is formulated in Sect. 3.4. Some important symbols used in this section are presented in Table 1.

3.1 Deterioration Model

For the planning of track maintenance activities, we divide a piece of railway track into N nonoverlapping sections, as shown in Fig. 2. The following discrete-time state-space model is used to describe the independent deterioration dynamics of each section $j \in \{1, \ldots, n\}$:

Table 1 Important symbols used in Sect. 3

Symbol	Meaning
$x_{j,k}$	State of section j at time step k
$u_{j,k}$	Maintenance option applied to section j at time step k
$\theta_{j,k}$	Realizations of all the uncertain parameters for section j at time step k
$v_{j,k}$	New binary and continuous decision variables in the transformed MLD model
N_P	Prediction horizon
$\hat{x}_{j,k+l\|k}$	Estimated state of section j at time step $k+l$, based on the information available at time step k
$\tilde{x}_{j,k}$	Estimated state of section j from time step $k+1$ to time step $k+N_\text{P}$
$\tilde{u}_{j,k}$	Maintenance option applied to section j from time step k to time step $k+N_\text{P}-1$; same notation applies to $\tilde{\theta}_{j,k}, \tilde{v}_{j,k}$.
$\tilde{x}_{j,k}^{(s)}$	Scenario s of $\tilde{x}_{j,k}$; similar notation applies to $\tilde{\theta}_{j,k}^{(s)}, \tilde{v}_{j,k}^{(s)}$

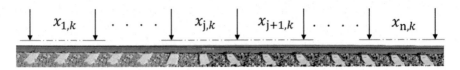

Fig. 2 Illustration of track sections for a single railway line

$$x_{j,k+1} = f_j(x_{j,k}, u_{j,k}, \theta_{j,k})$$
$$= \begin{cases} f_j^1(x_{j,k}, \theta_{j,k}) & \text{if } u_{j,k} = 1 \text{ (no maintenance)} \\ f_j^q(x_{j,k}, \theta_{j,k}) & \text{if } u_{j,k} = q \quad \forall q \in \{2, \ldots, N-1\} \\ f_j^N(\theta_{j,k}) & \text{if } u_{j,k} = N \text{ (full renewal),} \end{cases} \quad (1)$$

where the vector $x_{j,k} = \begin{bmatrix} x_{j,k}^{\text{con}} & x_{j,k}^{\text{aux}} \end{bmatrix}^{\text{T}} \in \mathcal{X}_j$ denotes the state of section j at time step k. In particular, $x_{j,k}^{\text{con}}$ indicates the "condition" of the track section, while $x_{j,k}^{\text{aux}}$ is an auxiliary state that can be viewed as the "memory" of the track section, e.g., the number of grindings that have been applied to this section since the last rail replacement. This auxiliary state is useful to capture the inefficiency of track maintenance activities. The discrete scalar $u_{j,k} \in \mathcal{U}_j = \{1, \ldots, N\}$ denotes the maintenance options, including maintenance activities and the "no maintenance" option, that is applied to section j. Finally, the vector $\theta_{j,k} \in \Theta_j$ contains the realizations of all the uncertain parameters for system j at time step k. Our only assumption on the uncertain parameters is that Θ_j is a bounded hyperbox.

We assume that for any $q \in \{1, \ldots, N\}$, the function f_j^q is either piecewise-affine or linear with respect to $x_{j,k}$. This is not a very restrictive assumption, as piecewise-affine functions can approximate any nonlinear function with arbitrary accuracy.

3.2 Local Chance-Constrained MPC Problem

Let N_p denote the prediction horizon. Define:

$$\tilde{x}_{j,k} = \begin{bmatrix} \hat{x}_{j,k+1|k}^{\text{T}} & \cdots & \hat{x}_{j,k+N_p|k}^{\text{T}} \end{bmatrix}^{\text{T}}$$
$$\tilde{u}_{j,k} = \begin{bmatrix} u_{j,k} & \cdots & u_{j,k+N_p-1} \end{bmatrix}^{\text{T}} \quad (2)$$
$$\tilde{\theta}_{j,k} = \begin{bmatrix} \theta_{j,k}^{\text{T}} & \cdots & \theta_{j,k+N_p-1}^{\text{T}} \end{bmatrix}^{\text{T}},$$

where $\hat{x}_{j,k+l|k}^{\mathrm{T}}$ denotes the estimated state of section j at time step $k+l$, based on the information available at time step k. Define

$$J_j^{\mathrm{Deg}}(\tilde{x}_{j,k}) = \|\tilde{x}_{j,k}^{\mathrm{con}}\|_1, \tag{3}$$

where $\|\cdot\|_1$ denotes the 1-norm, and P is a nonnegative weighting matrix. This term calculates the accumulated condition deterioration within the prediction window. Define the indicator function I_X, which takes value 1 if the statement X is true, and 0 otherwise. We then define

$$J_j^{\mathrm{Maint}}(\tilde{u}_{j,k}) = \sum_{l=0}^{N_{\mathrm{P}}-1} \sum_{q=1}^{N} c_{q,j}^{\mathrm{Maint}} I_{u_{j,k+l}=q}, \tag{4}$$

which computes the total maintenance costs for section j within the entire prediction window. The objective function for each local MPC controller can then be expressed as:

$$J_j(\tilde{x}_{j,k}, \tilde{u}_{j,k}) = J_j^{\mathrm{Deg}}(\tilde{x}_{j,k}) + \phi_j J_j^{\mathrm{Maint}}(\tilde{u}_{j,k}), \tag{5}$$

where the weighting parameter ϕ_j captures the trade-off between condition deterioration and maintenance costs. Finally, the chance-constrained MPC problem for section j can then be formulated as:

$$\min_{\tilde{u}_{j,k}} \mathbb{E}_{\tilde{\theta}_{j,k}}[J_j(\tilde{x}_{j,k}, \tilde{u}_{j,k})] \tag{6}$$

$$\text{subject to: } \tilde{x}_{j,k} = \tilde{f}_j(\tilde{u}_{j,k}, \tilde{\theta}_{j,k}; x_{j,k}) \tag{7}$$

$$\mathbb{P}_{\tilde{\theta}_{j,k}}\left[\max_{l=1,\ldots,N_{\mathrm{P}}} \hat{x}_{j,k+l|k}^{\mathrm{con}}(\tilde{u}_{j,k}, \tilde{\theta}_{j,k}; x_{j,k}) \le x_{\max}^{\mathrm{con}}\right] \ge 1 - \epsilon_j, \tag{8}$$

where the objective (6) is to minimize the expected condition deterioration and maintenance costs. The N_{P}-step prediction model (7) can be computed by recursive substitution of (1). Constraint (8) is the chance constraint, stating that the probability that the maximal degradation level within the prediction horizon is no more than the maintenance threshold $x_{\max}^{\mathrm{con}}$ is at least $1-\epsilon_j$, where the violation level ϵ_j is a small positive value, e.g., 0.05.

3.3 Two-Stage Robust Scenario-Based Approach

We apply the two-stage approach developed in [46] to approximate the chance-constrained problem (6)–(8) with a confidence level β_j indicating that the optimal solution of the resulting deterministic problem is also an ϵ-level solution of the originate chance-constrained problem with a probability at least $1 - \beta_j$, where β_j is a small positive value.

First, we generate the scenario set $\mathcal{H}_j$ satisfying the following condition [47]:

$$|\mathcal{H}_j| \geq \left\lceil \frac{1}{\epsilon_j} \cdot \frac{e}{e-1} \left(2 \dim(\tilde{\Theta}_j) - 1 + \ln \frac{1}{\beta_j} \right) \right\rceil \tag{9}$$

and solve the following convex scenario-based optimization problem

$$\min_{\{(\underline{\tau}_i, \overline{\tau}_i)\}_{i=1}^{\dim(\tilde{\Theta}_j)}} \sum_{i=1}^{\dim(\tilde{\Theta}_j)} \overline{\tau}_i - \underline{\tau}_i \tag{10}$$

$$\text{subject to: } (\tilde{\theta}_{j,k})_i^{(h)} \in [\underline{\tau}_i, \overline{\tau}_i] \quad \forall h \in \mathcal{H}, \forall i \in \{1, \ldots, \dim(\tilde{\Theta}_j)\} \tag{11}$$

to obtain the smallest hyperbox $\mathcal{B}_j^*$ covering all scenarios in $\mathcal{H}_j$. The notation $(\tilde{\theta}_{j,k})^{(h)}$ denotes the realization of $\tilde{\theta}_{j,k}$ for scenario h, and the symbol $(v)_i$ denotes the i-th entry of vector v.

Then we solve the robust optimization problem

$$\min_{\tilde{u}_{j,k}} \frac{1}{|\mathcal{H}_j|} \sum_{h \in \mathcal{H}_j} J_j\left(\tilde{x}_{j,k}^{(h)}, \tilde{u}_{j,k}\right) \tag{12}$$

$$\text{subject to: } \tilde{x}_{j,k}^{(h)} = \tilde{f}_j\left(\tilde{u}_{j,k}, \tilde{\theta}_{j,k}^{(h)}; x_{j,k}\right) \quad \forall h \in \mathcal{H}_j \tag{13}$$

$$\max_{\tilde{\theta}_{j,k} \in \mathcal{B}_j^* \cap \tilde{\Theta}_j} \max_{l=1,\ldots,N_\mathrm{P}} \hat{x}_{j,k+l|k}^{\mathrm{con}}(\tilde{u}_{j,k}, \tilde{\theta}_{j,k}; x_{j,k}) \leq x_{\max}^{\mathrm{con}}. \tag{14}$$

Furthermore, define the worst-case scenario w as

$$\tilde{\theta}_{j,k}^{(w)} \in \underset{\tilde{\theta}_{j,k} \in \mathcal{B}_j^* \cap \tilde{\Theta}_j}{\operatorname{argmax}} \max_{l=1,\ldots,N_\mathrm{P}} \hat{x}_{j,k+l|k}^{\mathrm{con}}(\tilde{u}_{j,k}, \tilde{\theta}_{j,k}; x_{j,k}), \tag{15}$$

and replace the robust constraint (14) by the following linear constraint:

$$P_j \tilde{x}_{j,k}^{(w)}\left(\tilde{u}_{j,k}, \tilde{\theta}_{j,k}^{(w)}; x_{j,k}\right) \leq x_{\max}^{\mathrm{con}}, \tag{16}$$

where the matrix P_j satisfies $P_j \tilde{x}_{j,k} = \tilde{x}_{j,k}^{\mathrm{con}}$. The local chance-constrained MPC problem (6)–(8) can then be approximated by the deterministic optimization problem (12), (13), (16) with the local scenario set $\mathcal{S}_j = |\mathcal{H}_j| \cup \{w\}$.

3.4 MLD-MPC Problem

For each scenario $s \in \mathscr{S}_j$, we can transform the local deterioration model (1) into the following standard MLD model [25]:

$$x_{j,k+1}^{(s)} = A_j^{(s)} x_{j,k}^{(s)} + B_j^{(s)} v_{j,k}^{(s)} \tag{17}$$

$$E_{j,1}^{(s)} v_{j,k}^{(s)} \leq E_{j,2}^{(s)} x_{j,k}^{(s)} + E_{j,3}^{(s)}, \tag{18}$$

where the new decision variable $v_{j,k}^{(s)}$ contains all the binary and continuous decision variables in the transformed MLD model. An example of how to transform the deterioration dynamics of a generic railway asset can be found in [48].

Define $\tilde{v}_{j,k}$ similar to $\tilde{u}_{j,k}$ as in (2). Furthermore, define $\tilde{v}_k = [(\tilde{v}_{j,k}^{(1)})^\mathrm{T} \ldots, (\tilde{v}_{j,k}^{(|\mathscr{S}_j|)})^\mathrm{T}]^\mathrm{T} \in \tilde{\mathscr{V}}_j$. Let $\tilde{v}_k = [\tilde{v}_{1,k}^\mathrm{T} \ldots \tilde{v}_{n,k}^\mathrm{T}]^\mathrm{T}$. The MPC optimization problem for the whole systems can then be expressed in the following compact MILP formulation:

$$\min_{\tilde{v}_k} \sum_{j=1}^{n} c_j \tilde{v}_{j,k} \tag{19}$$

$$\text{subject to:} \sum_{j=1}^{n} R_j \tilde{v}_{j,k} \leq r \tag{20}$$

$$F_j \tilde{v}_{j,k} \leq l_j \quad \forall j \in \{1, \ldots, n\}. \tag{21}$$

The objective function (19) is obtained by substituting (17) into the local objective function (12) for every section j. The linear constraint (20) is the global coupling constraint on resources, e.g., available track possession time for maintenance. Constraints (21) are the local constraints for each track section, including the deterministic approximation of the local chance constraint, and all the linear constraints from the transformation of the hybrid dynamics into an MLD model.

4 Distributed Optimization

The centralized MPC problem (19)–(21) is an NP-hard MILP problem, where the number of binary decision variables is proportional to the number of sections and the dimension of uncertain parameters. It becomes intractable for a railway infrastructure divided into a large number of sections, or for high-dimensional uncertainties. To improve the scalability of the proposed approach, we investigate two distributed optimization schemes. We call the first one the DWD algorithm, as it is based on Dantzig-Wolfe decomposition [49]. The second one is a modified version of the DRSBK algorithm [31] that uses a constraint tightening technique [50].

4.1 Dantzig-Wolfe Decomposition

Define the polyhedron $\mathscr{P}_{j,k} = \{\tilde{v}_{j,k} \in \tilde{\mathscr{V}}_j : F_j \tilde{v}_{j,k} \leq l_j\}$, which is the feasible region of the j-th local MPC problem. The set $\mathscr{G}_{j,k}$ that contains all the extreme points, i.e., columns, of the convex hull of $\mathscr{P}_{j,k}$, is called the *generating set* of the j-th subproblem. According to Minkowski's theorem [51], every point in a compact polyhedron can be represented by a convex combination of the extreme points. For each column $g \in \mathscr{G}_{j,k}$, let $\tilde{v}_{j,k}^{[g]}$ denote the value of $\tilde{v}_{j,k}$ at column g, and let $\mu_{j,g}$ denote the weight assigned to column g. Furthermore, define $\mu_j = [\mu_{j,1} \ldots \mu_{j,|\mathscr{G}_{j,k}|}]^\mathrm{T}$ and $\mu = [\mu_1^\mathrm{T} \ldots \mu_n^\mathrm{T}]^\mathrm{T}$. The *master problem* can then be defined as:

$$\min_{\mu} \sum_{j=1}^{n} \sum_{g \in \mathscr{G}_{j,k}} c_j \tilde{v}_{j,k}^{[g]} \mu_{j,g} \tag{22}$$

$$\text{subject to:} \sum_{j=1}^{n} \sum_{g \in \mathscr{G}_j} \left(R_j \tilde{v}_{j,k}^{[g]} \right) \mu_{j,g} \leq r \tag{23}$$

$$\sum_{g \in \mathscr{G}_j} \mu_{j,g} = 1 \quad \forall j \in \{1, \ldots, n\} \tag{24}$$

$$\mu_{j,g} \geq 0 \quad \forall g \in \mathscr{G}_{j,k}, \forall j \in \{1, \ldots, n\}. \tag{25}$$

This master problem is a reformulation of the LP-relaxation of the centralized MPC problem (19)–(21).

As the size of the generating set $\mathscr{G}_{j,k}$ is usually large, column generation [52], which starts with an initial partial generating set $\mathscr{G}_{j,k}^\mathrm{s} \subset \mathscr{G}_{j,k}$, is usually used to improve computational efficiency. Instead of solving the master problem, a *restricted master problem* that can be obtained by simply replacing $\mathscr{G}_{j,k}$ by $\mathscr{G}_{j,k}^\mathrm{s}$ in (22)–(25) is solved. The dual of this restricted master problem can be written as:

$$\max_{\lambda, \pi} -r\lambda + \sum_{j=1}^{n} \pi_j \tag{26}$$

$$\text{subject to: } \lambda \left(-R_j \tilde{v}_{j,k}^{[g]} \right) + \pi_j \leq c_j \tilde{v}_{j,k}^{[g]} \quad \forall g \in \mathscr{G}_{j,k}^\mathrm{s}, \forall j \in \{1, \ldots, n\} \tag{27}$$

$$\lambda \geq 0 \tag{28}$$

$$\pi \in \mathbb{R}^n. \tag{29}$$

Let μ^* and (λ^*, π^*) denote the optimal solutions of the restricted master problem and its dual, respectively. The *reduced cost* of section j can then be obtained by solving the following pricing subproblem:

$$\rho_j = \min_{g \in \mathscr{G}_{j,k}} c_j \tilde{v}_{j,k}^{[g]} + \lambda^* \left(R_j \tilde{v}_{j,k}^{[g]} \right) - \pi_j^*$$

$$= \min_{\tilde{v}_{j,k} \in \mathscr{P}_{j,k}} c_j \tilde{v}_{j,k} + \lambda^* (R_j \tilde{v}_{j,k}) - \pi_j^*, \tag{30}$$

which is an MILP. We only add the new column, i.e., the optimal solution of (30), into the partial generating set $\mathscr{G}_{j,k}^s$, when the reduced cost ρ_j is negative. Furthermore, an upper bound on the objective function value of the centralized MPC problem is obtained whenever μ^* is binary, and a lower bound is given by:

$$q(\lambda^*) = \inf_{\tilde{v}_k \in \times_{j=1}^n \mathscr{P}_{j,k}} \sum_{j=1}^n c_j \tilde{v}_{j,k} + \lambda^* \left(\sum_{i=1}^n R_j \tilde{v}_{j,k} - r \right)$$

$$= -\lambda^* r + \sum_{j=1}^n (\rho_j + \pi_j^*), \tag{31}$$

which is the Lagrangian dual of the centralized MPC problem.

The column generation procedure terminates when all the reduced costs are 0, or when the upper bound meets the lower bound. In particular, if the procedure ends with a binary μ^*, then we have also found the global optimal solution for the centralized MPC problem. If not, then a suboptimal solution of the centralized MPC problem can be found by solving the restricted master problem using the partial generating sets obtained at the end of the column generation procedure [35].

4.2 Constraint Tightening

We modify the DRSBK algorithm [31], which is based on a constraint tightening technique. First, we generate a random sequence s that is a permutation of the set $\{1, \ldots, n\}$. This sequence specifies the order of solving the subproblems. Then for each section j, we define the following subproblem:

$$\min_{\tilde{v}_{j,k} \in \mathscr{P}_{j,k}} c_j \tilde{v}_{j,k} \tag{32}$$

$$\text{subject to: } R_j \tilde{v}_{j,k} \leq r - \sum_{i=1, i \neq j}^n R_i \tilde{v}_{i,k}^\dagger, \tag{33}$$

where the local feasible region $\mathscr{P}_{j,k}$ is defined the same way as in Sect. 4.1. The left-hand side of constraint (33) is the resource allocated to section j, while the right-hand side represents the global resource reduced by the resource allocated to all the other sections. If the i-th subproblem is already solved before the j-th problem, then $\tilde{v}_{i,k}^\dagger$ denotes its optimum at time step k, otherwise $\tilde{v}_{i,k}^\dagger$ denotes the optimal solution of the i-th problem at time step $k - 1$.

If the subproblem (32)–(33) is infeasible for any section j, a new sequence s is generated, and the subproblems are solved in a new order. The iteration terminates when all the subproblems are feasible, and the difference of global objective function values between the current iteration and the previous iteration is less than the optimality tolerance. Unlike column generation, where the solution improves over each iteration, this random algorithm might need a large number of iterations for convergence. However, in practice this random algorithm works surprisingly well for MILP problems with a relatively small number of coupling constraints.

5 Case Studies

5.1 Settings

A numerical case study is performed on the optimal treatment of squats, a type of rolling contact fatigue. The evolution of a squat depends on the dynamic wheel-rail contact. A severe squat is shown in Fig. 3. The severity of a squat is determined by its visual length, which can be measured by techniques like axle box acceleration [53, 54], eddy current testing [55], or ultrasonic surface waves [56]. The degradation level, i.e., condition, of each section can be computed by aggregating the individual squat measurements within the section, as in [41]. For convenience we normalize the degradation level to [0, 1].

We consider three maintenance options, no maintenance, grinding, and replacing, to be applied to each track section. The deterioration model of section j can then be expressed as:

Fig. 3 A severe squat on the rail surface

MPC for Condition-Based Railway Track Maintenance Planning

$$x_{j,k+1}^{\text{con}} = f_j^{\text{con}}(x_{j,k}^{\text{con}}, u_{j,k}, \theta_{j,k})$$
$$= \begin{cases} f_j^{\text{Deg}}(x_{j,k}^{\text{con}}, \theta_{j,k}) & \text{if } u_{j,k} = 1 \text{ (no maintenance)} \\ f_j^{\text{Gr}}(x_{j,k}^{\text{con}}, \theta_{j,k}) & \text{if } u_{j,k} = 2 \text{ (grinding)} \\ 0 & \text{if } u_{j,k} = 3 \text{ (replacing)} \end{cases} \quad (34)$$

$$x_{j,k+1}^{\text{aux}} = f_j^{\text{aux}}(x_{j,k}^{\text{aux}}, u_{j,k})$$
$$= \begin{cases} x_{j,k}^{\text{aux}} & \text{if } u_{j,k} = 1 \text{ (no maintenance)} \\ x_{j,k}^{\text{aux}} + 1 & \text{if } u_{j,k} = 2 \text{ (grinding)} \\ 0 & \text{if } u_{j,k} = 3 \text{ (replacing)}. \end{cases} \quad (35)$$

The auxiliary state $x_{j,k}^{\text{aux}}$ counts the number of grindings on section j since the last renewal. The functions f_j^{Deg} and f_j^{Gr} in (34) are both piecewise-affine in the current condition $x_{j,k}^{\text{con}}$, i.e.

$$f_j^{\text{Deg}}(x_{j,k}^{\text{con}}) = \begin{cases} y_{j,1}^{\text{int}} + \dfrac{y_{j,2}^{\text{int}} - y_{j,1}^{\text{int}}}{x_{j,1}^{\text{swi}}} x_{j,k}^{\text{con}} & \text{if } x_{j,k}^{\text{con}} \in [0, x_{j,1}^{\text{swi}}) \\ y_{j,2}^{\text{int}} + \dfrac{y_{j,3}^{\text{int}} - y_{j,2}^{\text{int}}}{x_{j,2}^{\text{swi}} - x_{j,1}^{\text{swi}}} \left(x_{j,k}^{\text{con}} - x_{j,1}^{\text{swi}}\right) & \text{if } x_{j,k}^{\text{con}} \in [x_{j,1}^{\text{swi}}, x_{j,2}^{\text{swi}}) \\ y_{j,3}^{\text{int}} + \dfrac{y_{j,4}^{\text{int}} - y_{j,3}^{\text{int}}}{1 - x_{j,2}^{\text{swi}}} \left(x_{j,k}^{\text{con}} - x_{j,2}^{\text{swi}}\right) & \text{if } x_{j,k}^{\text{con}} \in [x_{j,2}^{\text{swi}}, 1], \end{cases} \quad (36)$$

$$f_j^{\text{Gr}}(x_{j,k}^{\text{con}}) = \begin{cases} 0 & \text{if } x_{j,k}^{\text{con}} \leq x_j^{\text{eff}} \\ \dfrac{y_j^{\text{sev}}}{x_j^{\text{sev}} - x_j^{\text{eff}}} \left(x_{j,k}^{\text{con}} - x_j^{\text{eff}}\right) & \text{if } x_j^{\text{eff}} < x_{j,k}^{\text{con}} \leq x_j^{\text{sev}} \\ y_j^{\text{sev}} + \dfrac{y_j^{\text{max}} - y_j^{\text{sev}}}{1 - x_j^{\text{sev}}} \left(x_{j,k}^{\text{con}} - x_j^{\text{sev}}\right) & \text{if } x_{j,k}^{\text{con}} > x_j^{\text{sev}}. \end{cases} \quad (37)$$

Five different deterioration models are used, and the model parameters are given in Table 3. The maintenance threshold $x_{\text{max}}^{\text{con}}$ is 0.95, and the following deterministic constraints are imposed on the auxiliary state:

$$x_{j,k+l}^{\text{aux}} \leq x_{\text{max}}^{\text{aux}} \quad \forall j \in \{1, \ldots, n\}, \forall l \in \{1, \ldots, N_{\text{p}}\}, \quad (38)$$

to bound the maximal number of consecutive grindings on one track section. We set $x_{\text{max}}^{\text{aux}} = 10$ in the case study.

Finally, we have the following global constraint:

$$\sum_{j=1}^{n} I_{u_{j,k}=1} \leq n_{\max}^{\text{Gr}} \quad \forall l \in \{1, \ldots, N_{\text{P}}\} \tag{39}$$

to bound the maximal number of sections that can be ground at each time step.

The proposed approach is implemented in Matlab R2016b, on a desktop computer with an Intel Xeon E5-1620 eight-core CPU and 64 GB of RAM, running a 64-bit version of SUSE Linux Enterprise Desktop 12. All the MILP and LP problems are solved by CPLEX V12.7.0.

5.2 Representative Run

A representative run with 53 track sections is performed to illustrate the proposed MPC approach. The length of each track section can range from 200 m to 5 km. Note that the size of the MPC optimization problem depends on the number of track sections in the network, rather than the length of each track section. For the same physical network, a finer partition captures the condition of a section more accurately, at the cost of heavier computational demand. The sampling time is 3 months, and the planning horizon is 20 steps, i.e., 5 years. The prediction horizon $N_{\text{P}} = 3$, and the maximal number of sections that can be ground is 15. The maximum number of section that can be ground at each time step is determined by multiple practical factors like the sampling time step (the larger the sampling time step, the more available track possession time for maintenance) and section length (longer section indicates more maintenance time to treat each section, thus less sections that can be ground). The realizations of the uncertain parameters within the planning window are randomly generated by Gaussian distribution. The simulation results of one of the 53 sections are shown in Fig. 4. From Fig. 4a we can see that the degradation level of this track section is kept below the maintenance threshold for the entire planning horizon. Due to the high maintenance cost, maintenance interventions, including grinding and replacing, are suggested when the degradation level is relatively high (above 0.8). Replacing is suggested when the degradation level almost hits the threshold, and there is a long interval (7 time steps) of no maintenance after rail replacement.

An overview of the simulation results of the whole network at one time step is shown in Fig. 5. In total 11 grindings and 2 replacements are suggested at the current time step, keeping the degradation levels of the whole network under the maintenance threshold at the next time step.

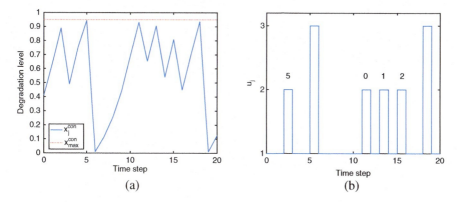

Fig. 4 Simulation results for section 24 by the chance-constrained MPC based on column generation. The number above each grinding action is the number of previous grindings on section 24 since the last replacement. (**a**) Simulated degradation levels within the planning horizon. (**b**) Interventions suggested by the MPC controller

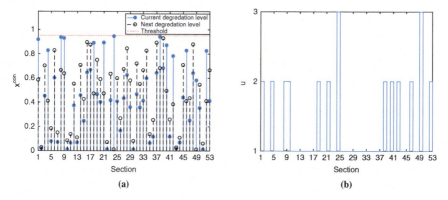

Fig. 5 Simulation results for the whole railway network at representative time step ($k = 6$). (**a**) Degradation levels of the whole railway network at time step 6 (current time step) and time step 7 (next time step). (**b**) Interventions suggested by the high-level MPC controller at time step 6 for the whole railway network

5.3 Computational Comparisons

We test the performance of the two distributed optimization algorithms on 12 randomly generated chance-constrained MPC optimization problems with the number of sections ranging from 10 to 120. The centralized approach becomes intractable (out of memory) when the number of sections reaches 130. The comparison of the 12 test problems is shown in Fig. 6. The DWD algorithm is the fastest one in all the 12 test problems. Moreover, the CPU time increases almost linearly as the size of the problem grows. The DRSBK algorithm does not show much advantage over the centralized method for small problems with no more than

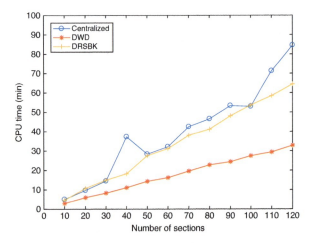

Fig. 6 CPU time of the centralized approach and two distributed approaches

30 sections. However, as the computation time of the DRSBK algorithm also grows linearly, the reduction in CPU time becomes more obvious for larger problems, especially those with more than 100 sections. The centralized approach is the slowest one in most of the test problems.

Neither of the two distributed algorithms provides theoretical guarantee on convergence to global optimum. However, the DWD algorithm is able to obtain global optimum in all the test problems. DRSBK algorithm converges to the global optimum in all the test problems except the one with 80 sections. It converges to a local optimum 70% away from the global optimum.

In summary, the DWD algorithm performs the best among the three solution methods. The centralized approach always provides global optimal solution, but its scalability is poor. The DRSBK algorithm is faster and more scalable than centralized approach. However, it might converge to a local optimum very far away from the global optimum. The DWD algorithm is the fastest among the three algorithms, and it converges to the global optimum in all the test cases. Moreover, due to its distributed nature, it is suitable for large-scale railway networks divided into many sections, as tractability of the DWD algorithm mainly depends on the tractability of the local pricing problem (30), which is an MILP of the same size as the centralized MPC problem for one single section.

5.4 Comparison with Alternative Approaches

We compare the results of the proposed chance-constrained MPC (solved by the DWD algorithm) with two alternative maintenance planning approaches. The first one is the nominal MPC approach, which uses a deterministic deterioration model that considers only the mean values of the uncertain parameters. The other one is the cyclic approach following a time-based maintenance strategy, and performing grinding and replacing at a predetermined optimal interval. The formulation of the cyclic approach is given in Appendix.

We compare robustness, optimality, and computational efficiency of the three maintenance planning approaches. Robustness is measured by maximal constraint violation v defined as:

$$v = \max\left(\frac{x_{\text{worst}}^{\text{con}} - x_{\text{max}}^{\text{con}}}{x_{\text{max}}^{\text{con}}}, 0\right), \quad (40)$$

where $x_{\text{worst}}^{\text{con}}$ is the highest degradation level of all sections within the entire planning horizon. Optimality is measured by the closed-loop objective function value, which is obtained by evaluating all the local objective function values (5) for the entire planning horizon and summing them up. Computational efficiency is measured by the CPU time needed for solving all the MPC optimization problems for all the 20 time steps. Since the cyclic approach is an offline optimization approach, i.e., it solves only one optimization problem for the entire planning horizon, we only compare the computational efficiency of the two MPC approaches.

We create 10 test runs where the realizations of the uncertain parameters within the planning horizon are randomly generated by a Gaussian distribution. The comparison of the three approaches for the 10 test runs is shown in Table 2. Both the chance-constrained MPC approach and the cyclic maintenance approach are robust, as neither of them has constraint violations for the 10 test runs. However, the cyclic approach shows much worse closed-loop performance. It is very conservative and tends to plan more maintenance than necessary. The nominal MPC approach has a slightly lower closed-loop objective function value than the chance-constrained

Table 2 Comparison between the proposed chance-constrained MPC approach (with subscript "CC") solved by the DWD algorithm, the nominal approach (with subscript "Nom"), and the cyclic approach (with subscript "Cyc")

Run	Constraint violation			Closed-loop performance			CPU time (h)	
	v_{CC} (%)	v_{Nom} (%)	v_{Cyc} (%)	$\frac{J_{\text{CC}}}{J_{\text{Cyc}}}$ (%)	$\frac{J_{\text{Nom}}}{J_{\text{Cyc}}}$ (%)	J_{Cyc}	T_{CC}	T_{Nom}
1	0	0.063	0	39.335	34.148	670,502	5.671	0.003
2	0	0.006	0	38.127	36.577	670,504	5.075	0.003
3	0	0.353	0	37.635	35.043	670,503	5.062	0.003
4	0	0.129	0	37.606	33.344	670,502	5.703	0.003
5	0	0	0	36.354	34.536	670,502	5.141	0.003
6	0	0.082	0	36.413	35.803	670,502	5.802	0.003
7	0	0.021	0	39.425	36.250	670,503	5.134	0.003
8	0	0.053	0	38.440	35.028	670,500	5.126	0.003
9	0	0.0344	0	40.244	33.359	670,503	5.088	0.003
10	0	0.172	0	38.902	34.656	670,503	5.082	0.003

MPC approach, and a much shorter CPU time. However, it is not robust, as it has constraint violations in 9 out of the 10 test runs. So in comparison, the proposed chance-constrained MPC provides an excellent balance between robustness and optimality, despite its high computational demand.

6 Conclusions and Future Work

In this paper we have developed a chance-constrained MPC approach for optimal condition-based maintenance planning for railway infrastructures. Two distributed optimization algorithms, the DWD algorithm based on Dantzig-Wolfe decomposition, and the modified Distributed Robust Safe But Knowledgeable (DRSBK) algorithm [31], have been investigated to improve the scalability of the proposed MPC approach. Computational experiments have shown that column generation is able to obtain the global optimum with a much shorter CPU time. Comparison with two alternative maintenance planning approaches has shown that the proposed chance-constrained MPC approach is robust and cost-effective.

In the future, it is interesting to consider heterogeneous components, e.g., rail and switches, in maintenance planning. Another interesting extension would be joint condition-based maintenance planning and train scheduling. Furthermore, a business case study with historical measurement data and actual maintenance costs can be performed to demonstrate the applicability of the proposed MPC approach for real-world railway track maintenance planning problems. For this purpose, a suitable key performance indicator should be chosen to evaluate the condition of each track section, and sufficient data should be used to identify the deterioration model.

Acknowledgements Research sponsored by the NWO/ProRail project "Multi-party risk management and key performance indicator design at the whole system level (PYRAMIDS)," project 438-12-300, which is partly financed by the Netherlands Organisation for Scientific Research (NWO).

Appendix

Parameters for Case Study

See Table 3.

Table 3 Parameters of the functions f_j^{Deg} and f_j^{Gr} for five different models. Both the nominal values and the 95% nonsimultaneous confidence bounds (given in the square brackets) are provided for all uncertain parameters

Parameter	Model 1	2	3	4	5
$x_{j,1}^{\text{swi}}$	0.512	0.526	0.543	0.363	0.563
$x_{j,2}^{\text{swi}}$	0.683	0.784	0.781	0.621	0.798
$y_{j,1}^{\text{int}}$	0.107 [0.086, 0.128]	0 [0,0]	0.051 [0.040, 0.063]	0.076 [0.036, 0.115]	0.058 [0.049,0.068]
$y_{j,2}^{\text{int}}$	0.783 [0.776, 0.790]	0.849 [0.845, 0.853]	0.815 [0.809, 0.821]	0.624 [0.615, 0.633]	0.805 [0.900, 0.809]
$y_{j,3}^{\text{int}}$	0.929 [0.924, 0.934]	0.975 [0.967, 0.983]	0.972 [0.966, 0.977]	0.859 [0.853, 0.865]	0.963 [0.958, 0.968]
$y_{j,4}^{\text{int}}$	1 [0.997, 1.003]	1 [0.997, 1.004]	1 [0.998, 1.002]	1 [0.994, 1.006]	1 [0.998, 1.002]
x_j^{eff}	0.156	0.177	0.172	0.141	0.106
x_j^{sev}	0.899	0.810	0.880	0.938	0.882
y_j^{sev}	0.506 [0.494, 0.518]	0.516 [0.505, 0.527]	0.502 [0.490, 0.514]	0.506 [0.490, 0.521]	0.443 [0.432, 0.455]
y_j^{max}	0.957 [0.944, 0.970]	0.991 [0.981, 1]	0.977 [0.965, 0.990]	0.922 [0.905, 0.939]	0.944 [0.931, 0.956]

Cyclic Approach

Let $t_{0,j}$ denote the time instant of the first replacement on section j. Grinding is performed every $T_{\text{Gr},j}$ after the first replacement for section j. Furthermore, we assume that replacement is performed after r consecutive grindings since the last replacement on section j. Let k_{end} denote the planning horizon. Then the offline optimization problem of the cyclic maintenance approach can be formulated as:

$$\min_{t_0, T_{\text{Gr}}, r} \sum_{k=1}^{k_{\text{end}}} \sum_{j=1}^{n} x_{j,k}^{\text{con}} + \lambda \sum_{q=2}^{3} c_{q,j}^{\text{Maint}} I_{u_{j,k}=q} \tag{41}$$

subject to

$$x_{j,k+1} = f_j(x_{j,k}, u_{j,k}; \mathbb{E}(\theta_{j,k})) \quad \forall j \in \{1, \ldots, n\}, \forall k \in \{0, \ldots, k_{\text{end}} - 1\} \tag{42}$$

$$x_{j,k}^{\text{con}} \leq x_{\max}^{\text{con}}, \quad x_{j,k}^{\text{aux}} \leq x_{\max}^{\text{aux}} \quad \forall j \in \{1, \ldots, n\}, \forall k \in \{1, \ldots, k_{\text{end}}\} \tag{43}$$

$$u_{j,k} = \begin{cases} 2, & \text{if } (k - t_{0,j}) \bmod \text{round}(T_{\text{Gr},j}) = 0 \\ 3, & \text{if } k = t_{0,j} \text{ or } (k - t_{0,j}) \bmod \text{round}(rT_{\text{Gr},j}) = 0 \\ 1, & \text{otherwise} \end{cases} \qquad (44)$$

$$\forall j \in \{1, \ldots, n\}, \ \forall k \in \{1, \ldots, k_{\text{end}}\}$$

$$1 \leq t_{0,j} \leq T_{\max} \quad \forall j \in \{1, \ldots, n\} \qquad (45)$$

$$1 \leq T_{j,\text{Gr}} \leq T_{\max} \quad \forall j \in \{1, \ldots, n\} \qquad (46)$$

$$1 \leq \mu_j \leq \mu_{\max} \quad \forall j \in \{1, \ldots, n\}. \qquad (47)$$

References

1. Zoeteman, A., Li, Z., Dollevoet, R.: Dutch research results in wheel rail interface management: 2001–2013 and beyond. Proc. Inst. Mech. Eng. F J. Rail Rapid Transit **228**(6), 642–651 (2014)
2. Sandström, J., Ekberg, A.: Predicting crack growth and risks of rail breaks due to wheel flat impacts in heavy haul operations. Proc. Inst. Mech. Eng. F J. Rail Rapid Transit **223**(2), 153–161 (2009)
3. Jamshidi, A., Núñez, A., Li, Z., Dollevoet, R.: Maintenance decision indicators for treating squats in railway infrastructures. In: 94th Annual Meeting Transportation Research Board, Washington, 11–15 January 2015. TRB (2015)
4. Zoeteman, A.: Life cycle cost analysis for managing rail infrastructure. Eur. J. Transp. Infrastruct. Res. **1**(4) (2001)
5. Al-Douri, Y., Tretten, P., Karim, R.: Improvement of railway performance: a study of Swedish railway infrastructure. Int. J. Mod. Transport. **24**(1), 22–37 (2016)
6. Kobbacy, K., Murthy, D.: Complex System Maintenance Handbook. Springer Science & Business Media, London (2008)
7. Ben-Daya, M., Kumar, U., Murthy, D.: Condition-based maintenance. In: Introduction to Maintenance Engineering: Modeling, Optimization, and Management, pp. 355–387. Wiley, Chichester (2016)
8. Jardine, A., Lin, D., Banjevic, D.: A review on machinery diagnostics and prognostics implementing condition-based maintenance. Mech. Syst. Sig. Process. **20**(7), 1483–1510 (2006)
9. Fararooy, S., Allan, J.: Condition-based maintenance of railway signalling equipment. In: International Conference on Electric Railways in a United Europe, pp. 33–37. IET, Amsterdam (1995)
10. Ahmad, R., Kamaruddin, S.: An overview of time-based and condition-based maintenance in industrial application. Comput. Ind. Eng. **63**(1), 135–149 (2012)
11. Dekker, R.: Applications of maintenance optimization models: a review and analysis. Reliab. Eng. Syst. Saf. **51**(3), 229–240 (1996)
12. Scarf, P.: On the application of mathematical models in maintenance. Eur. J. Oper. Res. **99**(3), 493–506 (1997)
13. Wen, M., Li, R., Salling, K.: Optimization of preventive condition-based tamping for railway tracks. Eur. J. Oper. Res. **252**(2), 455–465 (2016)
14. Famurewa, S., Xin, T., Rantatalo, M., Kumar, U.: Optimisation of maintenance track possession time: a tamping case study. Proc. Inst. Mech. Eng. F J. Rail Rapid Transit **229**(1), 12–22 (2015)
15. Mercier, S., Meier-Hirmer, C., Roussignol, M.: Bivariate Gamma wear processes for track geometry modelling, with application to intervention scheduling. Struct. Infrastruct. Eng. **8**(4), 357–366 (2012)

16. Quiroga, L., Schnieder, E.: Monte Carlo simulation of railway track geometry deterioration and restoration. Proc. Inst. Mech. Eng. O J. Risk Reliab. **226**,(3), 274–282 (2012)
17. Vale, C., Ribeiro, I.: Railway condition-based maintenance model with stochastic deterioration. J. Civ. Eng. Manag. **20**(5), 686–692 (2014)
18. Jamshidi, A., Faghih-Roohi, S., Hajizadeh, S., Núñez, A., Dollevoet, R., Li, Z., De Schutter, B.: A big data analysis approach for rail failure risk assessment. Risk Anal. **37**(8), 1495–1507 (2017)
19. Camacho, E., Alba, C.: Model Predictive Control. Springer Science & Business Media, London (2013)
20. Rawlings, J., Mayne, D.: Model Predictive Control: Theory and Design. Nob Hill Publishing, Madison (2009)
21. Nikolaou, M.: Model predictive controllers: a critical synthesis of theory and industrial needs. Adv. Chem. Eng. **26**, 131–204 (2001)
22. Zafra-Cabeza, A., Maestre, J., Ridao, M., Camacho, E., Sánchez, L.: Hierarchical distributed model predictive control for risk mitigation: an irrigation canal case study. J. Process Control **21**(5), 787–799 (2011)
23. Schildbach, G., Morari, M.: Scenario-based model predictive control for multi-echelon supply chain management. Eur. J. Oper. Res. **252**(2), 540–549 (2016)
24. Nandola, N., Rivera, D.: An improved formulation of hybrid model predictive control with application to production-inventory systems. IEEE Trans. Control Syst. Technol. **21**(1), 121–135 (2013)
25. Bemporad, A., Morari, M.: Control of systems integrating logic, dynamics, and constraints. Automatica **35**(3), 407–427 (1999)
26. De Schutter, B., De Moor, B.: Optimal traffic light control for a single intersection. Eur. J. Control. **4**(3), 260–276 (1998)
27. Boyd, S., Parikh, N., Chu, E., Peleato, B., Eckstein, J.: Distributed optimization and statistical learning via the alternating direction method of multipliers. Found. Trends® Mach. Learn. **3**(1), 1–122 (2011)
28. Feizollahi, M., Costley, M., Ahmed, S., Grijalva, S.: Large-scale decentralized unit commitment. Int. J. Electr. Power Energy Syst. **73**, 97–106 (2015)
29. Sebastio, S., Gnecco, G., Bemporad, A.: Optimal distributed task scheduling in volunteer clouds. Comput. Oper. Res. **81**, 231–246 (2017)
30. Mendes, P., Maestre, J., Bordons, C., Normey-Rico, J.: A practical approach for hybrid distributed mpc. J. Process Control **55**, 30–41 (2017)
31. Kuwata, Y., Richards, A., Schouwenaars, T., How, J.: Distributed robust receding horizon control for multivehicle guidance. IEEE Trans. Control Syst. Technol. **15**(4), 627–641 (2007)
32. Benders, J.: Partitioning procedures for solving mixed-variables programming problems. Numer. Math. **4**(1), 238–252 (1962)
33. Dantzig, G., Wolfe, P.: Decomposition principle for linear programs. Oper. Res. **8**(1), 101–111 (1960)
34. Rahmaniani, R., Crainic, T., Gendreau, M., Rei, W.: The Benders decomposition algorithm: a literature review. Eur. J. Oper. Res. **259**(3), 801–817 (2016)
35. Gunnerud, V., Foss, B.: Oil production optimization-a piecewise linear model, solved with two decomposition strategies. Comput. Chem. Eng. **34**(11), 1803–1812 (2010)
36. Morari, M., Zafiriou, E.: Robust Process Control, vol. 488. Prentice Hall, Englewood Cliffs (1989)
37. Gruber, J., Ramirez, D., Limon, D., Alamo, T.: Computationally efficient nonlinear min-max model predictive control based on Volterra series models – application to a pilot plant. J. Process Control **23**(4), 543–560 (2013)
38. Prekopa, A.: On probabilistic constrained programming. In: Proceedings of the Princeton Symposium on Mathematical Programming, pp. 113–138. Princeton University Press, Princeton (1970)
39. Grosso, J., Ocampo-Martínez, C., Puig, V., Joseph, B.: Chance-constrained model predictive control for drinking water networks. J. Process Control **24**(5), 504–516 (2014)

40. Jurado, I., Maestre, J., Velarde, P., Ocampo-Martinez, C., Fernández, I., Tejera, B.I., del Prado, J.: Stock management in hospital pharmacy using chance-constrained model predictive control. Comput. Biol. Med. **72**, 248–255 (2016)
41. Su, Z., Jamshidi, A., Núñez, A., Baldi, S., De Schutter, B.: Multi-level condition-based maintenance planning for railway infrastructures – a scenario-based chance-constrained approach. Transp. Res. C Emerg. Tech. **84**, 92–123 (2017)
42. Pintér, J.: Deterministic approximations of probability inequalities. Z. Oper. Res. **33**(4), 219–239 (1989)
43. Calafiore, G., Campi, M.: The scenario approach to robust control design. IEEE Trans. Autom. Control **51**(5), 742–753 (2006)
44. Shapiro, A.: Sample average approximation. In: Encyclopedia of Operations Research and Management Science, pp. 1350–1355. Springer, Boston (2013)
45. Esfahani, P., Sutter, T., Lygeros, J.: Performance bounds for the scenario approach and an extension to a class of non-convex programs. IEEE Trans. Autom. Control **60**(1), 46–58 (2015)
46. Margellos, K., Goulart, P., Lygeros, J.: On the road between robust optimization and the scenario approach for chance constrained optimization problems. IEEE Trans. Autom. Control **59**(8), 2258–2263 (2014)
47. Alamo, T., Tempo, R., Luque, A.: On the sample complexity of randomized approaches to the analysis and design under uncertainty. In: American Control Conference (ACC), 2010, pp. 4671–4676. IEEE, Baltimore (2010)
48. Su, Z., Núñez, A., Jamshidi, A., Baldi, S., Li, Z., Dollevoet, R., De Schutter, B.: Model predictive control for maintenance operations planning of railway infrastructures: In: Computational Logistics (Proceedings of the 6th International Conference on Computational Logistics (ICCL'15), Delft, Sept. 2015), pp. 673–688 (2015)
49. Dantzig, G., Ramser, J.: The truck dispatching problem. Manag. Sci. **6**(1), 80–91 (1959)
50. Chisci, L., Rossiter, J., Zappa, G.: Systems with persistent disturbances: predictive control with restricted constraints. Automatica **37**(7), 1019–1028 (2001)
51. Cassels, J.: An Introduction to the Geometry of Numbers. Springer Science & Business Media, Berlin (2012)
52. Vanderbeck, F., Wolsey, L.: Reformulation and decomposition of integer programs. In: Jünger, M., Liebling, T., Naddef, D., Nemhauser, G., Pulleyblank, W., Reinelt, G., Rinaldi, G., Wolsey, L. (eds.) 50 Years of Integer Programming 1958–2008, pp. 431–502. Springer, Berlin (2010)
53. Li, Z., Molodova, M., Núñez, A., Dollevoet, R.: Improvements in axle box acceleration measurements for the detection of light squats in railway infrastructure. IEEE Trans. Ind. Electron. **62**(7), 4385–4397 (2015)
54. Molodova, M., Li, Z., Núñez, A., Dollevoet, R.: Automatic detection of squats in railway infrastructure. IEEE Trans. Intell. Transp. Syst. **15**(5), 1980–1990 (2014)
55. Song, Z., Yamada, T., Shitara, H., Takemura, Y.: Detection of damage and crack in railhead by using eddy current testing. J. Electromagn. Anal. Appl. **3**(12), 546 (2011)
56. Fan, Y., Dixon, S., Edwards, R., Jian, X.: Ultrasonic surface wave propagation and interaction with surface defects on rail track head. NDT & E Int. **40**(6), 471–477 (2007)

Index

A

Abductive diagnosis system
  causes imply effects, 447
  definition, 439–441
  FMEA, 446
  health states, 446
  minimal abductive diagnoses, 441–442
  Modelica, 446, 447
  problem, 440
Adaptive cruise control (ACC) system, 456
Adaptive Random Forest (ARF), 113, 122–123
Adaptive synthetic sampling method (ADASYN), 73–74
Aeration system blowers
  blowers diffuser position and current consumption, 396–397
  description, 395
  experimental setup, 395–396
  residual values analysis, 397
  RME metric, 397, 398
  smoothed air temperature, 398, 399
Analytical redundancy relations (ARRs), 18, 463, 469
Anomaly/fault localization, 9
AnYa system, 271
Arousal index, 411
Arrow–Hurwitz–Uzawa procedure, 138
Automated model-based reasoning, 17–18
Automated process optimization, 18–19
  characteristics, 486
  chip production process, 485–486
    application, 512–513
    data collection, 513–515
    experimental setup, 513–515
    data-driven models, 486–487
    DoE, 487
    off-line, 488
    on-line, 488
    predictive mapping construction, 487
    process parameters, 486, 488–490
    process values, 486
      MOOP, 492–493
      pre-optimized production process, 493–494
      quality criteria, 491, 492
      time-series trends, 490–491
      un-optimized production process, 492, 493
  self-adaptive forecast models, 487
  static predictive mappings
    construction, 497–498
    cross-correlation analysis, 516–517
    default parametrization, 519–521
    expert knowledge, 487, 494, 495
    HDoE, 495–497
    linear and best-performing MLP models, 517, 518
    MOOP, 507–508
    multi-objective EA, 508–509
    multi-way factor analysis, 516
    NSGA-II and DECMO2, 519
    QC criteria, 506
    time-series-based forecast models (*see* Time-series-based forecast models)
Automatization in predictive maintenance (APM), 2
Autoregressive integrated moving average (ARIMA), 416–417
Auto regressive moving average (ARMA), 10

B

Backpropagation (BP) algorithm, 232
Barely visible impact damage (BVID)
　　criterion, 323
Battery discharge model, 368
Bayesian Cramér–Rao Lower Bounds
　　(BCRLBs)
　　CPBIM, 362
　　discrete-time dynamical systems, 357–359
　　JCP-BCRLB, 362–363
　　MCP-BCRLB, 363–364
Bayesian Information Matrix (BIM), 357
BCRLBs, *see* Bayesian Cramér–Rao Lower
　　Bounds
Benders decomposition, 536
Box–Cox models, 270

C

Cause–effect (CE) diagram, 446, 447, 495
Centralized diagnosis, 473–475
Chance-constrained optimization problems,
　　536–539
Chemical-mechanical planarization (CMP)
　　tool, 81
CHT, 225–228
CompAD algorithm, 442
CompMBD algorithm, 437–438
Computation complexity analysis, 475–476
Computerized maintenance management
　　system (CMMS), 335
Computer vision system (CVS), 453, 454
Concept drift
　　abrupt changes, 98
　　adaptive mode
　　　　components, 99–100
　　　　stream volatility and speed, 100
　　classical methods, 97
　　context-sensitive staged learning (*see*
　　　　Context-sensitive staged learning)
　　definition, 98
　　deployment, 98–99
　　detection problem, 98
　　EP approach, 113
　　feature variables and outcome, 98
　　individual learners, 99
　　neural networks, 128
　　parameter values, 117
　　recurrence pattern, 99
　　Recurrent Classifier, 113
　　sensitivity analysis, 125–127
　　spectrum learning, 128
　　synthetic data
　　　　drift intensities, 116

　　　　oscillating drift pattern, 116
　　　　preparation of, 114
　　　　RBF dataset, 115
　　　　real-world data, 117
　　　　Rotating Hyperplane dataset, 115
　　　　types, 114
Conditional gradient method, 235
Conditional predictive Bayesian information
　　matrix (CPBIM), 362
Condition-based maintenance (CbM), 314
Condition monitoring (CM) system, 315, 316
Constraint tightening technique, 543–544
Context-sensitive framework, 12–13
Context-sensitive staged learning, 100
　　decision trees, 101
　　DFT
　　　　aggregation of spectra, 105
　　　　binary-valued features, 104
　　　　classification outcomes, 104
　　　　coefficient order, 104–105
　　　　decision tree, 103
　　　　Fourier basis functions, 102–103
　　　　Fourier spectrum, 101–104
　　　　mapping, 102
　　drift detectors, 101
　　Fourier spectra, 101
　　incremental classifiers, 101
　　repository management, 106–107
　　SOL (*see* Staged online learning approach)
　　spectral learning, 112–113
　　trade-off, 101
Continuous components (Ccs), 464
Controlled undersampling technique, 74
Corrective actions, 10
Correlation-based feature selection, 69
Cramér–Rao Lower Bound (CRLB), *see*
　　Bayesian Cramér–Rao Lower
　　Bounds (BCRLBs)
Cyclic maintenance approach, 549, 551–552

D

Dantzig-Wolfe decomposition (DWD), 536,
　　542–543, 547–548
Data acquisition, 7
Data-driven models, 6, 15–16
Data preprocessing, 7, 12
　　accuracy and robustness, 54
　　characteristics, 79
　　classification algorithms, 77–78
　　data cleansing
　　　　example, 56
　　　　feature engineering step, 55
　　　　Mahalanobis distance, 57–61

Index

outlier detection accuracy, 84–85
Q and T^2 statistics, 61–63
data engineering, 56
  feature extraction, 70–71
  feature selection, 68–69
  results, 85–86
data normalization
  mean centering, 63
  performance degradation, 55
  scaling, 64
  standardization, 63–64
datasets
  artificial modifications, 82
  imbalanced data treatment, 83
  Mann–Whitney–Wilcoxon test, 82
  PHM challenge 2014, 80
  PHM challenge 2016, 81–83
data transformation
  chemical composition, 55–56
  statistical transformations, 61, 64–65
  feature discretization, 71–72, 86
  imbalanced data treatment, 56
  faulty and fault-free samples, 72
  mixed sampling, 76
  oversampling, 73–74
  results, 86–87
  undersampling, 74–76
missing values treatment, 56, 84–85
multi-and many-objective solutions, 54
regression algorithms, 78–79
signal processing, 65–67
Decentralized diagnosis, 473–475, 477
Decision maker (DM), 508
DECMO2, 508–510
Decoupled extended Kalman filter (DEKF) algorithm, 289
Deep learning (DL), 78
Degradation-based predictions, 333
Degrees of freedom, 61
Design of experiments (DoE) approaches, 486, 487
Deterioration model, 534, 537–538
Diebold–Mariano test, 415, 418
Discrete components (Dcs), 464, 467–468, 470
Discrete Fourier transform (DFT)
  aggregation of spectra, 105
  binary-valued features, 104
  classification outcomes, 104
  coefficient order, 104–105
  decision tree, 103
  Fourier basis functions, 102–103
  Fourier spectrum, 101–104
  mapping, 102

Discretely controlled continuous systems (DCCS), 462, 464, 470
Discrete-time dynamical systems, 357–359
Distance-weighted K-nearest neighbor algorithm, 77
Distributed Robust Safe But Knowledgeable (DRSBK) algorithm, 536, 543
Double-model strategy, 68
Dutch railway network, 533
DWD, *see* Dantzig-Wolfe decomposition
Dynamic production process values (DPVs), 493, 494
Dynamic time-series-based forecast models, 487

E
EANN, *see* Evolutionary neural network
EGNN, *see* Evolving fuzzy granular neural network
eHT technique, *see* Enhanced Hilbert–Huang transform technique
Electrical machine fault detection
  acquisition and data treatment module, 233
  advantages, 231–232
  backpropagation algorithm, 232
  10-class balanced classification problem, 255, 256
  condition monitoring, 231, 237
  EANN, 232
    acquisition and data treatment, 234
    classification rate, 263
    database, 236
    diagnosis report, 234, 237
    fault simulator module, 235
    GA (*see* Genetic algorithm)
    inter-turns fault detection, 257–259
    MLP neural network, 260, 264
    optimization module, 235–236
    parameters estimation module, 235
  EGNN, 232
    acquisition and data treatment, 234
    basic processing elements, 246
    connection weights, 254
    database, 236
    diagnosis report, 234, 237
    fault simulator module, 235
    fuzzy neuron model, 249–251
    fuzzy rules, 233
    granularity adaptation, 252, 265
    granules, 251–253
    learning algorithm, 254–255
    network architecture, 246–249

Electrical machine fault detection (*cont.*)
　　neurofuzzy EGNN classifier, 260
　　neurofuzzy structure, 234, 236
　　nonstationary decision boundaries, 262–264
　　numerical and fuzzy data, 247
　　optimization module, 235–236
　　parameters estimation module, 235
　　$T_{prod}-S_{max}$ EGNN model, 261, 263
　　instrumental setup, 255, 256
　　internal faults, 231
　　predictive maintenance, 231
　　schematic diagram, 233, 234
　　trial-and-error approach, 232
Electricity (ELEC) dataset, 117, 119
Embedded methods, 69
Empirical mode decomposition, 67
End-of-Discharge (EoD) time prognosis
　　discharge current profile, 369
　　efficiency criterion, 372
　　hyper-parameter candidates, 370, 371
　　ℓ^1 distances, 374–375
　　MCP-BCRLBs, 369–371
　　Monte Carlo simulations, 368
　　particle-filtering, 367–368
　　SoC, 365
　　state-space model, 365–367
　　statistical characterizations, 375–377
　　ToF PMFs, 372–374
Enhanced Hilbert–Huang transform (eHT) technique
　　closing and opening operation, 216–217
　　D'Agostino-Pearson normality measure, 222–224
　　dilation operation, 215–216
　　erosion operation, 215–216
　　experimental setup, 220–222
　　health condition monitoring, 225–226
　　IMF, 219–220
　　morphology-based filtering technique, 222–223
　　processing procedures, 214
　　processing results, 226–227
　　structural element (*see* Structural element (SE))
Enriched local model, 471–472
Ensemble Pool (EP), 113
EoD time prognosis, *see* End-of-Discharge time prognosis
Epanechnikov kernel, 370
eT2QFNN, *see* Evolving type-2 quantum fuzzy neural network
Evolutionary algorithms (EAs), 507

Evolutionary neural network (EANN), 232
　　acquisition and data treatment, 234
　　classification rate, 263
　　database, 236
　　diagnosis report, 234, 237
　　fault simulator module, 235
　　GA (*see* Genetic algorithm)
　　inter-turns fault detection, 257–259
　　MLP neural network, 260, 264
　　optimization module, 235–236
　　parameters estimation module, 235
Evolving fuzzy granular neural network (EGNN), 232
　　acquisition and data treatment, 234
　　basic processing elements, 246
　　connection weights, 254
　　database, 236
　　diagnosis report, 234, 237
　　fault simulator module, 235
　　fuzzy neuron model, 249–251
　　fuzzy rules, 233
　　granularity adaptation, 252, 265
　　granules, 251–253
　　learning algorithm, 254–255
　　network architecture, 246–249
　　neurofuzzy EGNN classifier, 260
　　neurofuzzy structure, 234, 236
　　nonstationary decision boundaries, 262–264
　　numerical and fuzzy data, 247
　　optimization module, 235–236
　　parameters estimation module, 235
　　$T_{prod}-S_{max}$ EGNN model, 261, 263
Evolving intelligent system (EIS), 288
Evolving tree (ET), 109
Evolving type-2 quantum fuzzy neural network (eT2QFNN)
　　architecture of, 290
　　cross-validation experiment, 306
　　DEKF algorithm, 289
　　direct partition experiment, 307
　　experiment setup, 304–305
　　fuzzy rule initialization, 299–300
　　input layer, 293
　　interval type-2 quantum membership function, 292, 293
　　learning policies, 289, 295
　　learning processes, 291
　　mathematical formulation, 289
　　membership function, 292
　　multi-model classifier, 294–295
　　network architecture, 291, 292
　　online learning mechanism, 289, 295

Index 559

output processing layer, 294
quantum layer, 293
RSS information, 290–291
rule growing mechanism, 289, 296–298
rule layer, 293–294
winning rule update, 300–304
Experience-based predictions, 333
Expert knowledge, 487, 494, 495
External recirculation pumping system
  description, 387
  experimental setup, 388
  linear regression, 388
  observed flow-rate *vs.* applied pump
    frequency, 388, 389
  relative mean error value, 389, 391
  residual values analysis, 388, 390
Extrapolation methods, 68

F
Failure mode and effect analysis (FMEA), 446
False positive rate (FPR), 279, 283
Fast Fourier transform (FFT), 66
Fault detection
  AnYa system, 271
  Box–Cox models, 270
  Chebyshev inequality, 271
  data-based statistical methods, 270
  data-driven monitoring, 269
  evolving-based methods, 270
  evolving mechanism, 273
  incremental unsupervised clustering
    algorithm, 271
  knowledge-based methods, 270
  learning/training phase, 274
  local densities, 274, 275
  model-based methods, 270
  RDE, 271
  real-time monitoring, 269
  residual-based approach, 270
  rule-based form, 271–273
  signal-based methods, 270
  subspace aided approach, 270
  Takagi–Sugeno fuzzy model, 270–271
  TEDA, 271
Fault identification phase, 275
Fault-tolerant control system (FTC), 10
Feature selection (FS), 68–69
First order logic (FOL), 431
Fisher information matrix, 357, 496
Forbus's qualitative process theory, 444
Forecast models/methods, 5
Fourier coefficient (FC), 103

Frank–Wolfe algorithm, 235
FRB models, *see* Fuzzy rule-based models
Freedman–Diaconis rule, 71
Frequency domain, 66
Fuzzy clustering
  adaptive fuzzy online algorithms, 132–133
  analysis, 154, 156
  artificial data sample, 147–148
  artificial neuron, 141
  attribute vectors, 147
  batch mode, 132
  Bezdek clustering procedure, 151, 153
  cluster centers, 151–152
  data sampling, 149–152
  distance parameter, 144
  goal function, 142
  Gustafson–Kessel clustering procedure,
    154–155
  Kuhn–Tucker equations, 143
  Lagrange function, 142–143
  local modification, 143
  location of cluster prototypes, 149
  multidimensional observations, 132–133
  n-dimensional attribute vectors, 133
  normal distribution, 141
  online combined approach, 138–141
  possibilistic approach, 136–137, 148
  preprocessing stage, 151
  probabilistic approach, 134–136, 148
  processed inputs, 133
  quality assessment, 141
  recurrent cluster analysis, 143–146
  results of classification, 148–149
  robust recurrent clustering procedures, 147
  target function, 142, 144–145
  time sequences, 146–147
  web-mining, 145
Fuzzy C-means, 135–136
Fuzzyfier, 134
Fuzzy modeling approach, 14–15
Fuzzy neuron model, 250–251
  aggregation neurons, 249, 260
  triangular norm and conorm, 250
Fuzzy rule-based (FRB) models
  ARIMA and VECM models, 416–417
  coverage and efficiency rates, 419, 420
  Diebold–Mariano test, 418
  IBOVESPA, 417, 420–421
  iMLP, 417
  interval arithmetic, 407–408
  interval-valued data, 404, 406, 407
  ITS, 403–405
  MDE, 419

Fuzzy rule-based (FRB) models (*cont.*)
    participatory learning, 406
    performance assignment, 414–416
    RMSE and SMAPE, 417, 418
    univariate models, 413

G
GA, *see* Genetic algorithm
Gamma process, 7
Gated recurrent unit (GRU), 79
Gaussian distribution, 63
Gaussian membership function (GMF), 288–289
Gaussian mutation, 243
Generalized linear models (GLMnet), 78
Genetic algorithm (GA), 14
    codified parameters, 233
    fitness function, 244–245
    flowchart, 237–238
    initialization and parameterization, 238–239
    mutation operator
        Gaussian mutation, 243, 257, 258
        local random mutation, 244
        random mutation, 243–244, 257, 258
    phenotype representation, 239–241
    recombination operator, 257
        arithmetic crossover, 241
        local intermediate crossover, 242–243
        multipoint crossover, 241–242
    selection operator, 245–246
    stopping criteria, 246
    trial-and-error approach, 233
Gen-Smart-EFS algorithm, 501–502

H
HDS, *see* Hybrid dynamic systems
Health and usage monitoring system (HUMS), 329
Health indicators, 8
Heating, ventilation, and air condition (HVAC) process model
    ACC, 279, 283
    control algorithm, 276
    description, 276
    fault detection results, 280–282
    fault types, 278, 279
    FPR, 279, 283
    Mollier diagram principles, 276
    PI controllers, 276
    schematic diagram, 276, 277
    tested faults, 280
    TPR, 279, 283
Hilbert–Huang transform (HHT), 14, 67. *See also* Rolling element bearings
Holistic approach for automated process optimization (HAPO), 487
Hotelling's T^2 statistic, 61
HVAC process model, *see* Heating, ventilation, and air condition process model
Hybrid Design of Experiments (HDoE), 495–497
Hybrid dynamic systems (HDS), 18
    ARR, 463
    centralized and decentralized diagnosis structures, 473–475, 477
    component, 464, 465
    computation complexity analysis, 475–476
    continuous dynamics, residual generation, 469–471
    DCCS, 462
    definitions and motivation, 461–462
    discrete component, 467–468
    discrete mode, 462
    enriched local model, 471–472
    local diagnoser construction, 472–473
    local diagnosers' decisions, 479, 482
    Matlab-Simulink^{TM}, 476
    output measured voltage, 476, 477, 479
    power converters, 464
    PWM, 476, 478
    real state outputs, 477, 480
    residuals, 478, 481
    single and multiple faults, 476, 478
    Stateflow^{TM} toolbox, 476
    stuck-open fault, 479
    system decomposition, 464, 466–467
    TCG, 462–463

I
iFRB, *see* Interval fuzzy rule-based model
Imbalanced data treatment, 83
Incremental feed-forward neural networks (NNs), 14
Inertial measurement unit (IMU), 453, 454
Instance hardness, 75–76
Instantaneous center of curvature (ICC), 429
Interpolation methods, 68
Interval-based evolving modeling (IBeM), 406
Interval fuzzy rule-based model (iFRB), 17. *See also* Fuzzy rule-based (FRB) models
    consequent parameters, 411–412
    identification procedure, 412–414

Index 561

iPL, 409–411
Takagi–Sugeno model, 409
Interval multilayer perceptron neural network (iMLP), 413, 417
Interval participatory learning (iPL), 409–411
Interval time series (ITS), 403–405, 407, 415, 416
Interval type-2 Gaussian membership function (IT2GMF), 297
Interval type-2 QMF (IT2QMF), 289, 291–294, 297
Intrinsic mode functions (IMFs), 219–220

J
Joint conditional predictive BCRLB (JCP-BCRLB), 362–363
Joint probability distribution, 98

K
Key performance indicator (KPI), 386
K-nearest neighbors algorithm, 77
Knowledge-based models, 6
Kullback–Leibler divergence, 503

L
Labelled transition system (LTS), 456
Latent variables, 65
LeveragingBag (LB), 113
Linear programming (LP), 536
Lithium-ion (Li-Ion) batteries, *see* End-of-Discharge (EoD) time prognosis
Local diagnoser, 472–473
Locality preserving projections (LPP), 71
Locally linear embedding (LLE), 65
Logit transformation, 64
Long short term memory networks (LSTM), 79
Long-term prediction, 16

M
Mann–Whitney–Wilcoxon test, 82
Manufacturing shopfloor, object location, *see* Radio frequency identification (RFID) localization technology
Many/multi-objective optimization problem (MOOP), 490, 492–493, 507–508
Marginal conditional predictive BCRLB (MCP-BCRLB), 363–365
MBD, *see* Model-based diagnosis
Mealy machine (MM) model, 471

Mean distance error (MDE), 415, 419
Mean square error (MSE), 356–358, 376
Mean time between failures (MTBF), 318
Measurement model, 366–367
Mechanism-based failure analysis (MBFA), 336–337
MED filter, *see* Minimum entropy deconvolution filter
MEMS-based accelerometers, 321
Minimal abductive diagnosis, 441–442
Minimal diagnosis, 436–439
Minimum covariance determinant, 57
Minimum description length principle (MDLP), 72
Minimum entropy deconvolution (MED) filter
  vs. convergence, 213–214
  defect-related impulses, 211
  entropy minimization, 211
  filter coefficients, 213
  flowchart, 211
  inverse filter, 212
  machinery system condition monitoring, 211
  optimal filter coefficient vector, 212
  signal denoising operation, 211–212
  vs. test signal response, 213–214
  Toeplitz autocorrelation matrix, 213
Minkowski's theorem, 542
Missing value imputation, 67
Mixed integer linear programming (MILP) problems, 536
Mixed integer programming (MIP), 535
Mixed integer quadratic programming (MIQP) problem, 536
Mixed logical dynamical (MLD) system, 535, 541
Model-based diagnosis (MBD), 442–443
  component type, 445–446
  definitions, 434–436
  minimal diagnosis, 436–439
  problem, 436
Model-based predictions, 333
Modelica programs, 446, 447
Model predictive control (MPC), 10, 19
  centralized approach, 547, 548
  chance-constrained MPC, 536–539, 549
  computational efficiency, 549
  cyclic approach, 549, 551–552
  Dantzig-Wolfe decomposition, 542–543, 547–548
  hybrid and distributed MPC, 535–536
  MLD model, 541
  violation, 549

Modified generalized type-2 Datum significance (mGT2DS), 289
Monitor, Analyze, Plan, Execute, Knowledge (MAPE-K), 448
Monte Carlo (MC) simulation, 355
Motivation, requirements, and challenges
  APM, 2
  classical quality control, 2
  condition monitoring, 2
  development of, 2–3
  fuzzy transition, 4
  system checks, 1
  system models, 2–4
  zero-defect manufacturing, 1
MPC, *see* Model predictive control
Multi-degradation process, 5
Multi-objective evolutionary algorithms (MOEAs), 507
Multistage processes, 5
Multivariate adaptive regression splines (MARS) algorithm, 392–393

N
Naïve Bayes (NB) algorithm, 77
Naval ship system, 315–316
Neuro-fuzzy approach, 13
  fuzzy clustering (*see* Fuzzy clustering)
  fuzzy segmentation, 132
  nonstationary time series, fault detection
    combined optimization criteria, 158
    computational intelligence, 156
    double wavelet-neuron, 160–163
    emulation results, 163–164
    explosion of parameters, 165
    forecasting error, 165
    function of influence, 158
    Hampel function, 158
    Huber function, 157
    least squares criterion, 157
    logistic function, 157
    nonlinear dynamic object, 163
    optimization criterion, 156
    POLYWOG-wavelets, 158–160
    RASP-wavelets, 158–159
    robust training algorithm, 161–163
    wavelet analysis, 156
Non-dominated sorting genetic algorithm II (NSGA-II), 508
Nonlinear regression modeling, 498
Non-smooth optimization problems, 535
Normalized symmetric difference (NSD), 415
N_P-step prediction model, 539

O
Off-line process optimization, 488
On-line model
  forgetting factor, 503
  functionalities, 502
  QC measurements, 502
  rule splitting, 503–504
  updating process, 505–506
On-line process optimization, 488
Open circuit voltage (OCV) curve, 366
Orthogonal distance, 62
Overall equipment effectiveness (OEE), 42

P
Page-Hinkley (PH) test, 503
Palmgren-Miner rule, 324
Partial least squares (PLS), 63–65
Partial least squares regression (PLSR), 499–502, 523
Participatory learning (PL), 406
Particle-filtering-based prognostic algorithm, 367
Performance assignment, 414–416
Personal protective equipment (PPE), 48
Physical-based models, 5–7, 15–16
Piecewise-affine model, 534
Piezo-electric transducers (PZT), 321
Piezo-electric wafer active sensor (PWAS), 343, 344, 348, 349
Plant input pumping system
  description, 391–392
  experimental setup, 392
  MARS algorithm, 392–393
  observed flow-rate *vs.* applied pump frequency, 392, 393
  residual values analysis, 393, 394
  RME value, 393, 395
  time interval, 392
Predictive maintenance
  CbM, 314
  CM system, 315, 316
  component level prognostic methods, 315
  data-driven approach, 317–318, 331–332
  data quality, 319
  decision support tools
    critical part selection, 335–337
    selection guidelines, 332–335
  diagnostic approach, 314
  health and condition monitoring techniques, 318
  maritime systems, 338–340
  model-based prognostics, 317–318

MTBF, 318
physical model-based prognostics
　associated physical model, 328–329
　critical part selection, 327–328
　failure mechanism, 328–329
　governing loads determination, 329
　HUMS, 329
　load-to-life relation, 326–327
　operational usage, 329
　time to failure prediction, 329–330
　usage-to-load relation, 326–327
　validate model, 330–331
qualitative tools, 318
railway infrastructures, 341–342
structural health and condition monitoring, 316
　sensors, 320–321
　vibration-based monitoring, 322–325
system and component level, 315–316
vibration monitoring of bearings, 317
wind turbines
　damage accumulation monitoring, 343
　damage index values, 346
　damage intensity probability, 345
　failure mechanisms, 343
　LabVIEW program, 345
　maximum damage probability value, 347, 348
　operational and maintenance costs, 342
　physics-based methodology, 342–343
　probability function, 344
　PWAS, 343, 344, 348, 349
　RAPID algorithm, 343
　RAPID maps, 347
　SAPS, 344
　vibration-based health monitoring approach, 343
　WMC, 345
Predictive mapping construction, 487
Pre-optimized production process, 493–494
Preventive maintenance, 4
Principal components regression (PCR), 63
Principle component analysis (PCA), 64–65
Probabilistic clustering method, 134–136
Process analytic technology (PAT), 78
Prognostic algorithm design
　Acuña's failure probability mass function, 361
　CRLB, 357–359
　EoD (*see* End-of-Discharge time prognosis)
　hyper-parameters, 359–360
　JCP-BCRLB, 362–363
　MCP-BCRLB, 363–365

PHM community, 355–356
step-by-step design, 360
system failure, 361
ToF PMF, 360–361
Prognostic and Health Management (PHM) community, 355–356
Prognostics and forecasting, 9–10
Prototype generation methods, 74
Prototype selection methods, 74
Pulse width modulation (PWM), 476, 478
PWAS, *see* Piezo-electric wafer active sensor

Q
Quality control (QC), 485–486
Quadratic programming (QP) problems, 536
Qualitative reasoning (QR), 444
Quality criteria (QCs), 490
Quantum membership function (QMF), 289

R
Radio frequency identification (RFID) localization technology, 15
　data processing subsystem, 287
　description, 287
　EIS, 288
　eT2ELM, 305–307
　eT2QFNN
　　architecture of, 290
　　cross-validation experiment, 306
　　DEKF algorithm, 289
　　direct partition experiment, 307
　　experiment setup, 304–305
　　fuzzy rule initialization, 299–300
　　input layer, 293
　　interval type-2 quantum membership function, 292, 293
　　learning policies, 289, 295
　　learning processes, 291
　　mathematical formulation, 289
　　membership function, 292
　　multi-model classifier, 294–295
　　network architecture, 291, 292
　　online learning mechanism, 289, 295
　　output processing layer, 294
　　quantum layer, 293
　　RSS information, 290–291
　　rule growing mechanism, 289, 296–298
　　rule layer, 293–294
　　winning rule update, 300–304
　Gaussian membership function, 288–289
　IT2GMF, 297

Radio frequency identification (RFID)
    localization technology (*cont.*)
  IT2QMF, 289, 291–294, 297
  LANDMARC, 288
  mGT2DS, 289
  quantum membership function, 289
  radar equation, 288
  RFID reader, 288, 290
  RFID tags, 288, 290
  SVR, 288
Railway infrastructures
  components, 533, 534
  constraint tightening, 543–544
  deterioration model, 534, 537–538
  grey-box model, 534
  parameters, 550–551
  piecewise-affine model, 534
  representative run, 546–547
  settings, 544–546
  two-stage robust scenario-based approach,
    539–540
Random forests (RF) algorithm, 77–78
Reciprocal transformation, 64
Reciprocating compressors, 13
  automatic compressor valves, 171
  automatic spring-loaded valves, 171
  broken valves, 167
  capacity/load control, 171
  components, 169
  compressor test bench, 193–194
  condition-based maintenance, 167
  cyclostationary modeling, 168
  double-acting cylinder, 169
  fault detection, 168
  monitoring system, 173
  operation, 169–170
  pressure pulsation, 168
  pV diagram, 168
    classification accuracy, 201
    classification problem, 191
    classifier training, 190–191
    cost function, 192, 193
    of faultless valves, 184, 185
    of faulty discharge valve, 184
    feature extraction, 185–188
    feature space, 188–190, 192, 199
    observations of valve type, 199
    piston at top dead centre, 170
    piston towards head end, 171, 172
    self-adapting approach, 193
    spike sorting, 192
    two-class SVM classification, 200
    validation accuracies, 200, 201
  reliable performance, 167
  sealing element
    baseline, 194
    broken, 194
    broken sealing element detection,
      172–173
    crack, 194, 195
    of plate valve, 195, 196
  time–frequency analysis, 168
  vibration analysis, 168
    compression cycles, spectrogram, 174,
      175
    confusion matrix, 198
    feature extraction, 180–182
    feature space, 182–183
    leave-one-valve-out approach, 198
    plastic valves, 198
    pointwise spectrogram, faultless valve,
      175–177
    raw accelerometer data, 174
    statistical analysis, 174
    steel valves, 197, 198
    test observations, 196, 197
    (time–)frequency space analysis, 174
    two-dimensional autocorrelation,
      177–179
    validation accuracies, 198
Reconstruction algorithm for probabilistic
    inspection of damage (RAPID)
    algorithm, 343
Recurrent Classifier (RC), 113
Recurrent networks (RNN), 79
Recursive density estimation (RDE), 271
Regression imputation, 68
Reliability statistics prediction, 333
Remaining useful life (RUL), 9–10, 78–79
REPAIR algorithm, 451
Restricted master problem, 542
RFID localization technology, *see* Radio
    frequency identification localization
    technology
Robust location estimator, 57
Rolling element bearings
  detection of fault, 227–228
  eHT technique
    closing and opening operation, 216–217
    D'Agostino-Pearson normality
      measure, 222–224
    dilation operation, 215–216
    erosion operation, 215–216
    experimental setup, 220–222
    health condition monitoring, 225–226
    IMF, 219–220
    morphology-based filtering technique,
      222–223

Index 565

processing procedures, 214
processing results, 226–227
structural element (*see* Structural element)
envelope analysis, 209
frequency analysis, 208–209
geometry of, 207–208
incipient bearing defect detection, 210
inner race defect, 209, 226–227
mathematical morphology-based analysis, 210
MED filter
   vs. convergence, 213–214
   defect-related impulses, 211
   entropy minimization, 211
   filter coefficients, 213
   flowchart, 211
   inverse filter, 212
   machinery system condition monitoring, 211
   optimal filter coefficient vector, 212
   signal denoising operation, 211–212
   vs. test signal response, 213–214
   Toeplitz autocorrelation matrix, 213
outer race defect, 209, 226
shaft frequency, 209
signal processing techniques, 208
time–frequency domain techniques, 209–210
vibration-based monitoring, 208
Root mean squared error (RMSE), 68, 414
Rule growing mechanism, 296–298

S
Score distance, 61
SE, *see* Structural element
Self-adaptive and autonomous systems
   ACC, 456
   control system, 448–449
   FOL, 431
   ICC, 429, 430
   internet of things, 427
   MAPE-K, 448
   mobile robot driving, 431, 432
   model-based reasoning, 428
      abductive diagnosis, 439–443
      abstraction, 444, 445
      components, 432, 443, 445
      deviation models, 445
      differential drive robot, 433
      electrical characteristics, 444
      MBD (*see* Model-based diagnosis)
      motors and wheel encoder, 433
      qualitative reasoning, 444
      system's architecture, 443
   repair, 447, 448
   system health state
      actuator, 455
      control block, 451–453
      REPAIR algorithm, 451
      sensor block, 453–454
      time step t, 449, 450
Self-adaptive forecast models, 487
Self-healing strategies, 4–5
SeqDrift2 detector, 111–112
Sherman–Morrison formula, 60–61
Short-time Fourier transform (STFT), 66–67
Signal amplitude peak squared (SAPS), 344
Smart devices, 11
   application-related limitations, 43–44
   approach of, 26
   assembly instructions, 27
   assisted reality, 29
   augmented reality, 29–30
   condition monitoring
      control-specific raw data, 39
      data exchange, 38
      direct socket communication, 38
      information system, 38
      irregularities/errors, 36
      machine-related operator support, 38–39
      predefined error libraries and codes, 38
      real-time worker information system, 37
   device selection and potentials
      climate-controlled assembly shop, 35–36
      construction environment, 35–36
      environment conditions, 34
      evaluation tool, 34–35
      hands-free operation, 34
      implementation approach, 33
      knowledge-and experience-based matching algorithm, 34
      productivity gains, 33
      use case's boundary conditions, 33
   environment-related limitations, 44
   equipment as built, 27–28
   equipment as planned, 27–28
   equipment as serviced, 27–28
   human-related limitations, 43
   information compression, 46–47
   Internet of production, 26–27
   legal aspects, 47–48
   logistics, 26–27
   mixed reality, 29

Smart devices (cont.)
  PricewaterhouseCoopers, 25
  process monitoring
    glass molding process, 42
    integrated menu structure, 41–42
    long-term application, 43
    production's software environment, 41
    short-term perspective, 43
  quality control, 27
  remote expert solutions
    cloud-based functions, 39
    device-integrated wireless module, 40
    downtimes and resulting costs, 40
    functions and options, 39–40
    machine experts, 39
    maintenance process, 36–37
    mobile devices, 39
  smartglasses, 30–32
  smartphones, 30, 32
  smartwatches, 31–33
  stand-alone solution, 36
  tablets, 30, 32
  user acceptance, 44–46
  virtual reality, 29
  visual tracking, 29
Square root transformation, 64
Staged online learning (SOL) approach, 113–114
  accuracy vs. throughput trade-off, 124
  accurate classifiers, 120–121
  vs. ARF, 122–123
  components, 108
  Covertype dataset, 117, 119
  data stream, 108
  deployment state, 107
  ELEC dataset, 117, 119
  evolving tree, 109
  Flight dataset, 117–121
  Friedman test statistic, 121–122
  historical properties, 108
  Hoeffding tree, 120
  internal parameters, 120
  Kleinberg's modelling, 108
  learning state, 107
  memory consumption, 124–125
  noisy RBF dataset, 118–119
  noisy RH dataset, 117–119, 121
  Occupancy dataset, 121
  performance measurement, 113–114
  throughput evaluation, 123–124
  transition, 109–112
  tree induction algorithm, 107
  winner classifier, 108–109
State-of-Charge (SoC), 365, 368

State-space model, 365–367
State transition model, 366, 368
Stochastic neighborhood embedding (SNE), 70–71
Stochastic regression imputation, 68
Stressor-based predictions, 333
Structural element (SE)
  impulse extraction and demodulation, 215
  morphological signal analysis, 215
  proposed morphological filter
    bearing vibration signal analysis, 218
    example, 218–219
    finer search spacing, 218
    kurtosis, 218–220
    proposed indicator, 219–220
    Renyi entropy, 217–220
Structural health and condition monitoring, 316
  sensors, 320–321
  vibration-based monitoring, 322–325
Subspace aided approach (SAP), 270
Supervisory control and data acquisition (SCADA) systems, 386
Support vector machines (SVM), 77
Support vector regression (SVR), 288
Symmetric mean absolute percentage error (SMAPE), 414, 415
Synthetic minority oversampling technique (SMOTE), 73–74
System decomposition, 464, 466–467
System health state
  actuator, 455
  control block, 451–453
  REPAIR algorithm, 451
  sensor block, 453–454
System models, 2–4

T
Taguchi L12 method, 495
Takagi–Sugeno (TS) fuzzy systems, 270–271, 409, 499–502, 521, 522
t-distributed SNE (t-SNE), 70–71
Temporal causal graphs (TCG), 462–463
Three-cellular converter, 466, 467, 469
Threshold autoregressive (TAR), 406, 407, 418
Time domain, 66
Time–frequency domain analysis, 66
Time instant optimization (TIO), 535
Time-of-failure PMFs, 372–374
Time-series-based forecast models, 487
  batch-processing, 498, 499
  dimensionality reduction, 511–512

influence analysis, flatness criteria, 524–526
initial batch model construction, 521–522
model adaptation, 522–524
multi-stage production, 526, 527
on-line model adaptation
　forgetting factor, 503
　functionalities, 502
　QC measurements, 502
　rule splitting, 503–504
　updating process, 505–506
PLS and TS fuzzy systems, 499–502, 521, 522
prediction horizon, 498
pre-optimized production process, 510–511
strategies, 509–511
three-dimensional matrix and flattening, 498
Tomek's links, 75
True positive rate (TPR), 279, 283
Typicality and eccentricity data analytics (TEDA), 271

U
Un-optimized production process, 492, 493
Update gates, 84

V
Vector error correction model (VECM), 404, 413, 416–417
Vibration-based monitoring, 322–325

W
Wastewater treatment plants (WWTPs), 16–17
　aeration system blowers
　　blowers diffuser position and current consumption, 396–397
　　description, 395
　　experimental setup, 395–396
　　residual values analysis, 397
　　RME metric, 397, 398
　　smoothed air temperature, 398, 399
　data mining modelling, 386
　dynamic and non-linear scenarios, 385
　energy consumption, 382–383
　energy savings, 383–385
　equipment, 382
　external recirculation pumping system
　　description, 387
　　experimental setup, 388
　　linear regression, 388
　　observed flow-rate *vs.* applied pump frequency, 388, 389
　　relative mean error value, 389, 391
　　residual values analysis, 388, 390
　KPI, 386
　modelling approach, 385
　overview, 381–382
　plant input pumping system
　　description, 391–392
　　experimental setup, 392
　　MARS algorithm, 392–393
　　observed flow-rate *vs.* applied pump frequency, 392, 393
　　residual values analysis, 393, 394
　　RME value, 393, 395
　　time interval, 392
　pumping/aeration processes, 385
　regression algorithm, 386
　relative mean error, 386
　SCADA systems, 386
　schematic diagram, 387
Wavelet transform (WT), 209–210
Weighted recursive least squares (wRLS) algorithm, 412
Whittaker's smoother, 187
Wiener Process, 7
Wigner–Ville distribution, 67, 209–210
Windows Mixed Reality, 32
Wind turbine materials and constructions (WMC), 345
Wind turbines
　damage accumulation monitoring, 343
　damage index values, 346
　damage intensity probability, 345
　failure mechanisms, 343
　LabVIEW program, 345
　maximum damage probability value, 347, 348
　operational and maintenance costs, 342
　physics-based methodology, 342–343
　probability function, 344
　PWAS, 343, 344, 348, 349
　RAPID algorithm, 343
　RAPID maps, 347
　SAPS, 344
　vibration-based health monitoring approach, 343
　WMC, 345
WWTPs, *see* Wastewater treatment plants

Z
Zero-defect manufacturing, 1

Printed by Printforce, the Netherlands